Design and Deployment of Small Cell Networks

This comprehensive resource covers everything you need to know about small cell networks, from design, to analysis, optimization, and deployment.

Detailing fundamental concepts as well as more advanced topics, and describing emerging trends, challenges, and recent research results, in this book experts explain how you can improve performance, decision making, resource management, and energy efficiency in next-generation wireless networks.

Key topics covered include green small cell networks and associated tradeoffs, optimized design and performance analysis, backhauling and traffic overloading, context-aware self-organizing networks, deployment strategies, and mobility management in large-scale heterogeneous networks (HetNets).

Written by leading experts in academia and industry, and including tools and techniques for small cell network design and deployment, this is an ideal resource for graduate students, researchers, and industry practitioners working in communications and networking.

Alagan Anpalagan is a Professor in the Department of Electrical and Computer Engineering at Ryerson University where he is the recipient of the Dean's Teaching Award; Faculty Scholastic, Research and Creativity Award; and Faculty Service Award. He is a registered Professional Engineer in the province of Ontario, Canada and a Fellow of the Institution of Engineering and Technology.

Mehdi Bennis is a Senior Research Fellow at the Centre for Wireless Communications (CWC), University of Oulu, Finland. Previously he worked as a research engineer at IMRA-EUROPE and was a visiting researcher at the Alcatel-Lucent Chair on Flexible Radio, SUPELEC.

Rath Vannithamby leads a team responsible for 5G and Internet of Things research at Intel Labs and was previously a researcher at Ericsson. He is currently a Senior Member of the IEEE and an IEEE Communications Society Distinguished Lecturer.

Design and Deployment of Small Cell Networks

ALAGAN ANPALAGAN
Ryerson University

MEHDI BENNIS
University of Oulu

RATH VANNITHAMBY
Intel Corporation

CAMBRIDGE
UNIVERSITY PRESS

University Printing House, Cambridge CB2 8BS, United Kingdom

Cambridge University Press is part of the University of Cambridge.

It furthers the University's mission by disseminating knowledge in the pursuit of education, learning and research at the highest international levels of excellence.

www.cambridge.org
Information on this title: www.cambridge.org/9781107056718

© Cambridge University Press 2016

This publication is in copyright. Subject to statutory exception
and to the provisions of relevant collective licensing agreements,
no reproduction of any part may take place without the written
permission of Cambridge University Press.

First published 2016

Printed in the United Kingdom by TJ International Ltd. Padstow Cornwall

A catalog record for this publication is available from the British Library

ISBN 978-1-107-05671-8 Hardback

Additional resources for this publication at www.cambridge.org/9781107056718

Cambridge University Press has no responsibility for the persistence or accuracy of URLs for external or third-party internet websites referred to in this publication, and does not guarantee that any content on such websites is, or will remain, accurate or appropriate.

Contents

	List of contributors	page x
	Preface	xv
1	**Mobility performance optimization for 3GPP LTE HetNets**	1
	Kathiravetpillai Sivanesan, Jialin Zou, Subramanian Vasudevan, and Sudeep Palat	
	1.1 Introduction	1
	1.2 Radio link monitoring and failure recovery process	4
	1.3 Handover process	7
	1.4 Mobility performance and challenges	11
	1.5 Mobility performance results	16
	1.6 RLM and RLF recovery enhancement in HetNets	21
	1.7 Discovery of small cells	23
	1.8 Summary	29
2	**Design and performance analysis of multi-radio small cell networks**	31
	Nageen Himayat, Shu-ping Yeh, Shilpa Talwar, Mikhail Gerasimenko, Sergey Andreev, and Yevgeni Koucheryavy	
	2.1 Introduction	31
	2.2 Integrated multi-RAT HetNet architectures	34
	2.3 Radio resource management in multi-RAT HetNets	37
	2.4 Analytical frameworks for radio resource management in integrated multi-RAT access network	41
	2.5 System evaluation methodology and assumptions	46
	2.6 Performance evaluation	48
	2.7 Summary and next steps	54
3	**Dynamic TDD small cell management**	58
	Cheng-Chih Chao, Yi-Ting Lin, and Hung-Yu Wei	
	3.1 Dynamic TDD system overview	58
	3.2 D-TDD deployment and issues	62
	3.3 Simulation methodology	64
	3.4 Interference mitigation method for D-TDD	69
	3.5 Conclusion	72

4	**3GPP RAN standards for small cells**	75
	Weimin Xiao, Jialing Liu, and Anthony C. K. Soong	
	4.1 Introduction	75
	4.2 Interference management	76
	4.3 Mobility management	83
	4.4 Spectrum utilization	85
	4.5 Dense network adaptation	87
	4.6 Other related 3GPP RAN items	90
	4.7 Future 3GPP features for small cell	93
	4.8 Appendix: brief description of LTE channels	94
5	**Dense networks of small cells**	96
	Jialing Liu, Weimin Xiao, and Anthony C. K. Soong	
	5.1 Introduction	96
	5.2 Evaluation and performance of dense networks of small cells	98
	5.3 Characterizing dense cellular networks	107
	5.4 Technologies for dense networks	111
	5.5 Summary and future directions	118
	5.6 Appendix	120
6	**Traffic offloading scenarios for heterogeneous networks**	122
	Adrian Kliks, Nikos Dimitriou, Andreas Zalonis, and Oliver Holland	
	6.1 Introduction	122
	6.2 The role of small cells in traffic offloading	124
	6.3 Technological solutions for HetNets	127
	6.4 HetNets offloading: benefits analysis	137
	6.5 HetNets offloading assessment	139
	6.6 Conclusions	144
7	**Required number of small cell access points in heterogeneous wireless networks**	148
	S. Alireza Banani, Andrew Eckford, and Raviraj Adve	
	7.1 Introduction	148
	7.2 Finite-area network with uniformly distributed SC APs	150
	7.3 Dependent placement of SCs in a hexagonal grid network	158
8	**Small cell deployments: system scenarios, performance, and analysis**	169
	Mark C. Reed and He Wang	
	8.1 System scenarios	171
	8.2 Analytical model and performance analysis	180
	8.3 Summary	187

9	**Temporary cognitive small cell networks for rapid and emergency deployments**	191
	Akram Al-Hourani, Sithamparanathan Kandeepan, and Senthuran Arunthavanathan	
	9.1 Introduction	191
	9.2 The concept of temporary cognitive small cell networks	192
	9.3 Network model	196
	9.4 Deployment process	198
	9.5 Cognitive interference mitigation techniques	199
	9.6 Spectrum sensing and radio environment learning	203
	9.7 Simulation of temporary cognitive small cell networks	208
	9.8 Conclusion	210
10	**Long-term evolution (LTE) and LTE-Advanced activities in small cell networks**	213
	Qi Jiang, Jinsong Wu, Lu Zhang, and Shengjie Zhao	
	10.1 Introduction	213
	10.2 Relay eNodeB in LTE-Advanced	215
	10.3 Pico eNodeB in LTE-Advanced	223
	10.4 Home eNodeB in LTE-Advanced	230
	10.5 Small cell enhancement in Release 12	235
	10.6 Summary	239
11	**Game theory and learning techniques for self-organization in small cell networks**	242
	Prabodini Semasinghe, Kun Zhu, Ekram Hossain, and Alagan Anpalagan	
	11.1 Small cell networks	242
	11.2 Self-organization	244
	11.3 Issues and challenges in self-organizing small cell networks	248
	11.4 Game theory for self-organizing small cell networks	250
	11.5 Game theory-based resource management for self-organizing small cells	258
	11.6 Learning techniques for self-organizing small cell networks	270
	11.7 Conclusion	278
12	**Energy efficient strategies with BS sleep mode in green small cell networks**	284
	Hong Zhang and Jun Cai	
	12.1 Introduction	284
	12.2 System model and problem formulation	288
	12.3 Methodologies	291
	12.4 Simulation results	302
	12.5 Conclusions	306
13	**Mobility management in small cell heterogeneous networks**	309
	Peter Legg and Xavier Gelabert	
	13.1 Introduction	309
	13.2 Mobility in LTE small cell HetNets	311

	13.3	Mobility robustness optimization (MRO)	328
	13.4	Inter-system mobility: LTE to WiFi	333
	13.5	Summary	335

14 The art of deploying small cells: field trial experiments, system design, performance prediction, and deployment feasibility — 338
Doru Calin, Aliye Özge Kaya, Amine Abouliatim, Gonçalo Ferrada, and Ionel Petrut

	14.1	Introduction	338
	14.2	LTE small cell field trials	339
	14.3	LTE performance prediction framework validation with measurements	349
	14.4	High density small cells' design for stadiums using the LTE performance prediction framework	354
	14.5	Summary	360

15 Centralized self-optimization of interference management in LTE-A HetNets — 363
Yasir Khan, Berna Sayrac, and Eric Moulines

	15.1	Introduction	363
	15.2	Interference management in HetNets	365
	15.3	Surrogate-based optimization (SBO)	370
	15.4	Centralized self-optimization for interference mitigation	383
	15.5	Concluding remarks and open issues	388

16 Self-organized ICIC for SCN — 393
Lorenza Giupponi, Ali Imran, and Ana Maria Galindo

	16.1	Femto–macro interference control: a time-difference learning approach	394
	16.2	Macro–femto interference minimization through self-organization of macro cell azimuth angles	407
	16.3	Summary	421

17 Large-scale deployment and scalability — 425
Iris Barcia, Simon Chapman, and Chris Beale

	17.1	Introduction	425
	17.2	L-SND for modern wireless networks	431
	17.3	Large-scale challenges	439

18 Energy efficient heterogeneous networks — 462
Y. Qi, M. A. Imran, M. Z. Shakir, and K. A. Qaraqe

	18.1	Introduction	462
	18.2	Conventional HetNet	465
	18.3	HetNet based on cloud architecture	467
	18.4	Multi-point coordination	473
	18.5	Conclusions	480

19   Time- and frequency-domain e-ICIC with single- and multi-flow carrier aggregation in HetNets   484
Meryem Simsek, Mehdi Bennis, and Ismail Guvenc

- 19.1   Inter-cell interference in HetNets   485
- 19.2   Time-domain e-ICIC in HetNets   486
- 19.3   Frequency-domain e-ICIC in HetNets   488
- 19.4   Single-flow and multi-flow transmission   490

Index   502

Contributors

Amine Abouliatim
Alcatel-Lucent

Raviraj Adve
University of Toronto

Akram Al-Hourani
RMIT University

Sergey Andreev
Tampere University of Technology

Alagan Anpalagan
Ryerson University

Senthuran Arunthavanathan
RMIT University

S. Alireza Banani
University of Toronto

Iris Barcia
Keima Limited

Chris Beale
Keima Limited

Mehdi Bennis
University of Oulu

Jun Cai
University of Manitoba

Doru Calin
Alcatel-Lucent

List of contributors

Cheng-Chih Chao
National Taiwan University

Simon Chapman
Keima Limited

Nikos Dimitriou
National Kapodistrian University of Athens

Andrew Eckford
York University

Gonçalo Ferrada
Alcatel-Lucent

Ana Maria Galindo
Orange Labs

Xavier Gelabert
Huawei Technologies

Mikhail Gerasimenko
Tampere University of Technology

Lorenza Giupponi
Telecommunications Technology Centre of Catalonia

Ismail Guvenc
Florida International University

Nageen Himayat
Intel Corporation

Oliver Holland
Kings College London

Ekram Hossain
University of Manitoba

Ali Imran
Qatar Mobility Innovations Centre

M. A. Imran
University of Surrey

List of contributors

Qi Jiang
Alcatel-Lucent

Sithamparanathan Kandeepan
RMIT University

Aliye Özge Kaya
Alcatel-Lucent

Yasir Khan
Orange Labs

Adrian Kliks
Poznan University of Technology

Yevgeni Koucheryavy
Tampere University of Technology

Peter Legg
Huawei Technologies

Yi-Ting Lin
National Taiwan University

Jialing Liu
Huawei R&D

Eric Moulines
Telecom ParisTech

Sudeep Palat
Alcatel-lucent

Ionel Petrut
Alcatel-Lucent

K. A. Qaraqe
Texas A&M University at Qatar

Y. Qi
University of Surrey

Mark C. Reed
NICTA, Australia

List of contributors

Berna Sayrac
Orange Labs

Prabodini Semasinghe
University of Manitoba

M. Z. Shakir
Texas A&M University at Qatar

Meryem Simsek
Dresden University of Technology

Kathiravetpillai Sivanesan
Intel Corporation

Anthony C. K. Soong
Huawei R&D

Shilpa Talwar
Intel Corporation

Rath Vannithamby
Intel Corporation

Subramanian Vasudevan
Alcatel-lucent

He Wang
NICTA, Australia

Hung-Yu Wei
National Taiwan University

Jinsong Wu
Alcatel-Lucent

Weimin Xiao
Huawei R&D

Shu-ping Yeh
Intel Corporation

Andreas Zalonis
National Kapodistrian University of Athens

Hong Zhang
University of Manitoba

Lu Zhang
Alcatel-Lucent

Shengjie Zhao
Tongji University

Kun Zhu
University of Manitoba

Jialin Zou
Alcatel-Lucent

Preface

The ever-increasing use of smart phone devices, multimedia applications, and social networking, along with the demand for higher data rates, ubiquitous coverage, and better quality of service, pose new challenges to the traditional mobile wireless network paradigm that depends on macro cells for service delivery. Small cell networks (SCNs) have emerged as an attractive paradigm and hold great promise for future wireless communication systems (5G systems). SCNs encompass a broad variety of cell types, such as micro, pico, and femto cells, as well as advanced wireless relays, and distributed antenna systems. SCNs co-exist with the macro cellular network and bring the network closer to the user equipment. SCNs require low power, incur low cost, and provide increased spatial reuse. Data traffic offloading eases the load on the expensive macro cells with significant savings expected to the network operators using small cells.

As the demand for increased bandwidth rages on, SCNs emerged in dense urban areas mainly to provide coverage and capacity. They have now gained momentum and are expected to dominate in the coming years, with the rollout in large scale – either planned or in ad-hoc manner – and the development of 5G systems with many small cell components. Already, the number of "small cells" in the world exceeds the total number of traditional mobile base stations. SCNs are also envisioned to pave the way for new services. However, there are many challenges in the design and deployment of small cell networks, which have to be addressed in order to be technically and commercially successful. This book provides various concepts in the design, analysis, optimization, and deployment of small cell networks, using a treatment approach suitable for pedagogical and practical purposes.

This book is an excellent source for understanding small cell network concepts, associated problems, and potential solutions in next-generation wireless networks. It covers from fundamentals to advanced topics, deployment issues, environmental concerns, optimized solutions, and standards activities in emerging small cell networks. New trends, challenges, and research results are also provided. Written by leading experts in the field from academia and industry around the world, it is a valuable resource dealing with both the important, core, and specialized issues in these areas. It offers a wide coverage of topics, while balancing the treatment to suit the needs of first-time learners of the concepts and specialists in the field. It serves as a one-stop reference book for students, instructors, researchers, and industry practitioners who are working in the design and deployment of small cell networks. Some highlights are:

- dense networking and multi-radio networking
- green small cell networks and tradeoffs
- optimized design and performance analysis
- backhauling and traffic overloading
- small cell network management
- deployment strategies
- latest standard activities
- context-aware self-organizing networks
- mobility management in large-scale HetNets
- coverage centric deployment of small cells
- enhanced inter-node carrier aggregation in small cells

The editors would like to thank all the chapter authors for their excellent and timely contribution. Special thanks go to the staff at Cambridge University Press for their professional and dedicated service. Last, but not least, we want to thank our families for their support, encouragement, and sacrifice.

Alagan Anpalagan
Mehdi Bennis
Rath Vannithamby

1 Mobility performance optimization for 3GPP LTE HetNets

Kathiravetpillai Sivanesan, Jialin Zou, Subramanian Vasudevan, and Sudeep Palat

Heterogeneous networks (HetNets) are being deployed as a feasible and cost-effective solution to address the recent data explosion caused by smart phones and tablets. In a co-channel HetNet deployment, several low-power small cells are overlaid on the same carrier as the existing macro network. While this is the most spectrally efficient approach, coverage areas of the small cells can be significantly smaller due to their lower transmit powers, which can limit the volume of data offload. Extending the range of pico cells to increase traffic offload via increased number of associated users to these cells is known as cell range extension (CRE). On the flip side, CRE results in interference issues that have been resolved via standards based solutions in 3GPP, known as the Release 10 enhanced inter-cell interference coordination (eICIC) capability. In this chapter, we address the problem of ensuring connected state mobility or handover performance in co-channel HetNets. HetNets with and without range extension are considered. We show how the aforesaid interference coordination techniques can also be leveraged to improve mobility performance. Furthermore, we discuss how the handover decisions and handover parameters can be further optimized based on user speed. We show that the handover failure rate can be significantly reduced using mobile speed dependent handover parameter adaptation and CRE with subframe blanking, although at the cost of an increase in the short time-of-stay (SToS) rate. Finally, other aspects such as radio link failure recovery, small cell discovery, and related enhancements are discussed.

1.1 Introduction

As a result of rapid penetration of smart phones and tablets, mobile users have started to use more and more data services, in addition to the conventional voice service, on their devices. Due to this trend, demand for network capacity has been growing significantly. It is observed that the capacity demand normally originates unevenly in the cellular coverage area. In other words, the demand is concentrated in some smaller geographical areas, for example shopping malls, stadiums, and high-rise buildings. The conventional homogeneous cellular networks are intended to provide uniform coverage and services with base stations having the same transmit powers, antenna parameters, backhaul connectivity, etc., across a wide geographical area. To serve spatially concentrated data demand, HetNets are a viable and cost-effective solution.

A HetNet is the result of embedding (or overlaying) a group of smaller cells with lower transmit power within a conventional macro cellular network. The smaller cells can be pico, micro, femto, or relay nodes, and can be located either indoors or outdoors. The deployments can be for capacity (in high-demand areas) or for coverage enhancement (deep penetration into buildings). Pico cells deployed with some awareness of the spatial traffic distribution or deployed at hotspots are targeted toward capacity augmentation. Deploying these cells at such hotspots will achieve substantial data offload with and without having to resort to range extension.

In this chapter, we focus on the typical HetNet deployment scenario where pico cells employ the same radio access technology (LTE) as that of the macro cellular network. Both co-channel and multi-carrier deployments are possible. In co-channel deployments, the small cells and the macro cells share the same frequency spectrum. This, from a spectral efficiency standpoint, is a preferable option for operators even though it results in several challenges due to mutual interference between the macro and pico cells. A multi-carrier deployment requires dedicating a new carrier frequency for the pico cells. The possibility of dedicating a carrier for use by pico is subject to spectrum availability and can often put limits on such deployments.

The conventional cellular radio interface technologies such as 3GPP LTE, CDMA $1\times$ and EVDO were designed from the perspective of a homogeneous network deployment. Their procedures and functions were intended to work for a macro cellular network. When a group of small cells with different transmit power and access privileges are introduced, the main techniques that need to be investigated are random access, best cell selection and reselection in idle mode, connected mode mobility or handover, interference coordination or mitigation between cells, and load balancing.

Approach to handover in LTE: Although the conventional LTE system has an OFDM-based air interface with no macro diversity on either uplink or downlink, it is still intended to employ a frequency reuse of 1. Thus cell-edge mobiles experience significant interference from surrounding cells. The handover (HO) process in the cellular network is intended to transfer the service of a mobile user to another base station without service interruption. The HO process is typically initiated by the mobile sending reports of signal power or quality (reference signal received power (RSRP) or referenced signal received quality (RSRQ)), of the current serving cell and the candidate target cells [1]. The measurement and reporting is based on event triggers that can be configured by the network. The serving base station directs the mobile to handover to the intended target cell after confirming availability of resources at the target cell via backhaul message exchanges between the source and target eNBs. The mobile then reconnects to the target cell. The HO process, including event triggers, and messages exchanged between mobile, serving, and target cells, are summarized in [2].

Challenges in HetNet mobility: In the co-channel macro–pico environment, the RSRP profile is very different from the macro-only system. A typical user RSRP profile in co-channel macro–pico HetNets is shown in Section 1.3.3, where measurement and signaling events are illustrated for the hand-in and hand-out between a macro and a pico. It is apparent that the time available for these handovers is reduced due to the smaller coverage area of the pico and that furthermore the steep rate at which the pico RSRP decreases has an impact on the reliability with which the handover messages can be delivered in this short interval of time. The mobility process in HetNets is also vulnerable to the co-channel interference between macro and pico cells. The large number of additional coverage boundaries created by embedding pico cells also increases the number of HOs in the system thereby increasing the load of network mobility management resources and the overall number of handover failures.

Handover optimization: In the current LTE system, the mechanisms available to optimize the HO process are: hysteresis (Hys), the offsets that are applied between source and target cell, time-to-trigger (TTT), and Layer 3 filtering [1]. Hysteresis refers to the path-loss difference between inbound and outbound handovers between a pair of cells and is designed to prevent frequent switching between cells. The offsets are used to change the relative signal strength levels at which mobility events are triggered at the mobile, and the TTT refers to the time that the mobile waits before sending the corresponding measurement reports to the serving base station. Layer 3 filtering is network-configurable filtering that is applied to implementation-specific RSRP measurements made by the mobiles at Layer 1. These HO parameters can be intelligently chosen to minimize HO failures and reduce the ping-pong effect (frequent HOs to maintain the connection and the service flow), while providing connection through the best available base station.

Mobility studies: The HO performance of the LTE macro networks, based on the received signal strength at the mobile, has been studied in [3–6]. The simulations were performed using the 3GPP LTE parameters and assumptions. In [3], the effects of the L3 filtering and other HO parameters (TTT, HO offset) on the tradeoff between the HO rates and the radio link failures are investigated. The effects of the measurement bandwidth, HO margin (A3 offset), and measurement period have been studied in [4]. The impact of the linear and dB domain L3 filtering on the HO performance is investigated in [5]. The HO failure rates, and the delay for the entire HO process, have been studied in [6] for various HO parameters and Layer 1 control-channel errors. In [7], improvement of the HO performance using interference coordination was studied. It is shown that significant HO performance gains were obtained using inter-cell interference coordination (ICIC). The HO between the overlaid macro and micro cells is also discussed in [8] in the TDMA, FDMA, and CDMA contexts. Velocity-based HO decision was suggested for the micro overlay situations.

Cell range extension refers here to the expansion of pico coverage beyond the contour at which its RSRP equals that of the macro, and almost blank subframes (ABS) to the partially muted transmissions from the macro to reduce interference in this expanded pico coverage region. In the eICIC-based study, 3 dB and 6 dB cell selection biases toward the pico have been considered. Early deployments of HetNets with R10 UEs will be constrained to such moderate bias values and moderate range extension since they will not be equipped with interference cancellation capability to combat residual interference from macro cells that increases with pico range extension.

Presentation summary: In this chapter, we study the connected state handover performance in co-channel 3GPP LTE HetNets. We have investigated the HO failure, and SToS rates for various mobile speeds. Two techniques to optimize the handover performance, namely, (a) adaptation of the TTT, HO offset and Layer 3 filter based on the

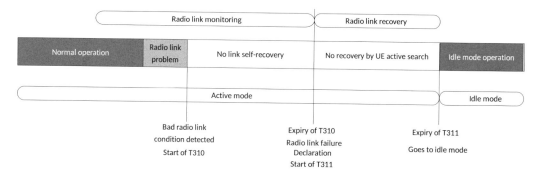

Figure 1.1 Radio link monitor (RLM) and radio link failure (RLF) recovery process.

mobile speed, and (b) cell range extension with ABS, which are the two key elements of enhanced ICIC (eICIC), are presented. To increase the modeling accuracy, aspects such as RSRP measurement errors, and the Layer 3 filtering, have been modeled.

The rest of the chapter is organized as follows. The 3GPP LTE radio link monitoring process and HO are described in Sections 1.2 and 1.3, respectively. The aspects of the handover process that are captured in the model for the mobility performance are described in Section 1.4. In Section 1.5, the performance results and observations are presented. Recommendations for further performance improvement, additional open questions in HetNet mobility, and a summary, are presented in Sections 1.6, 1.7, and 1.8, respectively.

1.2 Radio link monitoring and failure recovery process

In order to support the user equiment (UE) mobility and minimize the frequency and duration of service interruption, radio link monitor (RLM) and radio link failure (RLF) recovery functions are specified in LTE standards as an independent mobility process. In the HetNets, when co-channel small cells are deployed, they can have significant impact on the RLF performance [11]. In this section, we first describe the legacy RLM and RLF recovery process for a conventional macro cell only network.

Figure 1.1 illustrates the entire RLM and RLF recovery process. There are five important parameters defined in the standards [1], which are associated with the RLM, RLF detection, and RLF recovery procedures: T310, N310, N311, T311, and T301 (Table 1.1).

The parameters are configured by the network at the UE through dedicated signaling.

1.2.1 Out-of-sync detection

In RRC-connected, a UE continuously monitors the downlink link quality based on the signal-to-noise ratio (SNR) of the cell-specific reference signal (wide band CQI). The UE then compares it to the thresholds Q_{out} and Q_{in} for the purpose of evaluating the downlink radio link quality of the serving cell.

Table 1.1 Radio link monitor related parameters.

Parameters	Usage
T310	RLF timer, the maximum time allowed for the radio link self-recovery.
N310	Number of consecutive "out-of-sync" indications received from lower layers before the UE starts radio link self-recovery process.
N311	Number of consecutive "in-sync" indications received from lower layers before the UE considers the link has recovered.
T311	RLF recovery timer, the maximum time allowed for the recovery of the radio link actively performed by the UE.
T301	Maximum time allowed from the time when radio resource control (RRC) connection re-establishment request message is sent till when the response from the target evolved Node B (eNB) is received by the UE.

In non-DRX mode operation, the physical layer in the UE assesses the radio link quality every radio frame, evaluated over the previous 200 ms period, against thresholds (Q_{out} and Q_{in}) defined in [10]. When the downlink radio link quality estimated over the last 200 ms period becomes worse than the threshold Q_{out}, Layer 1 of the UE sends an out-of-sync indication to the higher layers.

When the downlink radio link quality estimated over the last 100 ms period becomes better than the threshold Q_{in}, Layer 1 of the UE sends an in-sync indication to the higher layers.

In DRX mode operation, the physical layer in the UE assesses the radio link quality at least once every DRX period, calculated over the previous time period, against thresholds (Q_{out} and Q_{in}) defined in [10].

The setting of Q_{out} and Q_{in} is dependent on the UE receiver implementation. The threshold Q_{out} is defined as the level at which the downlink radio link cannot be reliably received and corresponds to 10% block error rate of a hypothetical physical downlink control channel (PDCCH) transmission taking into account the physical control format indicator channel (PCFICH) errors [11]. The threshold Q_{in} is defined as the level corresponding to 2% block error rate of a hypothetical PDCCH transmission taking into account the PCFICH errors such that the downlink radio signals can be significantly more reliably received than at Q_{out}.

1.2.2 Radio link monitoring

As shown in Figure 1.1, when a UE is in normal RRC_CONNECTED mode, the lower layer of the UE continuously performs the downlink reference signal measurement of the serving cell. When a bad link condition is detected, an "out-of-sync" indication will be reported to the upper layer as discussed above. If N310 consecutive "out-of-sync" indications are received from lower layers, the RLF timer (T310) will be started.

Before the T310 timer expires if the in-sync indication were detected N311 times as reported from the lower layer, then the T310 is stopped. In this case, the radio link is considered to be self-recovered.

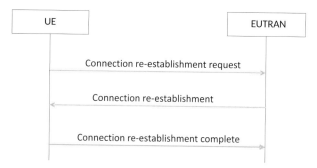

Figure 1.2 Radio resource control (RRC) connection re-establishment is successful.

Figure 1.3 Radio resource control (RRC) connection re-establishment request is rejected.

Otherwise, while T310 is running if not enough in-sync indication(s) are received from the lower layer, the RLF is declared by the UE upon the expiry of the T310 timer.

In addition to determining the RLF based on the physical layer measurement, a UE also detects the RLFs based on other criteria and indications including the random access problem indication from medium access control (MAC) and the indication from radio link control (RLC) that the maximum number of retransmissions has been reached.

When the RLF is determined by a UE, its RRC connection recovery process is triggered and timer T311 is started.

1.2.3 Radio link failure recovery

When a UE declares an RLF, it first starts the T311 RLF recovery timer. While T311 is running, the UE searches for a suitable cell to re-establish the new radio link. If a suitable cell is found, upon selecting the cell, the T311 timer is stopped and T301 timer is started. An RRC connection re-establishment request message is sent by the UE to the target cell eNB. In the case of successful connection establishment (Figure 1.2), the target eNB sends RRC connection re-establishment before the expiry of T301 at the UE. The UE acknowledges to the target eNB by sending back the connection re-establishment completion message. If no response is received from the target eNB before T301 expiry or if an RRC re-establishment reject is received (Figure 1.3), the UE transits to the RRC-idle mode.

If a suitable cell cannot be identified until the expiry of T311, the UE will go into the RRC_IDLE mode.

In the RLF recovery phase, in order to resume activity and avoid going into the RRC_IDLE mode when the UE returns to the same cell or a different cell from a different eNB, the following principles are followed.

- The UE stays in RRC_CONNECTED mode during the RLF recovery process.
- After identifying the suitable cell, the UE accesses the cell through the random access procedure.
- At the selected target cell eNB, the UE identifier used in the random access procedure for contention resolution is used to authenticate the UE and check whether it has a context stored for that UE.
  - If the eNB finds a context that matches the identity of the UE, it indicates to the UE (sends the confirmation message to the UE) that its connection can be resumed.
  - If the context is not found, the eNB sends the connection re-establishment rejection message to the UE. Then RRC connection is released and the UE initiates the procedure to establish a new RRC connection. In this case the UE is required to go via RRC_IDLE.

1.3 Handover process

The HO process in RRC_CONNECTED is one of the most important functions in mobility management. The legacy HO process was originally designed for macro only systems. The macro cells have the same transmit power, coverage area, and relatively wide border area between the cells. In HetNets, especially when the co-channel small cells are deployed with lower transmit powers and thus are smaller, the coverage area and more sharply changing signal strength at the border area of small–macro cells pose several challenges for mobility performance. Another major factor affecting the HO performance in co-channel overlay is the co-channel interference. The co-channel interference is quite severe in HetNets compared to the conventional macro network. More specifically, around the small cell transmit antenna the interference from the small cell is quite high for the macro link. Thus there is a high possibility for RLF for a UE served by the macro cell while the UE is deep inside the pico coverage. Therefore handover performance in HetNets is an important aspect to be studied and improved.

1.3.1 Triggering of UE handover

The HO process is typically started by triggering the measurement from the UE to the serving eNB. The LTE measurement for HO is normally triggered by the Event A3 as specified in [1]. The received signal strength (reference signal received power, RSRP) or the received signal quality (reference signal received quality, RSRQ) is used as a metric to make the HO decisions. The logic of Event A3 and HO parameters used in Event A3 for making the decision are shown as follows.

Event A3 (neighbor becomes better than serving after a preset offset)
The Event A3 is used for the handover measurement triggering as specified in [1].

Entering condition:

$$Mn + Ofn + Ocn - Hys > Ms + Ofs + Ocs + Off$$

Leaving condition:

$$Mn + Ofn + Ocn + Hys < Ms + Ofs + Ocs + Off$$

Mn is the measurement result of the neighboring cell, not taking into account any offsets.
Ofn is the frequency-specific offset of the frequency of the neighbor cell
Ocn is the cell-specific offset of the neighbor cell, and set to zero if not configured for the neighbor cell.
Ms is the measurement result of the serving cell, not taking into account any offsets.
Ofs is the frequency-specific offset of the serving frequency.
Ocs is the cell-specific offset of the serving cell and is set to zero if not configured for the serving cell.
Hys is the hysteresis parameter for this event.
Off is the system-wide common offset parameter for this event.
Mn and **Ms** are expressed in dBm in the case of RSRP, or in dB in the case of RSRQ.
Ofn, **Ocn**, **Ofs**, **Ocs**, **Hys**, and **Off** are expressed in dB.

When a mobile is moving toward another cell, if the target cell satisfies the entering condition it is included in the neighbor cell measurement report list for the HO. If a mobile is moving away from a cell and the leaving condition is satisfied, that cell is removed from the neighbor cell report list for the HO.

For the conventional co-channel macro-to-macro handover the parameters **Ofn**, **Ocn**, **Ofs**, and **Ocs** are normally set to zero. The **Hys** is a positive quantity and used to prevent ping-pong and unnecessary HOs. The **Off** is used to specify a handover threshold or margin and generally is a positive quantity and common for all target cells and is commonly referred to as A3 offset.

When co-channel small cells are deployed on top of the macro cells, the handover behavior in different scenarios, such as macro to macro, macro to pico, pico to macro, and pico to pico, can be very different. The HO performance especially in the scenarios involving pico cells is significantly impacted by the UE speed.

1.3.2 The Layer 3 RSRP filtering

As is specified in [1], the RSRP (or RSRQ) is measured at the physical layer (Layer 1) and periodically measured RSRP (or RSRQ) is passed through an IIR filter at the Layer 3. The filtered RSRP (or RSRQ) is used to make the accurate HO decisions. The filter is denoted as:

$$Fn = (1 - a)Fn - 1 + a\, Mn \qquad (1.1)$$

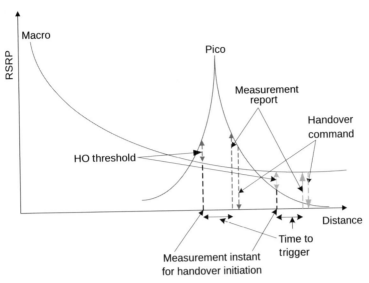

Figure 1.4 Handover triggering based on measurement in the macro and small cell co-channel environment.

where a = ½ ^(k/4), k is the filter coefficient, Mn is the current measurement, Fn − 1 and Fn are the previous and current filtered RSRP values, respectively.

It is noted in the standard [1] that the k value is specified based on the Layer 3 sampling rate of once every 200 ms. For a different sampling rate a different filter coefficient is required to preserve the same time characteristic of the filter.

1.3.3 Lower layer handover behavior in HetNets

Figure 1.4 illustrates the HO decision process based on RSRP measurement in HetNets. Consider that a mobile is moving through the center of the small cell; the connected mode mobility with HO triggering based on Event A3 is demonstrated here. The RSRP profile of the macro and the small cell is depicted in Figure 1.4.

When the small cell RSRP is greater than the macro RSRP the mobile should be handed over to the small cells. In practice, due to shadowing the RSRP profiles are fluctuated curves. Thus there could be several crossovers of the macro and small cell RSRP profiles. There are several parameters that can be tuned to reduce unnecessary handovers or ping-pongs, such as time to trigger (TTT), handover threshold or offset (i.e., target versus serving measurement offset), entering/exiting hysteresis, and Layer 3 filtering.

1.3.4 Handover procedures and signaling

Figure 1.5 shows the details of the major HO procedures involved and signaling at the radio access network (RAN).

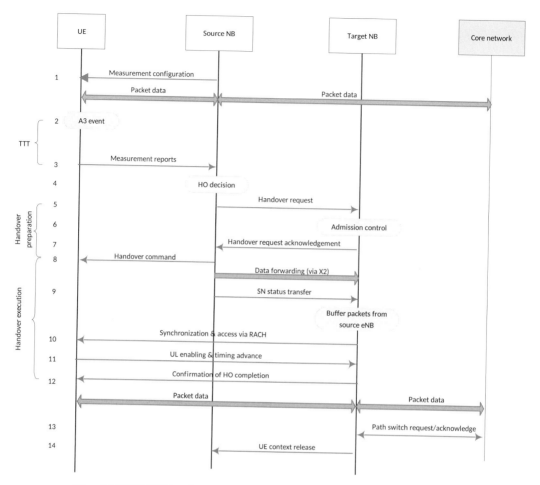

Figure 1.5 3GPP LTE handover procedures and signaling.

1. The source eNB configures the UE measurement procedures according to the roaming and access restriction information. Measurements provided by the source eNB may assist the function controlling the UE's connection mobility.
2. UE measurements meet the A3 event triggering condition. The A3 event is deemed to have occurred. Time-to-trigger (TTT) timer is started.
3. After TTT timer is expired, a measurement report is triggered and sent to the source eNB.
4. The source eNB makes a decision based on the measurement report and radio resource management (RRM) information to hand off the UE.
5. The source eNB issues an HO request message to the target eNB passing necessary information to prepare the HO at the target side.
6. Admission control may be performed by the target eNB dependent on the received E-RAB QoS information to increase the likelihood of a successful HO, if the resources can be granted by the target eNB.

7. The target eNB prepares HO with L1/L2 and sends the HO request acknowledgement to the source eNB. The handover request acknowledgement message includes a new cell radio network temporary identifier (C-RNTI), target eNB security algorithm identifiers, may include a dedicated random access channel (RACH) preamble, and possibly some other parameters.
8. When the source eNB receives the HO request acknowledgement, it sends the handover command to the UE. As soon as the transmission of the handover command is initiated in the downlink, data forwarding may be initiated.
9. The source eNB sends the sequence number (SN) status transfer message to the target eNB to convey the uplink packet data convergence protocol (PDCP) SN receiver status and the downlink PDCP SN transmitter status of E-RABs for which PDCP status preservation applies (i.e., for radio link control acknowledged mode RLC-AM).
10. After receiving the HO command with the mobility control information, the UE performs synchronization to target the eNB and accesses the target cell via RACH. The UE derives target eNB specific keys and configures the selected security algorithms to be used in the target cell.
11. The target eNB responds with uplink assignment (resource allocation) and timing advance.
12. When the UE has successfully accessed the target cell, the UE sends the connection configuration complete message, along with an uplink buffer status report, to confirm that the HO procedure is completed for the UE. After verifyimg the feedback information, the target eNB can now begin sending data to the UE.
13. The target eNB sends a path switch request message to the core network to inform it that the UE has changed cell. The core network switches the path and acknowledges to the target eNB the completion of the path switch.
14. The target eNB sends the UE context release message to the source eNB to inform success of HO and triggers the release of resources by the source eNB. Upon reception of the release message, the source eNB can release radio and control-plane related resources associated to the UE context.

1.4 Mobility performance and challenges

As shown in Section 1.3.4, the HO measurement configuration, uplink (UL) grant for the measurement report, measurement report, HO command, and random access with the target cell, are the signaling required for a successful handover. The mobility performance generally depends on the successful transmission and reception of the associated signaling and prevention of unnecessary HOs. Suboptimal selection of the mobility parameters leads to increased frequency of HOs, leading to greater probability of loss of signaling messages and hence HO failures.

In order to evaluate the mobility performance, several metrics are defined [11] including: HO failure (HOF) rate, number of HOFs/UE/s, time of stay (ToS), SToS rate, and ping-pong rate [11], which allows us to capture not just HO failures but also frequent mobility events.

1.4.1 Handover performance modeling

1.4.1.1 Handover states

The modeling approach replicates the actual HO procedures that are well defined in the LTE specifications [1, 10]. In this model, the focus is on downlink simulations, and uplink messages that are required to support HO are assumed to be delivered reliably. This choice is driven by observations in macro LTE networks that failures in scheduling of measurement reports or sending HO commands to mobiles are a dominant contributor to overall HO failure performance. For link quality measurement, instantaneous user RSRP is first measured periodically in Layer 1 at the UE, and averaged over 200 ms. This average is passed to a Layer 3 RSRP filter, which is an ARMA filter with a forgetting factor parameterized by "k" in Section 1.3.2. Reference signal received power measurement error modeling is adopted as in [11].

Two independent processes govern mobility in the LTE network: the HO process and the RLM process [1, 10]. The Event A3 based HO process is divided into three states, designated as states 1, 2, and 3 [11]: covering the interval prior to the Event A3 trigger, the subsequent interval up until the HO command is received, and finally the remaining interval till the receipt of HO complete message. The HO failure is modeled in states 2 and 3 as follows:

1. In state 2 when the UE is attached to the source cell, a HOF is counted if one of the following criteria is met:
   (a) Timer T310 has been triggered or is running when the HO command is received by the UE (indicating PDCCH failure)
   (b) RLF is declared in state 2.
2. In state 3, target cell downlink filtered average (the filtering/averaging here is the same as that used for starting T310) wideband channel quality indicator (CQI) is less than the threshold, Q_{out} at the end of the handover execution time (indicating RACH failure with target cell).

The two possibilities for the HO failure in state 2 are shown in Figures 1.6 and 1.7 [11].
The RLM is based on periodic wideband CQI measurement and is initiated by a test against "Q_{out}" and terminated by a test against "Q_{in}" as described in Section 1.2.

1.4.1.2 Handover frequency and ping-pong modeling

The handover frequency and ping-pong behavior is another important aspect of the mobility performance. A UE's time of stay (ToS) with a cell is used as a basic metric for HO frequency evaluation. The ToS in cell A is the duration from when the UE successfully sends a "handover complete" message to cell A, to when the UE successfully sends a "handover complete" message to another cell B. A UE-stay with a cell where the condition ToS < minimum time-of-stay (MTS) is met regardless of the cell from which UE was handed out or the cell to which it was handed in, is considered a short

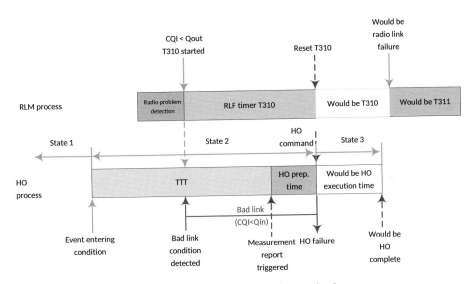

Figure 1.6 Timer T310 is running when the HO command is received.

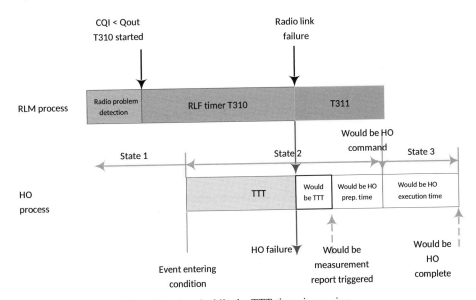

Figure 1.7 Radio link failure is declared while the TTT timer is running.

time of stay (SToS). The generic SToS metrics including the SToS rate and number of SToSs/UE/second are defined in [11].

A handover from cell B to cell A then handover back to cell B is considered as a ping-pong if the ToS connected in cell A is less than a pre-determined MTS [11].

Figure 1.8 demonstrates examples of ping-pong and SToS [11].

1.4.2 HetNet mobility performance evaluation metrics

There are several important HO performance metrics defined in [11].

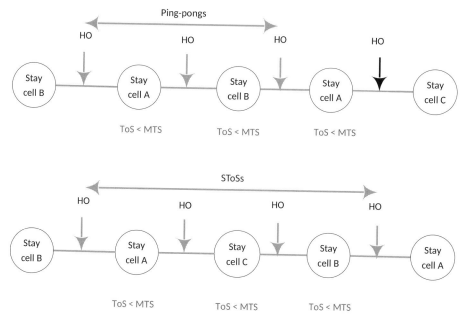

Figure 1.8 Examples of ping-pong and short time of stay.

Handover failure (HOF) rate

Handover failure rate = (number of handover failures)/(total number of handover attempts), where the total number of HO attempts is the sum of number of HO failures and number of successful HOs.

Number of HOFs/UE/second

Number of HOFs/UE/second = (the total number of handover failures)/(total number of the UEs observed)/(the total observation time).

Another metric is *number of successful HOs per UE per second*, which can be calculated similarly.

For handover frequency evaluation, a SToS is counted when a UE's ToS in a cell is less than a predetermined MTS parameter, i.e., a UE with ToS < MTS.

Short time of stay (SToS) rate

SToS rate = (number of SToS occurrences)/(total number of successful handovers).

Number of SToSs/UE/second

Number of SToSs/UE/second = (total number of short ToS occurrences)/(total number of the UEs observed)/(the total observation time).

Ping-pong rate

Ping-pong rate = (number of ping-pongs)/(total number of successful handovers).

As is shown in Subsection 1.4.1.1, in HetNets most of the RLFs are closely associated with HOFs. The metrics for RLF performance evaluation are also specified in [11].

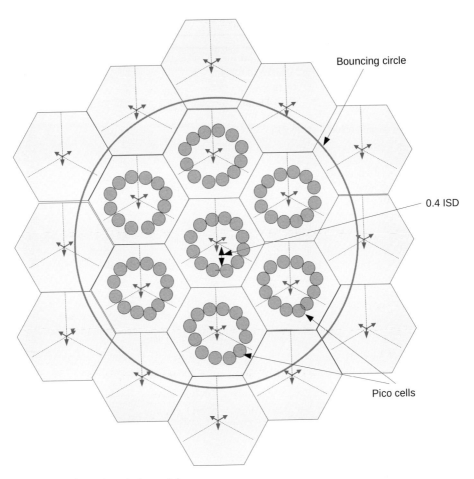

Figure 1.9 A bouncing-circle model.

Number of RLFs/UE/second

Number of RLFs per UE per second = (total number of RLF occurrences)/(total number of the UEs observed)/(the total observation time).

In order to observe the HetNet mobility behavior, HO performance results are logged separately for macro to macro, macro to pico, pico to macro, and pico to pico HOs.

1.4.3 Simulation layout and parameters

Figure 1.9 depicts the simulation layout considered for performance evaluation in this chapter. A 19-cell macro cellular site with 12 picos placed per eNB at equal distance on a circle of radius 0.4 ISD (inter-site-distance) is assumed. A bouncing circle radius of 1.7 ISD is imposed and 30 mobiles are randomly dropped within the bouncing circle. These mobiles then travel in straight lines at constant speeds in random directions. On encountering the bouncing circle, they bounce back in random directions. This motion continues for the duration of the entire simulation. The choice of mobility

Table 1.2 Assumptions and parameter values used in the simulations.

Parameter	Macro	Pico
Carrier frequency/bandwidth	2.0 GHz/10 MHz	2.0 GHz/10 MHz
ISD	500 m	
Macro distance-dependent path loss	3GPP TR 36.814 macro-cell model 1	3GPP TR 36.814 pico-cell model 1
BS antenna gain (incl. cable loss)	15 dBi	5 dBi
MS antenna gain	0 dBi	0 dBi
Shadowing standard deviation	8 dB	10 dB
Shadowing correlation distance	50 m	50 m
Shadow correlation	0.5 between eNBs/1 between cells	0.5
Antenna pattern	3D (TR 36.814, Table A.2.1.1–2)	Omni (TR 36.814, Table A.2.1.1.2–3)
BS total TX power	46 dBm	30 dBm
Penetration loss	20 dB	20 dB
Antenna configuration	2×2	2×2
Cell loading	100%	
UE speed	30 km/h, 60 km/h, 120 km/h	
Channel model	ETU	
Time to trigger	160 ms	
HO threshold (A3-offset)	2 dB	
$T_{\text{Measurement Period,Intra}}$	200 ms (TS36.133)	
Layer 3 filter parameter k	2	
Layer 1 filtering (sampling)	10 ms over 200 ms window	
RSRP measurement error modeling	$N(0, 1.216\text{ dB})$ (TR 36.839)	
Handover preparation delay	50 ms	
Handover execution time	40 ms	
$Q_{\text{out}}/Q_{\text{in}}$	−8 dB/−6 dB	
T310/N310	1 s/1	
$Q_{\text{out}}/Q_{\text{in}}$ testing period	200 ms/100 ms	
Minimum time of stay (MTS)	1 s	

model is intended to force UE traversal across the entire coverage area and have a spatially representative performance characterization. Key simulation parameters and assumptions are listed in Table 1.2. Unless otherwise noted, simulation assumptions and parameter values follow those given in 3GPP TR 36.839 [11] and general LTE simulation parameters and assumptions are given in [9].

1.5 Mobility performance results

1.5.1 HO and SToS vs. speed

In the performance study, all mobiles traveled at the same constant speed; speeds of 30, 60, and 120 km/h were simulated. Figure 1.10 depicts the overall HOF rate, and the SToS

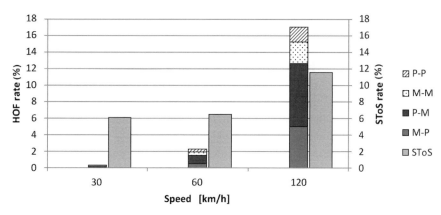

Figure 1.10 Overall HOF rate and SToS rate performance comparison for different speeds.

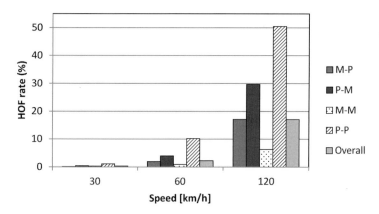

Figure 1.11 Handover failure rate performance comparison for different scenarios and speeds.

rate, for different mobile speeds. As expected, the HOF rate increases with speed. The pico–macro HO failure is the largest contributor of the overall HO failure rates. Since high-speed mobiles stay in the picos for a short period of time, the SToS rate increases with speed.

The contributions of pico–pico (P–P), macro–macro (M–M), pico–macro (P–M), and macro–pico (M–P) toward the overall HOF rate are shown in Figure 1.10.

The individual HOF rates for different HO scenarios are presented in Figure 1.11. The results indicate that the pico–pico HO has a higher failure rate than other HO scenarios; failures increase with higher mobile speeds. However, Figure 1.10 shows that the contribution of pico–pico HO failure is very small compared with macro–macro, macro–pico, and pico–macro HO failures but this is because the number of pico–pico handovers is small in the simulation model relative to other types of HOs. Thus the effect of pico–pico HOs on the overall HOF rate is minimal. The pico–macro HOF rate is always higher than that of macro–pico, and the gap widens with the speed. This indicates that the handing-out of a mobile to a macro from a pico is more susceptible to interference than handing-in. These two scenarios need to be optimized differently to lower the HOF rate.

1.5.1.1 Handover performance optimization for high-speed UEs

Handover performance of high-speed UEs deteriorates significantly due to the smaller coverage area and the lower transmit power of the picos. Thus it can be concluded that high-speed UEs are under high risk of HO problems and service interruption in HetNets. In addition, frequent handover of high-speed UEs to small cells will incur a large signaling overhead between eNBs, as well as between the UE and eNB. High-speed mobiles typically stay for very short period of time in the picos. In this situation, pico cells do not efficiently carry traffic load from high-speed UEs, while suffering from more signaling overhead.

There are several solutions to address HO performance degradation in high-speed mobiles. Adaptation of HO parameters based on speed, keeping the high-speed UEs to stay at the macro cell layer, employing eICIC for mobility management, and scenario dependent extension of the T310 timer (expiry of which causes RLF) are some of the possible solutions discussed in the subsequent sections. User equipment speed-based solutions depend on the speed estimation with reasonably good accuracy and reliability. There are many UE-speed estimation methods under study. In general, several techniques may be used for UE-speed estimation including HO counting, UE location estimation, neighboring eNB location collecting, and Doppler estimation methods. They can be conducted at the network or at the UE. In addition to the UE-based mobility state estimation (MSE), 3GPP also included (in Release 12) the feature of UE reporting the history of its visited cells including the global cell ID and ToS with the cell. This will assist a better MSE conducted at the network side.

1.5.2 Adaptation of HO parameters based on mobile speed

For the adaptation of HO parameters, mobile speed is based on TTT, HO offset, and the "k" value of the Layer 3 filter. Some heuristics that we have developed are:

- Since high-speed mobiles spend less time in the HO region, smaller TTT and HO offset should help to improve HO performance. The coherence time is smaller for high-speed mobiles. Thus averaging over a shorter period should remove the effect of fast fading. A smaller "k" value should be better for fast-moving mobiles.
- In addition, for low-speed mobiles, ping-pong is a major concern, and relatively high (to serving cell measurement) HO offset should be applied to reduce ping-pong, at the expense of slightly increased HO failure rate.
- For medium-speed mobiles, the HO failure rate is of more concern and less ping-pong is observed, and relatively low (to serving cell measurement) HO offset should be used.

Note: here the HO offset is in terms of the target cell RSRP (or RSRQ) measurement minus the serving cell measurement.

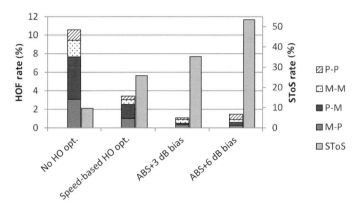

Figure 1.12 Mobility performance comparisons for different HO optimization techniques for mixed speed.

Large-area simulations were conducted to optimize the mobility parameters based on the UE speed. As discussed in the assumptions in Section 1.2, 30 mobiles were placed uniformly, and moved randomly within the bouncing circle. Each mobile randomly selected a speed from 30 km/h, 60 km/h, and 120 km/h. For mobile speeds greater than 60 km/h, TTT, HO offset, and "k" value were scaled by a factor 0.5; for mobile speeds less than 60 km/h, those parameters were scaled by a factor 1.5. A case with no parameter optimization was also simulated for performance comparison. Figure 1.12 shows the overall HOF rate and SToS rate performances with and without HO parameter optimization. Using speed-dependent parameter optimization, the overall HOF rate was reduced by 7% for a slight increase in SToS rate. Figure 1.6 presents the HOF rate of the different HO scenarios. It is evident that the speed-dependent parameter optimization improved the HOF rate of all HO scenarios.

1.5.3 Leveraging enhanced inter-cell interference coordination (eICIC) for mobility management

3GPP has standardized ABS to protect UEs in the cell extended range of the picos from macro interference. From an interference-mitigation perspective therefore, the use of ABS is coupled with cell-range extension. However, as we see below, there is value in treating ABS as an available tool that can potentially be used to improve mobility performance. By decoupling its use from cell-range extension, or use only at the macro, HO performance can be improved by blanking at least one subframe each at macro and pico(s) to create an interference-free subframe at either cell to improve pico–macro and macro–pico HO respectively. The pico and macro use their knowledge of these interference-free subframes to schedule their respective HO related messages with very high reliability to the user. There is a penalty paid in terms of spectral efficiency due to this blanking and so the benefits to mobility performance have to be weighed against the

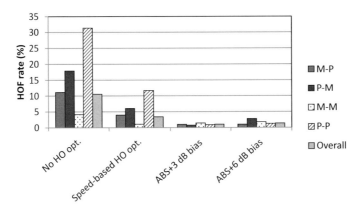

Figure 1.13 Handover failure rate comparison for different scenarios and different HO optimization techniques for mixed speed.

loss to sector throughput (although the fraction of such blanked subframes will be very small and on the order of a couple of subframes every 200 ms). Moving onto the cell-range extension case, we note first that mobility performance worsens dramatically if the pico cells' range is extended while maintaining full re-use of air–interface resources at the macro. This is expected, since PDCCH SINR, and therefore reception, degrades at the range-extended cell boundary. It is unlikely that even speed-based optimizations can remedy this problem.

In the simulations here, we have modeled ABS in the macros only, and the cell IDs of the umbrella macro and picos are selected in such a way to avoid the collision of common reference signals (CRS) of the umbrella macro with pico transmission in the ABS. The same mix of mobiles at different speeds was considered as in Section 1.5.1. No CRS interference cancellation at the mobiles is considered in this study.

Figure 1.13 shows a significant HOF rate reduction of around 9.5%, while increasing SToS rate by 25% for the 3 dB cell selection bias with ABS. The HOF rate reduction of 9% was observed for a 6 dB bias. The HOF performance gain in 6 dB bias softens due to greater CRS interference from the surrounding macros in the ABS. As expected, the SToS rate also increases with the bias values. In general, a higher SToS rate is better tolerated than a higher HOF rate in a network.

1.5.4 Per scenario based HO parameter optimization

As shown in [11], the HO behavior in different scenarios such as macro–macro, macro–pico, pico–macro, and pico–pico can be very different. Therefore if the HO parameters can be set at a per cell basis, they may be optimized for different HO scenarios to achieve a tradeoff between the HOF rate and SToS. In addition to per-cell based HO offset parameter, 3GPP allows in Release 12 the target cell specific TTT. Time-to-trigger can also be optimized based on the target cell type now.

1.6 RLM and RLF recovery enhancement in HetNets

Details of the existing RLM and RLF recovery process are described in Section 1.1. In HetNets the large number of small cells leads to much more frequent HO activities. Co-channel small cells can be coverage holes of the overlaid macro cell if a UE (especially with high speed) is not able to HO to the small cell properly. It has been observed that the HO performance of medium- to high-speed UEs deteriorates significantly due to the small size of the "low-transmission power node"; most of those HO failures eventually lead to RLF. Due to the cost of the RLF (even if it may be recovered, it will still cause service interruption over a long time and an additional signaling overhead), it is desired to further enhance the RLM and RLF recovery process to reduce the chance of RLF occurrence, and if RLF occurs to reduce the overall service interruption time.

1.6.1 T310 adaptation

The decision of RLF in the UEs is highly dependent on the setting of T310. If T310 is set too short, the RLF could be often triggered in a UE, even by insignificant measurement fluctuation. It may cause the UE to unnecessarily get into the high-cost RLF recovery process and miss the chance of radio link self-recovery. On the other hand, if T310 is set too long, in the situation of un-recoverable radio link in the serving cell such as when the UE is experiencing an HO failure, the necessary RLF recovery process can be delayed and overall service interruption time will be increased. In a conventional macro-only system, the setting of T310 is really for avoiding RLF recovery procedure during the real radio coverage hole. The value of T310 is optimized for this purpose. However, in a HetNet system the situation becomes very different. The simulations indicated that most RLFs occuring while the UE is in a pico in HetNets are not because of radio coverage holes but due to HO failure. The T310 setting for the legacy macro-only system may not be suitable for the different scenarios in HetNets.

1.6.1.1 The serving cell of the UE is the pico cell

This includes the mobility scenarios of pico–macro and pico–pico HOs. In this case, the radio link quality of the high-speed UEs with the serving pico cell deteriorates rapidly due to the small size of the pico cell. Since the link with the pico cell is likely to deteriorate in a short period of time, and the radio link with the pico is unlikely to be recovered with further movement of the UE, the T310 in a pico can be set to a shorter value than the normal value setting for the macro-only system. Using this setting, it can avoid any unnecessary delay in conducting RLF recovery by triggering the re-establishment procedure early, and thereby shortening the overall service interrupt time.

1.6.1.2 The serving cell of the UE is the macro cell

For this scenario, there are two cases: (1) high-speed UEs are forced to stay with the macro layer in HetNets; or (2) they can hand over to the pico cell. We will discuss these separately below.

> **Case 1.1** If the UE is high speed and HO from macro cell to pico cell is not allowed
>
> In the real system deployment, it is possible that ABS at pico cell is not available. In this situation, if a high-speed macro-cell-connected UE (MUE) is forced to stay at the macro layer when it moves in the pico cell coverage, there will be strong interference to the MUE from the pico cell. The RLF and the re-establishment process are likely to be triggered. Due to short time of stay with the pico coverage, the re-establishment may also likely fail and call drop will occur.
>
> In fact if the RLF is not triggered, with the further moving of the UE its radio link with the macro cell will be recovered. Therefore for this case if the T310 can be extended to a longer value it will benefit to keep the MUE connected with the macro cell. If we extend T310 when the pico cell is discovered, it will not incur RLF and will still keep connected with the macro cell even if the link condition is bad. This will allow enough time before the extended T310 expired. At this point the link with the macro cell has resumed.

> **Case 1.2** If the UE is high speed and HO from macro cell to pico cell is allowed
>
> In this case the UE often just moves across the pico tangentially and a high macro–pico HO failure rate is observed. Again due to the short time of stay of a high-speed UE with pico coverage, RLF and call drop are likely occur. On the other hand, since the UE speed is high, the UE will pass the pico coverage in a short period of time. It will also be beneficial to use a longer T310 value. With a long T310 setting, if the UE trys to hand in to pico but fails, it will still keep connected with the macro cell, even if the radio link with the macro is very bad. If the UE could pass the pico cell before the T310 expires, the radio link with the macro cell will be resumed again. Then the T310 should be set back to the normal default value with the macro-only case.

1.6.1.3 Possible enhancement considerations

Based on the analysis in previous sections, a straightforward solution is that the network configures different T310 settings to the UE based on the mobility scenarios that the UE is currently in. However, in HetNets, a serving cell may have multiple neighboring cells of different types (macro or pico). In this case, a cell type sensitive update of parameters may need to wait till the type of the target cell is determined. Too late parameter update may impact the mobility performance. In addition, more HO activities and, as a result, more signaling overhead for T310 update is expected in HetNets. In order to minimize the signaling overhead and minimize the delay for T310 update, an alternative approach has been adopted by 3GPP as follows:

a. Allow a UE maintaining two timers: T310 and T310b. The usage of T310 is the same as for a macro-only system, a value for the normal RLF case is set to T310, and a short time value for the RLF cases caused by HOFs is set to the timer T310b.

b. T310b is started upon TTT expiry if T310 is already running. RLF is declared when T310b expires.

The RLFs caused by HOFs more likely cannot be self-recovered over longer time and most HOFs involve small cells in HetNets. By doing this, an unnecessary long T310 waiting time can be avoided in most cases. The overall service interruption time can be reduced.

1.6.2 Context fetch

There are two conditions that should be met to recover UE's connection by the RRC connection re-establishment procedure: one is that the UE can select a suitable target cell based on some predefined criteria, the other is that the selected target cell has the UE context. If one of these two conditions cannot be satisfied, the UE will lose its connection and return to idle mode.

When an RLF is caused by HOF, the failures may have occurred at different stages of the HO process. If a failure occurs at the early stage of the HO process, e.g., in the measurement reporting phase, the source cell may not have enough information for preparing the target cell(s). In this case, the UE context may be unavailable in the target cell selected by the UE based on some cell selection criteria (e.g., s-Measure criteria), which will lead to re-establishment failure and the UE will return to idle mode accordingly.

One of the known solutions is to pre-deliver the UE's context to all the candidate neighboring cells that are possible handover targets. This approach may waste a lot of network resource and there is still a chance of missing context at the target cell.

An alternative approach can be considered if a UE performs handover to a target cell and the accessing UE's context is not available there: let the target eNB get it from the source eNB via the X2 interface or from the MME via S1 interface. This procedure is known as *context fetch*. Supported by X2 between the eNBs now, the delay due to context fetch is small enough to be practical. With respect to the RRC connection re-establishment in HetNet scenarios, the context fetch approach is one of the possible solutions for RLF recovery enhancement.

1.7 Discovery of small cells

The deployment of pico cells is generally for two different reasons: capacity enhancement and coverage extension. Therefore there are two typical scenarios that should be treated differently for inter-frequency pico cell discovery.

Scenario 1: Capacity enhancement. Most of the pico cell deployments are for capacity enhancement by offloading UEs and consequently their traffic from the macro cellular network. Therefore this section will focus more on the discussion of inter-frequency small cell discovery in the scenario where at least one macro frequency layer provides full coverage and overlaid pico cells are provided on a different frequency layer. For this scenario, the UE power consumption would be the major concern. The impact of pico discovery delay would be secondary, mainly in terms of how much offloading opportunity and QoS benefit will be lost due to delayed detection of the small cell.

Scenario 2: To cover the macro cell coverage holes. In order to enhance the coverage of the umbrella macro cell, pico cells can be deployed at the edge of the macro cell to fill the coverage hole(s). In this case, if the pico cells cannot be discovered early enough, RLF and HOF will be increased significantly. Therefore when the pico cells are deployed for coverage purpose, the requirement for pico cell discovery will be to ensure the robustness of UE mobility, i.e., to minimize the RLF and HOF rates.

1.7.1 Capability of existing mechanism

1.7.1.1 UEs in the RRC_CONNECTED mode

When a UE is in the RRC_CONNECTED mode, the existing measurement configuration mechanism could be used by the network to request when to search for the pico cell. The knowledge of the UE location at the eNB may not be accurate. S_measure and DRX could be used to balance measurements and power with faster pico discovery in certain scenarios.

Unnecessary measurements and measurement reporting can be avoided if the network could request the UE to search the pico cell only when it is necessary (for example when the UE is nearby a pico cell and the cell is not overloaded). Therefore the key is based on different scenarios, a UE is able to perform the pico search or measurement differently based on the network instruction or its own knowledge.

1.7.1.2 Discovery of CSG cell versus pico cell

In HetNets, tens of hundreds of public pico cells could be deployed with the overlaid macro cells. The pico cells could share the same carrier with the macro cell or use different carriers. The UE power consumption will be a big concern if the UEs have to exhaustively search for the pico cells when the pico cells' carrier frequency is different from that of the macro cells. Therefore it is desirable to develop a much more efficient method for the UE to discover the pico cells in HetNets.

The known CSG cell discovery scheme includes the autonomous search function (ASF) and proximity indication mechanism, with the primary goal of reducing UE power consumption. ASF depends on UE implementation to decide when and where to search the CSG cells in the UE's white list. A UE will store a white list of its associated CSGs. The radio and location information associated with the CSGs in the white list are also stored in the UE. Together with other UE dynamically measured/obtained information, the pre-stored information can be used by the UE to determine whether it is nearby a CSG in its white list and then triggering the ASF or sending proximity indication to the serving macro cell.

The CSG cell discovery methods are based on the fact that a UE is only associated with a limited number of CSG cells. Therefore the UE is able to memorize the information of all its associated CSG cells, such as location information and radio fingerprints. For public pico cells, however, it is not possible for a UE to store the information of all the pico cells. Therefore we may not be able to simply reuse the CSG cell discovery methods for the discovery of the public pico cells.

On the other hand, pico cells are normally deployed only at hotspots. Not all the macro cells are overlaid with the pico cells. Even if there are pico cells in a macro cell, most of the macro cell coverage area may not be also covered by pico cells. It is not necessary for a UE to search the pico cells when it is in a macro cell without any pico cell, or in a pico overlaid macro cell but far away from the pico cells. It is possible to find a more efficient method for pico cell discovery other than the exhaustive search to minimize the UE power consumption.

1.7.2 Factors impacting inter-frequency pico cell discovery

In the process of pico cell discovery, three elements are involved: the umbrella macro cell, the target pico cell(s), and the UE. With the three elements, there are several potential factors that could be involved for improving performance.

- From the perspective of the umbrella macro cell, the load of the macro cell is an important factor to be considered in pico cell discovery – especially for the scenario of traffic offloading.
- From the perspective of target pico cells to be discovered, the following factors might be considered for improving the performance:
  - the location of the pico cells
  - the distribution/density of the pico cells
  - the types of the pico cells (i.e., for capacity boosting or for coverage enhancement)
  - the load of the target pico cells, especially for the scenario of traffic offloading
  - GoS/QoS requirement
- From the perspective of triggering the UE to perform pico cell discovery, the following factors might be considered for improving the performance:
  - the position of the UE
  - the speed of the UE
  - the trajectory of the UE
  - the traffic of the UE
  - the DRX configuration of the UE
  - the measurement criteria of the UE

Considering the above-listed factors could improve the performance of UE power consumption and the efficiency of inter-frequency pico cell discovery. The impacts of different factors are different for different scenarios. Therefore the factors should be counted in differently when we investigate inter-frequency pico cell discovery solutions for different scenarios. In the following sections, we show several examples considering the impact of some important factors for pico cell discovery in different scenarios.

1.7.3 Network controlled pico discovery for different scenarios

According to the above analyses, additional guidance/control from the network is desirable for inter-frequency pico cell discovery. The example of network-assisted solutions for inter-frequency pico cell discovery is based on the assumption that pico cells are generally deployed in a planned fashion and the information regarding their location and density is available to the macro eNB. The macro eNB is also able to obtain a reasonable estimation of UE location. The impacts of involving some potential factors are analyzed for different scenarios in the following sub-sections.

1.7.3.1 Scenarios of offloading macro UEs to pico cells

For this scenario, the UE inter-frequency measurement for pico cell discovery is generally encouraged as long as the load of the macro cell is higher than a predefined level. The details of the major factors to be considered for triggering the pico discovery process are discussed as follows.

A. **Location of each UE**. Since the location of each pico cell is available to the macro eNB, if the location of each UE is also available to the macro eNB, the macro eNB could judge whether a UE is approaching a pico cell in advance and decide whether to trigger the UE to perform pico cell discovery in time.

B. **Load information of the target pico cell**. Since the objective of this scenario is to offload macro UEs to the target pico cells, the load of the target pico cells should be considered first. If the load of the target pico cell is too high to accept new traffic, it is unnecessary to be discovered by the macro UEs. For load balance purposes, the macro eNB can use its knowledge of the load of the pico cells proactively. The macro eNB will decide whether to instruct the UEs to start the inter-frequency pico cell discovery and how often to perform the search based on its load knowledge of the involved picos. The instruction could be delivered to the UE via broadcast or unicast signaling.

C. **Density of the target pico cells**. The impact of UE power consumption depends on how often and for how long the UE performs inter-frequency measurements. If the network only has a very rough idea of a UE's location, the density of the target cells to be discovered could be considered as a factor to determine the inter-frequency configuration. More specifically, if the macro eNB knows that the UE is approaching an area with sparse pico cell deployment, it could configure the inter-frequency measurement gap sparsely with more relaxed inter-frequency measurement performance requirement. If the macro eNB knows that the UE is approaching a hotspot with high pico cell density, it could configure the inter-frequency measurement more frequently.

D. **Mobility state of each UE**. Considering the fact that even if a suitable inter-frequency pico cell is discovered, the high-speed UE is likely to pass through its coverage so quickly that a handover to the pico cell is more likely to fail or at least not be beneficial for the service. In addition, for high-speed UEs frequent inter-frequency measurements of pico cells may cause unnecessary power consumption. Therefore it would make sense to avoid unnecessary inter-frequency measurements of pico cells for very high speed UEs.

For the medium- or low-speed UEs, their handover to the pico cells should be supported. In order to provide a balance between UE power consumption and efficiency of pico cell discovery to meet the handover performance requirement, the measurement gap pattern for them could be configured based on the UE speed.

1.7.3.2 Scenario of macro cell coverage extension

For this scenario, the pico cells are often deployed at the edge of the umbrella macro cell for coverage enhancement. Thus pico cell discovery is generally triggered by the serving cell measurement degradation or proximity information when a UE is moving into the proximity of a pico cell or a hotspot with multiple picos. In this case, more responsive inter-frequency measurements are necessary since infrequent inter-frequency measurements by UEs with higher speed may cause RLF. If the UE location, speed information and the target pico cell location are available to the macro eNB, the macro eNB will be able to configure the inter-frequency measurement gap accordingly for the high-speed UE to perform pico cell discovery quickly.

In this scenario, even high-speed UEs will have to switch to the pico cells. A speed-dependent measurement pattern is desirable.

1.7.4 Pico proximity determination by UEs with network assistance

Due to the limitation of network knowledge about individual UEs (e.g., UE location and speed), it will be beneficial if a UE is able to determine whether it is in the proximity of a pico cell. In this section we discuss an example where the UE determines the proximity of the pico cells with the assistance of the network.

In this example, a "small cell alert zone" is introduced as shown in Figure 1.14. It is larger than the normal coverage area of pico cell(s). There could be multiple alert zones under an umbrella macro cell. The umbrella macro cell broadcasts or unicasts the latitude and longitude range of the "alert zone" of the pico cells to the UEs. When the UE knows it is moving into a small cell alert zone, it will start to search the small cell in the neighborhood. After the UE entering the alert zone, based on the measurement, the UE will send a small cell proximity to the eNB of the umbrella macro cell. This solution is good for both connected and idle mode UEs. This solution requires good UE location estimation method at the UEs. If a UE doesn't have the capability to estimate its location with low power consumption, the sub-optimum solution would be that the search of the small cells is started only after a UE enters the macro cell that is overlaid with the small cells. The indication of the presence of the pico cells overlaid with the macro cell can be delivered to the UE via broadcast or unicast signaling.

1.7.5 Pico proximity determination by UEs with history information

Another possible UE-based pico discovery and proximity determination is to take a similar approach for CSG cell discovery. When a UE approaches Pico cell A, the UE determines the proximity of the pico based on its prior knowledge, such as the fingerprint of this pico cell. Similar to the white list in the UEs for the CSG cell discovery, for those frequently visited pico cells such as home, office, and coffee shops a *"frequently visited pico list"* may be maintained in a UE for storing their cell IDs and associated fingerprint information. When the UE approaches one of the frequently visited pico cells, based on the stored information it will decide when and where to search the pico cells in its list autonomously. The UE will send proximity indication to the network based on

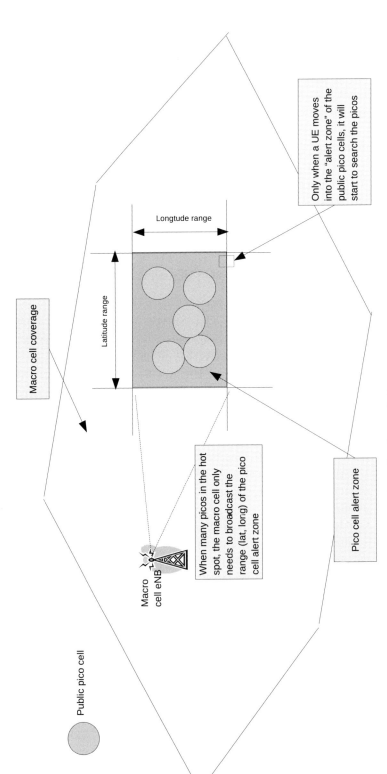

Figure 1.14 Pico cell alert zone instruction for assisting UE to determine the proximity.

the measurement of the pico cell. The network will pre-prepare and get the UE pre-prepared for handover to the pico cell. In this way, those frequently visited pico cells can be discovered much more efficiently with lower signaling overhead and UE power consumption.

1.8 Summary

The connected mode user mobility in 3GPP LTE HetNets with and without range extension has been considered. Due to low-power co-channel small cell deployment in HetNets the connected mode user mobility becomes one of the most critical aspects. The legacy 3GPP LTE handover and the radio link monitoring processes were described. The system-level handover performance modeling and a few mobility performance enhancement techniques were presented. Speed-based handover parameter optimization, and ABS with cell range extension, were considered and the performance gains were compared. We showed that the handover failure rate can be significantly reduced using mobile speed dependent handover parameter adaptation and cell range extension with subframe blanking, although at the cost of an increase in the short time-of-stay rate. It was shown that the inter-cell interference coordination techniques can also be leveraged to improve mobility performance. Finally, other aspects such as radio-link failure recovery, small cell discovery, and related enhancements were also discussed.

References

[1] TS36.331 v.8.8.0 "3GPP Technical Specification Group Radio Access Network; E-UTRARadio Resource Control (RRC); Protocol specification".
[2] D. Lopez-Perez, I. Guvenc, and X. Chu, "Mobility management challenges in 3GPP heterogeneous networks", *IEEE Commun. Mag.*, Dec. 2012.
[3] D. Aziz, R. Sigle, H. Bakker, H. Grob-Lipski, and S. Kaminski, "Design criteria for optimum handover parameters in LTE", *CCC Conference* 2009.
[4] M. Anas, F. D. Calabrese, P. E. Mogensen, C. Rosa, and K. I. Pedersen, "Performance evaluation of received signal strength based hard handover for UTRAN LTE", *IEEE Vehicular Technology Conference (VTC)*, Spring 2007.
[5] M. Anas, F. D. Calabrese, P. Ostling, K. I. Pedersen, and P. E. Mogensen, "Performance analysis of handover measurements and Layer 3 filtering for UTRAN LTE", *IEEE Symposium on Personal, Indoor and Mobile Radio Communications (PIMRC)*, 2007.
[6] K. Dimou, M. Wang, Y. Yang, M. Kazmi, *et al.*, "Handover within 3GPP LTE: design principles and performance", *IEEE Vehicular Technology Conference (VTC)*, Fall 2009.
[7] D. Aziz and R. Sigle, "Improvement of LTE handover performance through interference coordination", *IEEE Vehicular Technology Conference (VTC)*, Spring 2009.
[8] G. P. Pollini, "Trends in handover design", *IEEE Commun. Magazine*, pp. 82–90, March 1996.
[9] TS36.814 v9.0.0 "3GPP Technical Specification Group Radio Access Network; further advancements for E-UTRA physical layer aspects".

[10] TS36.133 v.9.3.0 "3GPP Technical Specification Group Radio Access Network; E-UTRA requirements for support of radio resource management".
[11] TR36.839 v 11.1.0 "3GPP LTE mobility enhancements in heterogeneous networks", Rapporteur Alcatel-Lucent.
[12] R. N. Pupala, S. Vasudevan, K. Sivanesan, G. Sundaram and A. Rudrapatna "Improving cell-edge user performance with multi streaming", *IEEE International Conference on Communications 2011*. 2011.

2 Design and performance analysis of multi-radio small cell networks

Nageen Himayat, Shu-ping Yeh, Shilpa Talwar, Mikhail Gerasimenko, Sergey Andreev, and Yevgeni Koucheryavy

Multi-tier, heterogeneous networks (HetNets) using small cells (e.g., pico and femto cells) are an important part of operators' strategy to add low-cost network capacity through aggressive reuse of the cellular spectrum. In the near term, a number of operators have also relied on un-licensed WiFi networks as a readily available means to offload traffic demand. However, the use of WiFi is expected to remain an integral part of operators' long-term strategy to address future capacity needs, as licensed spectrum continues to be scarce and expensive. Efficient integration of cellular HetNets with alternate radio access technologies (RATs), such as WiFi, is therefore essential for next-generation networks.

This chapter describes several WiFi-based multi-RAT HetNet deployments and architectures, and evaluates the associated performance benefits. In particular, we consider deployments featuring integrated multi-RAT small cells with co-located WiFi and LTE interfaces, where tighter coordination across the two radio links becomes feasible. Integrated multi-RAT small cells are an emerging industry trend toward leveraging common infrastructure and lowering deployment costs when the footprints of WiFi and cellular networks overlap. Several techniques for cross-RAT coordination and radio resource management are reviewed and system performance results showing significant capacity and quality service gains are presented.

2.1 Introduction

Multi-tier HetNets based on small cells (e.g. pico cells, femto cells, relay cells, WiFi APs, etc.) are considered to be a fundamental technology for cellular operators to address capacity and coverage demands of future 5G networks. Typical HetNet deployment architectures comprise an overlay of a macro cell network with additional tiers of densely deployed cells with smaller footprints, such as picos, femtos, relay nodes, WiFi access points, etc. Figure 2.1 illustrates the various deployment options in a multi-radio HetNet.

HetNets allow for greater flexibility in adapting the network infrastructure according to the capacity, coverage, and cost needs of a given deployment. As shown, the macro base station tier may be used for providing wide area coverage and seamless mobility, across large geographic areas, while smaller inexpensive low-powered small cells may be deployed, as needed, to improve coverage by moving infrastructure closer to the

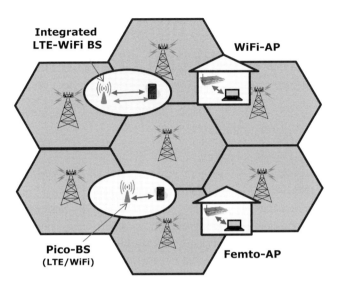

Figure 2.1 Multi-radio HetNet.

clients (such as for indoor deployments), as well as to add capacity in areas with higher traffic demand. Conceptually, mobile clients with direct client-to-client communication may also be considered as one of the tiers within this hierarchical deployment, wherein the clients can cooperate with other clients to *locally* improve access in an inexpensive manner. Several papers have characterized the significant performance gains achieved with HetNets [1–4]. In particular, references [2, 3] characterize the linear growth in network capacity with increasing infrastructure even with aggressive reuse of the same spectrum across the multiple tiers in the network. A comprehensive overview of multi-tier, multi-radio HetNets is provided in [4].

While the main focus of HetNets has been on cellular deployments operating on licensed bands, operators have also aggressively focused on harnessing low-cost spectrum in unlicensed bands through WiFi offload mechanisms [5, 6]. Although the trend toward the use of WiFi in conjunction with cellular networks grew from the near-term need of operators to relieve congestion on cellular networks, the use of WiFi is expected to remain an integral part of operators' long-term strategy to address future capacity needs, as licensed spectrum continues to be scarce and expensive.

In fact, in our view, the capacity and connectivity limitations faced by future networks will continue to drive the need for utilization and integration of different radio access technologies (RATs), encompassing other technologies beyond WiFi and additional use cases beyond simple aggregation of capacity across unlicensed bands. For instance, even when new spectrum allocations are available, they are likely to be fragmented and may require different transmission technologies (air interface schemes) for use, thereby driving the need to operate and integrate multiple RATs in a single network. Shrinking cell sizes in HetNet deployments have also resulted in an increasing overlap between the footprints of cellular wide area networks (WANs), local area networks (LANs), and personal area networks (PANs), creating an opportunity to simultaneously utilize the distinct RATs designed for these hitherto distinct networks. The Internet of Things (IoT) will also require networks to connect a diverse range of devices requiring connectivity at different scales. Inter-working across the multiple RATs that are optimized for each scale of connectivity will therefore be essential. Hence we expect that HetNets supporting different radio access technologies and the associated device/system intelligence for their efficient use will be a fundamental characteristic of future networks (also see [7, 8]).

This chapter therefore focuses exclusively on integration of multiple RATs within HetNet architectures, focusing on multi-radio small cell deployments. In particular, the integration of WiFi-based small cells within operator managed HetNet deployments is used as a case study to illustrate the various architectural options for integration and their associated performance benefits. We note that inter-working between WiFi and cellular networks has been considered in the past, but largely from the point of view of inter-network (vertical) handoff perspective. The cellular standards (3GPP: Third Generation Partnership Project) community has also been engaged in developing specifications that consider inter-working between the cellular as well as WLAN (wireless local area networks) for a number of years [9–11]. Several new study and work items are currently in progress, which now are developing specifications toward increased integration of WiFi with cellular networks [12, 13]. These efforts treat WiFi as an independent network and focus on loose inter-working solutions, which inter-work with WiFi within the packet core network. Specific changes within the core network focus on improving security over connections using WiFi access, and inter-RAT mobility with WiFi networks. However, there has been a recent shift to address tighter inter-working options, which inter-work with WiFi at the radio access network (RAN) layer [13]. This shift is guided by the need to support better QoS on unlicensed spectrum as prescribed by a consortium of network operators who have put together aggressive requirements for carrier grade WiFi. Hence it is timely to investigate RAN-based integration solutions, which assume increased cooperation between 3GPP and WiFi RATs.

Tighter cross-RAT integration and cooperation is also facilitated by recent trends toward the deployment of integrated multi-RAT small cells with co-located WiFi and 3GPP interfaces. Multi-RAT small cells have recently emerged as a popular industry trend, as they help reduce deployment costs by leveraging common infrastructure and site locations across multiple RATs. The co-located radios on the base station allow for tighter coordination across the multiple radio links, when used together with multi-radio client devices. Multi-RAT small cells also allow for a common network deployment across both WiFi and 3GPP RATs, using the same 3GPP core network to serve both RATs. Here, WiFi no longer needs to be treated as a separate radio network, but may be considered as an additional "extension" or "secondary" carrier anchored within the 3GPP radio network; thus creating a single unified "integrated multi-radio access network," encompassing both licensed and unlicensed spectrum. The concept of an integrated multi-radio network allows for increased coupling and coordination across multiple RATs, which can exploit the rich multi-dimensional diversity (e.g., spatial, temporal, frequency, interference, load, etc.) to achieve beyond-additive gains in network capacity and user connectivity experience. The integrated RAN architecture considered here may be extended to include other RATs and generalized HetNet architectures under consideration [14].

This chapter emphasizes integrated multi-radio network architectures that exploit tighter radio network based integration and coordination between WiFi and 3GPP networks, and highlights their associated benefits. The chapter presents several joint radio resource management techniques across WiFi and 3GPP LTE RATs, encompassing RAT assignment, selection, and scheduling algorithms, which provide significant gains in overall system performance and user quality of service (QoS). We address both network-centric and UE-centric approaches, wherein the control of how different radio technologies are utilized rests with the network or the UE respectively. Our focus is on performance metrics such as user and system throughput, assuming best effort traffic, as well as QoS metrics suited for delay-sensitive traffic. Some results presented are based on material covered in [15–18], but several results are presented here for the first time.

This chapter is organized as follows. Section 2.2 describes architectural options for integrating multiple RATs within an operator-managed 3GPP heterogeneous network, emphasizing integrated multi-radio network architectures. In Section 2.3 we describe several approaches for improving the utilization of multi-RAT resources in the network, in particular focusing on schemes that allow for their joint management within the integrated RAN. Section 2.4 covers the analytical framework based on network utility optimization, which facilitates development of schemes allowing for joint radio resource management. Section 2.5 covers the simulation assumptions and framework used for evaluating performance of the proposed schemes. Detailed performance results are covered in Section 2.6. The final summary and a discussion on future work are covered in Section 2.7.

2.2 Integrated multi-RAT HetNet architectures

Figure 2.2 illustrates several different architectural choices for integrating WiFi and LTE networks. These architectures provide different mechanisms to control key operations required for multi-RAT integration: (a) RAT discovery, (b) mechanisms for RAT selection or assignment, (c) control of multi-RAT radio resource management, and (d) protocols for inter-RAT mobility or session transfers, etc. Here we briefly review these options.

Option (A) in Figure 2.2 describes the application layer or higher layer integration architecture, wherein the user device and the content server can communicate directly, over a proprietary or a higher layer interface, and exchange data over multiple RATs without any coordination between the two networks. Here, distinct transport connections over the two RATs may be established between the server and the UE (for instance, different IP connections may be established for each RAT), and the transport addresses of these connections are known between the server and the UE through proprietary or higher layer signaling. Such solutions have been explored in the context of improving the performance of OTT (over the top) applications, relying on higher layer protocol interfaces between the client and the server (e.g. [19] and the references therein).

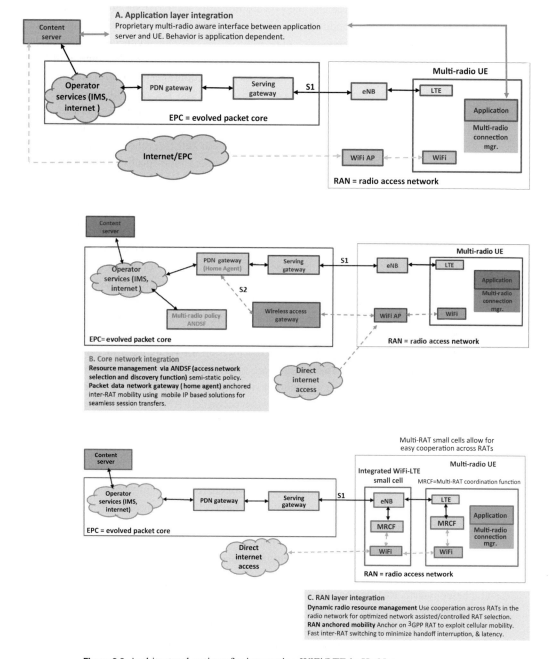

Figure 2.2 Architectural options for integrating WiFi/LTE in HetNets.

Option (A) may be beneficial for supporting user quality of experience (QoE), as end-to-end application QoE-aware metrics may easily be incorporated into RAT selection decisions. However, this option requires compatible application or higher layer support, at both the server and the UE, which may not be always available. Additionally, as the higher layers at the server and the UE are not fully aware of the underlying dynamically changing radio conditions, their choice of RATs may adversely affect network performance.

Option (B) in Figure 2.2 illustrates the current solutions developed by 3GPP for WiFi/3GPP integration, which are based on integration/interworking in the core network [10–12]. Here, an access network discovery and selection function (ANDSF) assists the user in discovering WiFi APs as well as specifies policies for network selection. The UE controls the overall network selection decision. The UE can combine knowledge of local radio-link conditions, operator policies, and user preferences to make a decision that maximizes user QoE. Seamless inter-RAT session transfers are accomplished via higher layer signaling protocols. The packet data network gateway (PDN-GW) acts as the mobility anchor, wherein connections over different radio links are always routed through the PDN-GW, and the GW seamlessly transfers traffic to and from these connections during inter-RAT handoff, without tearing down the session between the server and the UE. Several protocols, based on variants of mobile-IP have been developed for this purpose [11, 12]. There are several advantages of this approach as it can properly account for both operator policies and user preferences. However, the performance of such schemes may still be limited due to the fact that UEs only have local knowledge of their own radio-link conditions, but are unaware of other users' use of shared radio resources, and therefore will make greedy decisions that may ultimately harm overall performance across users [20]. While the UEs can report their varying radio-link information to the core network, such information exchange cannot be updated dynamically due to the high signaling overhead involved. Here, local radio resource management within the radio access network (RAN) may offer better QoS, especially to address the dynamically changing channel conditions encountered for wireless access networks. Therefore architectures that allow for multi-RAT integration within the RAN, where network-wide knowledge of radio-link conditions is available, are of interest.

Option (C) in Figure 2.2 illustrates RAN-based WiFi/LTE integration architectures, where integrated multi-RAT small cells (or where cells are connected through near-ideal backhaul links) allow for full cooperation and more dynamic radio resource management for improved system and user performance. Further, the 3GPP RAT can serve as the primary RAT to be used as a mobility and control anchor, and WiFi may be used as a "secondary" or "extension" RAT for data offload. Here, the UE maintains a single 3GPP-based interface to the operator's network. WiFi is no longer treated as a separate network but rather is integrated as an additional radio resource within the 3GPP RAN. Duplication of control functions such as authentication, security paging, etc., may be eliminated through this approach. Additionally, with integrated architectures WiFi connections may be anchored within the RAN (at the eNB), allowing the user to use local RAN-based switching to move sessions between WiFi and LTE. By anchoring on the 3GPP RAT, the UE may also leverage mobility-optimized 3GPP protocols for transferring sessions between multi-radio cells as the UE moves across the small cells' coverage areas. Such approach allows for faster inter-RAT session transfers reducing session interruptions and packet loss, as well as reducing the handoff signaling generated toward the core network. Operator and user preferences may be made available within the RAN through feedback from the UE or through suitable configuration of the radio network by the operators. Even when integrated cells are not available, some RAN-based coordination across the RATs may be feasible through UE assistance in reporting measurements across the RATs or through a suitable interface between the distributed 3GPP and WiFi radio infrastructure.

The degree of cooperation within the RAN can take the form of simple assistance information (such as network load information) from the radio network or can be based on tight coupling and joint/centralized radio resource management across the radio network. In this chapter, we describe the various levels of cross-RAT cooperation options available across a multi-RAT radio network and characterize the associated performance benefits. We pay specific attention to the integrated RAN architecture where WiFi serves as a "secondary" data carrier in the 3GPP network, and present several results illustrating the performance benefits of this approach.

2.3 Radio resource management in multi-RAT HetNets

This section details various options available for utilizing and managing multi-RAT radio resource available in the network. Both UE and network-controlled or assisted radio resource management (RRM) may be considered for the range of architectural options described in Section 2.2. For application or core network-based integration options, only UE-based RRM schemes may be feasible. A richer set of options are available for RAN-based multi-RAT integration, which depend on the degree of inter-RAT cooperation achieved with different RAN topologies. Here, RAN can play a larger role in determining how multi-RAT resources are managed in the radio network. Even if RAN does not directly control the RRM decisions, it may provide optimized network assistance to enable better decisions by the UE. In integrated RAN architectures, where the mobility and control anchor is moved from the core network to the RAN, more dynamic radio resource management with fast session transfers between RATs (dynamic switching) may be feasible. For co-located multi-RAT small cells or where the delay between the backhaul interfaces between multi-RAT interfaces is negligible, tighter cooperation involving joint RAT scheduling may also be enabled.

In the following, we detail specific RRM schemes that are investigated in this chapter. They range from typical implementations used by UEs today, where the UE always prefers to connect to the free WiFi network if it is in its coverage ("WiFi-preferred"), to more intelligent network cross-RAT selection/assignment, dynamic RAT switching, and joint scheduling schemes.

2.3.1 UE-centric approaches

Received signal strength based (RSRP). This policy compares the received signal power from the cellular and WiFi cells and selects the RAT with the maximum received signal power. Ideal estimation is assumed in our analysis.

WiFi preferred. This scheme is based on preferring WiFi if certain minimum performance is available. Minimum performance may be based on coverage or QoS. A coverage-based definition is used in this chapter.

- *Minimum coverage* is determined by selecting the WiFi link with the best rate predicted by the preferred modulation and coding scheme (MCS), if the best rate equals or exceeds the WiFi link's lowest supported data rate for transmission (e.g., BPSK, rate ½). The preferred MCS is determined by comparing the average of the measured SNR over a certain window with the SNR threshold required to achieve the target error rate of the lowest modulation and coding scheme. Higher targets for coverage may also be used by employing a higher SNR threshold.

Simultaneous transmission (Sim-Op). This scheme is based on the UE being able to transmit on both RATs, without any intelligent coordination across them. This allows for utilization of resources across all RATs providing simple spectrum aggregation gains. However, as we show later, intelligent use of radio resources across RATs can achieve beyond-additive gains in system performance and user QoS.

2.3.2 RAN-assisted approaches

These approaches rely on network assistance from the RAN to improve UE-based RAT selection decisions. Network assistance can be very simple in that the RAN may transmit certain assistance parameters (e.g., network load, utilization, and expected resource allocation), which allow the UE to improve its RAT selection decisions. With increased cross-RAT cooperation in the radio network, RAN assistance may also be improved.

Max throughput (Max-TP) scheme. This scheme is based on a greedy approach, where the UE selects a RAT that can offer the best throughput to the user. The network provides assistance information (for example, the current network load), which allows the UE to estimate its expected throughput on a given RAT. The analytical performance and convergence behavior of this scheme has been extensively analyzed via game theoretic analysis in [20]. In this chapter we investigate its behavior in a 3GPP HetNet deployment.

- This scheme selects the network/RAT providing the highest estimated throughput.
- Throughput is estimated by dividing the rate predicted by the preferred MCS with the "actual loading" (e.g., the number of users sharing the radio). The preferred MCS is determined based on mapping the measured SNR/SINR to the best rate that meets that target error-rate requirement. As described earlier, improved rate estimation for WiFi rate, which relies on knowledge of the WiFi access delay for each UE, may also be used. Ideal knowledge of this access delay is assumed in this chapter. The load information may be available from the WiFi link in the BSS_load information element [21], which indicates the channel utilization as well as the number of users associated with WiFi. The throughput on the 3GPP link may be estimated if assistance information from the RAN in the form of network load or "expected resource allocation" is made available. Note that such information is currently not available in cellular networks.

As shown in [20, 22], the Max-TP is susceptible to ping-pong behavior given the dynamically varying load in the network. Hence suitable hysteresis mechanisms are needed to control its dynamic behavior. An example hysteresis method is described below. More sophisticated hysteresis mechanisms are also feasible.

- Hysteresis mechanism for Max-TP scheme is based on applying a switching threshold, i.e., the user switches to a new RAT only if

$$Throughput_{RAT-new} > TH_{NS} x\, Throughput_{RAT-current} \quad \text{(Condition 1)}$$

The hysteresis threshold for network selection TH_{NS} can be set to a high value to discourage excessive switching and can be adjusted based on the uncertainty in parameters affecting the reliability of throughput estimates. Typical thresholds are set such that the new throughput is two to three times the current throughput. To avoid mass switching, randomization is also applied: at each decision instance, the RAT selection criterion is evaluated and if Condition 1 is met, then a coin toss is performed to make a decision to switch. The randomization probability of the coin toss may also be adjusted dynamically.

Biased received signal strength (RSRP ± bias) scheme: steers the user to prefer certain RATs by adding a bias value to the received signal strength measured at the UE. In current multi-tier network architecture, a network-wide bias is applied to "received signal strength" from the small cell tier to steer users from the macro cell tier to small cells. This cell-range extension for small cells may also be generalized for multi-RAT scenarios [23]. The optimal bias value depends on particular deployments and UE distribution and therefore assumes network-wide cooperation.

2.3.3 RAN-controlled approaches

RAN-controlled approaches place the control of the radio resource management in the radio network. Here the base station controls and assigns the UE to use certain RATs. Such network control may be distributed across base stations, or may utilize a central radio resource control entity that manages radio resources across several base stations/RATs. Network-controlled schemes may utilize proprietary or standardized interfaces between cells/RATs. Distributed network-controlled schemes are currently being discussed as part of the 3GPP study on WLAN/3GPP RAN interworking [13]. Here, the network establishes certain triggers for the UEs to report measurement on their local radio environment. The final RAT selection decisions are then made by the 3GPP base station based on UE measurement reports. An example of centralized network-controlled architecture is the emerging dual connectivity, or "anchor/booster" architecture, where the UE always maintains a control link to the macro cell tier and the macro cell centrally manages the user offload to smaller cells [14]. Here, the macro cell can centrally determine the optimal offload mechanisms.

2.3.3.1 Radio resource management in integrated multi-RAT access networks

Much tighter multi-RAT integration is feasible through network-controlled mechanisms provided that the multi-RAT interfaces are co-located through the use of integrated multi-RAT small cells or through equivalent low-latency interfaces between cells. Additionally, with integrated architectures, network-controlled schemes may be implemented without requiring extensive UE feedback. We study these tighter integration options in the context of the integrated multi-RAT access network. However, the findings may be generalized to non co-located architectures that are connected through near-ideal low-latency interfaces.

In an integrated RAN architecture, 3GPP RAN is used as the primary RAT serving as the control and mobility anchor for WiFi. The UE always associates with the 3GPP network, which controls the usage of WiFi RATs through network-assisted or controlled approaches. The mobility anchor resides within the RAN and 3GPP protocols are used for inter-cell session transfers between multi-radio cells, when the UE moves across these cells. Local switching at each cell is used to transfer sessions to and from co-located WiFi with low latency.

Several techniques for exploiting the cross-RAT coordination within integrated RAN architecture may be considered. In one approach, a multi-RAT interworking function can be used to monitor the quality of radio links across users, and the users (or data flows) can be better assigned across RATs to satisfy user QoS requirements. Typically, such assignments will be made in a semi static fashion, wherein they are updated infrequently during the course of a session. More dynamic updates are also feasible with fast inter-RAT switching available. Tighter cooperation involving joint RAT scheduling is another option, which requires MAC layer coordination across RATs. In this case, a MAC scheduler selects across all associated users based on its scheduling rule and a user may simultaneously transmit over multiple radios if scheduled to do so by the cross-RAT MAC scheduler. Dynamic scheduling may provide benefits, especially in tracking fast-fading and rapidly varying interference conditions.

Coordinated RAT assignment

A RAT assignment framework based on network utility maximization is used for intelligent RAT assignment. The utility framework considers utility metrics such as proportional fair throughput and on-time throughput [24, 25]. The proportional fair metrics are indicative of performance for best effort traffic, while the on-time throughput metric is reflective of performance for delay-sensitive traffic. The main idea behind this framework is to assign users to cells or RATs such that proportional fair or product of user utilities is maximized. This framework is described in detail in the next section (more details are covered in [15]). The assignment problem can be simplified greatly if conventional methods are used to associate users between the macro cell tier and the multi-RAT small cells. Once the user is associated with the integrated cell, the assignment problem simply becomes a problem of partitioning users across the WiFi and 3GPP RATs. When the number of users associated with the small cell is low, this partitioning of users may be solved optimally through exhaustive search. A dynamic-assignment scheme may be developed if regular adjustments to user RAT assignments are made based on a periodic update period.

Cross-RAT scheduling
With integrated WiFi-LTE small cells, cooperation at the MAC layer and joint scheduling of LTE and WiFi resources also becomes feasible. Here, the joint scheduler manages and schedules radio resources of both LTE and WiFi links. Cross-RAT scheduling can be particularly helpful when dealing with delay-sensitive traffic that requires tighter management of packets to avoid excessive latency. Hence in this chapter we consider cross-RAT scheduling in the context of maximizing on-time throughput for users to illustrate the QoS benefits with tighter cross-RAT coordination. To make a fair assessment of the gains in on-time throughput achievable with cross-RAT scheduling, we also develop a scheduling algorithm specifically to optimize on-time throughput [16].

2.4 Analytical frameworks for radio resource management in integrated multi-RAT access network

In this section, we describe the network utility based optimization framework, which may be used for more centralized network-based radio-link assignment of users to different base stations and RATs within the network. This framework offers a network-centric alternative to the UE-based game theoretic approach covered in [20]. While the framework can consider several utility metrics of interest, we focus below on the proportional fair and the on-time throughput utility metrics, as they may be used to investigate performance of the important classes of best-effort and delay-sensitive traffic respectively. We also describe how joint cross-RAT scheduling is performed for maximization of these metrics.

We note that while the proposed utility maximization framework with proportional fair and on-time throughput metrics can provide a useful approach toward RAT assignment problems, it does not specifically account for various factors that can impact actual system performance. Hence we use simulations to determine the effectiveness of this approach across various deployments of interest.

2.4.1 Utility optimization framework

For HetNets based on integrated multi-RAT small cells, user association addresses both the decision to connect each user to a macro or pico base station (cell association) and the decision on the radio utilized by each user once base-station association has been made (RAT association). We formulate the user-association problem from a network utility point of view. While maximization of system utility across the network requires joint optimization across all cells in the network, for simplicity we focus on utility maximization across a single macro cell (coverage area with a single macro base station) and P small cells (pico base stations). The cell provides downlink service to K^{sys} mobile users, constituting the set S^{sys}. At any given time, each user is either associated with the macro base station (MBS) or a pico base station (PBS), i.e., users are divided into two groups:

(a) Macro users: constituting the set S^{macro} with the number of users associated with the macro $K^{macro} = |S^{macro}|$ (the cardinality of the set S^{macro})
(b) Pico users: constituting the set S^{pico} with $K^{pico} = |S^{pico}|$.

The division is such that $S^{sys} = S^{macro} \cup S^{pico}$ with $K^{sys} = K^{macro} + K^{pico}$. The pico user-set is further divided between P pico base stations such that

$$S^{pico} = \bigcup_{p=1}^{P} S_p^{pico}, \quad \text{and} \quad K^{pico} = \sum_{p=1}^{P} K_p^{pico}.$$

Note that the sets are all mutually exclusive.

For the case of integrated multi-RAT small cells (e.g., integrated WiFi and LTE PBSs), the users associated with the PBS may be further partitioned between WiFi and LTE. Hence for PBS p, the associated users are further sub-divided into $S_p^{pico} = S_{p,LTE}^{pico} \cup S_{p,WiFi}^{pico}$ and $K_p^{pico} = K_{p,LTE}^{pico} + K_{p,WiFi}^{pico}$.

Assume that the service delivered by each base station results in a certain "utility," u_k for the kth user, and consider utilities that are additive across users, such as proportional fair throughput described in the next subsection. The overall system utility summed across all users within the cell may be defined as

$$U^{sys} = \sum_{k=1}^{K^{sys}} u_k. \tag{2.1}$$

Equivalently, by summing the user utilities per base station we have

$$U^{sys} = U^{macro} + \sum_{p=1}^{P} U_p^{pico} = U^{macro} + \sum_{p=1}^{P} \left(U_{p,LTE}^{pico} + U_{p,WiFi}^{pico} \right). \tag{2.2}$$

Given this formulation, the user association problem may be stated as follows.

User association problem. Find the optimal user association (partitioning) between MBSs and PBSs (LTE and WiFi interfaces) for all the users in the cell such that the total cell utility is maximized.

$$\max_{S_{macro}, \left(S_{p,LTE}^{pico}, S_{p,WiFi}^{pico} \right) \forall p} U^{sys} \tag{2.3}$$

As a simplification, we assume that cell association is performed using conventional techniques, although this assumption may not be used if more centralized solutions for cell association are implemented. Conventional cell association rules for offloading traffic to small cells can be found in [23]. The cell range extension (CRE) method may also be employed. Conventionally, by measuring downlink reference (pilot) signal powers, each user in the cell determines the downlink carrier signal strength corresponding to the MBS and each PBS in the cell. A positive bias value is added to the received signal power from the small cell to steer users to prefer the small cell over a macro cell. The user simply associates with the base station with the highest value of biased received signal power. Once all the users have associated with their base stations, the user-set S^{macro} is fixed and may be dropped from the optimization. The problem remaining is that of RAT association for the user-set of each PBS. We call this the user-mapping problem.

User RAT assignment problem. Once cell association is fixed, the original optimization problem can be simplified as P parallel user RAT assignment problems. For each PBS, the user RAT assignment problem may be stated as finding the optimal user mapping (partitioning) between the LTE and WiFi radio interfaces. The total cell utility is maximized when optimal RAT assignments are performed for all PBSs.

$$\max_{\left(S_{p,LTE}^{pico}, S_{p,WiFi}^{pico}\right)} \left(\sum_{k \in S_{p,LTE}^{pico}} u_k + \sum_{k \in S_{p,WiFi}^{pico}} u_k \right), \forall p. \tag{2.4}$$

The user RAT assignment problem at each small cell can be solved using exhaustive search across all possible RAT assignments of users associated with the small cell, if the number of associated users is small. However, for the case of proportional fair utility, the RAT assignment problem per small cell may be optimally solved by using an approximation to the proportional fair throughput.

2.4.2 Cross-RAT coordination for proportional fair utility maximization

2.4.2.1 Cross-RAT assignment for maximizing proportional fair throughput

The overall system utility may be defined as a "proportionally fair" product of the per user utilities, or equivalently $U^{sys} = \sum_{k=1}^{K^{sys}} \log u_k$. The utilities per base station may be defined as proportionally fair utilities across the users associated with each base station. Specifically,

$$U^{sys} = \sum_{k \in S^{macro}} \log u_k + \sum_{k \in S_{1,LTE}^{pico}} \log u_k + \sum_{k \in S_{1,WiFi}^{pico}} \log u_k + \cdots + \sum_{k \in S_{P,LTE}^{pico}} \log u_k$$

$$+ \sum_{k \in S_{P,WiFi}^{pico}} \log u_k \tag{2.5}$$

If we assume that proportional fair scheduling is used at the LTE and the WiFi interfaces of each pico cell, the utility per user can then be represented by the resulting long-term throughput achieved with the proportionally fair scheduler (PFS). The long-term throughput is proportional to the average achievable rate when the total number of users sharing the same radio is fixed. In the special case of static channel, the long-term throughput approximation for the kth user over a total of K users with proportional fair scheduler is, $\bar{T}_k \sim \frac{\bar{R}_k}{K}$. For the case of dynamic Rayleigh fading, the throughput may be approximated by $\bar{T}_k \sim \frac{\log K}{K} \bar{R}_k$, where ($\log K/K$) is the multi-user diversity gain [15]. More generally, $\bar{T}_k \sim c(K) \bar{R}_k$, where the "$c$" depends on the exact approximation used. Here the average rate may be approximated by the Shannon's formula, $\bar{R}_k = E[B \log(1 + SINR_k(t))]$, which is the expected value of the rate achieved by the user, given its SINR, resulting from associating with a particular base station. We note that for

WiFi, the Shannon rate approximation does not hold and more accurate rate-estimation models may be substituted.

Using the PFS approximation, it is shown in [15] that the user RAT assignment algorithm is optimally solved by rank ordering the ratios of LTE and WiFi rates and comparing them to a threshold that is based on the optimal split of users between the WiFi and LTE RATs. In particular, if we assume that (m) users are associated with the WiFi RAT and the remaining $(K - m)$ users are associated with the LTE RAT, then the rank-ordered rate ratios may be expressed as follows.

$$\frac{\bar{R}_{1,LTE}}{\bar{R}_{1,WiFi}} < \frac{\bar{R}_{2,LTE}}{\bar{R}_{2,WiFi}} <, \ldots, \frac{\bar{R}_{m,LTE}}{\bar{R}_{m,WiFi}} < \frac{c(K-m)}{c(m)} < \frac{\bar{R}_{m+1,LTE}}{\bar{R}_{m+1,WiFi}}, \ldots, \frac{\bar{R}_{K,LTE}}{\bar{R}_{K,WiFi}}. \quad (2.6)$$

As the rate ratios are monotonically increasing and $c(K - m)/c(m)$ is monotonically decreasing, the optimum may easily be found such that

$$\frac{\bar{R}_{m,LTE}}{\bar{R}_{m,WiFi}} < \frac{c(K-m)}{c(m)} < \frac{\bar{R}_{m+1,LTE}}{\bar{R}_{m+1,WiFi}}. \quad (2.7)$$

2.4.2.2 Cross-RAT scheduling for maximizing proportional fair throughput

The proportional fair metric ($\frac{\bar{R}}{T}$) is used to jointly assign cross-RAT resources across users. This joint scheduling method may be equivalently implemented by sharing the per-user average proportional fair throughput across the WiFi and LTE RATs.

2.4.3 QoS-aware cross-RAT coordination for timely throughput maximization

This subsection describes cross-RAT coordination scheme focused on optimizing the on-time throughput as the QoS metric. We first cover the QoS-aware scheduling algorithm designed specifically for on-time throughput optimization. Then we explore different options for multi-RAT coordination in integrated LTE-WiFi small cells, which include RAT assignment and cross-RAT scheduling procedures.

2.4.3.1 Scheduling for maximizing on-time throughput

The metric for on-time or timely throughput is defined in [25, 26]. The following briefly summarizes the concept. Assume that transmission time is divided into time slots of size τ where the nth time slot represents the time duration between $(n - 1)\tau$ and $n\tau$. Define $T_k(n\tau)$ as the total data throughput of user k at the nth time slot. The average on-time throughput, λ, of user k is then defined as the probability that a target packet of size T_0 is successfully received within every time-slot of length τ,

$$\lambda_k = \liminf_{N \to \infty} \frac{\sum_{n=1}^{N} I(n\tau)}{N} \quad (2.8)$$

where, $I_k(n\tau)$ is an indicator function,

$$I_k(n\tau) = \begin{cases} 0, & T_k(n\tau) < T_0 \\ 1, & T_k(n\tau) \geq T_0 \end{cases} \quad (2.9)$$

The effective goodput g_k of user k is the throughput weighted by the proportion of time slots where the target data throughput T_0 was received on time.

$$g_k = \lambda_k T_0 \qquad (2.10)$$

From the above formulation, we can observe that higher effective goodput per user can be achieved by increasing the allocation of time slots that support data throughput in excess of the target throughput T_0, i.e., increasing the number of (n, k) pairs where $T_k(n\tau) \geq T_0$. Hence our scheduling algorithm for maximizing on-time throughput is based on increasing the number of users receiving data throughput in excess of T_0 within the current time slot. The instantaneous data rate and data throughput of users within the current slot is considered when selecting the user to be scheduled. No additional resources are allocated to users who have already received more than T_0 bits within the current time slot, as the additional data bits over T_0 do not increase their average timely throughput. The scheduler thus only scans through users receiving less than T_0 bits within the current time slot and selects the one who is most likely to achieve the target throughput T_0 while using the minimal amount of resources. If a user can achieve T_0 using the least amount of resources, there will be more time-frequency resources available for other users. Thus the probability that more users achieve their targeted T_0 within the current time slot will increase.

The amount of resources, B_k, required to achieve throughput T_0 for user k is estimated via the following equation:

$$B_k = \begin{cases} \dfrac{T_0 - T_k(t)}{R_k(t)}, & T_k(t) < T_0 \\ 0, & T_k(t) \geq T_0 \end{cases} \qquad (2.11)$$

which applies for $(n-1)\tau < t < n\tau$. $T_k(t)$ is defined as the accumulated throughput for user k from time instant $(n-1)\tau$ to t, and $R_k(t)$ is the instantaneous rate of user k at time instant t. The scheduler will select the user with the minimum positive B_k for transmission.

2.4.3.2 Cross-RAT assignment for on-time throughput maximization

The same utility-based framework developed in Section 2.4.1 may be used to perform utility-based RAT assignment across users. Here the utility metric u_k can be thought of as the on-time throughput metric. A measurement period is helpful to estimate the on-time throughput performance on the radio links for each user. During the measurement period, there is no coordination between LTE and WiFi radios. With each radio adopting QoS-aware scheduling, the effective goodput collected over the measurement period provides a good indicator of how likely a user is to achieve on-time throughput on each radio link. We can thus use the effective goodput as the utility metric and the RAT mapping can be done through an exhaustive search across assignments achieving the highest effective goodput. We name the above approach *Max Timely-TP Mapping*. Note that exhaustive search across users is feasible if the number of users associated with each pico cell is small.

The measurement method described provides an approximation to the on-time throughput. Accurate estimation is difficult because the on-time throughput measurement actually depends on the set of users being scheduled, which implies that it requires measuring the effective goodput of the exact set of users being assigned to the radio link to obtain the actual on-time throughput performance. As this may not always be feasible, a simpler "*Min-Resource Mapping*" algorithm, based on measuring the average rate per user on both links, is also investigated. The average rate measurement provides a good estimate of the proportion of resources required for a user to achieve its target throughput T_0 within a given time slot. As discussed, scheduling the user requiring the least amount of resources to achieve its target, T_0, can improve the chances of more users being able to achieve their target timely throughput in upcoming transmissions. Hence for the *Min-Resource Mapping* technique, we partition the users such that minimum amount of resources be required across both the LTE and the WiFi links to support associated users in achieving T_0. Both the loadings on the LTE and WiFi links should be minimized and we use the maximum of the two as the optimizing metric. Our RAT assignment strategy ensures that the spectral resources across the radio links are economically utilized to increase the probability of users achieving their target QoS requirements for on-time traffic delivery. More details appear in [16].

2.4.3.3 Cross-RAT scheduling for timely throughput maximization

With joint RAT scheduling, radio resources of both LTE and WiFi links are managed by the same scheduler. The scheduler monitors both radio links and keeps track of the aggregated throughput, ongoing transmissions that are incomplete, and the unacknowledged packets. We also combine cross-RAT scheduling with the QoS-aware scheduling algorithm designed for optimizing on-time throughput, [16].

2.5 System evaluation methodology and assumptions

In this section, we briefly describe the multi-RAT system evaluation methodology to investigate the performance of the various multi-RAT resource management schemes discussed earlier. There is limited coverage on methodology for evaluation of multi-RAT networks in the literature. Hence our evaluation is based on extending the 3GPP evaluation methodology for heterogeneous deployments, [26–28], to also model the contention-based WiFi MAC. Specifically, we focus on outdoor small cell deployments, as this scenario results in the most challenging interference scenario. The small cells are randomly placed throughout the coverage area, and users are either uniformly distributed across the cell or clustered around the cell according to the procedure described in [26]. We use a dynamic simulation model that utilizes fast-link adaptation and frequency-selective scheduling to react to fast channel variations. The channel models and the network topology are kept constant across the WiFi and LTE-based small cell deployments. The WiFi state MAC state machine is accurately modeled wherein the various state transitions of the WiFi are triggered asynchronously based on event timing. The LTE processing occurs periodically at events generated at regular

Table 2.1 Simulation assumptions.

LTE	
Topology	4 AP, 30 UEs/sector (1 pico/sector, 9 UEs/sector for timely throughput analysis), 19 or 7 cell wrap-arounds
RSRP bias, ABS	0 dB, 0%
UE dropping	Clustered or uniform UE distribution
Channel/UE speed	[IMT] UMa Macro, UMi Pico, UE speed = 3 km/h
LTE mode	Downlink FDD @ 10 MHz
No. antennas (macro, pico, UE)	(2, 2, 2), SISO uplink
Antenna configuration	macro, pico: co-polarized, UE: co-polarized ($\|\rightarrow\|$)
Max rank per UE	2 (SU-MIMO)
UE channel estimation	Ideal
Feedback/control channel errors	No error
Scheduler	Prop. fair, max. On-time throughput maximizing
Scheduling granularity	5 PRBs
Traffic load	Full buffer for both WiFi and LTE
Receiver type	Interference-unaware MMSE
Feedback periodicity	10 ms
CQI and PMI feedback granularity in frequency	5 PRBs
PMI feedback	3GPP Rel.-10 LTE codebook (per sub-band)
Outer loop for target FER control	10% PER for 1st transmission
Link adaptation	MCSs based on LTE transport format
HARQ scheme	CC
WiFi	
WiFi parameters	802.11g, same network deployment as LTE
WiFi frequency/channel	2.4 GHz band/20 MHz
Number of frequency bands	3
MPDU size	1,500 bytes
Rogue interferers	3 rogue APs/sector

frame boundaries. We use the WiFi standards based on 802.11g, and consider a 1 × 2 system in our evaluation. The WiFi link budget used in our simulations assumes a 20 dB maximum EIRP limit based on meeting the requirements imposed in Europe and Japan.

Unless otherwise noted, results are presented for the deployment scenarios of 4 APs and 30 UEs per macro cell sector for either clustered or uniform user distribution. Full-buffer traffic, consistent with 3GPP methodology, is considered across the results presented, which is useful to assess performance under high-load conditions. However, we expect that the full-buffer assumption will introduce some limitations to assess performance gains available under different loading conditions. Detailed simulation assumptions are covered in Table 2.1.

To evaluate the on-time throughput performance, we use a full buffer traffic model, but we set target on-time throughput requirements, which are induced by a video application stream running in real time with $\tau = 30$ ms and $T_0 = 0.20787$ Mbits. The maximum achievable effective goodput for this stream is g_k is 6.93 Mbps.

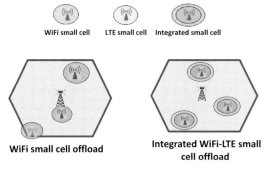

Figure 2.3 Multi-RAT HetNet deployment scenarios.

2.6 Performance evaluation

This section compares both throughput and QoS (on-time throughput) performance across the range of radio-resource management schemes described in the previous section. In particular, we compare performance of UE versus network-based schemes across distributed and integrated multi-RAT deployments. Results are presented for different deployment scenarios shown in Figure 2.3. HetNet topologies based on WiFi-only small cell deployments and integrated WiFi plus LTE small cell deployments are considered.

The chapter provides more coverage of downlink system performance, but several results comparing the uplink throughput performance of UE-centric schemes with selected network-based optimization schemes are presented. Results that illustrate the benefits of dynamic RAT switching for avoiding WiFi interference are also given.

2.6.1 Downlink throughput performance enhancements

Table 2.2 compares the throughput performance of conventional UE-centric schemes, network-assisted schemes, and centralized RAT assignment and scheduling schemes. Simultaneous transmission on LTE and WiFi interfaces without any coordination between the two interfaces is also included for reference.

Joint scheduling for macro plus WiFi-only deployments is considered for completeness, and joint operation between the macro cell and the serving WiFi cell providing the best coverage is assumed. In practice such tight coordination for the distributed deployments may not always be feasible, unless low-latency backhaul is available for coordination. For macro plus integrated WiFi/LTE small cell deployments, an integrated multi-RAT access network model is assumed, in that users associate with the small cell based on best LTE coverage without the use of any cell-range extension mechanisms. Centralized RAT assignment and joint scheduling is then carried out across the co-located WiFi and LTE interfaces. Unless otherwise noted we use ideal estimation of WiFi rate, to get the best comparison across all schemes. For the Max-TP scheme shown, network selection decisions are assessed every 200 milliseconds and hysteresis

Table 2.2 Comparing system performance across UE and network-centric approaches for cell-edge and average user throughput performance assuming full-buffer traffic.

(Mbps)	WiFi preferred (MCS based)	Max TP with hysteresis (ideal rate estimates)	Sim-Op (no cooperation)	Sim-Op joint scheduling at macro cell
Macro plus WLAN small cell deployments				
5% users	0.35	0.76 (+117%)	0.74	0.89 (+17%)
50% users	2.2	3.53 (+60%)	2.73	3.42 (−3.1%)
Average rate	2.34	3.44 (+47%)	2.86	3.33 (−3.1%)

(Mbps)	WiFi preferred (MCS based)	Max TP with hysteresis (ideal rate estimates)	Integrated RAT assignment at small cell	Sim-Op (no coordination)	Sim-Op (joint scheduling)
Macro plus integrated WiFi/LTE small cell deployments					
5% users	0.49	0.37	0.52 (+6%)	0.54	0.54 (+3.8%)
50% users	3.14	4.96	5.9 (+88%)	4.74	5.78 (−2%)
Average rate	3.44	6.02	6.39 (+86%)	5.52	6.35 (−0.4%)

and randomization schemes described in Section 2.3 are applied. A clustered UE distribution was used for our results, but the comparison is similar for the uniform scenario as well.

As expected, Max-TP schemes based on network assistance in terms of network load and knowledge of ideal WiFi rate provide substantial gains over existing WiFi-preferred (based on minimum coverage) mechanisms, as WiFi-preferred schemes tend to overload the WiFi capacity and do not take advantage of the available LTE capacity. Better load balancing may be achieved through load-aware Max-TP schemes. As shown, approximately 117% and 47% gains in cell-edge and average system performance are observed respectively.

The gains in average throughput performance increase when additional LTE capacity is available with integrated multi-RAT small cells, where 75% gains are observed in average system throughput. Some loss in cell-edge performance is observed due to restricting use of WiFi to be on the co-located interface of the integrated cells. Such loss in performance can typically be addressed by improving macro to LTE small-cell offload via optimized cell-range extension. Further performance gains may be obtained by using the optimal user RAT partitioning algorithm based on maximizing proportional fair throughput, which was described in Section 2.4. The gains in average throughput performance over conventional schemes may be improved by approximately 88%, whereas modest cell-edge improvements are also observed. When comparing with Max-TP schemes, over 40% gains are observed in cell-edge results with modest gains in average system performance.

When compared to RAT assignment and selection schemes, joint scheduling over RATs does not provide significant benefits. Only modest gains are observed in cell-edge results (17% and ~4% for WiFi only and integrated WiFi/LTE small cell deployments).

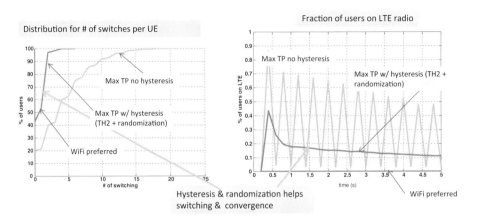

Figure 2.4 Convergence behavior of UE-centric schemes.

This is mostly due to the full-buffer traffic model used in our simulation, which precludes realizing the benefits of traffic-aware scheduling. We expect improved results when the full-buffer assumption is relaxed. Also note that simultaneous operation across WiFi and LTE interfaces with no cross-RAT cooperation results in poor performance compared to intelligent RAT assignment and selection schemes, as weaker users may end up consuming substantial resources across both links, thus degrading overall performance.

Note that proper hysteresis design is essential for the network-assisted Max-TP scheme to achieve performance that compares well with network-based schemes. Otherwise, Max-TP is susceptible to ping-pong behavior as it tracks dynamically varying load conditions in the network. Figure 2.4 illustrates how proper hysteresis design is essential for controlling the convergence behavior of Max-TP schemes. While the Max-TP scheme can be used with conventional RAN-anchored mobility protocols, the latency involved in performing session transfers may preclude frequent updates to network selection decisions, thereby substantially increasing the convergence time. Further 3GPP load information is considered proprietary by operators and is not available in current standards, which also makes the adoption of such schemes difficult in practice. Hence the integrated RAT assignment schemes may be preferred for providing stable and improved performance.

2.6.2 Uplink throughput performance enhancements

This subsection focuses on uplink throughput performance enhancements achieved with network-assisted RAT selection. The uplink environment is more challenging compared to downlink, as the RAT selection and assignment decisions change the interference environment in the network. Results are again presented for a WiFi-only (macro plus WiFi deployment) and integrated (macro plus WiFi/LTE integrated cells) small cell deployments. The performance of the coverage-based WiFi-preferred scheme is compared with the Max-TP scheme with hysteresis. The SNR threshold for coverage-based WiFi preferred scheme is set to be 40 dB, to ensure high QoS. A hysteresis threshold of

Table 2.3 Uplink performance of RAN-assisted Max-TP scheme: comparison with WiFi-preferred (SNR threshold scheme).

(Mbps)	WiFi-preferred (SNR threshold) with hysteresis	Max-TP With hysteresis
Macro plus WLAN small cell deployments		
5% users	0.94	1.54 (+64%)
50% users	3.49	3.84 (+10%)
Average rate	4.10	4.41 (+7.6%)
Macro plus integrated WLAN/LTE small cell deployments		
5% users	0.73	1.36 (+86%)
50% users	4.04	4.81 (+19%)
Average rate	5.56	5.71 (+2.6%)

3 dB is applied to both schemes. The performance is further compared with network-assisted schemes relying on biased signal strength association, where the offload bias is optimized across the network.

The simulation assumptions are consistent with Table 2.1, with a few exceptions:

- 1 AP and 10 UEs (clustered user distribution) per macro cell sector are assumed.
- Uplink power control consistent with LTE methodology is applied.
- Frequency reuse 3 is assumed in the macro network to mitigate more severe interference in the uplink. Round-robin scheduling is used for LTE, whereas WiFi is based on random access.
- RAT selection decisions may be updated every 100 milliseconds. LTE throughput estimates are updated every 10 milliseconds, while WLAN throughput estimates are updated on 100-millisecond granularity. The throughput estimates are averaged over several measurement periods as well as predicted measurements are refined based on actual measured throughput once the UE connects to a given RAT.

As shown in Table 2.3, the load-aware Max-TP scheme with hysteresis provides improved cell-edge performance when compared to the WiFi-preferred scheme based on SNR threshold. A 64% and 86% improvement is observed in cell-edge performance for WLAN based and integrated WLAN/LTE small cell deployments respectively.

Figure 2.5 also shows that hysteresis design continues to be important for achieving performance gains with Max-TP schemes. Results show that the number of reconnections per UE can be substantially reduced by increasing the hysteresis threshold.

The performance of the Max-TP scheme is further compared with RSRP-based schemes with optimum offload bias. Figure 2.6 shows a comparison of average system throughput and cell-edge performance across these schemes as the offload bias is changed. Results are shown for both WiFi-only and integrated small cell deployments. It is seen that load-aware Max-TP schemes typically show better performance when compared with RSRP plus bias schemes, in particular for cell-edge results. This is due to poor WiFi coverage in sparse WiFi deployments as well as the cross-tier interference in integrated deployments. Adding QoS criteria (such as SNR threshold used earlier)

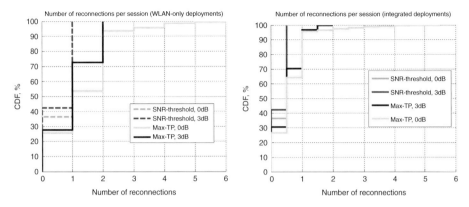

Figure 2.5 Role of hysteresis in improving the convergence performance of load-aware Max-TP schemes.

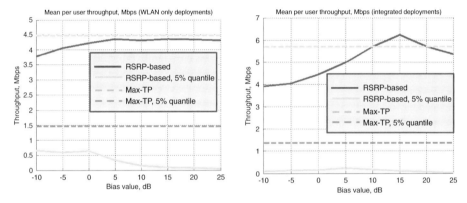

Figure 2.6 Uplink performance of RAN-assisted Max-TP scheme: comparison with RSRP-based scheme with optimized bias.

therefore seems essential for associating with WiFi APs. Proper management of cross-tier interference will also further improve performance of both Max-TP and "RSRP plus bias" based schemes. Results for integrated deployment further indicate that load-aware scheme may be combined with optimal bias setting to further improve performance. Note that the optimal bias value is dependent on a particular deployment scenario, therefore requires network-wide knowledge to determine the best setting.

Further improvement of uplink performance with tighter cross-RAT cooperation is an area for further investigation. Also a comparison with improved small cell offloading mechanisms, such as those based on reference signal received quality (RSRQ) may also be performed.

2.6.3 Performance of dynamic interference avoidance

An important advantage of integrated RAN architectures is their ability to provide fast session transfers, allowing for fast fall back to LTE in case of poor WiFi radio

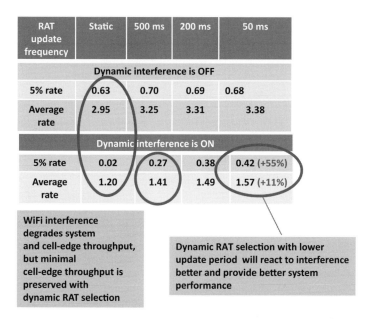

Figure 2.7 Performance of dynamic RAT switching in the presence of rogue WiFi interference.

conditions due to rogue or uncoordinated WiFi interference. In this subsection, we explore the effectiveness of dynamic RAT switching in providing robust performance in the presence of uncoordinated WiFi interferers. Figure 2.7 compares the system performance of static RAT assignment schemes with schemes that allow assignments to be updated at an update rate of 50, 200, and 500 milliseconds. A five-second simulation period is assumed. The WiFi interference turns on and off at two and four seconds respectively. Again a 4 AP/30 users/sector uniform UE deployment is considered. It can be seen that system performance across both schemes degrades in the presence of rogue WiFi interference as the WiFi capacity gets shared across the interferers as prescribed by WiFi's random-access mechanism. However, dynamic RAT assignment can preserve a minimal level of performance for users heavily impacted by the WiFi interference. A faster update rate shows more robust performance, with a gap of 55% being observed for cell-edge users when the update rate is changed from 50 to 500 milliseconds.

2.6.4 Downlink performance enhancement for on-time throughput

In this subsection, we consider the effectiveness of joint scheduling techniques in improving user QoS (on-time throughput metric). Figure 2.8 compares the performance of assignment and scheduling algorithms designed for maximizing on-time throughput. It shows the percentage of users that can achieve a given target timely throughput. Results given are based on single RAT schedulers optimized for maximizing on-time throughput. Several options are shown in the figure and include the single radio case using only LTE or WiFi radios as a reference. The other options assume co-located LTE and WiFi radios and cover techniques for different levels of cooperation between LTE and

Percentage users achieving "on-time" throughput for delay sensitive (real-time video) traffic				
Multi-RAT coordination scheme	% Users for target throughput of 4 Mbps	% Users for target throughput of 5 Mbps	% Users for target throughput of 6 Mbps	% Users for target throughput of 6.93 Mbps
LTE-only	42	35	22	4.2
WiFi-only	69	41	19	9.3
Uncoordinated Sim-Op	85	71	53	23.3
Low-complexity RAT assignment (minimum resource mapping)	89	81	59	13.6
Optimal RAT assignment (maximum timely TP)	93	89	73	18.9
Multi-RAT scheduling	94	90	82	56.7

Figure 2.8 QoS performance gains (number of users achieving target on-time throughput) with cross-RAT coordination.

WiFi radios: (a) simultaneous operation without assuming any cross-RAT coordination, (b) cross-RAT assignment schemes, and (c) joint scheduling across RATs.

It can be seen from Figure 2.8 that increased cooperation across RATs can improve the number of users achieving their target on-time throughput, with substantial improvements at higher target timely throughput rates. In particular, cross-RAT scheduling can service up to three times more users at the target on-time rate of 6.93 Mbps, for the real-time video application considered. These results are in contrast to the throughput enhancement gains observed with joint scheduling, which offered marginal improvements. Here, the ability to send data simultaneously on both RATs can help improve performance of users with tighter delay deadlines.

2.7 Summary and next steps

Multi-RAT heterogeneous networks with technologies for efficient integration of multiple RATs are expected to become a fundamental characteristic of future 5G networks. Using the integration of WiFi within 3GPP cellular HetNets as a case study, we describe several architectures for multi-RAT integration and argue the importance of radio layer integration across RATs. In particular, we emphasize the need for a unified RAN architecture where radio resources are jointly managed locally with the RAN.

Performance evaluation of several RAN-assisted and controlled RRM schemes shows significant performance gains in system and user throughput performance, when compared with conventional WiFi-preferred schemes. In particular, we show that a load-aware UE-centric RAT selection scheme (Max-TP), based on RAN assistance, can achieve near-comparable performance to network-controlled RAT assignment. However, this scheme requires 3GPP load information to be available, as well as requiring carefully designed hysteresis mechanisms to achieve convergence with dynamic

UE-based RAT selection. Faster session transfers available with RAN-anchored mobility may be used to improve the speed of convergence for such schemes. In comparison, network-controlled RAT assignment designed for an integrated RAN network can work without network-load information and can provide further improvement in performance without the need for careful management of convergence behavior. We also observe that intelligent RAT selection and assignment performs better than uncoordinated simultaneous use of both RATs.

Tighter joint RRM mechanisms, such as cross-RAT scheduling, do not provide significant throughput improvements for the full-buffer traffic considered in our evaluation. However, significant QoS performance gains are observed with joint scheduling when evaluating performance for delay-constrained traffic. Hence we expect that tighter cross-RAT integration will be important for delivering application layer QoS, in particular for supporting real-time traffic such as voice and video. Again, such coordination mechanisms are more easily implemented using integrated RAN architectures.

Our results also show that integrated RAN architectures supporting dynamic RAT switching for interference avoidance can improve system robustness in the presence of rogue or uncoordinated interference, thereby allowing operators to offer improved QoS over the unlicensed spectrum.

While the chapter provides performance evaluation results across a range of radio resource management options, several areas remain for future study.

Effect of practical impairments. Our evaluations assume ideal estimation of measurements and do not account for practical effects such as channel estimation errors etc. These effects may be better addressed in future studies.

Additional deployment scenarios with advanced WiFi and alternative standards for unlicensed spectrum. Our evaluation focused on outdoor deployment scenarios assuming basic capability of the WiFi air interface with 2.4 GHz operation. The results should further improve with more advanced WiFi standards. Results in [29] show that WiFi at 5 GHz can also play an important role in improving indoor coverage. Further investigation of cross-RAT coordination in indoor deployments, especially at 5 GHz bands, is therefore of interest.

Further, alternative approaches to the use of unlicensed spectrum have also been recently proposed, [30], and comparison of such approaches with the integrated multi-radio RAN architectures is also of great interest.

Application-aware multi-RAT optimization. We expect further gains in performance with joint RRM if traffic characteristics are explicitly exploited in RRM decisions. In particular, application-aware cross-RAT optimization remains an important area of investigation.

Energy efficiency. While energy efficiency was not explicitly covered in this chapter, it remains an important consideration for multi-RAT design. Our preliminary results show that intelligent RAT selection performs better than simultaneous RAT operation when energy efficiency is considered as a metric. Hence the power efficiency impact of advanced joint RRM techniques must be evaluated.

Extension to next-generation RAN architectures. As noted, advanced 3GPP HetNet architectures based on dual connectivity anchor-booster architectures have recently emerged [14]. These architectures are also based on the "integrated RAN" concept, wherein the entire tier of small cells is used for traffic offload, and the macro-cell tier acts as the control and mobility anchor. Extension of such architectures to include WiFi operating on unlicensed bands is another area of future interest.

Acknowledgments

We acknowledge V. Gupta, K. Etemad, H. Yin, S. Sirotkin, M. Fong, H. Niu, J. Zhu, R. Vannithamby, and K. Johnsson at Intel for helpful discussions. We also thank W. C. Wong, Y. Zhu, T. Papathanassiou, H. Sherani-Meher, and E. Perahia for their help with the multi-RAT system-level simulator.

Part of this work has been supported by the Internet of Things program of Digile (funded by Tekes).

References

[1] H. Claussen, "Performance of macro-and co-channel femto-cells in a hierarchical cell structure," *IEEE PIMRC*, 2007.

[2] J. Andrews, *Can Cellular Networks Handle 1000x the Data?* http://users.ece.utexas.edu/~bevans/courses/realtime/lectures/Andrews_Cellular1000x_Nov2011.pdf, 2011.

[3] S. Yeh, S. Talwar, S.-C. Lee, and H. Kim, "WiMAX femto-cells: a perspective on network architecture, capacity and coverage," *IEEE Communications Magazine*, pp. 58–65, June, 2008.

[4] S. Yeh, S. Talwar, G. Wu, N. Himayat, and K. Johnsson, "Capacity and coverage enhancement in heterogeneous networks," *IEEE Wireless Communications*, 18(3), 32–38, 2011.

[5] RWS 120045, "Summary of 3GPP TSG RAN workshop on Release 12 and beyond," June, 2012.

[6] "Carrier WiFi offload and hotspot strategies: global service provider survey," Infonetics Research, 2012.

[7] IEEE C80216-10_0016r1, "Future 802.16 networks: challenges and possibilities," March, 2010.

[8] "5G Radio access: research & vision," Ericsson White Paper, June, 2013.

[9] Technical specification (TS) 24.302, "Access to the 3GPP Evolved Packet Core (EPC) via non-3GPP access networks," 3GPP, 2013.

[10] 3GPP Technical specification (TS) 23.402, "Architecture enhancements for non-3GPP accesses," 2013.

[11] 3GPP Technical Report (TR) 23.861, "Network based IP flow mobility," 2012.

[12] 3GPP Technical Report TR 23.852, "Study on S2a Mobility based on GTP and WLAN access to the EPC network (SaMOG)," 2013.

[13] 3GPP Technical Report (TR) 37.834, "Study on WLAN/3GPP radio interworking," 2013.

[14] 3GPP Technical Report (TR) 36.842, "Study on small cell enhancements for EUTRA and EUTRAN: higher layer aspects," 2012.

[15] A. Y. Panah, S. Yeh, N. Himayat, and S. Talwar, "Utility-based radio link assignment in multi-radio heterogeneous networks," *Proc. of International Workshop on Emerging Technologies for LTE-Advanced and Beyond-4G on IEEE Globecom*, pp. 618–623, 2012.

[16] S. Yeh, A. Y. Panah, N. Himayat, and S. Talwar, "QoS aware scheduling and cross-radio coordination in multi-radio heterogeneous networks," *Proc. of IEEE VTC*, Fall, 2013.

[17] N. Himayat, S. Yeh, A. Y. Panah, *et al.*, "Multi-radio heterogeneous networks: architecture and performance," *to appear in Proc. of ICNC, 2014*, Feb. 2014.

[18] M. Gerasimenko, N. Himayat, S. Yeh, S. Talwar, S. Andreev, and Y. Koucheryavy, "Characterizing performance of load aware network selection in multi-radio (WiFi/LTE) heterogeneous networks," *Proc. of IEEE Globecom workshop on Broadband Wireless Access*, Dec. 2013.

[19] V. Gupta, S. Somayazulu, N. Himayat, H. Verma, M. Bisht, and V. Nandwani, "Design challenges in transmitting scalable video over multi-radio networks," *Proc. of IEEE Globecom workshop on Broadband Wireless Access*, Dec. 2012.

[20] E. Aryafar, A. Keshavarz-Haddad, M. Wang, and M. Chiang, "RAT selection games in Het-Nets," *Proc. of IEEE INFOCOM*, pp. 1–11, 2013.

[21] IEEE standard 802.11-2012.

[22] 3GPP R2-133479, "Performance benefits of RAN level enhancements for WLAN/3GPP interworking, 3GPP RAN WG2 #83bis", October, 2013.

[23] S. Singh and J. Andrews, "Joint radio resource partitioning and offloading in heterogeneous networks," *IEEE Transactions on Wireless Communications*, 2013.

[24] I.-H. Hou, V. Borkar, and P. R. Kumar, "A theory of QoS for wireless," *Proc. IEEE INFOCOM 2009*, pp. 486–494, Apr. 2009.

[25] S. Lashgari and A. S. Avestimehr, "Timely throughput of heterogeneous wireless networks: fundamental limits and algorithms," (arXiv:1201.5173v1) *IEEE Trans. on Information Theory* (forthcoming).

[26] 3GPP Technical Report (TR) 36.814, "Further advancements for E-UTRA physical layer aspects," 2012.

[27] 3GPP Technical Report (TR) 36.819, "Coordinated multi-point operation for LTE physical layer aspects," 2011.

[28] "Guidelines for evaluation of radio interface technologies for IMT-Advanced," ITU, 2009.

[29] L. Hu, L. L. Sanchez, M. Maternia, *et al.*, "Modeling of WiFi IEEE 802.11ac offloading performance for 1000x capacity expansion of LTE Advanced, *Proc. IEEE VTC*, Fall, 2013.

[30] "Extending LTE-Advanced to un-licensed spectrum," Qualcomm white paper, December, 2013.

3 Dynamic TDD small cell management

Cheng-Chih Chao, Yi-Ting Lin, and Hung-Yu Wei

In contrast to voice traffic, wireless data traffic is mostly asymmetric and time-variant with a requirement for a dynamically adjusting technique to divide the uplink (UL) and downlink (DL) resource. In typical cellular systems, the length of UL resource and the length of DL resource are predetermined. In a typical frequency-division duplex (FDD) system, the UL and DL transmission use distinctive frequency bands, which is especially efficient in cases of symmetric traffic due to the avoidance of possible interference between UL and DL transmission. However the FDD system has difficulty in adjusting its UL and DL resource in asymmetric traffic since the resource division is operated by the duplexer in the hardware. A typical time-division duplex (TDD) system is capable of adjusting the UL and DL transmission in time domain. However, due to the requirement of synchronization in order to eliminate the interference, the UL and DL resource is still fixed. To support asymmetric and time-variant traffic, LTE provides small cell base stations (BSs) with dynamic TDD by supporting seven TDD UL/DL configurations, enabling the BSs dynamically to change the ratio of UP and DL resource to handle the time-variant traffic. Nevertheless, such a scheme also induces two type of interference: BS–BS interference and MS–MS interference. In this chapter the interference issues and several interference mitigation methods will be extensively discussed.

3.1 Dynamic TDD system overview

3.1.1 Introduction

To divide the UL and DL traffic resource, some typical communication systems apply FDD, where different frequency bands are used for transmitting and receiving, the benefit of which is that no interference will be incurred between UL and DL signals. For the symmetrical traffic on UL and DL (e.g., voice service), the FDD system is suitable since the BS is assigned the same amount of radio resource in the UL and DL. Whereas for wireless data services, FDD is not flexible enough to handle this type of dynamic UL/DL traffic due to the character of the UL and DL traffic being asymmetric and time-variant in these cases.

Compared to FDD, TDD is different in that the UL and DL resource is divided in time domain and can be easily adjusted. It possesses an advantage of greater flexibility

Uplink-downlink configuration	Downlink-to-uplink switch-point periodicity	Subframe number									
		0	1	2	3	4	5	6	7	8	9
0	5 ms	D	S	U	U	U	D	S	U	U	U
1	5 ms	D	S	U	U	D	D	S	U	U	D
2	5 ms	D	S	U	D	D	D	S	U	D	D
3	10 ms	D	S	U	U	U	D	D	D	D	D
4	10 ms	D	S	U	U	D	D	D	D	D	D
5	10 ms	D	S	U	D	D	D	D	D	D	D
6	5 ms	D	S	U	U	U	D	S	U	U	D

Figure 3.1 Seven configurations supported by 3GPP in TS 36.211.

in handling the dynamic UL/DL traffic. In the TDD system, the boundary between the UL and DL duty cycle is adaptively adjustable according to service requirements.

Dynamic-TDD (D-TDD) is also supported in the 3GPP standard. In TS 36.211 [1], seven configurations are provided for LTE TDD with respect to different UL and DL traffic ratios. The supported UL/DL configurations are listed in Figure 3.1. Each radio frame consists of ten subframes, and for each subframe in a radio frame, "D" denotes the subframe reserved for DL transmissions, "U" denotes the subframe reserved for UL transmissions, and "S" denotes a special subframe with the three fields: downlink pilot time slot (DwPTS), guard period (GP), and uplink pilot time slot (UpPTS). Every femto cell can select the configuration according to its traffic load of UL and DL.

As shown in Figure 3.1, there are seven configurations: from Configuration 0 to 6. In each configuration, the ratio of UL/DL subframes is different; for example, the UL/DL ratio in Configuration 0 is 6:2, which supports higher UL traffic. The BS can select the configuration according to the ratio of its UL/DL traffic. If one femto cell demands more UL resources, it can choose Configuration 0, with six UL subframes for UL transmission. When one femto cell demands more DL resources, it chooses Configuration 5, which has eight subframes in the support of DL transmission. Other configuration sets can also support different ratios of UL/DL traffic. The femto cell can choose the configuration set that is closest to its DL/UL traffic ratio. By dynamically changing the configuration, the BS can achieve the dynamic UL/DL ratio adaptation. The cycle period of Configurations 3 to 5 is 10 ms, while the cycle period of other configurations is 5 ms.

In addition to the support on the D-TDD technique, 3GPP also studies some issues related to D-TDD. The issues are discussed in a study item called "Further Enhancements to LTE TDD for DL-UL Interference Management and Traffic Adaptation (eIMTA)" in 3GPP. The study item was generated from the 3GPP meeting of RAN1 68 in February, 2012. At that meeting, the discussion on the evaluation of performance gain from dynamic adaptation to the UL/DL traffic and the simulation scenario for evaluation were firstly addressed. The first proposed scenario is the isolated outdoor pico cell.

During the simulation, they found that by dynamically adapting the TDD configuration on a pico cell, the packet throughput and resource utilization is significantly improved. Furthermore, it provides better user experience and power-saving capability on UEs. At the next meeting, 3GPP discussed eIMTA on the multiple cells scenario. There are two types of simulations. The first type is to adapt the traffic without interference mitigation between UL and DL. The second is to sacrifice some flexibility to mitigate the interference. The simulation results indicate that in multiple cells, when the traffic load is light, the traffic adaptation benefits for both UL and DL, and the interference between cells, is slight. However, when the traffic load is heavy, the performance gain of the traffic adaptation is limited under the effect of the interference. Therefore they suggest that the employment of interference mitigation methods is required. At this meeting, they also discussed the adaptation period for eIMTA. At the next meeting, they discussed the scenario of multiple cells with a fixed TDD macro cell and proposed some interference mitigation methods. The conclusion of the technical specification group of D-TDD can be found in TR 36.828 document [2].

In this chapter, an overview of current D-TDD research on small cells and the deployment supported in the standard is given first, and the interference issue on D-TDD and the mitigation methods are illustrated accordingly.

3.1.2 Related work

The first work considering asymmetric traffic and applying D-TDD appeared in a macro cell deployment in 2000. Li *et al.* [3] propose D-TDD on a macro cell BS. In their scheme, each TDD frame is divided into 48 time slots. They assume the system is fully buffered, and analyze its SINR performance. For interference mitigation, they utilize smart antennas with an 8-element, 16-element, and 26-element array to mitigate the interference between macro cell BSs. In the stage of D-TDD implemented on macro cell BSs, smart antennas are the mostly adopted interference mitigation technique for the intense interference between macro cell BSs. However, this smart antennas method may be hard to apply in the D-TDD small cell system, since the small cell is simply deployed and uses omni-directional antennas. Such type of interference is also addressed earlier in a TDD asynchronous problem [4].

In addition to the general TDD macro cell system, D-TDD is also discussed in a TDD/CDMA system. A paper addressing dynamic TDD in a TDD/CDMA system applies the space-time method on multiple antennas and resource allocation on CDMA domain to mitigate the interference [5].

For D-TDD in small cells, a survey [6] illustrates the benefits of the implementation of D-TDD within small cells. In a small cell, since the coverage of a small cell BS is limited compared to a macro cell BS, the required transmission power of a small cell BS is smaller than that of a macro cell BS. Therefore the interference level is lower in small cells. Moreover, the channel attenuation between small cells is higher than the path loss between macro cells, since the transmission between small cells is often non-line-of-sight.

In small cell TDD systems, the majority of the papers discuss the TD-LTE system since the D-TDD is supported in TR 36.828 [2] and the frame structure is supported in TS 36.211 [1]. In this work, most of the papers are in the stage of performance analysis to see the performance of D-TDD in the TD-LTE system. Among these papers, several papers apply some interference mitigation methods on D-TDD. For instance, Khoryaev *et al.* [7] demonstrate the feasibility of D-TDD in outdoor pico cells and evaluate the performance of D-TDD in outdoor pico cells. As a result, the D-TDD system has throughput gain compared to the typical TDD system. By taking into consideration the different conditions of buffer status and conducting reconfiguration more efficiently on the resource utilization, Wang *et al.* [8] evaluate D-TDD in the LTE system, and show that through the implementation of D-TDD, the throughput on both UL and DL are improved. Other than the TD-LTE system, Chiang *et al.* [9] propose the D-TDD scheme in the IEEE 802.16 system, perform the performance analysis on D-TDD, and prove that D-TDD obtains performance gain as well in the IEEE 802.16 system. There are also some studies that discuss how to determine the TDD configuration. Kim *et al.* [10] propose a mechanism to determine the TDD configuration considering both traffic and channel quality.

Some papers investigate the D-TDD system from different aspects in system assumptions. Yu *et al.* [11] utilize stochastic geometry to obtain the distribution of DL and UL SINR, assuming that the traffic in this system is fully buffered and there is no coordination between the subframes in the same small cells. The transmission direction of the subframes is determined by probability, and they analyze the optimal probability. Lassila *et al.* [12] analyze D-TDD in the flow level by using queueing analysis and no full-buffer traffic model to propose the optimal policy of dynamically selecting the configurations.

Some of the works apply interference-mitigation methods on D-TDD to improve the performance degraded by new type of interference. The authors in [11] apply open-loop power control for interference mitigation, and the throughput in UL is greatly enhanced with little sacrifice on the throughput in the DL. Al-Rawi and Jantti [13] propose inter-cell interference coordination to mitigate the interference. By coordinating the transmission time and mode of users in neighboring cells, they reduce the interference between neighboring cells. Another work proposed by Ji *et al.* [14] also applies inter-cell interference coordination on the D-TDD system. Song and Jin [15] apply the beamforming technique for interference mitigation. The combination of resource allocation with the beamforming technique separates them from the previous work. The throughput improvement by dual-layer beamforming is shown in their simulation. Dowhuszko *et al.* [16] propose a distributed cooperative scheme to select the TDD configuration to enhance system throughput. Janis *et al.* [17] analyze the D-TDD system performance within two femto cell BSs and propose decentralized radio resource management for multiple femto cell BSs to improve the throughput.

Concerning the issues of deployment scenarios, Khoryaev *et al.* [18] analyze adjacent macro–femto deployment and co-channel macro–femto deployment in their simulation. The results show that the femto cell is able to potentially change the transmission direction when the macro cell is in UL while the femto cell can only be in DL when the macro cell is in DL considering the strong interference from the macro cell.

In the following sections, we will illustrate the D-TDD interference problem in detail and categorize the current interference mitigation methods.

3.2 D-TDD deployment and issues

The first issue in deployment is the adaptation period. Due to the characteristic of the data traffic in a femto cell being bursty, the UL/DL configuration should change periodically. Depending on the required adaptation time scales, different methods are therefore considered in applications for TDD UL/DL reconfiguration. There are three different adaptation time scales: 640 ms, 200 ms, and 10 ms, each of which is supported by one or two different methods. The four possible supporting methods, which use current signaling to indicate the change of UL/DL configuration, are introduced in the following paragraph; however, some issues related to the four methods need further study.

The first method is to support TDD UL/DL reconfiguration via system information signaling. In this method, the TDD UL/DL configuration is indicated by SIB. With the Rel-8 system information change procedure, it can support 640 ms [19] as the time scale for TDD UL/DL reconfiguration. Not only new UEs can use the new configuration, but the legacy UEs are capable of receiving the benefits of TDD UL/DL reconfiguration because the method to adapt the TDD UL/DL configuration is backward compatible. However, this method always affects all UEs connected to the cell, even those that do not have data to transmit or receive. The other disadvantage of this method is that an ambiguity exists between the eNB and the UE on the TDD UL/DL configuration. Owing to the inability in perceiving the exact time when the UE correctly decodes the updated SI, the eNB may apply a scheduling restriction during the uncertain period to maintain the communications between the eNB and the UE properly. Further study is required to assess its impact on performance.

The second method is to support TDD UL/DL reconfiguration by RRC signaling. The corresponding time scale supported by this method is in the order of 200 ms, a smaller time scale for the reconfiguration, which is better than the first method. However, this method is not applicable for legacy UEs. A possible solution is to stop the usage of the legacy UE on the changed subframe in the new configuration, but such solution is unreasonable in that it seriously restricts the usage of legacy UEs. It needs another solution for this situation. This method also has the problem of ambiguity existing between eNB and UE on the TDD UL/DL configuration.

The third method is to support TDD UL/DL reconfiguration by MAC control element signaling in the MAC header with the ability to accommodate the time scale of adaptation on the order of a few tens of milliseconds. It provides an even smaller time scale than the former two methods to support the reconfiguration period of 200 ms. Nevertheless, as with the second method, it is inapplicable to legacy UEs and the ambiguity existing between eNB and UE on the TDD UL/DL configuration remains if the eNB does not know the exact time that updated UL/DL configuration is applied on the UE.

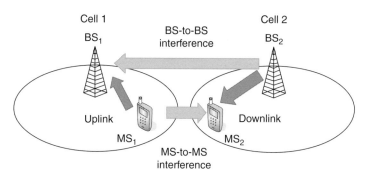

Figure 3.2 Cross-slot interference.

The fourth method is to support TDD UL/DL reconfiguration by physical layer design with a time scale of adaptation in the order of 10 ms. This method provides the shortest adaptation period. The transmission direction of a subframe can be explicitly indicated by a physical channel or signal. However, this method is unable to be applied to legacy UEs as with the second and third methods. Further study is also required for this method.

3.2.1 Interference problem

Dynamic TDD provides system throughput gain thanks to its adaptation ability in UL/DL configuration, but it also generates some other issues. The most severe is the interference within small cells. Typical types of interference in small cells are co-tier interference and inter-tier interference: co-tier interference is between small cells while inter-tier interference is between the macro cell BS and small cells, both including DL–DL interference and UL–UL interference. In D-TDD, since the UL/DL configuration is different between each cell, there exists interference between different transmission directions, i.e., DL–UL interference and UL–DL interference.

As shown in Figure 3.2, femto cell 1 is receiving the UL signal from its MS whereas femto cell 2 is transmitting the DL signal to its MS. The DL–UL interference, also called BS–BS interference, is caused by the DL signal from femto cell BS 2 to the UL signal from MS 1 to femto cell BS 1. The UL–DL interference, also called MS–MS interference, is caused by the UL signal from MS 1 to the DL signal from femto cell BS 2 to MS 2. Moreover, since the power level of the femto cell BS exceeds that of the MS, the BS–BS interference results in significant decrease in UL SINR.

Furthermore, the induced interference differs from different D-TDD deployment scenarios within small cells. Dynamic TDD deployment scenarios of small cells include: (1) small cells without a macro cell, (2) small cells with a macro cell on a different frequency, (3) small cells with a macro cell on the same frequency. In scenario 1 of small cells only, the BS–BS interference and the MS–MS interference are only between small cells. In scenario 2, since the macro cell and the small cells are on a different frequency, the interference is also between small cells. In scenario 3, since the macro cell and the small cells are on the same frequency, interference is between the macro

Table 3.1 Parameters settings in simulation of dynamic TDD.

Thermal noise	−174 dBm/Hz
Bandwidth	10 MHz
Dual-stripe model	
Number of femto BSs	24
Number of UEs	48 (2 UE per femto BS)
Number of rows per floor	4
Maximum number of cells per row	10
Number of blocks per cell	1
Number of floors per block	6
FTP traffic model 2 in 3GPP	
File size	0.2 M bytes
UL power control	Femto UE: P0 = −75 dBm; alpha = 0.8
Minimum distance between UE and cell	3 m
Shadowing standard deviation between UE and UE	12 dB
UE power class	23 dBm (200 mW)
UE noise figure	9 dB
UE antenna gain	0 dBi
Femto modeling parameters	
Femto antenna pattern	Omni-directional
Femto antenna gain	0 dBi
Shadowing standard deviation between UE and femto	4 dB
Femto noise figure	13 dB
UE power class	20 dBm

cell and small cells. There are macro cell BS–femto cell BS interference, femtocell BS–macro cell BS interference, macrocell MS–femto cell MS interference, and femto cell MS–macro cell MS interference. Interference between femto cells also exists in this scenario.

In scenario 3, where the macro cell and the femto cells are on the same frequency, the strong interference from the macro cell causes severe decrease in UL SINR of femto cells. Therefore the deployment of scenario 2 is more preferred than scenario 3, or interference mitigation methods on the macro cell should be applied in scenario 3 for the elimination of such significant interference.

3.3 Simulation methodology

Aiming at analyzing the D-TDD system, we perform the simulation under the femto cell BSs deployment scenario. In this scenario, we employ the dual-stripe model [2] as the deployment scenario and path-loss model. For MSs, we assume each femto cell BS has one MS. In this model femto cell BSs are uniformly distributed in the area of buildings with the associated MSs located in the same buildings with the femto cell BSs.

The remaining parameters used in the simulation are listed in Table 3.1.

For the path-loss model in the dual-stripe model, we categorize the path-loss model into three cases. The first case considers that femto cell BSs are inside a different

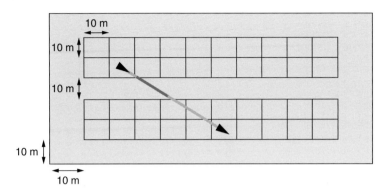

Figure 3.3 Path-loss model where HeNBs are inside a different apartment stripe.

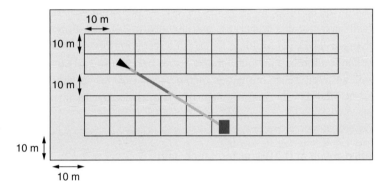

Figure 3.4 Path-loss model where UE is inside a different apartment stripe.

apartment stripe, as shown in Figure 3.3. In this case, the path-loss model is

$$PL(dB) = \max(2.7 + 42.8 log10R, 38.46 + 20log10R) + 0.7d_{2D,indoor}$$
$$+ 18.3n((n+2)/(n+1) - 0.46) + q * L_{iw} + L_{ow,1} + L_{ow,2}.$$

The distance R is composed of a lighter line and a darker line in Figure 3.3 and the distance $d_{2D,indoor}$ is the lighter line in the same figure. The notation q is the number of walls separating apartments between femto cell BSs. L_{iw}, $L_{ow,1}$, and $L_{ow,2}$ are the shadowing factors in the dual-stripe model.

As shown in Figure 3.4, the second case considering the MS is inside a different apartment stripe of femto cell BSs. In this case, the path-loss model is

$$PL(dB) = \max(2.7 + 42.8 log10R, 38.46 + 20log10R) + 0.7d_{2D,indoor}$$
$$+ 18.3n((n+2)/(+1) - 0.46) + q * L_{iw} + L_{ow,1} + L_{ow,2}.$$

The third case considers that the MS is inside the same apartment stripe as the femto cell BSs are, as shown in Figure 3.5. The path-loss model in this case is

$$PL(dB) = 38.46 + 20log10R + 0.7d_{2D,indoor}$$
$$+ 18.3n((n+2)/(n+1) - 0.46) + q * L_{iw}.$$

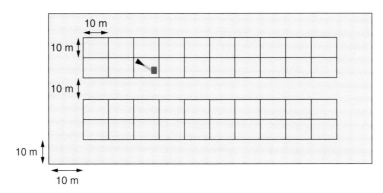

Figure 3.5 Path-loss model where UE is inside the same apartment stripe as HeNB.

In the following sections, we will present several simulation scenarios. (1) The isolated femto cell scenario: this scenario is not included in TR 36.828, but shows the benefit of applying D-TDD. (2) Multiple femto cells scenario: this scenario is included in TR 36.828, and it addresses the interference issues between D-TDD femtocells. (3) Femto cells with an interference-mitigation method scenario: this scenario is not included in TR 36.828. It shows the improvement of performance by the interference-mitigation method.

3.3.1 Simulation result of D-TDD performance gain

In the first simulation with two schemes, fixed TDD scheme and dynamic TDD scheme, respectively being performed on an isolated femto cell, we show the benefits of using D-TDD. The traffic model in this simulation is FTP traffic model 2, which is standardized in 3GPP, TR 36.814 [20]. The file size is 0.2 M bytes, and the UL file size is 0.05 M bytes. The packet arrival rate of UL is two packets/subframe, and the packet arrival rate of the DL is changed from one, two, four, to eight packets/subframe.

After that, set the configuration for both fixed TDD scheme and D-TDD scheme. In the fixed TDD scheme, the configuration settings for different UL and DL arrival rate are as follows.

1. Use Configuration 0 when the arrival rate of UL is two packets/subframe and the arrival rate of DL is one packet/subframe.
2. Use Configuration 1 when the arrival rate of UL is two packets/subframe and the arrival rate of DL is two and four packets/subframe.
3. Use Configuration 2 when the arrival rate of UL is two packets/subframe and the arrival rate of DL is eight packets/subframe.

In the D-TDD scheme, the configuration is dynamically adjusted according to the traffic.

The result is shown in Figure 3.6. We can observe that the throughput of D-TDD outperforms the throughput of fixed TDD in a higher traffic arrival rate. The reason is that when the arrival rate is higher, the UL/DL resource is not fully utilized in fixed

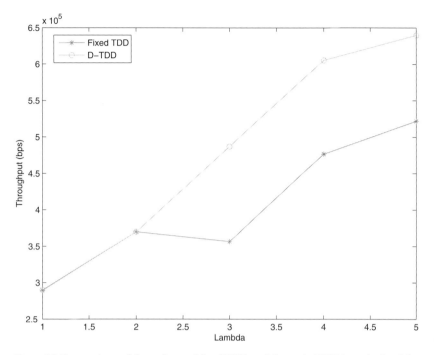

Figure 3.6 Comparison of throughput of fixed TDD and dynamic TDD in an isolated femto cell.

TDD. In contrast, in D-TDD the ratio of UL/DL is able to be adjusted according to the instantaneous change of UL and DL traffic load. Therefore D-TDD provides higher throughput in comparison to fixed TDD.

3.3.2 Simulation result of interference problem in D-TDD

In order to present the interference issue between small cells in D-TDD, we perform the simulation to show the effect of interference on D-TDD throughput. In this simulation, the scenario of multiple femto cell BSs is operated. We generate 24 femto cell BSs in the dual-stripe model. There are six floors in the dual-stripe model, and 40 rooms on each floor, so the total number of rooms is 240 in the dual-stripe model. The femto cell BSs are generated within 24 rooms randomly chosen from the 240 rooms. We set two MSs per femtocell BS. The example of the generated femto cell BSs location is shown in Figure 3.7. We compare both UL SINR on femto cell BSs and DL SINR on MSs in the fixed TDD scheme and the dynamic TDD scheme in this scenario.

For the traffic load, each femtocell BS is randomly assigned a traffic pattern. We also used FTP traffic model 2 in the specification as in the previous simulation. The DL file size is 0.2 M bytes, and the UL file size is 0.05 M bytes. The packet arrival rate of UL is two packets/subframe. We change the packet arrival rate of DL from one, two, four, to eight packets/subframe.

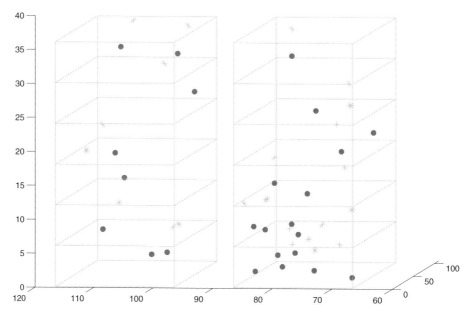

Figure 3.7 Femto cell BSs generated in dual-stripe model.

After that, we set the configuration for both the fixed TDD scheme and the D-TDD scheme. In the fixed TDD scheme, the configuration settings for different UL and DL arrival rate are as follows.

1. Use Configuration 0 when the arrival rate of UL is two packets/subframe and the arrival rate of DL is one packet/subframe.
2. Use Configuration 1 when the arrival rate of UL is two packets/subframe and the arrival rate of DL is two and four packets/subframe.
3. Use Configuration 2 when the arrival rate of UL is two packets/subframe and the arrival rate of DL is eight packets/subframe.

In the D-TDD scheme, the configuration is dynamic according to the traffic.

The result of the UL SINR value is shown in Figure 3.8. It shows that the range of SINR of fixed TDD is from −15 to 15 dB and the range of SINR of D-TDD is from −40 to 10 dB. We can see that the interference of UL in D-TDD is more severe. This is because that in fixed TDD, an UL femto cell is only interfered by the other UL transmission, which has lower transmission power. However, in D-TDD, an UL femto cell can be interfered with by some DL transmissions, which have higher transmission power.

In DL SINR, the result in Figure 3.9 shows that the SINR are almost the same in fixed TDD and dynamic TDD since the effect of interference in the DL is smaller. A DL femto cell is interfered with by the power of both the UL and DL femto cell BSs, which are a little bit lower than the fixed TDD.

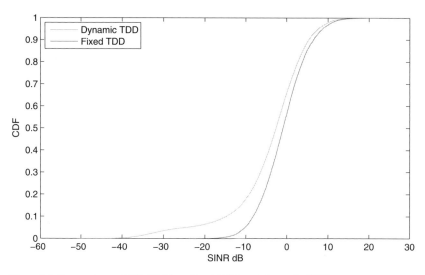

Figure 3.8 Comparison of UL SINR on fixed TDD and dynamic TDD.

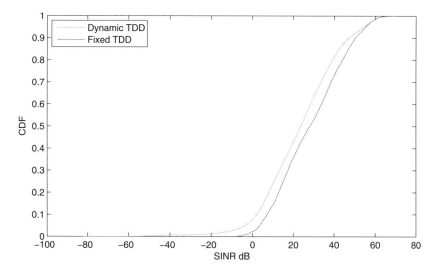

Figure 3.9 Comparison of DL SINR on fixed TDD and dynamic TDD.

The strong interference on the UL affects the performance of D-TDD. It reduces the throughput gain of D-TDD. Therefore interference mitigation is needed to improve the performance of D-TDD.

3.4 Interference mitigation method for D-TDD

The first interference mitigation method is to eliminate the interference between the tightly coupled femto cell base stations, which is called the cell clustering. The operation of it is to figure out the femto cells that affect each other the most and group them into a cluster. The effect can be quantized as a factor, namely the mutual coupling loss (MCL). The MCL between femto cell BS1 and femto cell BS2 is the sum of the transmit antenna gains and the receive antenna gains minus the path-loss factor. For the femto cell BSs in the same cluster, they will use the same configuration, so the transmission direction in each subframe will be the same. As a result, the BS–BS interference and MS–MS interference can be mitigated within the cell cluster. Since the femto cell BSs in a cluster use the same configuration, some flexibility of UL/DL reconfiguration is inevitably sacrificed, which is the tradeoff of cell clustering.

The cell-clustering method is also proposed in [21]. In this paper, the authors grouped the femto cells that highly interfered with each other into the same cluster. The specialty distinguishing their clustering from other typical clustering schemes is that the femto cells in a cluster can use different configurations in their scheme. With the application of appropriate DL/UL configurations, they optimize the system throughput taking in consideration interference and traffic load.

The second scheme is scheduling dependent interference mitigation (SDIM). In this scheme, the eNB adjusts the scheduling strategies through taking some situations into consideration. The scheduling strategies include link adaptation, resource allocation, transmit power, and transmission direction of a subframe, while the considered situations contain the DL and UL channel quality, the eNB–eNB and UE–UE interference, traffic load, etc. In this scheme, the eNB should measure the interference level from/to another eNB.

There are some ideas of interference mitigation method design in SDIM. To reduce the interference, the transmission power of the femto cell BSs is determined based on the path loss to the neighboring cells such that the pico cell's DL transmission causes interference no higher than a predetermined target interference over thermal (IoT) level in UL reception at the eNB that is the closest to the pico eNB.

Moreover, in SDIM, the femto cell BSs can coordinate the usage of each time/frequency resource in order to prevent severe interference. For example, if an eNB is able to inform the neighboring cells of the set of subframes where it intends to change the communication direction (e.g., from UL to DL), the neighboring cells are enabled to utilize this information to avoid UL transmissions vulnerable to the interference from that eNB.

3.4.1 Simulation of clustering

In this simulation, we apply clustering for interference mitigation. In our clustering algorithm, we randomly pick a femto cell BS as the anchor BS, and check the mutual coupling loss between the chosen femto cell BS and other femto cell BSs. The mutual coupling loss threshold is –70 dB. When the mutual coupling loss of the femto cell BS is above the threshold, the femto cell is categorized into the same cluster with the anchor BS.

The simulation result is shown in Figure 3.10 and Figure 3.11. We can see that the difference of DL SINR between cluster and no cluster is quite little for the interference problem is minor on DL in D-TDD. However, in UL, the SINR of the cluster is higher

Dynamic TDD small cell management 71

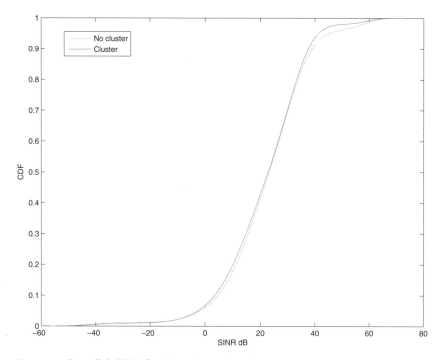

Figure 3.10 Downlink SINR for clustering and no clustering.

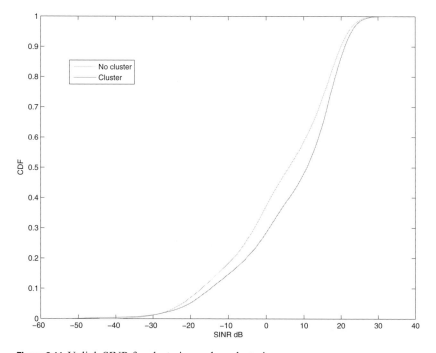

Figure 3.11 Uplink SINR for clustering and no clustering.

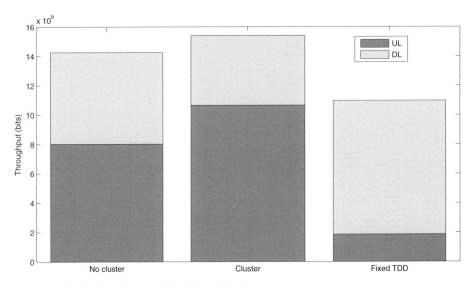

Figure 3.12 Total throughput for clustering and no clustering.

than that of no cluster. This suggests that the interference problem is mitigated through the employment of the clustering method. By using the same configuration in highly interfered with femto cell BSs, the interference is reduced on the adjacent femto cell BSs.

The throughput of using cluster and no cluster is shown in Figure 3.12. We can observe that the UL throughput is significantly improved due to the higher UL SINR in the clustering scheme. Despite the sacrifice of some flexibility of reconfiguration, the overall throughput of the clustering scheme still outperforms the scheme of no clustering. This proves that clustering is an efficient interference mitigation method in this case.

The progress in interference mitigation methods is still at an embryonic stage with many details awaiting to be determined; therefore there exist innumerable challenges and opportunities for researchers in interference mitigation design to further enhance the performance of D-TDD in small cells.

3.5 Conclusion

Dynamic TDD is beneficial for asymmetric and time-variant traffic in the improvement of the utilization in frame resource through adjusting the UL and DL configurations. However, it results in a new interference issue. BS–BS interference and MS–MS interference occur when the adjacent small cells are in different transmission directions. The two interference types, especially the BS–BS interference, cause significant degradation in the throughput gain from D-TDD. The effect of interference is shown in the simulation. In order to alleviate interference, interference mitigation methods are proposed to enhance the system throughput. In the simulation, we apply the clustering method on D-TDD. The result indicates that the SINR and system throughput are improved. The optimal interference mitigation methods for specific traffic models are still under investigation.

References

[1] "Radio frequency (RF) system scenario (release 10)," *3GPP TS 36.211*, June 2012.
[2] "Further enhancements to LTE time division duplex (TDD) for downlink-uplink (DL–UL) interference management and traffic adaptation," *3GPP TR 36.828*, June 2012.
[3] J. Li, S. Farahvash, M. Kavehrad, and R. Valenzuela. "Dynamic TDD and fixed cellular networks," *IEEE Communication Letters*, 2000.
[4] J. C.-I. Chuang. "Performance limitations of TDD wireless personal communications with asynchronous radio ports," *Electron. Lett.*, 28(6): 532–534, March 1992.
[5] I. Spyropoulos and J. R. Zeidler. "Supporting asymmetric traffic in a TDD/CDMA cellular network via interference-aware dynamic channel allocation and space v time LMMSE joint detection," *IEEE Trans. on Vehicular Technology*, February 2009.
[6] Z. Shen, A. Khoryaev, E. Eriksson, and X. Pan. "Dynamic uplink–downlink configuration and interference management in TD-LTE," *IEEE Communications Magazine*, November 2012.
[7] A. Khoryaev, A. Chervyakov, M. Shilov, S. Panteleev, and A. Lomayev. "Performance analysis of dynamic adjustment of TDD uplink–downlink configurations in outdoor picocell LTE networks," *3rd International Workshop on Recent Advances in Broadband Access Networks 2012*, pp. 914–921, 2012.
[8] Y. Wang, K. Valkealahti, K. Shu, R. Sankar, and S. Morgera. "Performance evaluation of flexible TDD Switching in 3GPP LTE system," *2012 IEEE Sarnoff Symposium*, pp. 1–4, 2012.
[9] C. H. Chiang, W. Liao, T. Liu, I. K. Chan, and H. L. Chao. "Adaptive downlink and uplink channel split ratio determination for TCP-based best effort traffic in TDD-based WiMAX networks," *IEEE Journal on Selected Areas in Communications*, 27(2):182–190, February.
[10] J. S. Kim, B.-G. Choi, S. J. Bae, I. Doh, and M. Y. Chung. "Adaptive time division duplexing configuration mode selection mechanism for accommodating asymmetric traffic in time division long term evolution-advanced systems," *Trans. Emerging Tel. Tech.*, March 2013.
[11] B. Yu, S. Mukherjee, H. Ishii, and L. Yang. "Dynamic TDD support in the LTE-B enhanced local area architecture," *GC'12 Workshop: The 4th IEEE International Workshop on Heterogeneous and Small Cell Networks (HetSNets)*, pp. 585–591, 2012.
[12] P. Lassila, A. Penttinen, and S. Aalto. "Flow-level modeling and analysis of dynamic TDD in LTE," *2012 8th Euro-NF Conference on Next Generation Internet (NGI)*, pp. 33–40, 2012.
[13] M. Al-Rawi and R. Jantti. "A dynamic TDD inter-cell interference coordination scheme for long term evolution networks," *2011 IEEE 22nd International Symposium on Personal, Indoor and Mobile Radio Communications*, pp. 1590–1594, 2011.
[14] H. Ji, Y. Kim, S. Choi, J. Cho, and J. Lee. "Dynamic resource adaptation in beyond LTE-A TDD heterogeneous networks," *IEEE International Conference on Communications 2013: IEEE ICC'13*, pp. 133–137, 2013.
[15] P. Song and S. Jin. "Performance Evaluation on dynamic dual layer beamforming transmission in TDD LTE system," *The 3rd International Conference on Communications and Information Technology (ICCIT-2013)*, pp. 269–274, 2013.

[16] A. A. Dowhuszko, O. Tirkkonen, J. Karjalainen, T. Henttonen, and J. Pirskanen. "A decentralized cooperative uplink/downlink adaptation scheme for TDD small cell networks," *International Symposium on Personal, Indoor and Mobile Radio Communications*, pp. 1682–1687, September 2013.

[17] P. Janis, C. B. Ribeiro, and V. Koivunen. "Flexible UL-DL switching point in TDD cellular local area wireless networks," *Mobile Network Applications*, 17: 695–707, October 2012.

[18] A. Khoryaev, M. Shilov, S. Panteleev, A. Chervyakov, and A. Lomayev. "Feasibility analysis of dynamic adjustment of TDD configurations in macro–femto heterogeneous LTE networks," *Internet of Things, Smart Spaces, and Next Generation Networking*, pp. 174–185, 2012.

[19] "Evolved universal terrestrial radio access (E-UTRA); radio resource control (RRC); protocol specification (Release 10)," *3GPP TR 36.331 v10.11.0*, September 2013.

[20] "Further advancements for E-UTRA physical layer aspects (Release 9)," *3GPP TR 36.814 v9.0.0*, March 2010.

[21] D. Zhu and M. Lei. "Cluster-based dynamic DL–UL reconfiguration method in centralized RAN TDD with trellis exploration algorithm," *2013 IEEE Wireless Communications and Networking Conference*, 50(10), April 2013.

4 3GPP RAN standards for small cells

Weimin Xiao, Jialing Liu, and Anthony C. K. Soong

4.1 Introduction

The so-called Third Generation Partnership Project (3GPP) Long Term Evolution (LTE) and its evolved version, LTE-Advanced, is currently the most prominent and advanced mobile communication system. When it was designed and standardized, starting in 2004 and first released in 2008 as Release 8, it was targeted mainly for macro base station networks with uniform, well-planned and deployed high-power nodes where coverage, mobility, and the provision of high throughput across large areas are at the heart of its requirements. As a consequence, small cell (or low-power node) deployments together with high-power macro base stations were not considered for the original designs. The specific issues with the introduction of small cells in the system therefore cannot be addressed sufficiently without additional features to deal with interference, mobility, traffic load management, etc., related to such deployments.

The major problems from deploying small cells, especially together with macro base stations, include:

1. **More severe interference conditions**. Although interference is always a key issue for cellular communication and handling interference is built into the core of LTE designs, coexistence of network nodes of different power levels, especially in the co-channel scenario where the same frequency channel (known as the component carrier) is used for both macro base station cells and small cells, results in much worse interference condition than before. The LTE system generally has a very robust physical layer design to ensure that each physical channel can be reliably received at fairly low signal to interference-plus-noise ratio (SINR) range. However, in order to fully utilize the potential of small cells to offload traffic, small cells sometimes need to serve users at even lower SINR, which requires a mechanism to either avoid or cope with strong interference.
2. **Mobility management and traffic load balancing**. The introduction of small cells with low transmission power basically creates cells with a small footprint within the system, which increases the frequency of handovers between cells due to user mobility. This then results in dramatically larger handover failure as well as associated backhaul signaling. Furthermore, also because of much smaller coverage area, the traffic loads between the cells are more likely to be unbalanced and time-varying, and a more efficient load-balancing and shifting mechanism is needed.

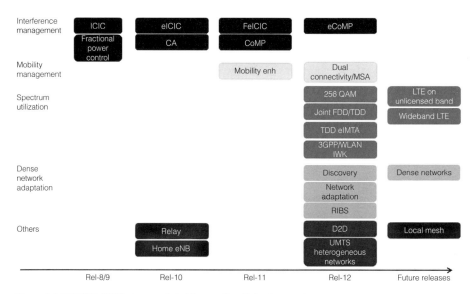

Figure 4.1 3GPP RAN features related to small cell deployments.

3. **Efficient utilization of spectrum resource**. Small cell deployments result in different channel conditions, levels of load variation, and feasible spectrum resource than macro base station networks. The efficiency of an LTE small cell network can be further improved to take advantage of these specific characteristics.
4. **Increasing network density**. The gain of cell splitting of small cell deployment comes from the increasing network density, which generates interesting behaviors and issues. Accordingly, standards efforts are taken to enhance LTE to cope with high network density in the areas of network adaptation, small cell discovery, etc.

Figure 4.1 summarizes the related features developed in LTE during releases 8 to 12 and potentially in the future releases to address the above issues. The rest of this chapter describes each of these issues and related features developed by the 3GPP radio access network (RAN) standards body, and concludes with future features for small cell networks. The reader may refer to the appendix to this chapter for a brief introduction to the channels involved in LTE. The acronyms of the channels are directly used in this chapter and are spelled out in the appendix.

4.2 Interference management

The heterogeneous network (Figure 4.2) is a network deployment supported in LTE that can effectively deliver high throughputs to user equipment (UE). The heterogeneous network is characterized by the overlay of low-power node (LPN) layers on top of the high-power node layer. The LPNs may include relays, micros, picos, femtos, and remote radio heads (RRHs), and the high-power nodes are typically macros. The coverage is mainly provided by the macro layer, whereas the high throughput is provided by the LPN

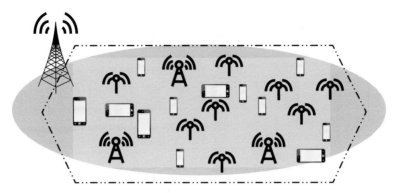

Figure 4.2 A diagrammatic representation of the deployment of a heterogeneous network.

layer. The main reasons for the high-throughput performance of the LPN layer are that the LPN layer can have significant cell-splitting gain thanks to the increased network resources available to serve the UEs, and that it allows more efficient communications of the network nodes and UEs via the reduced distances (and hence reduced path loss) between them. Therefore the heterogeneous network is considered as a key enabler for LTE-Advanced and future evolutions. Among the different types of LPNs, the picos are of the most interest in LTE-Advanced study.

There are some fundamental designs of LTE from its beginning that can be used to handle interference in heterogeneous networks. These include an inter-cell interference control (ICIC) feature developed during Release 8 with the specification of relative narrowband transmit power (RNTP), overload indication (OI), and high interference indication (HII) signaling between the eNodeBs.

Relative narrowband transmit power signaling is for coordination of downlink power control between cells to facilitate fractional frequency reuse (FFR) for downlink data channels. However, the limitation of such a feature is that it cannot deal with the interference to/from control channels and reference signals. Its usefulness is further limited as the reception side of these signals has no obligation of desired actions.

For LTE uplink, fractional power control [1, 2] as one of the fundamental features has proved to be very robust and useful even for heterogeneous networks as shown in related studies. Therefore no severe issue has been identified so far for the uplink part. In addition, OI and HII signaling can be used to coordinate uplink power control.

In the following, downlink interference for co-channel and non-co-channel deployments and the related 3GPP RAN features are described.

4.2.1 Co-channel scenario

4.2.1.1 ICIC/eICIC/FeICIC

There are a few challenges in heterogeneous network support, more specifically in the macro–pico scenario support. The most important challenge is the interference caused by sharing the same time-frequency resources for both the macro layer and pico layer. Especially the macro interference to the UEs attached to the picos may be strong due to

Figure 4.3 ABS-based interference management (eICIC) for heterogeneous networks.

the high power of the macros, and it may severely impact the pico UE PDCCH reception. Another challenging aspect is that the pico range is rather small due to its low power, and hence the likelihood of a UE selecting a pico over macro based on the reference signal received power (RSRP) may be small, leading to limited performance improvements.

One way to address the issue, related to the very limited range of pico cells, is to signal a cell selection bias to a UE so that the pico layer can be more effectively utilized. This method is generally termed cell range extension (CRE). Rel-8 and Rel-9 support range expansion up to about 3 dB bias, due to the control channel performance limitation.

However, CRE further increases the interference from the macro to pico UEs, and thus CRE should be small to moderate, at c. 3 to 6 dB, or CRE should be used together with additional interference management methods. These methods, which can be used to address both of the abovementioned challenges, are discussed below.

To alleviate the interference issue and also to support larger bias toward small cells, Rel-10 also introduced the enhanced ICIC, or eICIC, feature. In particular, eICIC is enabled by two techniques: the almost blank subframes (ABS) and resource-restricted measurements. In one ABS, the macro mutes its transmission of PDSCH and PDCCH used to schedule PDSCH. This reduces the interference from the macro to pico UEs; however, the macro interference cannot be completely eliminated in ABS since the macro CRS (and some other common channels as well) cannot be muted for legacy and measurement support. Nevertheless, ABS may be useful for a pico to transmit to its UEs, especially cell-edge UEs, with reduced interference (i.e., improved SINR), and the pico and pico UEs should exploit the instances with higher SINR by, for example, performing proper link adaptation differently in a macro ABS and macro non-ABS. This leads to the introduction of resource-restricted measurements, by which is meant that a pico signals measurement patterns to its UE such that the UE can measure and average its interference and hence report its channel quality indicator (CQI) feedback separately for the protected subset of subframes and non-protected subset of subframes. Then the pico can schedule the UE properly according to the CQI reports. In short, eICIC is a time division multiplex (TDM) based interference management employing ABS and resource-restricted measurements. An example ABS allocation is illustrated in Figure 4.3.

Subframe-specific measurement and reports are needed to support the time domain interference variations expected in the heterogeneous networks' eICIC operation. In

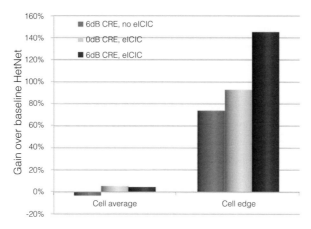

Figure 4.4 The potential system gain in LTE-A with four picos per macro cell. The baseline has 0 dB CRE and no eICIC.

particular, the radio link control (RLC), radio resource management (RRM), and channel state information (CSI) measurements for LTE Rel-10 UEs are restricted to certain subframes. Radio resource control signaling is used to inform the UE across which resources interference can be averaged for the measurement reports.

Figure 4.4 shows the possible system throughput gains by applying CRE and/or eICIC over baseline heterogeneous network performance. The baseline is obtained with four picos in each macro cell area, but without any CRE or eICIC. It is seen that moderate CRE and/or eICIC provide significant throughput benefits, especially at the cell edge, i.e., fifth percentile user throughput.

Rel-11 defined further enhancements to non-carrier aggregation based eICIC (FeICIC). 3GPP RAN1 considered and supports UE performance requirements for UE receiver based techniques for 9 dB cell range expansion bias. The UE can cancel interference on common control channels of ABS caused by interfering cells, such as CRS signals of high-power macro cells. The interference cancellation receiver fully handles colliding and non-colliding CRS scenarios and removes the need for cell planning of heterogeneous deployment. Without an IC-capable UE receiver, heterogeneous networks' eICIC can only work effectively for non-colliding CRS cases.

The performance gains from heterogeneous networks using eICIC in Rel-10 are about 25 to 50% [3]. FeICIC in Rel-11 provides additional gains. The gains vary and are in the range of 10 to 35% [4, 5].

4.2.1.2 CoMP and eCoMP

Coordinated multi-point transmission/reception (CoMP) can be utilized to improve coverage, cell-edge throughput, and/or system efficiency. The interested reader may refer to [6] for more detailed descriptions of the study and work done in 3GPP.

CoMP may be effective in the following scenarios. In an interference-limited network, a UE may be able to receive signals from more than one cell at possibly different locations, and by the same token, the UE's transmission may be received at more than one cell at possibly different locations. Thus if the transmissions from more than one cell at the possibly different locations can be coordinated to result in constructively superposed signals or less interfered signals at the UE receiver, the downlink performance may be improved. This is called DL-CoMP. Likewise, UL-CoMP may be done by the network taking advantage of receptions at more than one cell.

The following four scenarios were considered for 3GPP Rel-11 CoMP study and standardization:

Scenario 1: homogeneous network with intra-site CoMP
Scenario 2: homogeneous network with high transmission power RRHs
Scenario 3: heterogeneous network with low-power RRHs within the macro cell coverage where the transmission/reception points created by the RRHs have different cell IDs as the macro cell
Scenario 4: heterogeneous network with low-power RRHs within the macro cell coverage where the transmission/reception points created by the RRHs have the same cell IDs as the macro cell

In Scenario 1, CoMP is done for the cells at the same site location. The communication needed to support CoMP is limited at the site and does not involve backhaul connections, which link cells at different locations. However, the resultant interference management is also limited to cells at the same site location.

In Scenarios 2, 3, and 4, the cells at different locations are generally assumed to be connected via ideal backhaul with zero latency and sufficient capacities. Such assumptions may be approximately true in reality if there is a dedicated fast connection between any two sites involved in CoMP.

With the ideal backhaul assumption, three different DL CoMP approaches were studied: coordinated scheduling (CS) or coordinated beamforming (CB or CBF); dynamic point selection (DPS) or dynamic point blanking (DPB); and coherent or non-coherent joint transmission (JT). In the approach of CS/CB, the transmission to a UE is transmitted from its serving cell, similar to non-CoMP transmission, but the scheduling and beamforming is coordinated between the cells to reduce the interference between different transmissions. For example, when a cell is scheduling a cell-edge UE, neighboring cells that may cause strong interference to the UE will not perform any transmission toward the UE, by proper scheduling or beamforming (i.e., precoding).

For DPB and DPS, a UE may receive signal from more than one serving transmission point (TP), but only one single TP can transmit to the UE at each instant of time. For example, in DPS this single point can dynamically change from subframe to subframe, and other serving TPs that are not transmitting to this UE may be transmitting to other UEs at the same time. In DPB, on the other hand, this single point can dynamically change from subframe to subframe, and one or more other serving TPs that are not transmitting to this UE may be muting at the same time, thus reducing interference to the UE.

For JT, the transmission to a single UE is simultaneously transmitted from more than one TP. In coherent JT, joint precoding is applied to all TPs under coordination and the transmission is as if from a single transmitter, even though the antennas may be actually geographically separated. This requires UE feedback jointly for all antenna ports of the TPs. In non-coherent JT, however, UE can provide feedback for individual TPs and the separate precoding is done for each TP.

To support DL CoMP, LTE specifications provide a framework for CSI feedback and signaling that can enable JT, DPS, CBS, and CBF. A UE is allowed to be configured with up to three non-zero-power CSI-RS resources, up to four sets of zero-power CSI-RS resources, and up to three CSI interference measurement resources (CSI-IMR) via RRC signaling. A CSI process can be defined for a configured non-zero-power CSI-RS resource and a configured CSI-IMR, for a combination of signal-interference condition. Up to four CSI processes may be defined for a UE, and therefore the corresponding CoMP transmissions can be supported.

To enable standardized UL CoMP, two components were introduced. First, 3GPP defines a UE-specific PUSCH DMRS (demodulation reference signal) whose parameters can be configured via RRC signaling leading to reduced interference. Second, it allows the UL reception points to be decoupled from the DL TPs, by introducing a new dynamic ACK/NACK region for PUCCH.

The potential performance of CoMP was investigated in 3GPP RAN1 [6]. In low-load cases, DL CoMP may provide about 6% throughput gain for an average user and about 17% throughput gain for cell-edge users. In high-load cases, these gains become about 35% and 17%.

Finally, it should be emphasized that Rel-11 CoMP relies on several fundamental assumptions, such as the network synchronization, ideal/fast backhaul connections between the nodes involved in CoMP, etc. The synchronization and related issues are to be addressed later (see Section 4.5.3), while here we point out efforts have been made to extend CoMP concepts to networks without ideal/fast backhaul connections, or non-ideal backhaul (see Sec. 6.1.3 in [7] for classification of ideal/non-ideal backhaul connections). This is called CoMP with non-ideal backhaul, or simply eCoMP for short; see [8] for recent progress. Performance of networks with various backhaul latency values was evaluated. It is seen that the throughput performance gains are limited with eCoMP. Nevertheless, due to the wide presence of non-ideal backhaul and advantages of multi-cell access as enabled by CoMP/eCoMP, this problem has also been approached from another angle and significant standardization effort has been devoted; see Sections 4.3.1 and 4.3.2.

4.2.1.3 Other related features

Though eICIC may be effective in reducing the significant interference from macro to pico UEs in the data region (i.e., PDSCH), the control region of a pico cell may still be heavily impacted by the macro control region, since the macro cell transmits at least CRS/PCFICH/PHICH and some PDCCH in the region overlapping with pico cell control regions.

One approach to address this is the introduction of enhanced PDCCH (EPDCCH) in Rel-11. An EPDCCH is a downlink control channel with similar functionality as a PDCCH, but EPDCCH is located in a data region of the downlink frame instead of being limited to being located in a control region of the downlink frame. Therefore

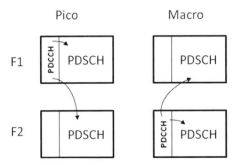

Figure 4.5 Carrier aggregation-based interference management for heterogeneous network.

frequency-domain interference coordination to ensure a more reliable control channel can be realized.

The demodulation of an EPDCCH is based on a demodulation reference signal (DMRS) embedded in the EPDCCH rather than a common reference signal (CRS) based demodulation used for a PDCCH. This feature can be used by a pico cell to communicate with a UE without CRS transmission, thus removing the persistent overhead due to CRS and reducing persistent interference. In CoMP Scenario 4, a pico may rely on EPDCCH to control its UEs without resorting to CRS. In general, EPDCCH may be used as the control channel for a non-standalone cell and it provides enhanced interference management capability for control channels in heterogeneous networks, CoMP, and so on.

Yet another approach to addressing co-channel interference is to enhance the intra-cell and inter-cell interference mitigation capabilities at the UE receiver side, with some degree of knowledge about interfering transmissions enabled by network coordination and assistance. This is referred to as network-assisted interference cancellation and suppression (NAICS), and requires advanced receivers be implemented at the UE and assistance information be sent from the network to the UE so that the UE can utilize the information to suppress or cancel the interference. The assistance information may include interfering cell resource allocation, MCS level, etc., and the signaling needs to be standardized. Refer to [9] for the status of the standardization work.

4.2.2 Non-co-channel scenario

4.2.2.1 Carrier aggregation (CA)

For the scenario where operators have multiple LTE carriers, a carrier aggregation-based interference-management approach is possible (Figure 4.5). This method was introduced in LTE Rel-10, for heterogeneous network interference management, namely CA-based heterogeneous network support. The pico PDCCH (as well as some other common channels) is located in a component carrier different from where the macro PDCCH (as well as some other common channels) is, and cross-carrier scheduling is used so that the PDCCH from a network node in one component carrier can schedule multiple component carriers. As a result, the pico PDCCH (and common channels) can be free of macro interference and is well protected. For PDSCH, various interference coordination and avoidance schemes may be implemented at carrier level.

4.2.3 Multi-stream aggregation (MSA)/dual connectivity

A typical application scenario of MSA/dual connectivity (see Section 4.3.2 for more details) is a heterogeneous network with non-co-channel macro and pico nodes, connected via non-ideal backhaul. This approach is mainly studied from a higher layer perspective and the main findings are summarized in [10].

4.3 Mobility management

4.3.1 Mobility enhancement

The introduction of small cells with much smaller coverage areas raised the question of how mobility is impacted. For example, a moving UE served by small cells may encounter more handovers than in the case where it is served by macro cells. In 3GPP, working groups looked at various mobility performance evaluations and improvements such as possible improvements to support seamless and robust mobility of users between LTE macro to pico cells in heterogeneous networks, better strategies to identify and evaluate small cells, handover performance with and without eICIC features, improvements to re-establishment procedures, etc. There is consensus that enhancements should be considered to improve the mobility performance of HetNets [11]. This includes UE and network-based mechanisms.

The heterogeneous network mobility enhancement study has concluded that handover performance degrades compared to a macro-only deployment, but the issue is not significant if the UE mobility is low (e.g., speed <30 km/h). However, the mobility performance depends on many factors, such as UE speed, system load, density of the pico nodes, cell range extension bias, UE behavior in discontinuous reception (DRX), etc. Nevertheless, several mobility enhancements have been proposed under various situations, the most prominent being dual connectivity as described below.

4.3.2 Multi-stream aggregation (MSA)/dual connectivity

Dual connectivity is considered as a fundamental aspect of small cell enhancements in 3GPP LTE Rel-12 standardization work [10]. In dual connectivity, or more broadly, multi-stream aggregation (MSA), a UE in RRC_CONNECTED state has connectivity with more than one eNB, with one (called the master eNB, or MeNB) mainly for control plane and mobility functionalities, and the other(s) (called secondary eNB, or SeNB) mainly for user plane data transmissions. The MeNB can be a macro eNB capable of providing wide area coverage and hence the mobility performance can be ensured, and the SeNB can be a small cell eNB capable of providing high data rate to nearby UEs.

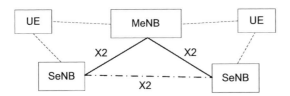

Figure 4.6 System diagram for MSA/dual connectivity.

At the network side, an MeNB is connected to one or more SeNBs via non-ideal backhaul using an X2 interface. Between the SeNBs, there may also be an X2 interface. A UE is connected wirelessly to MeNB and SeNB, with the MeNB as the mobility anchor. See Figure 4.6 for an illustration of the system diagram. Each MeNB and SeNB may have one or more component carriers, or cells, serving the UE. The cells associated with the MeNB form the MeNB cell group (MCG) and the cells associated with a SeNB form a SeNB cell group (SCG). For a UE, the cells within the MCG are aggregated according to CA, with one cell acting as the primary cell (Pcell) and others as the secondary cells (Scells). The cells within an SCG are also aggregated according to CA, except that the "Pcell" (or a Pcell-like cell) in SCG does not need to have all the functionalities of a Pcell.

In general, the MeNB can be a macro eNB and the SeNB a pico eNB. The dual connectivity architecture applies to either MeNB/SeNB co-channel deployment or MeNB/SeNB non-co-channel deployment, but likely the more typical scenario is the non-co-channel one. Therefore dual connectivity amounts to yet another way to deploy a heterogeneous network with non-co-channel macro and pico nodes.

Compared to a fast backhaul, a general backhaul connection (or any backhaul, or a non-ideal backhaul) generally imposes constraints on macro–pico coordination and/or pico–pico coordination. For example, with any backhaul, the backhaul delay may be long enough to prevent macro–pico coordination on scheduling. MSA/dual connectivity can enable the macro and small cell to serve a UE simultaneously with non-ideal backhaul connecting them.

There are several major benefits for dual connectivity. First, significantly more radio resources can be used for a UE at possibly the same time, thus improving UE throughput (especially in the case of large file transmission) and user experience. This extends along the line of CoMP, but with non-ideal backhaul connecting the network nodes. Second, it provides robust mobility, as the UE is anchored to the MeNB, which has wide coverage. The UE can thus enjoy both the wide coverage and mobility support provided by the macro and large pipe provided by picos. Third, signaling overhead toward the core network can potentially be saved by keeping the mobility anchor in the macro cell.

The ongoing standardization of small cell on/off in 3GPP physical layer work raised some concerns regarding mobility. However, the mobility issue can be resolved if the macro cells provide the needed coverage and the macro cells do not perform on/off adaptation (or at least not too frequently or significantly). For a UE with dual connectivity capability, it always has its mobility anchor with the macro cell and hence small cell on/off has little impact on the mobility performance. For a low-mobility UE without dual connectivity capability, it may rely on legacy procedures such as handover to handle mobility; for example, it may be handed over to another small cell or the macro cell before its serving small cell is turned off. For a high-mobility UE, it may rely on the coverage layer and does not connect to the small cells.

To summarize mobility support given the small cell enhancements: typical deployment scenarios may include a macro layer whose cells cover wide areas and do not perform network adaptation, and another layer whose cells (mainly small cells) provide high-capacity pipes and may perform network adaptation. Coverage/mobility support is mainly ensured by the macro cells. UE may connect to cells in the macro layer first, and then connect to small cells when needed, and both connections (to the macro layer and small cell layer) can be maintained if dual connectivity is supported.

4.4 Spectrum utilization

Deployment of small cells renders specific characteristics that may be exploited to further improve the efficiency of small cell networks. In this section, such features, including higher order modulation taking advantage of channel conditions within local coverage especially in indoor environments; dynamic TDD configuration adaptation to time-varying UL/DL traffic asymmetry, and TDD/FDD joint operation; and 3GPP/WLAN interworking utilizing more spectrum resources, are discussed.

4.4.1 Higher order modulation

Higher order modulation allows more efficient utilization of spectrum resources, but it requires higher SINR. 256 quadrature amplitude modulation (QAM) is already standardized in IEEE 802.11 and has been considered by 3GPP for Rel-12 standardization.

The required minimum SINR to support 256QAM, or effectively the performance benefits resulting from 256QAM, depends on several factors, such as transmission rank, transmitter error vector magnitude (EVM), receiver impairments, and deployment scenarios. Under reasonable assumptions, it is found that minimum SINR for which a gain is observed is around 25 to 30 dB. Such SINR is possible for high-geometry UEs in some small-cell deployments in downlink transmissions. Specifically, the indoor sparse small-cell scenario is the main focus for the evaluations, in which UEs are of low mobility and have high SINRs. It is concluded that 256QAM is beneficial, with around 7 to 30% gain on cell average UE throughput if transmit power back-off is not applied, and about 5 to 19% gain otherwise. 3GPP is currently working on support for 256QAM in the Rel-12 specifications. New CQI and modulation/coding scheme (MCS) entries and new transport block size (TBS) entries are to be introduced.

4.4.2 TDD enhanced interference management and traffic adaptation (eIMTA)

Time-division duplex (TDD) is an alternative deployment to deployments with a pair of spectrum resources. As the operators move toward shrinking cell sizes, the suitable spectrum becomes more and more unpaired. Therefore optimization of a TDD network is useful in future small cell networks. Usually, the traffic loads and directions of the traffic loads (UL traffic demand or DL traffic demand) at neighboring cells may be rather different. In order better to match the traffic loads, it is desirable to adapt the UL–DL configurations of the cells. In 3GPP study it has been found that UL–DL reconfiguration according to traffic conditions provides packet throughput benefits for UL and DL transmission, if the system loading is not very high. In addition, faster adaptation provides higher throughput gains. However, existing mechanisms in LTE allow only semi-static reconfiguration of UL–DL allocations, via system information update procedure in the time scales of at least 640 ms but typically much longer. Thus the support for more dynamic, flexible allocation of UL–DL subframes is needed.

One issue that needs to be addressed is the interference issue when the UL–DL configurations of neighboring cells are not aligned. One cell operating in a DL subframe will cause strong interference to a neighboring cell operating in a UL subframe, which is called base station–base station interference. This type of interference can be detrimental. Likewise, there is also UE–UE interference, albeit the impact on throughput performance is less severe. Therefore 3GPP has set up a work item to support UL–DL configuration adaptive to traffic conditions with consideration of interference management [12].

Many aspects are under discussion. A new physical layer signaling is to be designed to convey the reconfiguration information, with reconfiguration between 5 ms and 10 ms switching point periodicity to be supported. Various mechanisms for interference management are being discussed, such as UL power control, CSI feedback enhancement, etc.

4.4.3 FDD/TDD joint operation

In the above-described small-cell scenarios, there are cases where the macro cells are operating in FDD mode but the small cell is operating in TDD mode at a higher frequency. To allow a UE to use both cells possibly at the same time, the joint operation of frequency-division duplex (FDD) and TDD may be needed. In general, there are cases where it is desired to have the LTE network to support duplex modes of both FDD and TDD to the same UE simultaneously. In a certain sense, this is along the line of carrier aggregation and dual connectivity to let a UE use more spectrum resources, regardless of duplex mode, to improve the user experience and hence system performance.

If the nodes operating in FDD mode and TDD mode are connected via ideal/fast backhaul, generally CA mechanisms can be applied. For example, the FDD cell may act as the Pcell and the TDD cell(s) may act as the Scell(s). On the other hand, if the nodes are connected via non-ideal backhaul, then the general framework for the joint operation should be dual connectivity. The interested reader may refer to [10, 12] for recent outcomes.

4.4.4 3GPP systems to wireless local area network (WLAN) interworking

An enhancement is to allow the interworking between 3GPP systems, such as LTE networks, and wireless local area networks (WLANs), such as 802.11 networks. This allows a UE to be offloaded to a WLAN if it is close to the WLAN access points, and to connect back to the LTE network for better QoS or mobility support, and the transitions between the LTE network and WLAN can be seamless. The enhancement is also applicable to other wireless technologies built upon IP-based access networks with similar capabilities as 802.11. The work fits into the scope of 3GPP System Architecture Working Group 2, which develops the Stage 2 specification of the 3GPP networks. With the enhancement, the services and functionalities available in 3GPP systems can be received in a WLAN access environment. To support this, bearer services are provided to allow a 3GPP subscriber to use a WLAN to access 3GPP PLMN services. Details can be found in [13].

4.5 Dense network adaptation

It is expected that a large portion of the data will be around offices, homes, and other hotspots. How to exploit these traffic characteristics in system design will be one of the areas focused on. For example, hyper-dense deployment of a large number of low-power nodes (picos, femtos, or relays) can be deployed around these traffic-concentrated areas to pick up most of the localized traffic, while macro nodes provide wide-area coverage and capacity.

4.5.1 Network adaptation

Unlike macro deployment, where the network is well planned and each macro cell provides coverage for a large area, deployment of low-power nodes is likely to be more ad hoc in nature and each individual low-power node typically has smaller footprint. How to cope with traffic mobility (for example, during office hours versus during night hours) in such deployments is a new challenge. One possible solution is to adapt the network topology based on the location of data demand. For example, instead of turning on all the low-power nodes, only those nodes with traffic are turned on (such as nodes around offices during working hours and nodes around residential areas during the night). Such an opportunistic node on/off deployment also helps reduce inter-cell interference as well as lower power consumption and operational expense (OPEX). For the TDD system, adaptive operation can also be realized as adaptive time allocation between downlink and uplink based on the DL/UL traffic loads.

Small cell on/off means the adaptive turning on and off of a small cell. A cell here may refer to a component carrier (CC). In more detail, the behaviors for on and off are as follows:

- When a small cell is on, it transmits signals necessary for a UE to receive data from the cell, such as reference signals used for measurements and demodulation.
- When a small cell is off, it does not transmit signals necessary for a UE to receive data from the cell. Therefore for a legacy UE, the cell is not perceivable when turned off. However, enhancements to assist turning on a turned-off cell may be introduced for

new UEs, for example, signals to detect and/or measure the turned-off small cell. The signal transmitted from a turned-off small cell for the UEs to detect and/or measure is the discovery reference signal (DRS) under current design of 3GPP Rel-12 in the small cell enhancements (SCE) work item.

The following schemes are relevant to small cell on/off study:

1. **Baseline schemes without any on/off**. In these schemes, the small cell is always on.
2. **Long-term on/off schemes for energy saving**. In these schemes, the small cells may be turned on/off in large time scales, at most a few times in a day. These schemes are studied in the Energy Saving study item and work item.
3. **Semi-static on/off schemes**. In these schemes, the small cells may be turned on/off semi-statically. With legacy procedures, the feasible time scales of semi-static on/off schemes are generally of seconds to hundreds of millisecond level; with possible enhancements, the transitions may reduce to tens of milliseconds. The criteria used for semi-static on/off are mainly the traffic load increase/decrease, UE arrival/departure (i.e., UE–cell association), and packet call arrival/completion. Examples of semi-static on/off schemes include:
    a. **Semi-static on/off scheme based on traffic load**. In this case, a turned-off small cell may be turned on if the traffic load in a neighborhood of the cell (including the cell itself) increases to a certain level. Conversely, a turned-on small cell may be turned off if the traffic load in a neighborhood of the cell decreases to a certain level.
    b. **Semi-static on/off scheme based on UE–cell association**. In this case, a turned-on small cell may be turned off if there is no UE associated to it, and a turned-off small cell may be turned on if the network decides a UE to be associated to it. The UE–cell association may be decided by the network taking into account of UE measurements (e.g., discovery measurements) and load balancing/shifting considerations.
    c. **Semi-static on/off scheme based on packet call arrival/completion**. In this case, a turned-off small cell may be turned on if a packet call arrives and needs to be transmitted, and the cell may be turned off after the packet call is completed.
4. **Ideal, dynamic on/off schemes**. In these schemes, the small cells may be turned on/off in subframe level, following criteria such as packet arrival/completion and the need for interference coordination/avoidance in subframe time scales. In other words, at the moment of a packet arrival, the small cell can be turned on immediately and transmit the packet to a UE, and it can be turned off at the moment of completion of the packet. Likewise the small cell can be turned on/off immediately based on the need for interference coordination/avoidance. Clearly, these schemes cannot be supported at least according to current standards, and they are studied in the 3GPP SCE study item and work item to provide performance gain upper bounds for on/off adaptation.

Network adaptation based on existing mechanisms may involve multiple procedures (e.g., PSS/SSS based cell detection, RSRP measurements, configuration signaling, handover, etc.) usually taking hundreds of milliseconds to complete a transition. Faster transitions may be desirable in some cases. Some or all of the involved procedures may be simplified, modified, or even eliminated, to shorten the transition. The methods for supporting small cell on/off transition time reduction generally include:

- **Utilizing discovery signals**. Discovery signals may be sent from a turned-off small cell and the UE can perform necessary measurements. The measurements may be utilized so that additional measurement duration after the cell is turned on can be significantly reduced (to, for example, tens of milliseconds or even shorter).
- **Utilizing procedures including handover, carrier aggregation, and dual connectivity/MSA**. The legacy handover procedures may be streamlined under the assumptions such as dual connectivity/MSA. Dual connectivity/MSA may allow a faster transition by reducing/eliminating the needs for handover to and from a small cell performing on/off. Once dual connectivity/MSA between a UE and a small cell is configured, the activation/deactivation of the cell based on a procedure similar to carrier aggregation may be used, and the time scale may be in the tens of milliseconds level or even less.

4.5.2 Discovery

From the above descriptions, it can be seen that the enhancements needed to support small cell on/off are mainly the DRS-based RRM measurements (e.g., for a UE to discover a turned-off small cell and connect to it when it is turned on) and dual connectivity configuration, activation, and deactivation. The dual connectivity-based operations are similar to the CA-based operations introduced in Rel-10/11 and can shorten on/off transitions.

Below we focus on DRS-based measurements. Generally speaking, during the time when a small cell is turned off, only DRS is transmitted, and UEs monitor only DRS from the cell. DRS is defined in 7.2.2 of TR36.872. The time/frequency resources need to be known by UEs via network assistance signaling to detect DRS. DRS can be used by the UEs for the following specific functionalities: small cell identification, coarse time/frequency synchronization with the small cell, and RRM measurements of the small cell.

These DRS-based operations, when configured by the network, may be performed by Rel-12 UEs regardless of the on/off of a small cell. Therefore the UE and network can function without the UE being aware of the on/off status of the cell. The measurements may be reported to the network (according to network configuration) and used by the network to make decision for on/off, load balancing/shifting, handover, or activation/deactivation, etc. Most importantly, as the DRS-based measurements are available even when the cell is off, network adaptation decision-making and on/off transitions do not need to rely on CRS-based measurements, which are available only during the small cell on stage. This can significantly shorten the transition durations since the CRS-based measurements dominate the transition times in most cases.

4.5.3 Radio-interface based synchronization (RIBS)

As mentioned multiple times in previous sections, network synchronization is necessary for many existing features, e.g., (F)eICIC, CoMP, some carrier aggregation functionalities, small cell discovery, network-assisted interference suppression and cancellation, as well other forms of advanced receivers, TDD system operations, etc. However, densification makes existing backhaul-based synchronization mechanisms difficult to be utilized; for example, in some dense heterogeneous networks, wireless backhaul is used. Therefore it is beneficial to have a radio-interface based synchronization mechanism when global navigation satellite system (GNSS) or backhaul-based synchronization are unavailable. The target synchronization accuracy is 3 μs [10].

Network listening and UE-assisted synchronization have been proposed for this purpose; see summary in [14]. A cell providing synchronization for another cell is called a source cell, and a cell acquiring synchronization from another cell is called a target cell. Multiple source cells may be used for a target cell, and multiple target cells may exist for a source cell. Multi-hop synchronization may be used if a target cell cannot synchronize directly to a source cell.

In network listening, the target cell monitors the network listening RS (e.g., CRS, CSI-RS, and PRS) of the source cell directly to maintain synchronization with the source cell. When the target cell monitors the source cell, the target cell mutes its own transmission at least when the target cell and the source cell are in the same frequency. To enable and facilitate the listening process, several pieces of information may be needed, for example, the indication of the synchronization stratum level, the maximum supported hop number, etc.

In UE-assisted synchronization, the target cell may use some information provided by assisting UEs or obtained from selected UEs to achieve synchronization to the source cell. Similar to the network listening method, the indication of the synchronization stratum level is also needed. Furthermore, whether an assisting UE is available or how the UEs are selected may impact the synchronization accuracy.

4.6 Other related 3GPP RAN items

There are several functionalities that were defined in the standards that have some impact on heterogeneous networks. In particular, the heterogeneous networks of the future may include relay nodes, home evolved NodeB (HeNB), and device–device communications. These functionalities as well as recent work to include heterogeneous network in UMTS will be discussed in this section.

4.6.1 Relaying

The relay node, as a type of network note in heterogeneous networks, was introduced in LTE-A Rel-10. It was mainly intended to improve the coverage of high data rates, cell-edge coverage, to extend coverage beyond the cell edge as well as in heavily shadowed

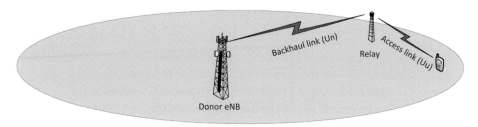

Figure 4.7 A schematic representation of a relay network showing the interface names.

areas, and to improve the overall capacity of the network. To enable this new node, a new link the Un link from the donor eNB to the relay was introduced (Figure 4.7).

The Un link is a wireless link and is known as an in band if the Un link shares the same frequency as the direct donor eNB to UE link within the donor eNB cell, and is known as an out band if the frequencies for Un link are different from the direct donor eNB to UE link. Several classes of relays were defined in 3GPP. A so-called "Type 1" relay node is characterized mainly as an in-band relaying node, which appears to the UE as a separate cell distinct from the donor cell. A "Type 1a" relay node is like a Type 1 relay node except that it operates out of band. A "Type 1b" relay node is a full duplex Type 1 relay. A "Type 2" relay node is an in-band relay node that does not have a separate physical cell ID. In LTE-A Rel-10, only the Type 1 and Type 1a relays are supported in the standard.

A new physical channel, the relay physical downlink control channel (R-PDCCH) was introduced. The R-PDCCH is frequency and time multiplexed within the subframe. The first slot of the R-PDCCH is used for the transmission of the downlink grant while the uplink grants are transmitted in the second slot. The resources used for the R-PDCH may be mapped to non-adjacent physical resources to allow for frequency diversity. Additionally the standard provides several features to improve the robustness of the R-PDCCH. The first is that the R-PDCCH transmissions from multiple users can be interleaved. The second is to obtain frequency diversity via an LTE Rel-8 concept known as distributed virtual resource block (DVRB) where the resources on the two slots of a DVRB pair use different slots. Consequently, frequency diversity is introduced when the R-PDCCH transmission uses multiple DVRB blocks.

The data transmission in the Un link reuses the PDSCH except that the carrier aggregation is not supported for the Un link.

4.6.2 Home evolved NodeB

Another fundamental node that will be found in the heterogeneous networks is the so called femto cells or as they are defined in the standard; Home NodeB (HNB) and Home Evolved NodeB (HeNB). An HeNB is suitable for indoor coverage of a few tens of meters, which has similar functionalities to a WLAN access point, but the HeNB operates on the licensed spectrum using LTE technologies for the benefits of higher spectrum efficiency. In Rel-8/Rel-9, the basic solution for inbound handover was defined for HNB. In Rel-10, support for both the HNB and HeNB were enhanced with the following features:

- intra-CSG/inter-HeNB handover optimization
- inter-CSG handover optimization
- inter-RAT handover (UTRA to HeNB and LTE to HeNB)
- enhancements to existing mobility mechanisms for HeNB cells
- footprint accuracy improvement
- signaling and UE battery minimization
- inbound handover performance enhancements
- IDLE mode improvements for inter-PLMN CSG reselection
- interference management between HeNB and macro eNB
- interference management among HeNB.

In Rel-11, the HNB support was enhanced by the introduction of the Iur interface between the HNB-GW and the macro RNC. Consequently, this enables the support of:

- hard handover between HNB and macro RNC using enhanced SRNS relocation, thus reducing CN load
- soft handover between HNB and macro RNC
- for UMTS, legacy UE mobility – support of non-CSG/legacy UEs.

For Rel-12, the standard is currently working on the following features for enhanced H(e)NB support:

- for LTE, X2-GW
- CELL_FACH support for HNBs
- for UMTS and LTE RAN sharing support.

4.6.3 Device to device communications

In the heterogeneous network, the localization of the traffic makes device–device (D2D) communication an interesting feature. In simple terms, D2D communications is a technology that allows multiple devices to communicate directly with each other, thus offloading the network controllers while achieving high data rates in the meantime. It is not a new technology but rather a technology that has been around for decades. However, with the rapid changes in social behavior and the rapid proliferation of wireless devices, there is a renewed interest in D2D in the wireless industry. The main reasons for the interest are: the potential to generate new revenue for both the operator and the content providers, extend the coverage for public safety applications, and also as a technology that can help mitigate the impending data crunch for cellular networks. Beyond this, D2D has the potential to evolve into an important technology that would nicely complement the existing cellular infrastructure.

Standardization work of D2D began in 2011 in 3GPP. TSG-SA1 devoted considerable effort in 2012 to study the service requirements for D2D, and identified a number of potential use cases for both commercial and public safety scenarios. This work was captured in technical report TR 22.803. The December 2012 RAN plenary meeting, TSG-RAN, agreed to start a D2D study item that addresses both commercial and public safety aspects.

New study items on D2D are currently underway for Rel-12 in the various SA groups, as well as in RAN groups. The current focus is on broadcast D2D communication for the public safety use case and on in-network discovery to support the localized advertising use case. It is anticipated that these study items will be converted to work items in 2014 so that the feature can be completed for Rel-12.

4.6.4 Heterogeneous network support for UMTS

With over 1.4 billion subscribers worldwide, 3G UMTS systems still remain the major technology for enabling broadband wireless access. Consequently with the anticipated data crunch, it was recognized that heterogeneous network technology also provides a powerful method for enhancing the capacity of 3G systems. Consequently, 3GPP initiated work on heterogeneous network support for UMTS in 2012 with the anticipation of support being completed for Rel-12. The current work focused on four topics: small cell discovery and identification, UE speed based mobility, mass small cell deployment, and further mobility enhancements. For small cell discovery and identification, solutions applied for both DCH and non-DCH states are currently under discussion. For UE speed-based mobility, different solutions, including network implementation based on reliable speed estimation, keeping the macro cell in the active set and cell-specific TTT, are being considered. For mass small cell deployment, the common understanding to extend the size of the neighbor cell list (NCL extension) was a feasible methodology reached. For further mobility enhancements, the work is not yet mature with many different divergent techniques being discussed.

4.7 Future 3GPP features for small cell

Small cell and dense network deployments are becoming an essential part of the mobile communication networks. 3GPP standards will continue evolving with small cell and dense network as a key focus point. The features to be considered for future 3GPP and specifically for future LTE development include:

1. **Ultra-dense networks**. As the network density keeps increasing, at least for areas of ultra-dense traffic demand, much denser deployment of small cells with new issues/problems is possible. Intriguing questions are where the limit of network density is and what the limiting factors are.
2. **LTE for unlicensed spectrum**. The interests in adapting LTE to unlicensed spectrum triggered a lot of discussions. Whether and how 3GPP may specify LTE for unlicensed spectrum can potentially change the landscape of the mobile communication industry.
3. **Wideband LTE**. To respond to the dramatic increase of traffic demand, utilizing more spectrums is as desirable as deploying more network nodes. However, radio

spectrum is a very scarce resource. In order to find a large amount of new spectrum, we have to turn to much higher carrier frequency where the path loss is much larger and feasible transmission power is lower. Therefore a wideband LTE system designed for higher frequency bands with small cell deployment is another potential direction.

4. **Cellular with local mesh**. With the development of D2D technology and the closely deployed small cells, forming a mesh network for a local area becomes feasible and may be a solution to the backhaul problem of dense networks.

As a last point, although the discussion for the fifth-generation (5G) mobile communication system has just started and it is yet to see what technologies may be developed there, small cell and dense network support should be a critical element for its fundamental design.

4.8 Appendix: brief description of LTE channels

In downlink transmission of LTE systems, there are some reference signal (RS) channels, e.g., cell-specific RS (CRS), dedicated/demodulation RS (DMRS), and channel state information RS (CSI-RS); channels designed for initial access, e.g., primary synchronization signals (PSS), secondary synchronization signals (SSS), and physical broadcast channels (PBCH); control channels, e.g., physical downlink control channels (PDCCH), and enhanced PDCCH (EPDCCH); and data channels, i.e., physical downlink shared channels (PDSCH). In uplink, there are mainly physical uplink control channels (PUCCH) and physical uplink shared channels (PUSCH).

CRS, with one, two, or four ports, is used for UEs to perform channel estimation for demodulation of PDCCH and other common channels as well as for radio resource management (RRM) and channel state information (CSI) measurements and feedback. DMRS, with up to eight ports, can be transmitted together with its associated PDSCH for channel estimation during PDSCH demodulation. DMRS can also be transmitted together with EPDCCH for the channel estimation of EPDCCH by the UE. CSI-RS, with one, two, four, or eight ports, is used for UEs to measure CSI, especially for MIMO/CoMP cases. CRS-RS are configured in a UE-specific way, and there may be multiple CSI-RS resources configured for a UE. Both CRS and CSI-RS are transmitted periodically, but DMRS is transmitted only with its associated PDSCH/EPDCCH.

Both the PSS and SSS are periodically transmitted, occupying the center six resource blocks of the downlink carrier. A UE first detects PSS, followed by detection of SSS. The UE thus gains information of frame configuration (FDD or TDD), cyclic prefix type, the cell ID, subframe timing, and coarse frequency and timing synchronization. The PBCH contains the master information block (MIB) and indicates to a UE the cell's system frame number (SFN), bandwidth, antenna ports, and information used for decode PDCCH.

A UE monitors the PDCCH to find out if there is any PDSCH transmitted for the UE to receive in downlink, or if there is any PUSCH resource scheduled for it to transmit in uplink. The UE can also obtain necessary information, such as the time and frequency resources, modulation, coding, etc., of the associated PDSCH/PUSCH. The EPDCCH has similar functionalities as a PDCCH, but EPDCCH may be located in a data region of a downlink frame instead of being limited to being located in a control region of a downlink frame. Demodulation of an EPDCCH may be based on a DMRS rather than a CRS used for a PDCCH.

The PDSCH carries UE-specific data or system information. It may be demodulated based on CRS or DMRS.

The PUCCH may contain channel quality indicator (CQI) report, acknowledgement/non-acknowledgement (ACK/NACK) of downlink reception, and scheduling request (SR). The PUSCH may carry UE uplink data or measurement feedback reports.

References

[1] 3GPP TS 36.213, "Evolved universal terrestrial radio access (E-UTRA); physical layer procedures," v12.0.0, Dec. 2013.

[2] S. Sesia, I. Toufik, and M. Baker, *LTE: The UMTS Long Term Evolution: From Theory to Practice*, John Wiley & Sons Ltd., 2009.

[3] R1-101083, Huawei, "Cell association analysis in outdoor hotzone of heterogeneous networks," available at www.3gpp.org/ftp/tsg_ran/WG1_RL1/TSGR1_60/Docs/R1-101083.zip, 2010.

[4] R1-113566, Qualcomm Inc., "eICIC evaluations for different handover biases," available at www.3gpp.org/ftp/tsg_ran/WG1_RL1/TSGR1_66b/Docs/R1-113566.zip, 2011.

[5] R1-112894, Huawei and HiSilicon, "Performance evaluation of cell range extension," available at www.3gpp.org/ftp/tsg_ran/WG1_RL1/TSGR1_66b/Docs/R1-112894.zip, 2011.

[6] 3GPP TR36.819 v11.2.0, "Coordinated multi-point operation for LTE physical layer aspects (Release 11), 3rd generation partnership project; technical specification group radio access network standards," 2013.

[7] 3GPP TR 36.932, "Study on scenarios and requirements of LTE small cell enhancements," v0.2.0, Nov. 2012.

[8] 3GPP TR36.874 v12.0.0, "Coordinated multi-point operation for LTE with non-ideal backhaul (Release 12), 3rd generation partnership project; technical specification group radio access network standards," 2013.

[9] 3GPP TR36.866 v12.1.0, "Network-assisted interference cancellation and suppression for LTE (Release 12), 3rd Generation Partnership Project; technical specification group radio access network standards," 2013.

[10] 3GPP TR 36.872, "Small cell enhancements for E-UTRA and E-UTRAN – higher layer aspects," v12.0.0, Dec. 2013.

[11] 3GPP TR 36.839, "Evolved universal terrestrial radio access (E-UTRA); mobility enhancements in heterogeneous networks," v11.1.0, Jan. 2013.

[12] 3GPP TR 36.828, "Further enhancements to LTE time division duplex (TDD) for downlink–uplink (DL–UL) interference management and traffic adaptation," v11.0.0, June 2012.

[13] 3GPP TS 23.234, "3GPP system to wireless local area network (WLAN) interworking; system description," v11.0.0, Sept. 2012.

[14] 3GPP TR 36.872, "Small cell enhancements for E-UTRA and E-UTRAN – physical layer aspects," v12.1.0, Dec. 2013.

5 Dense networks of small cells

Jialing Liu, Weimin Xiao, and Anthony C. K. Soong

5.1 Introduction

During the last few years, wireless data traffic has skyrocketed, driven mainly by a large penetration of smart phones and devices. In 2013, an exabyte of data traveled across the global mobile network monthly [1]. By 2020, data traffic served by such networks is expected to increase by up to a factor of 100, including traffic generated by the widespread adoption of device–device (D2D) and the Internet of Things (IoT) connected via machine–machine (M2M) communications. It is widely recognized that this general trend toward more explosive growth may accelerate even further in future, and how to meet such a demand has been one of the most active and rapidly growing areas in the wireless communication community in the past decade in terms of both academic and industrial research and development [2].

Facing the unprecedented challenge, the wireless communication community has considered many candidate solutions. A significant portion of these are focused on increasing the communication resources, e.g., deploying more network nodes, which leads to densification of existing networks, utilizing wider bandwidth, increasing antenna numbers, and employing additional resources to offload. Among them, the dense network approach stands out for its high scalability of providing magnitudes of capacity increase. Extensive research has been devoted to dense networks (see e.g., [3–8] and references therein).

Indeed, commercial wireless networks are already becoming denser and dense-network deployment will be a critical factor (together with other solutions) to meet the ever-increasing traffic demand. The trends for traffic and network-density growth over a 20-year span are illustrated in Figure 5.1.

In Figure 5.1, the network densities for the years 2000 to 2015 were estimated from 3GPP publications (e.g., [3, 9], etc.), whereas the network density for year 2020 is a projection based on historic data and recent trends. In more detail, around year 2000, sparse 3G network deployments of macro base stations covered wide areas with typical cell radii of several kilometers. Starting from around 2005, the network density increased to about 10 to 20 nodes/km^2 and cell radii shrunk to between one kilometer and several hundreds of meters, according to the study in 3GPP LTE Rel-8/9; however, macro eNBs (evolved NodeB, also known as base stations, BS, BTS, etc.) were still the main focus. Since 2009, heterogeneous networks with both macro and pico eNBs have been extensively researched (e.g., in 3GPP LTE Rel-10/11) and attract attention for deployments.

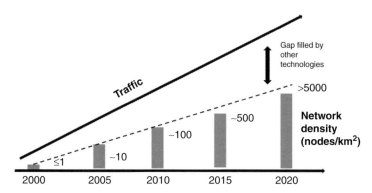

Figure 5.1 Illustration of trends for traffic and network-density growth.

A pico eNB typically covers one hundred meters in radius, and accordingly the studied density reached about 50 to 200 nodes/km². The pico eNBs in Rel-10/11 are generally isolated (i.e., not close to each other). Most recently, both the wireless communication industry and academic research community have introduced small cell networks, with a cluster of several small cells covering only tens of meters in radius. In late 2012, 3GPP started research on small cell enhancement in Rel-12, which considers the support for networks at least two to four times denser than the densest networks studied in Rel-10/11. Some operators are currently studying very high density networks, which are about ten times denser than the densest networks studied in Rel-11; in other words, the networks under their recent deployment considerations may have densities of 100 small cells in a macro area, or more than 1,000 nodes/km². It is expected that the shrinking of cell sizes and the densification of networks will extend into the future and continuously deployed dense networks will emerge, which may further evolve into ultra-dense networks.

However, it should be noted that dense networks of small cells are not meant to replace wide area networks, since small cells are mainly deployed for high-load local areas only (as opposed to providing wide-area coverage as macro eNBs do); rather, the dense networks of small cells and macro networks should coexist and complement each other.

5.1.1 Main topics and organization

As network density increases, we will first need to assess if existing modeling and evaluation methodologies are suitable for dense network research, and if the answer is "no," then it is essential to identify a set of appropriate modeling and evaluation methodologies for dense networks. Based on such methodologies, fundamental questions about the dense networks can be answered, including:

> **Q1: Why the dense network?** What types of benefits, and how much benefit, should be expected from the increasingly denser deployments?

Q2: What characterizes the dense network? What are the new, defining behaviors and critical issues that set the dense network apart from other forms of networks and call for fundamentally new solutions?

Q3: How to deploy and operate the dense network? How may the overall network efficiency be optimized through implementation and standardization?

The rest of this chapter is devoted to the discussion of modeling and evaluation methodologies and providing the answers to the fundamental questions. The organization is as follows. First, in Section 5.2 we discuss the system model, and evaluation methodologies and performance metrics. We show that based on simulation studies, existing modeling and evaluation methodologies do not capture some dense network behaviors well and are not able to provide insights for dense network studies, and thus new methodologies and performance metrics suitable for dense networks should be introduced. Then we present dense network capacity evaluation results, specifically we characterize numerically the growth trend of dense network capacity as the network density increases. Furthermore, in Section 5.3, we characterize and analyze new issues identified for dense networks. To address these issues, network adaptation and interference coordination techniques, as the primary enabler to dense network performance enhancements, are presented in Section 5.4. Finally, we summarize the dense network research work and provide the answers to the fundamental questions, followed by discussing possible future directions. We also call for more attention and more contributions from both academia and industry to develop suitable means, which possibly may include extensions of the stochastic geometry to model, evaluate, analyze, and optimize dense networks.

5.2   Evaluation and performance of dense networks of small cells

To fully benefit from a dense deployment, the uniqueness of dense networks should be well understood, and specific techniques should be adopted accordingly. However, we discover that, based on our simulation studies, existing modeling and evaluation methodologies do not capture dense network behaviors well, and thus new methodologies and performance metrics suitable for dense networks should be introduced. Then we present dense network capacity evaluation results in terms of the newly proposed framework.

5.2.1   Literature review

In studying the networks with a variety of densities, the wireless communication industry developed a set of methodologies to generate numerical simulation based results for evaluations. The methodologies and results are mainly summarized in [3, 7, 9, 10] and references therein. The (purely) simulation-based methodologies can provide fairly realistic results numerically but are generally not tractable mathematically to obtain any analytic expressions of network performance. As the networks become much denser and more complex, however, the existing simulation-based methodologies face considerable challenges due to the mounting computation resources required.

In academia, on the other hand, several approaches have been adopted, the most prominent being a stochastic geometry framework for analyzing coverage, signal to interference-plus-noise ratio (SINR) distribution, and ergodic capacity in heterogeneous networks [6, 11]. The Poisson point process models and Matérn hard-core process of type I models [12] are mainly used in the stochastic geometry-based analyses. This framework has attracted significant interest and has led to many useful insights that have significantly increased our understanding for dense networks, such as closed-form expressions for user-throughput distributions. However, such analytic models often sacrifice reality for mathematic tractability; namely, they heavily rely on unrealistic, oversimplified assumptions, and thus have limited usages in realistic situations. For example, this framework assumes a stationary set of users with full buffer traffic demands, which does not hold true in practice. For another example, traffic variations across cells are generally ignored. In a few references, different cell loads are considered, e.g., in [9], but the load at each eNB is assumed to be a random variable independent of the real user demands to allow stochastic geometry to be applicable. In the rest of the section, we point out that in some cases the stochastic geometry framework fails to capture important network behavior in a dense network. Therefore before employing the stochastic geometry-based framework for research and applying the results derived in this framework, care has to be taken and the underlying assumptions for the stochastic geometry approach have to correspond to realistic deployment scenarios. Another drawback of the stochastic geometry framework is that, in order to obtain analytic expressions, only the "averaging effects" or "ensemble averages" can be derived; for example, an SINR distribution may be derived by averaging over all possible eNB/user locations (following certain distributions) and averaging over time. If, instead, the eNB/user locations (i.e., a realization of the distribution) are given, the framework may not be applicable.

In addition to the stochastic geometry framework, other approaches are also proposed in the literature. For example, in [13][1], some more realistic situations are considered in the system modeling, such as the non-full buffer traffic loads at the eNBs. In this approach, aspects such as the channel models, system dynamics, interference calculations, rate computations, etc., are formulated as mathematic expressions, as opposed to that in the "purely" simulation-based methodologies, they are largely left for numerical evaluations. Based on these expressions, usually some optimization problems, such as maximizing the sum rate, minimizing the total energy consumption given some quality of service (QoS) constraints, etc., are formulated. These optimization problems generally cannot lead to any formulas that are readily evaluated, and numerical simulations are required to obtain network performance results. This approach may be viewed as a "semi-analytic" approach. As we will see in this section, the semi-analytic approach may turn out to be useful for dense network research.

---

[1] [12] was sponsored by the authors for this chapter and partially captures the views of the authors of this chapter.

5.2.2 System model and evaluation methodology

In this subsection, system models, evaluation methodology, and performance metrics tailored for dense networks will be provided.

5.2.2.1 Channel models

Among the various channel characteristics, the large-scale propagation loss, mainly the distance-dependent path loss, plays the most prominent role in evaluating system performance. The single-exponent path-loss model,

$$PL = \alpha \log_{10}(d/d_0), \tag{5.1}$$

where PL is the path loss in dB scale, α is the pathloss exponent, d is the distance, and d_0 is the reference distance, is widely used in the literature for its mathematic tractability, e.g. [2, 5, 6]. Based on this model, [6] shows an interesting result that, in a multi-tier interference-limited network, the user SINR distributions are generally invariant with densification (i.e., remain unchanged for different network densities). The equation that allows us to compute SIR (which approximates as geometry, generally defined as long-term SINR in a fully loaded network, if the effect of noise is small enough to be negligible) of a network is given by [6]:

$$\Pr(Geometry < \beta) = \frac{\alpha}{2\pi \csc(2\pi/\alpha)\beta^{2/\alpha}}, \quad \beta > 1 \tag{5.2}$$

where β is the SIR in linear scale. This equation tells us that the change in the density of the network nodes leads to the change in the received and interference powers with the same factor and hence the effects cancel. In other words, [6] implies that, as long as the network is interference limited, the network density does not change the SINR distribution and hence the sum rate scales linearly with the network density.

In addition to other assumptions leading to the linear sum rate scaling law, we point out the main underlying reason is the assumption of distance-invariant path-loss scaling law given by (5.1). Such a power law is widely known as "scale free," and hence densification can be viewed as altering nothing but the scale of the network and not the SINR of the network. However, realistic path loss does not exhibit the distance-invariant scaling property within the range of interest, and the invariance principle needs to be re-examined under more realistic assumptions. For example, in [5] it is shown for indoor networks whose path loss follows an exponential decay along distance, the capacity asymptotically scales with the square root of the network density.

In industry, WINNER-type models are widely accepted, such as urban micro (UMi) for small cells (e.g., picos) [9, 10]. Based on such a model, our evaluation in Figure 5.2 shows that the geometry distribution does not remain invariant with densification and therefore (5.2) is not applicable in a more pragmatic scenario. Furthermore, Figure 5.2 shows that the UE received SINR distribution also varies significantly for different densities. Since the UE rates and network sum rate are determined by UE received SINR, it is expected that a linear sum rate scaling law will not hold for practical dense networks.

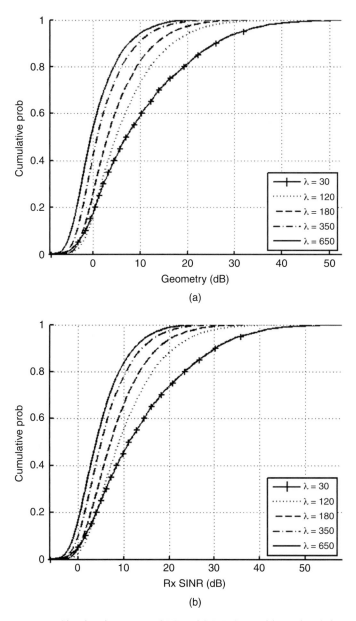

Figure 5.2 Simulated geometry CDF and SINR CDF with UMi path-loss model with different network node density λ (#/km^2). (Pico-only deployment, network stability criterion 1 (see Section 5.2.2.3), on average 50% time/frequency resource used for each pico, spatial Matern point process model for picos, Poisson point process model for UEs with minimum distance constraints to picos, per-UE Poisson traffic arrival and uniform arrival rate for all UEs over all the time, 10 MHz bandwidth at 2 GHz, 10% overhead.)

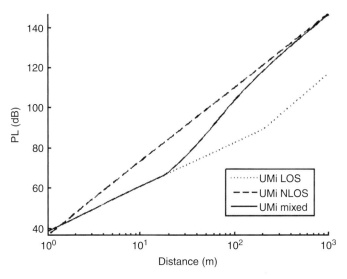

Figure 5.3 UMi path-loss model (at 2 GHz carrier frequency).

To better understand the reason behind this, we illustrate the UMi path loss in Figure 5.3. The UMi model has line-of-sight (LOS) and non-line-of-sight (NLOS) components (Table B.1.2.1-1 in [9]), and a distance-dependent LOS probability is also specified (Table B.1.2.1-1 in [9]). In the figure, "UMi mixed" is a mixture of LOS/NLOS components according to the LOS probability. One can see that the path loss is quite different from the single-exponent model given by (5.1). Roughly speaking, when the distance is far, the path loss increases much faster than it does when the distance is near, which affects the interference composition in a dense network. Therefore densely deployed neighboring nodes can contribute to an even higher portion of interference than a network with a single-exponent path loss. Densification thus leads to faster increase of interference than signal power, and hence the SINR distribution degrades with densification, as demonstrated in Figure 5.2.

A number of the WINNER-type models behave similarly to the UMi model. Realistic as they are, these models may be difficult to use other than resorting to numerical simulations. We thus propose a two-exponent approximation:

$$PL = \begin{cases} \alpha_1 \log_{10}(d/d_0), & d < D, \\ \alpha_2 \log_{10}(d/d_0), & \text{otherwise,} \end{cases} \quad (5.3)$$

where $\alpha_1 \leq \alpha_2$ and D is the turning point. This is illustrated in Figure 5.4. The two-exponent approximation with the distance-related path-loss exponents captures the main characteristics of practical path loss, but it is much simpler to analyze than, e.g., UMi models. Some studies based on the two-exponent approximation will be discussed later.

To summarize, WINNER-type path-loss models and two-exponent approximation can better capture interference conditions and hence the behavior of a dense network than single-exponent models and should be the focus for future analysis.

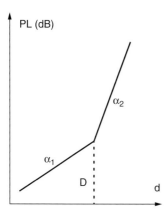

Figure 5.4 Two-exponent path-loss model.

5.2.2.2 System operation and dynamics modeling

In addition to channel models, a critical large-scale modeling factor includes network topology/configuration and system dynamics. From a high-level perspective, a dense network is modeled as an interconnected dynamic system with multiple tiers of components at possibly multiple carriers, externally driven by random traffic in shorter time scales (in the seconds scale) and by traffic load variations in larger time scales, and internally driven by the schedulers in shorter time scales and by network adaptation in larger time scales.

However, many static analyses, though providing fundamental insights in various scenarios, do not model some of the large-scale factors and hence may not capture some important aspects of dense network behaviors. For example, the existing literature assumes that SINR distributions (and hence throughput distributions) only depend on static network configurations, but in reality the interference depends on the traffic load (or traffic intensity) via a metric, the network resource utilization ratio (UR). For example, at high loads, a cell uses more time-frequency resources and emits more interference, generating a different interference profile than that at lower loads. In other words, the network performance may not be solely determined by static network settings; the interactions between UR and interference have to be considered and are generally missing in the literature.

On average, UR is the percentage of used resources over total resources, which is related to the *overhead channels* (mainly common channels, including control signals and reference signals) and each served UE's transmission time. More precisely, the UR of node i is

$$u_i = u_0 + (1 - u_0) \sum_{k=1}^{K} \tau_{ik} \qquad (5.4)$$

where u_0 is the fraction of time/frequency resources occupied by overhead channels, τ_{ik} is the fraction of time/frequency resources used by node i to transmit the packets for UE k. Note that τ_{ik} depends on I_{ik}, the total ICI received at UE k from all other transmitting nodes (namely, all nodes except for the serving node i),

$$I_{ik} = \sum_{j \neq i} p_j h_{jk} u_j, \qquad (5.5)$$

where p_j is the transmission power of node j, h_{jk} is the path loss between node j and UE k, and u_j is the UR of node j. In other words, the UR of node i depends on the URs of other nodes (through interference links), which further depend on the UR of node j. Therefore there is strong coupling between URs and interference.

The interplay between the UR and interference is widely ignored in system modeling and analysis of the existing literature. While such interplay makes an analytical approach more difficult, numerical or "semi-analytic" study may become the main tool. On the other hand, it points out new directions to understand network performance; for example, the impact of broadcast overhead u_0 becomes visible (more on this later), and UR-related performance metrics should be introduced.

5.2.2.3 Network stability and network efficiency

The UR of a network node measures how resources are utilized. High UR (e.g., >50% or >75%) generally implies that the node is overloaded, newly arrived packets will be queued for a long time, and some packets may be dropped, unfinished eventually, according to certain packet-drop criteria. If the UR is too low, then the node is underutilized (i.e., not yet reaching capacity). Based on this observation, one may define the capacity and network stability as follows.

DEFINITION:

i. A network is stable if $g(F_{UR}(x)) \leq g_0$, where $F_{UR}(x)$ is the CDF of UR over the network, $g(\bullet)$ is a non-decreasing function, and g_0 is a threshold.
ii. The network capacity is $C = \max_{s.t.\ g(F_{UR}(x)) \leq g_0} \lambda$, where λ is the aggregated traffic arrival rate (in bps/km^2).
iii. The normalized network capacity, or network efficiency, is $\bar{C} = C/N$, where N is the number of cells in the network.

In other words, (i) defines the stability of a network by requiring a statistics of the network UR not to exceed a threshold; (ii) defines the network capacity as the maximum traffic load that the network can afford without losing stability. Furthermore, (iii) defines the network efficiency as the per-node capacity rate. In a network with homogeneous spatial UE and pico distributions and uniform traffic arrival rate, the efficiency can be reduced to at most how many UEs a pico can support without losing stability. In such a case, the network efficiency may be simply (and intuitively) referred to as the "UE/pico ratio" evaluated at the boundary of stability region.

Various choices of $g(\bullet)$ may be used, and the above definition can also be extended to more generic cases. For example, one may choose either of the following:

$$\text{Network stability criterion 1:} \quad \frac{1}{N} \sum_{i=1}^{N} UR_i \leq 50\% \qquad (5.6)$$

$$\text{Network stability criterion 2:} \quad \Pr(UR > 75\%) \leq 10\% \qquad (5.7)$$

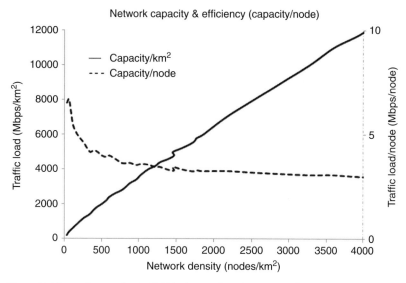

Figure 5.5 Network capacity (solid black curve) and network efficiency (dashed curve) over density. (Pico-only deployment, network load = #UE × per UE file arrival rate (0.5/s) × file size (0.5 MB), network stability criterion 1, UMi path-loss model, 20 MHz bandwidth with 10% overhead.)

Namely, with criterion 1, the network stability is measured according to the UR averaged across the network, that is, if the average UR is above 50%, then the network is seen as unstable. However, this stability criterion based on the average UR cannot reflect the distribution of URs in the network. Since packet drop is mainly associated with individual picos with high UR, from the network point of view, generally the percentage of nodes with high URs (e.g., UR > 75%) is desired not to exceed a fraction of the nodes (e.g., 10%). This may be especially useful for networks with non-uniform deployments and traffic loads, such as one in a large hotspot area with heavy traffic around it but light traffic far away from it.

5.2.3 Dense network performance evaluation

5.2.3.1 Sub-linear capacity growth with density

A dense network must be able to maintain high throughputs and low latency even with a very large number of users and bursty traffic demands. Using the pico-only deployment scenario (the simplest deployment scenario) as an example, with the increase of density, the network can support higher loads without losing stability, and the supportable traffic load (or capacity) grows sub-linearly; see Figure 5.5 for the network capacity and efficiency as functions of the network density. Intuitively speaking, sub-linearity means the growth is slightly slower than linear growth. The sub-linearity is also reflected in the trend of the network efficiency, which shows that as the density increases, moderate loss is observed but the loss almost flattens at high density.

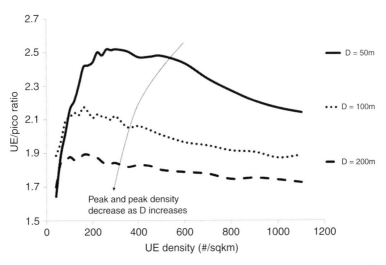

Figure 5.6 Simulated network efficiency with two-exponent path-loss model. (Pico-only deployment, efficiency normalized as #UE/#pico satisfying network stability criterion 1, $\alpha_1 = 2.2, \alpha_2 = 3.76$ as in (5.3), 10 MHz bandwidth with 0% overhead.)

The characterization of the capacity growth trends has significant consequences. First, a multi-fold increase of network density yields a multi-fold increase of capacity, which is the major premise of the dense cellular network. Second, sub-linearity implies that the growth rate is not constant and some degradation of the growth rate at a high density should be expected, as alternatively demonstrated in the network efficiency curve. Network efficiency is not density invariant as proven in the idealized case in [6]; more interesting behavior is expected for a dense network. The reason for the difference lies in the increased interference at higher density, which is caused by several factors, such as the UMi path-loss model, overhead channels, etc. All such factors will be further studied.

To see the impact of path-loss modeling, the two-exponent path-loss model in (5.3) is tested to see how the network efficiency (or equivalently in this case, the UE/pico ratio) changes. The results are presented in Figure 5.6. As the density increases, the network efficiency first improves, then peaks, and finally degrades. The three regimes roughly correspond to coverage limited (intuitively speaking, a sparse network has cell coverage issues and thus increasing the network density improves performance), transition between coverage limited and interference limited, and interference limited (and thus increasing the network density increases interference and degrades performance). The degradation of network efficiency at high density is consistent with the degradation of geometry/SINR at high density shown in Figure 5.2 for the UMi path-loss model; in contrast, we note that a single-exponent path-loss model has an almost flat network efficiency (a horizontal line) against network density (plot skipped for brevity). In addition, Figure 5.6 shows that spatial separation matters: the smaller the path-loss turning distance D, the larger the spatial separation and isolation, which further leads to smaller interference and better network efficiency. Note that this trend depends on the exact path-loss model and will differ for various deployment environments.

In summary, our study confirms the benefits of dense network deployment as a highly scalable solution to deliver multi-fold capacity gains and superior user experience. Therefore it is widely considered as a top candidate solution for future wireless communications. We have also identified more pragmatic and suitable methodologies to study dense network behaviors and showed that the increased ICI at higher density prevents the network capacity from scaling with network density. As a result, only sub-linear capacity growth is feasible in a dense network due to more severe interference conditions.

5.3 Characterizing dense cellular networks

With the preliminary understandings of how a dense cellular network may operate and perform, it is then clear that several critical issues exist and need to be addressed in order to take full advantage of the dense deployment.

5.3.1 Interference issues

Dense deployment scenarios will see different interference conditions from the existing, mostly sparse scenarios. By interference conditions we generally mean several aspects of interference, including the strength/magnitude of interference, profile/composition (i.e., relative strengths of contributing interferers in the interference experienced by UEs), and spatial/temporal variations of interference.

In the macro networks studied in 3GPP Releases 8 and 9 and co-channel heterogeneous network scenarios studied in 3GPP Releases 10 and 11, the dominant interferers are the strongest one or two macro cells, which account for at least 50% of the total interference received at UEs, and the dominant interferers are generally regularly located. In other words, the interference conditions in existing networks are characterized by one or two regularly located dominating interferers. Thus conventional interference management technologies can be adopted, such as spatial/frequency/time reuse, interference cancellation, etc., which have been shown as quite effective in dealing with a small number of dominating interferers.

In dense networks, however, interference conditions are characterized by several randomly located strong but less dominant interferers, generally with more severe interference with larger spatial/temporal fluctuations. The decreased dominance of the strongest interferers is illustrated in Figure 5.7, which shows that in the pico-only layer, as the network density increases with the increase of UE density (i.e., load, while maintaining the network stability), more and more interferers matter and the strongest interferers play lesser roles. As there is less planning in a typical dense network, these strong interferers are generally more randomly located in the spatial domain, and the traffic loads are distributed in the network in a more non-uniform way, leading to more significant fluctuations of interference in both the spatial domain and temporal domain.

This distinctive interference condition reflects the defining characteristic of a typical dense network, namely its irregularly deployed network nodes with the coverage areas

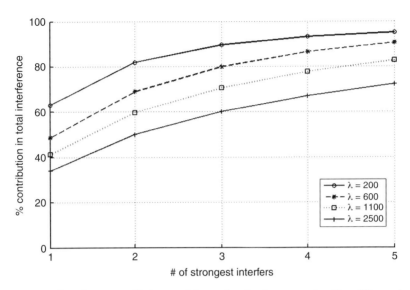

Figure 5.7 Interference profile/composition at the pico layer over densities. (Pico-only deployment, network stability criterion 1, UMi, 10M Hz bandwidth with 0% overhead.)

(defined according to received signal strength) overlapping between two or more nodes. Since a typical dense network is deployed with less planning than regular macro networks for apparent practical reasons, the neighboring nodes are generally close to each other with overlapping coverage areas, and the overlapping areas may have irregular shapes formed by a varying number of nodes. In contrast, if the network nodes have more regular spatial locations, then generally the overlapping areas have regular shapes formed by a constant number of nodes.

Therefore many existing interference management approaches may not apply to dense networks. In a typical dense network, eliminating the top four or five strongest interferers may lead to only a few decibel received SINR gain at high network density, namely capacity gains through existing interference avoidance will be limited and become increasingly difficult with density. For example, interference cancellation is not expected to work efficiently under the new interference conditions. For another example, it is found that frequency reuses with reuse factors ranging from two to six do not provide throughput benefit as shown in Figure 5.8, and the reason is indeed due to the randomized and less dominant interferer distribution.

From Figure 5.8(a), it is seen that frequency reuse in a dense network can significantly improve the geometry; note that geometry may be seen as SINR in a fully loaded system. As pointed out in Section 5.2.1, the interplay between the UR and interference should be taken into consideration for network performance. In fact, Figure 5.8(b) shows that the system average UR grows almost linearly with the reuse factor, to compensate the loss of frequency resources at each layer, and it largely offsets the benefits of reduced interferers due to frequency reuse. Consequently, as demonstrated in Figure 5.8(c), the throughput performance degrades significantly from reuse 1 to reuse 2 (the trend extends

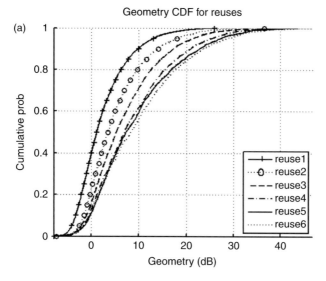
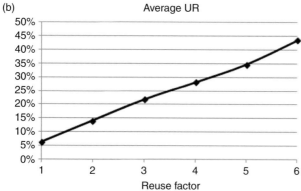
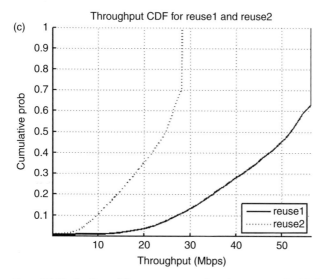

Figure 5.8 Evaluation of frequency reuses in a dense network. (a) Geometry CDFs for reuse 1 to reuse 6. (b) Average UR for reuse 1 to reuse 6. (c) UE throughput CDFs for reuse 1 and reuse 2. (Pico-only deployment with $\lambda = 400$ #/km^2, $\lambda_{UE} = 200$ #/km^2, UMi, 10 MHz bandwidth with 0% overhead.)

to higher reuses)[2], due to limited gains in SINR (or even losses in SINR) and significant losses of frequency resources. Note that the heuristic coloring algorithm mentioned in Section 5.1 is used here to separate the nodes into different frequency layers.

In addition, due to the more severe and randomized interference, some performance metrics, such as the network efficiency (see Figure 5.5) may not improve, or may even degrade, with straightforward densification (though enhancement techniques may be adopted to improve those performance metrics). The sub-linear, as opposed to linear, capacity growth of a dense deployment is a consequence of the new, worsened interference condition.

Furthermore, we observe that interference from common channel (such as CRS) overhead can substantially degrade performance. As plotted later in Figure 5.10, a 10% increase of the common channel overhead contributes to more than 25% network efficiency degradation. This high sensitivity of performance to overhead is also a characteristic and an issue associated with high density.

5.3.2 Other issues

The backhaul connections, including both the base station to core-network backhaul connections and the inter-base station backhaul connection but perhaps especially the former, may impose significant restrictions on dense networks. Due to the high density, direct and dedicated connections between the network nodes and the core network as well as among network nodes may incur prohibitive cost and are not realistic. Therefore network nodes are more likely to be connected to the core network and among each other by non-ideal backhaul, and instantaneous coordination among nodes may not be feasible.

The traffic demand in a dense network may have large variations in spatial/temporal domains. It may be desired that the network capacity can adaptively track the external traffic load. Other side effects of the dynamic variations of traffic loads include the fluctuations of ICI, the load imbalance among different nodes, and so on.

The dense network nodes are generally deployed much less regularly. Combined with the traffic-fluctuation issue mentioned above, this can enlarge the divergence of the network capability and the traffic load, and exacerbate the interference issue. This also makes it difficult to apply conventional interference avoidance, which is based on reuse. When the network is deployed regularly as three-sector hexagonal grids, reuse-3 leads to good performance, but reuse no longer delivers the performance gain with a less regularly deployed network, as shown in the companion technical report.

The energy consumptions in a dense network may increase significantly, despite the fact that the small cells may consume considerably less energy than macro cells.

The core network, known as the enhanced packet core (EPC) in LTE, also plays critical for the successful deployment of ultra-dense networks. In particular, the EPC was not

---

[2] We note that other views exist regarding the performance benefits of frequency reuse. Simulation calibrations and further study are needed for complete understanding on this issue.

designed for dense deployment of the access nodes and thus it will need to be evolved with the densification of the network.

In summary, we have identified a number of issues associated with dense networks, most prominently the issue with more severe interference from multiple strong but less dominant interferers, calling for new enhancement solutions to be employed for dense networks, which we now turn to.

5.4 Technologies for dense networks

In order to address the above-mentioned critical issues as well as to fully utilize a dense cellular network, solutions and suggestions are proposed along the following lines: how the dense network may be deployed and planned, how the interference may be managed, how the network may adapt its operations, and how the non-ideal backhaul may be dealt with.

The deployment strategies are commonly investigated with respect to spectral efficiency, coverage, or outage probability [14].

5.4.1 Deployment and planning of dense networks

A dense network should be deployed to fully take advantage of the available physical resources (e.g., bandwidth, backhaul), propagation environment, knowledge about traffic patterns, etc. The way that dense networks may be deployed depends on several *key factors*, including: (i) traffic demand; (ii) backhaul limitation; (iii) availability of macro coverage; and (iv) spectrum availability. The following scenarios (as typical deployments) are considered:

- **Scenario A:** high-density small cells with non-co-channel macro coverage, i.e., small cells and macro cell are on different frequency carriers
- **Scenario B:** high-density small cells without macro coverage
- **Scenario C:** high-density small cells with co-channel macro coverage.

For all these three scenarios, non-ideal backhaul connection between the small cells (and between macro cell and small cells) is assumed. The average user performance for these scenarios is shown in Figure 5.9 with different network densities and different bandwidths.

The results in Figure 5.9 indicate that, in general, Scenario B outperforms Scenario C, which outperforms Scenario A when the small cell layer occupies half of the available bandwidth. This is because small cells are capable of providing high capacity, the interference from co-channel macros may degrade the small cell capacity and hence overall network capacity, and restricting/reducing the small cells' usable bandwidth may not be beneficial. However, other aspects of the system design/operation also need to be considered. With Scenario A, there is a clear functionality split between the macro layer and the pico layer, which, for example, can ease the legacy UE support (done at

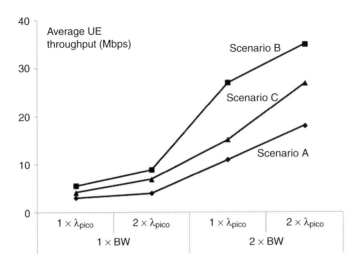

Figure 5.9 Average user performance for Scenarios A, B, and C, with different network densities ($1\times$ or $2\times$ λ_{pico}) and different bandwidths ($1\times$ or $2\times$ bandwidth). ($\lambda_{pico} = 125$ #/km^2, $\lambda_{UE} = 800$ #/km^2, BW = 10 MHz, UMi, 10% overhead.)

the macro layer) and enhance next-generation UE performance (done at the small cell layer).

The spectrum availability also plays an important role. If the available spectrum is abundant, then several methods may be adopted when deploying the dense network to better take advantage of the wide spectrum. As an example, the dense small cell layer may be deployed in a carrier separated from the macro carrier and hence simplify the design and operation while providing good support in all the aspects of coverage, mobility, and throughput (i.e., Scenario A); as another example, the operator can strike a *much* better balance between the bandwidth and network density (e.g., doubling the bandwidth requires lower than half of the network density to achieve the same or even better performance; see Figure 5.9. If, however, the operator has limited spectrum, the carrier to support high-density small cells may have to overlap with the carrier to support macro cells (if any) for which a variety of interference management/coordination techniques are to be applied for performance enhancement.

5.4.2 Network adaptation and interference management

The developments in previous sections have well established that interference is a top issue for a dense network. Network topology adaptation can be used to address several issues in a dense network, such as energy saving, spectrum efficiency improvement, load balancing, etc. The adaptation can include turning on/off a subset of network nodes, power adaptation, adaptive carrier selection, etc. For example, some redundant nodes may be turned off when the traffic loads are light, aiming to minimize the energy costs under certain QoS constraints. The on/off switching in these works is generally in large time scales (at least minutes). However, these works are not concerned about throughput performance enhancements (such as capacity enhancements, QoS enhancements, etc.) or interference management while improving energy efficiency in the meantime. Instead of viewing energy efficiency and throughput performance as conflicting goals, in [13] it is argued that interference management through network adaptation can significantly improve performance for dense networks while simultaneously saving energy (such as improved QoS with reduced power and/or bandwidth resources). A network utility maximization problem is formulated, which takes as input the network topology and aggregate demand information, and optimizes over the set of active nodes and the assignment of demands to the nodes.

To manage interference and to secure good performance gains in a dense network, one may be required to develop advanced interference-management techniques such as the adaptation of the network resources according to traffic variations, including turning on/off a channel and/or a low-power node flexibly, UE adaptively selects a serving layer (e.g., macro layer versus pico layer) and serving node under network instructions, downlink power control, the separation of common channel resources and data channel resources, etc.

5.4.2.1 Turning off transmissions of an inactive pico

Clearly, a pico without any UEs attached contributes nothing but interference (e.g., interference due to CRS) to the network, which reduces the performance. It is then desired to turn off this pico's downlink transmissions. The methods to support more dynamic on/off switching of pico transmissions (as well as more complicated pico on/off switching and load shifting) will be discussed in the network adaptation part.

5.4.2.2 Overhead reduction

One source of interference is due to overhead channels; in legacy deployments a pico serving no UE will still transmit these channels. Figure 5.10 shows that by reducing the overhead u_0 from 25% (typical for existing systems) to 10%, the capacity almost doubles. The network performance is very sensitive to the overhead. Therefore, turning off the transmissions of inactive picos can help reduce interference, which motivates (partially) further discussion of network adaptation.

5.4.2.3 Common channel/data channel separation

The current LTE system has a mixture of common channels and data channels, meaning they may interfere with each other, such as CRS transmission affecting data transmission and vice versa. Evaluations show that cross-interference can severely degrade the reliability of key common channels and data throughputs at high network density. To resolve this issue, separation between common channels and data channels is proposed and evaluated. With the same common channel overhead percentage, the separation demonstrates more than 50% gain in terms of network efficiency in the moderate and high overhead systems. Note that if frequency-domain interference coordination is used to avoid strong interference between PDSCH and EPDCCH, then it is effectively a type of separation of common channels and data channels, and hence our proposed approach may be considered as an extension along the lines of EPDCCH. See Figure 5.11.

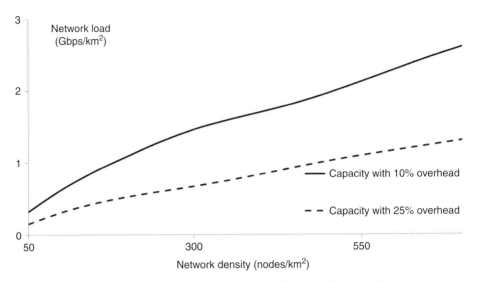

Figure 5.10 Increasing capacity by reducing overhead u_0. (Pico-only deployment, network stability criterion 1, UMi, 10 MHz bandwidth, network load = #UE × per UE file arrival rate (0.5/s) × file size (0.5 MB).)

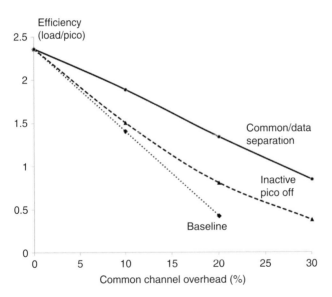

Figure 5.11 Network efficiencies for different interference management schemes. (Pico-only deployment, efficiency normalized as #UE/#pico satisfying network stability criterion 1, RSRQ cell selection, 10 MHz bandwidth. Baseline: all picos are always on and at least the 10% overhead channels are always transmitted. Inactive pico off: a pico serving no UE is turned off, including the overhead channels. Common/data separation: for each cell, common channel overhead and data are on orthogonal time/frequency resources, and overhead of all cells are on the same time/frequency resources.)

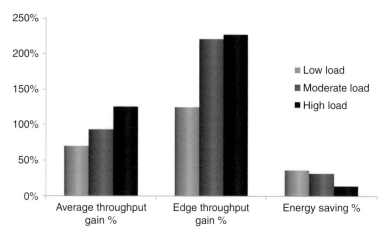

Figure 5.12 Performance improvements introduced by adaptive on/off and cell selection (clustered small cell deployment with macro layer at different carrier). (Baseline: pico-only network and RSRP association; enhancement: macro–pico network SCE Scenario 2a with inter-layer adaptive load balancing and pico on/off adaptive.)

5.4.2.4 Network adaptation

Network adaptation with on/off switching of network nodes, adaptive cell/carrier selection, power adaptation, etc., are prime candidates for dense network interference management. With the (arguably newly) introduced possibility of turning on/off network nodes and other adaptation mechanisms, new doors are open for further optimization of network performance. Two sets of network adaptation mechanisms are described and evaluated below.

Adaptive inter- and intra-layer load balancing and pico on/off adaptation (Example 1)

Consider a network with densely distributed pico nodes overlaid with sparsely deployed macro eNBs on a different frequency layer. To see the effectiveness of network adaptation, the following experiments are done. First, assume all traffic loads are supported by the pico layer only, and RSRP cell selection without any pico node on/off switching is evaluated as the baseline. Second, adaptive inter- and intra-layer load balancing is performed. A UE is decided to attach to a node in either the macro layer or the pico layer if the node maximizes its SINR plus a UR-dependent offset. The offset steers UEs away from nodes with high UR to avoid congestion. Clearly, any decision of cell selection affects the existing UR/interference distributions and hence existing cell selection results, and iterations need to be performed. Third, adaptive pico on/off switching is performed. A pico is turned off and its attached UEs are distributed to other nodes if this operation leads to an increase to a utility function (weighted sum throughputs).

As shown in Figure 5.12, compared with baseline results without adaptation, adaptive inter- and intra-layer load balancing and pico on/off adaptation lead to about two times average UE throughput and about three times cell-edge UE throughput, while turning off transmissions from a significant fraction of the picos.

Some observations are in order. A somewhat typical understanding is that the pico layer is the main throughput pipe and the macro layer should mainly provide mobility/coverage support and cannot contribute much to throughput performance. However, our evaluations have demonstrated that the macro layer can provide considerable throughput boost to the pico layer, as it can relieve the picos from covering UEs not too close to the picos, which reduces pico URs and interference. Moreover, adaptive load aggregation is identified as beneficial. Rather than attempting to distribute traffic loads as evenly as possible to different nodes as done in load balancing, in load aggregation certain traffic loads are aggregated to a subset of nodes so that other nodes may reduce their transmission activities, which leads to lower interference and improves overall throughput performance. The details are to be reported elsewhere.

Adaptive inter- and intra-layer load balancing and pico on/off adaptation (Example 2)

From the above analysis, network adaptation, which is an "unconventional" interference coordination technique, seems essential to dense networks. A dense network is deployed in an *overprovisioning* way, but some of the resources are not needed (or may even be harmful) to stay on all the time; rather, they should be adaptively put into use only if sufficient traffic loads are in their coverage areas. This issue related to overprovisioned network resources will be more exacerbated when small cells are equipped with wider and wider frequency resources, which may be termed as adaptive carrier selection and deserves some specific study.

For example, consider a case for evaluation where each small cell has K component carriers in a dense network with clustered and non-clustered small cell deployments. The baseline is to turn on all carriers and they all transmit at the same power level. In contrast, we strive to determine each carrier's on/off status and power level (between zero power and highest power) in order to maximize user experience (or equivalently, minimize sum user packet delay). More specifically, we aim to solve the following global optimization problem involving J UEs served by I picos on K component carriers:

$$\min_{\{p_i^k\}} \sum_{\substack{j=1,\ldots,J \\ i \text{ such that } j \in B_i}} m_j \left(\frac{F}{\sum_{k=1}^K R_j^k} + \bar{\tau}_i \right)$$

$$\text{s.t.} \begin{cases} R_j^k = W \log \left(1 + \frac{p_i^k h_{ij}}{\sum_{i' \neq i} p_{i'}^k u_{i'} h_{i'j} + N_0} \right) \\ \bar{\tau}_i = \frac{u_i - u_0}{(1 - u_0)\lambda((u_i - u_0)^{-1} - 1) \sum_{j \in B_i} m_j} \\ u_i = u_0 + (1 - u_0) \sum_{j \in B_i} \frac{m_j F}{\sum_{k=1}^K R_j^k} \end{cases} \quad (5.8)$$

where p_i^k is the transmission power of ith pico on kth carrier with bandwidth W, F is the packet size, m_j is the number of packets for UE j, which is a realization based on a given Poisson arrival rate λ_f, R_j^k is the transmission rate of UE j served by the ith pico on kth

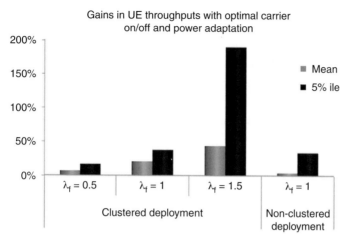

Figure 5.13 Adaptive carrier on/off status and power level selection gains in mean and 5th percentile UE throughputs, for clustered and non-clustered network deployments, and for different file arrival rates λ_f (#/UE/s). (Four of 10 MHz-bandwidth carriers available for each pico, SCE Scenario 2a for clustered deployment, uniformly pico dropping for non-clustered deployment (subject to minimum distance requirements to macros/picos), 30 UEs per macro area, throughput statistics do not include macro UEs.)

carrier, $j \in B_i$ denotes UE j served by pico i, h_{ij} is the channel gain between pico i and UE j, u_i denotes the UR of pico i, N_0 is the noise power, u_0 is the overhead, and $\bar{\tau}_i$ is the queuing time for pico i. For each pico, UE packets are served using the entire bandwidth on all active carriers based on a "first come first serve" rule, namely a packet is served upon its arrival if the pico is not serving another packet or put in the queue of the pico. In other words, in (5.8) the objective is to minimize the sum delays experienced by all UEs, where a delay contains two portions, the first being the queuing time of a packet (from the time when the packet arrives to the queue to the time when the packet starts to be transmitted), and the second being the transmission time. Moreover, the first equation in the constraints calculates the transmission rate based on the Shannon equation, the second estimates the queuing time assuming an M/M/1 model, and the third calculates the UR.

Figure 5.13 illustrates the performance gains of adaptive carrier selection. Only the pico-layer UEs are considered as we keep the macro layer the same for the baseline and the enhancement. It can be seen that, in a clustered network, optimal allocation leads to significant improvement of network capacity and user experience. Additionally, as load (measured by file arrival rate per UE) increases, the gain also increases; this is because at higher loads, the interference issue becomes more severe and hence it is more beneficial to manage the interference by proper carrier selection and power adaptation. However, the gains are lower (especially for the mean UE throughput metric) if the network is not deployed in a clustered way, as the dominant interferers become more randomly located and more challenging to avoid. Future study is needed for further enhancements, especially for the non-clustered case. It is noted that the clustered case follows the assumptions in Figure 5.13 with four small cells per macro cell area except otherwise mentioned, while the non-clustered case has the same assumptions and the same total number of small cells in the network except for that the small cells are uniformly distributed.

5.5 Summary and future directions

The development in this chapter has thus far established that high-density network deployment is a key for future wireless communications, based on the modeling and evaluation methodologies proposed in this chapter.

In more detail, we have seen that existing methodologies for studying dense networks are limited and sometimes fail to capture certain fundamental, defining behaviors of dense networks. As an initial attempt to rectify this issue, we identified more realistic and suitable methodologies to study dense network behaviors, proposing viable path-loss models; system dynamics models capturing the interplay of resource utilization, interference, and rates; and definition of network capacity taking into account network stability.

Based on these proposals, we have showed that the increased inter-cell interference at higher density prevents the network capacity from scaling with network density, and only sub-linear capacity growth is feasible in a dense network due to more severe interference conditions. Several solutions are proposed to further enhance the performance and operation efficiency in dense networks, most notably the network adaptation for interference coordination and load balancing/aggregation.

In summary, this study provides the following answers to the fundamental questions:

- Q1: Why the dense network?

  A1: The dense cellular network deployment is justified as a highly scalable solution to deliver magnitudes of capacity gains and superior user experience. The system capacity grows sub-linearly with the density of the network. A dense cellular network may maintain high user throughputs and low latency even in the presence of a very large number of users with heavy and bursty traffic demands.

- Q2: What characterizes the dense network?

  A2: Generally speaking, a network is dense when its efficiency degrades with network density. In other words, an X time increase in network density leads to less than X time increase of network capacity. Distinctive characteristics and challenges at high densities are identified, including more severe interference from multiple strong but less dominant interferers, performance sensitivity to overhead channels, and so on.

- Q3: How to deploy and operate the dense network?

  A3: Novel techniques are proposed and evaluated, such as network adaptation (of, e.g., nodes, carriers, power, etc.) for interference coordination and avoidance (which also reduces power consumption), common channel and data channel separation, etc., to sustain the high network efficiency at high densities.

Table 5.1 Simulation assumptions.

Items	Parameters	Comments
Simulation area	An area containing two rings of hexagonal macro cells (57 macro cells)	
Simulated scenario	1. SCE-2a; 2. pico-only scenario	
Macro dropping	Regular dropping, ISD = 500 m	Macros are not present in pico-only simulations
Pico dropping	1. SCE-2a; 2. Matérn hard-core process of type I models with density λ and minimum distance h [12]	In 2, the pico–pico minimum distance is min $(0.02, \sqrt{1/\lambda/\pi}/2)$ km
UE dropping	1. SCE-2a without indoor dropping; 2. uniformly dropping (subject to minimum distance requirements to macros/picos)	In 2, UE–macro minimum distance = 35 m, UE–pico minimum distance = 5 m
UE density	1. 30/60 per macro sector; 2. 50~5,000 per km^2	#/km^2
Pico density	1. Fixed 4/10 per cluster, 1/2/4 clusters per macro sector; 2. fixed 30 ~4,000 per km^2; 3. determined based on stability condition	
Path-loss model	1. UE–pico: UMi, with 2D distance; UE–macro: UMa, with 3D distance; 2. single- or double-exponent model for UE-pico path loss	
Shadow fading	SCE-2a	Shadowing correlation modeled for macro eNBs only
Traffic model	Packet size = 0.5 M bytes, arriving rate $\lambda = 0.5$	FTP2 model without considering reading time
Macro Tx power	46 dBm	
Pico Tx power	30 dBm	
Bandwidth	1. 1 carrier of 10 MHz for SCE-2a; 2. 1 carrier of 20 MHz for pico-only simulations	
Overhead channel resources	10% of total time/frequency resources	Overhead is accounted for in interference computation unless a cell is turned off
Cell selection criteria	RSRQ or adaptive association	

Based on these findings, a coherent paradigm built upon the dense deployment is emerging, which we expect to develop into a key, indispensable technology for future wireless networks to meet the traffic demands. Various aspects of the proposed methodologies and techniques for dense networks are shown to warrant further research and development. We thus call for more attention and more contributions from both academia and industry to develop suitable means to model, evaluate, analyze, and optimize dense networks.

Notwithstanding the various advantages of the dense LTE networks, we emphasize that no one technology can meet the multi-dimensional demands of future wireless networks. The dense-deployment based solution should be seamlessly integrated into the network with other solutions, such as the spectrum-based solutions, WiFi (or more generally, inter-RAT), massive MIMO, CoMP, next-generation antenna systems, and advanced coding/modulation/transmission/receiving schemes, etc. This may require a holistic approach to wireless communication system research, design, optimization, deployment, and operation, to integrate all necessary components.

To conclude the future directions, we point out that special efforts are needed for the following studies: first, further development of modeling and evaluation methodologies for dense networks; second, further research of network adaptation-based interference management for dense network performance enhancement; third, the evolution of dense networks toward ultra-dense networks; and fourth, the integration of dense networks with other wireless technologies.

5.6 Appendix

For the numerical evaluations presented in this chapter, the simulation assumptions in [7, 9] are followed unless otherwise specified. Table 5.1 summarizes some detailed simulation assumptions adopted in this chapter. The reader may refer to [7, 9] for terminologies used in the table. "SCE-2a" standards for SCE Scenario 2a specified in [7].

References

[1] 4G Americas, "Mobile broadband explosion: the 3GPP wireless evolution," 4G Americas white paper, www.4gamericas.org, Aug. 2013.

[2] D. Cavalcanti, D. Agrawal, C. Cordeiro, B. Xie, and A. Kumar, "Issues in integrating cellular networks WLANs, and MANETs: a futuristic heterogeneous wireless network," *IEEE Wireless Communications*, 12, 30–41, June 2005.

[3] 3GPP TR 36.932, "Study on scenarios and requirements of LTE small cell enhancements," v0.2.0, Nov. 2012.

[4] D. H. Kang, K. W. Sung, and J. Zander, "Cost efficient high capacity indoor wireless access: denser Wi-Fi or coordinated pico-cellular?" http://arxiv.org/pdf/1211.4392, Nov. 2012.

[5] J. Ling and D. Chizhik, "Capacity scaling of indoor pico-cellular networks via reuse," *IEEE Communications Letters*, 16:2, Feb. 2012.

[6] H. S. Dhillon, R. K. Ganti, F. Baccelli, and J. G. Andrew, "Modeling and analysis of K-Tier downlink heterogeneous cellular networks," *IEEE Journal on Selected Areas in Communications*, 30, 550–560, April 2012.

[7] 3GPP TR 36.872, "Small cell enhancements for E-UTRAN – physical layer aspects," v12.1.0, Dec. 2013.

[8] 4G Americas, "Developing and integrating a high performance Het-Net," 4G Americas white paper, www.4gamericas.org, Oct. 2012.

[9] 3GPP TR36.814, "Further advancements for E-UTRA physical layer aspects," 2010.
[10] IST-WINNER D1.1.2 P. Kyösti, *et al.*, "WINNER II Channel Models", ver 1.1, Sept. 2007. Available: www.ist-winner.org/WINNER2-Deliverables/D1.1.2v1.1.pdf.
[11] F. Baccelli and B. Blaszczyszyn, "Stochastic geometry and wireless networks," *Foundations and Trends in Networking*, 2010.
[12] M. Haenggi, *Stochastic Geometry for Wireless Networks*. Cambridge University Press, 2012.
[13] B. Zhuang, D. Guo, and M. L. Honig, "Energy management of dense wireless heterogeneous networks over slow timescales," In Communication, Control, and Computing (Allerton), 2012 50th Annual Allerton Conference, pp. 26–32, IEEE, 2012.
[14] H. S. Dhillon, R. K. Ganti, and J. G. Andrews, "Load-aware modeling and analysis of heterogeneous cellular networks," *CoRR*, vol. abs/1204.1091, 2012.

6 Traffic offloading scenarios for heterogeneous networks

Adrian Kliks, Nikos Dimitriou, Andreas Zalonis, and Oliver Holland

This chapter considers the challenges faced by network operators and service providers accommodating the increasing traffic demands in cellular networks in the most efficient yet inexpensive way. It proposes that these challenges are addressed by offloading part of the traffic to femto cell access points (FAPs) and WiFi access points (WiFiAPs). Whereas 4G micro and pico cell base stations are assumed to be managed by the network operator in terms of setup and maintenance, FAPs and WiFiAPs are normally bought and operated by the end-user. The main difference between these two solutions is that the FAPs operate on the frequency bands assigned to the network operator by national regulators, while WiFiAPs work on unlicensed spectrum. This chapter analyzes the pros and cons of such approaches, and the tradeoffs related to the different mechanisms employed in cellular and WiFi networks for interference management. Moreover, various methods for traffic offloading by the mobile network operator are discussed in detail, including local IP access (LIPA), selective IP traffic offloading (SIPTO), IP flow mobility (IFOM), access network discovery and selection function (ANDSF), and Hotspot 2.0, etc. In the experimental part of this chapter, means for traffic management from the network operator's perspective are discussed, taking into account costs and energy savings. Furthermore, a novel resource usage coordination concept in conjunction with the WiFi offloading concept is presented.

6.1 Introduction

The traffic served by cellular networks grows significantly on a yearly basis, making the efficient, fair, and inexpensive management of users' traffic demands a real challenge. Various forecasts provided both by legal bodies and commercial companies show that this trend will not change in the future. One can, for example, consider the expectations provided in Table 6.1, created based on the data delivered in the Cisco Visual Networking Index [1, 2] in 2012 and 2013. It is shown that the monthly average traffic generated by mobile devices is foreseen to grow exponentially. Based on [3], it can be stated that in 2017 IP traffic originating from mobile devices will comprise more than 13% of fixed IP traffic, while in 2013 this comprised only around 4% of fixed IP traffic. The dominant role will be played by mobile video, which is predicted to create around 66.5% of the total mobile traffic generated in 2017. The remaining part of the traffic is

Table 6.1 The growth of the traffic generated by mobile terminals, expressed in exabytes per month.

Year of forecast (below)	2011	2012	2013	2014	2015	2016	2017
2012	0.6	1.3	2.4	4.2	6.9	10.8	–
2013	–	0.9	1.6	2.8	4.7	7.4	11.2

anticipated to be split between mobile web/data services (24.9%), mobile machine–machine communications (5.1%), and mobile file sharing (3.5%).

The above data traffic demand projections force network operators to urgently find novel alternatives for better traffic management. Of course, one solution is to apply advanced radio interface and frequency reuse technologies guaranteeing the achievement of higher data rates and the reduction of interference, such as sophisticated beamforming solutions or cooperative radio-resource management algorithms. However, the capabilities of systems are getting very close to the Shannon limit, requiring excessive complexity to extract further performance increases causing increased energy consumption among other issues. Other proposals assume the possibility of traffic offloading from macro cells to smaller ones, such as micro, pico, femto, or metro cells. The reduction of cell sizes as well as of transmit power has the potential for significant improvement in area-capacity and efficiency, since the shorter the transmission distance and greater the number of base stations, the higher the frequency reuse hence overall network area-capacity. Among other solutions, the idea of data offloading, leading to a reduction in mobile network traffic, is particularly worth mentioning. Here, the traffic that can be managed either locally (two devices in the same local network communicating by means of a home base station H(e)NB[1]) or will terminate in the Internet and will not be conveyed by the core network of the mobile operator. It allows for effective management of the network congestion. Finally, the concept of WiFi traffic offloading has also become a vivid research topic and area of interest for network operators. Given that femto cell offloading refers rather to shifting the traffic from the macro cells to femto cells, thus reducing the load of the macro base station (MBS), the problem of mutual interference between macro and small cell users becomes immediately apparent. This is due to the fact that the macro base stations and femto access points (FAPs) operate in the same frequency bands belonging to the same mobile network operator (MNO), leading to interference management issues for the operator – a critical problem. On the other hand, under WiFi offloading the overall network load is reduced since a portion of the traffic is redirected to the WiFi network, which operates on separate unlicensed spectrum. Furthermore, it should be noted that in contrast to micro and pico cells, which are assumed to be managed by MNOs in terms of setup and maintenance, FAPs and WiFi access points (WiFiAP) are deployed by the end-user who incurs many of the

---

[1] Home base stations can be understood as either Home Node B (HNB) in UMTS-based systems, or Home eNode B (HeNB) in LTE-based solutions.

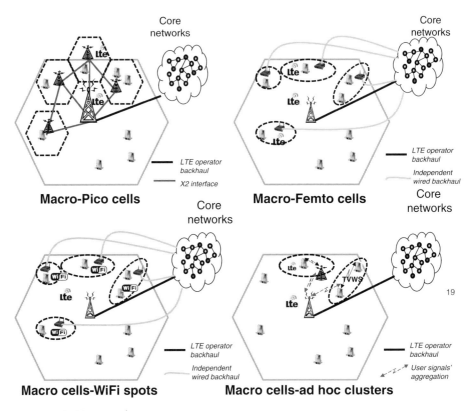

Figure 6.1 HetNet scenarios.

costs related to operation and maintenance [4, 5]. In this chapter we will discuss various aspects of possible traffic offloading through so-called small cells, e.g., through the usage of femto-, pico- and metro-access points, as well as WiFi hotspots.

6.2 The role of small cells in traffic offloading

Traditionally, in order to address the traffic growth, the cellular networks enhance their infrastructure with more cells by using various network planning mechanisms. The base stations (BSs) in that case work under similar characteristics, such as transmit powers, antenna patterns, number of serving users, and backhaul connections. In recent years, the focus in the research community and in the relevant standardization bodies has moved to new, more complicated and advanced network topologies, known as heterogeneous networks (HetNets) (Figure 6.1). A HetNet can be defined as a multi-tier network comprised of traditional large macro cells and smaller cells (sometimes called low power nodes – LPNs) including relays, micro cells, pico cells, and femto cells [6–9]. HetNets can also involve the combination of different radio access technologies, such as WiFi and WiMax, for traffic offloading. Small cells are usually classified according to their transmit power. Pico cells and micro cells are similar to the macro cell BSs, only using lower transmission powers and thus covering smaller areas. Femto cell access points (FAP), on the other hand, are designed for indoor scenarios, where their maintenance and installation is performed by the end users and the deployment is done mostly in an unplanned manner. Furthermore, not all of the small cells are connected to the backhaul of the cellular network in the same way. In LTE, pico cells connect to the core network with the X2 interface and thus they are able to use the standard's inter-cell interference coordination (ICIC) techniques developed for controlling interference and for proper resource allocation.[2] Femto cells, on the other hand, are intended to serve mainly indoor scenarios. They have low transmit power, and the installation and maintenance is carried out by the consumers in an unplanned manner. Thus operators are not aware of the changes that may occur to the network structure and this becomes one of the greatest challenges. Moreover, femto cells are not connected to the backhaul of the cellular network; instead, they utilize, e.g., the DSL connection or the cable modem of the end user. With that approach, however, in LTE, the X2 interface is not accessible, thus classic ICIC techniques cannot be implemented. This means that in order to control and/or mitigate the related interference alternative techniques should be developed. The 3GPP [10] and the Small Cell Forum [5] have worked to identify the critical problems of this multi-tier topology, and they have proposed specific procedures and techniques for proper interference management.

In this network enhancement with small cells operating in the cellular band, the benefits of offloading the traffic from the macro cell tier to smaller cells are only attributed to the reduced transmit power of the smaller cells that allows for the re-use of the available resources more frequently. Recently operators, in order to address the lack of available spectrum, have started to examine other options, including opportunistic access of other licensed (e.g., TVWS) or unlicensed (e.g., ISM) bands. The so-called integrated femto–WiFi (IFW) networks have been proposed for installation in residential buildings, in enterprise premises, as well as in metro stations [11]. The idea is to use both existing WiFi infrastructure and the new small cell deployments to offload traffic from the main macro cell network.

Figure 6.1 illustrates the different scenarios of HetNets that can be envisioned for traffic offloading from the traditional macro cell network.

A recent article [8] identifies the new topics for future research investigation caused by this multi-tier topology and highlights the need for further research in the area of network modeling, users' association to nearby cells, and handover mechanisms. Of course, the most critical challenge in this HetNet environment remains the proper interference management (IM) either between the small cells themselves (co-tier interference) or between the small cells and the overlaying network (cross-tier interference). In the last

---

[2] In the ICIC concept the assumed frequency reuse is set to unity. In order to achieve this goal and minimize interference, advanced resource assignment among users is applied, in which all available resource blocks in the time-frequency plane can be assigned with maximum power to every user located in the inner part of the cell. In contrast, for cell-edge users, a portion of the available resource can be assigned, and only with reduced power. In the case of LTE, almost blank subframes (ABS) are added for the purpose of further interference management enhancement.

few years the problem of IM in two-tier networks has been studied for the seamless integration of the small cells into existing macro cell cellular networks. An interference analysis and a classification of various IM approaches and techniques for selected OFDMA macro/femto cell scenarios is presented in [12]. Some characteristic works in the literature are references [13] and [14]. In [13] an optimum decentralized spectrum allocation policy for two-tier OFDMA networks is presented. The paper proposes orthogonal assignment between macro cell and femto cells to eliminate cross-tier interference. For co-tier interference the paper proposes a frequency ALOHA strategy where each femto cell accesses a random set of frequency subchannels. A spatial Poisson process is used to model the presence and position of femto cells instead of assuming a fixed femto cell deployment scenario. For co-tier interference, [14] proposes a distributed inter-cell power allocation algorithm where each cell computes by an iterative process its minimum power budget to meet its local QoS constraints. As it is presented in [12], and discussed above, the differences in HetNets from traditional cellular engineering create opportunities for novel ideas and approaches, including the incorporation of techniques that were originally developed for CRS, ad-hoc, and flexible networks. The exploitation of enhanced environmental awareness (above and beyond the classic channel-state information feedback) is a useful tool for enhancing the performance of IM. In that direction, various relevant studies have been published in the literature in which the most important part is the collection, the storage, and the accuracy of the information used by the proposed optimization techniques. The deployment of coexistence-enhancing technologies such as sensing and geo-location databases has been proposed as tools for collecting, accessing, and using this information. Some initial approaches toward the use of databases were proposed in [15, 16] where the notion of background interference matrix (BIM) was introduced, which maintains information on all the potential interfering cells. In [17] the focus was on generating an interference matrix by combining data generated by the users and the system operator respectively. In [18] and [19] the use of enhanced environmental information was proposed for the improvement of RRM and IM techniques in two-tier networks. In these studies the information was collected by various means (e.g., from the sensing capabilities of the network elements) and stored in a dedicated database, a radio environmental map (REM) [20]. In [21] the authors consider a method for determining the optimum user–AP association ensuring a network-wide maximum–minimum fair bandwidth allocation. They formulated the association control problem and presented approximation algorithms that provide guarantees on the quality of the solution. Additionally, in [22] the authors presented a distributed user association considering different degrees of load balancing that covered scenarios with different emphasis on user-related physical capacity (or SINR) and on network-related traffic conditions, that converged to the global optimum and was also robust to changes of traffic distributions. Another approach was followed in [23], where LTE HetNet load balancing is proposed by performing range offset optimization of low-power nodes, considering also cell load-coupling relation between adjacent or overlapping cells. In [24] the authors extended the scope of resource scheduling across a cluster of BSs to enhance the overall performance through interference avoidance and dynamic load balancing. Other related works that introduce also the concept of cell breathing to dynamically coordinate the scheduling and the coverage of neighboring cells are documented in [25, 26]. Apart from these techniques, other strategies that have been proposed include a network-wide utility maximization with interference avoidance that assumed partial frequency reuse [27], a distributed algorithm based on the relaxation of a network-wide utility maximization problem [28], and methods that rely on the metric of rate coverage, which captures the effect of both SINR and load distribution across the network [29].

6.3 Technological solutions for HetNets

Following the generic description of various offloading scenarios with the usage of small cells, let us now concentrate on the technological aspects of these solutions. In this section we will provide the brief analysis of the provisioning systems for femto–WiFi networks, particularly focusing on the offloading issues.

6.3.1 Provisioning systems for femto–WiFi networks

Integrated WiFi–femto (IWF) networks can be provisioned in various configurations, not only for home users (typical household scenario), but also for enterprise customers or metropolitan users, where the IWF devices are considered for the public environment. Let us now look at the provisioning system for the femto–WiFi solutions by focusing – without any loss of generality – on the residential case. The exemplary architecture of the residential network can be illustrated as follows. The end user equipped with a dedicated mobile terminal connects to the IWF access point, which is then connected to the network delivered by the local service provider through the digital subscriber line access multiplexer (DSLAM). The use of a multiplexer is evident since the access network provider will deliver the same common backhaul to any other IWF access point located in the vicinity. After that the DSLAM can be connected through the broadband remote access server (BRAS) to the network managed by the internet service provider. The traffic will be then scheduled through Internet routers to the global network and finally to the mobile core network (which could be 3G core network or LTE-Enhanced packet core network). Let us now look at this network from the provisioning and management perspectives, as illustrated in Figure 6.2, where the provisioning system of the 3G femto–WiFi network is shown. One can notice that the IFW cells (realized by the integrated FAPs and WiFiAPs) are connected to the so-called residential gateway (ReGW) and through the access network and broadband Internet to the security gateway (SeGW) using Iu-h interface with IPSec tunneling mechanism. Finally, again through the Iu-h interface, SeGW is connected to the femto gateway (FeGW) and then to the mobile core network (using Iu-PS and Iu-CS accordingly for packet-switched and circuit-switched data). Clearly the WiFi traffic will follow the same route to the point-of-presence (POP) in the internet service provider network. The possible incorporation of the dedicated provisioning servers is also shown in Figure 6.2; these are connected via TR-069 protocol over SLS/TLS (secure sockets layer/transport layer security protocols) or IPSec (regarding femto provisioning server) and via TR-069 protocol only (referring

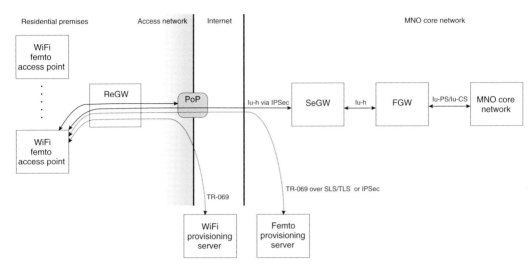

Figure 6.2 Provisioning system for 3G femto–WiFi network, based on [5, 11].

to the WiFi provisioning server). It is worth mentioning that in the case of enterprise or metro IWF networks, the WiFi traffic can be managed by the simple network management protocol (SNMP).

6.3.2 LIPA and SIPTO concepts

The problem of congestion in the network refers not only to the user data, but also to the backhaul traffic and the traffic in the core network. In order to deal with this problem two concepts have been recently proposed and standardized by 3GPP, which rely on the assumption that not all the traffic shall be routed via the MNO core network. These are briefly discussed below [30–33]:

- In the local IP traffic offload concept, known as LIPA, two devices located in the same local IP network can be connected directly without the use of the MNO core network; in such a case the traffic within the same LAN will be managed without the use of the core network architecture.
- In selective IP traffic offload, known as SIPTO, the traffic that originates or finishes at the public Internet does not need to traverse through the MNO core network.

The illustration of the LIPA and SIPTO proposals are shown in Figure 6.3. Let us stress that comparing to IFOM case the problem (described later in the chapter) is now more RAN-centric; the data offloading in the two cases considered now is mostly transparent to the user.

6.3.2.1 Local IP access (LIPA)

The idea of local IP access, LIPA [30], has been introduced in Release 9 of the 3GPP cellular network standards, and is continued in the further releases as well. As already

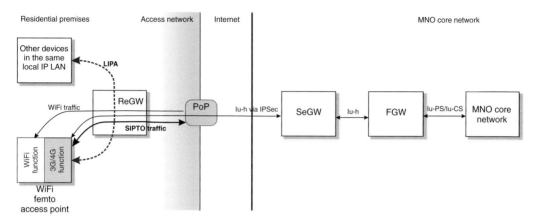

Figure 6.3 LIPA–SIPTO concept, based on [5, 11].

mentioned, the LIPA concept deals with the transferring of the data generated by the UE through the H(e)NB (it is not considered for macro stations), to local network also connected to that H(e)NB, without traversing the core network managed by the MNO. It is also feasible to allow the UE to connect to any external network that is also connected to the local network. Since LIPA can be applied to both residential or enterprise scenarios, one can imagine the following example, where a network printer could be reached by a mobile user through H(e)NB without traversing the MNO core network. Another example considered the wireless access of the mobile H(e)NB to the local computer with the purpose of photo searching and viewing. It suggests that various classes of traffic could be offloaded. In general, IP capable UE can benefit from offloading one of the following traffic types [32].

- Type I: where the whole traffic associated with the certain application protocol is offloaded.
- Type II: where all traffic related to the certain access point name (APN) is subject to offload.
- Type III: where the traffic is offloaded based on the IP address of the destination point of the network.

By assumption, the data flow classified as LIPA traffic should traverse the private network without involving any of the elements being possessed by the mobile network operator. It means that the traffic will be offloaded before the HNB-GW block in the 3G/4G network architecture. The agreed breakout point for the LIPA traffic will be always localized in the local gateway (L-GW), being either the standalone entity connected with the H(e)NB, or simply a part of the H(e)NB device. This situation is illustrated in Figure 6.4. The L-GW is responsible for implementing such functionality as IP address allocation of the UE or packet filtering. It connects to the private and public networks via the SGi interface (L-GW is seen as regular router by the external network), and communicates with the serving gateway (S-GW) by means of the L-S5 interface (based on GPRS tunneling protocol for control plane, GTP-c). It is worth mentioning that although the

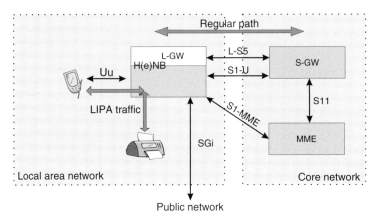

Figure 6.4 Local IP access (LIPA) traffic flow.

traffic will be offloaded at the H(e)NB level, the whole session management procedures to set up the LIPA connection with the packet data network (PDN) will be handled by the core network. For example the IP address is assigned by the H(e)NB to L-GW and then it is transferred via IPSec tunnel to S-GW and then via S11 interface to the mobile management entity (MME).

Separate PDN connection will be established if the UE will connect simultaneously to the core network. Thus since the dedicated PDN will be created the LIPA data flow can be then treated using separate APN or QoS requirements, but also different billing policies can be applied to users that utilize the LIPA functionality. Realization of the LIPA concept requires slight enhancements in the home subscriber server (HSS) in order to include the LIPA subscription information, e.g., it has to be provided if the selected user (or close subscriber group) is allowed to access selected APN through LIPA, etc.

It is also assumed that UE should have simultaneous access to the local and public network, as the second one will be managed by the MNO through the core network. Moreover, LIPA is also subject to all roaming agreements, meaning that the user can use LIPA even in the visited network. It is worth mentioning that in any of the previous cases all security aspects, defined by 3GPP for the cellular networks, cannot be compromised by application of the LIPA concept. Finally, it is said (starting from Release 11) that service continuity has to be guaranteed if the user moves between different H(e)NBs being part of the same local network (called inter-H(e)NB handover). Besides that one, handover between the macro base stations and H(e)NB are also considered [32].

Let us focus on the management aspects of the LIPA concept. It has been assumed that MNO, or any third party responsible for hosting H(e)NB, shall be able to activate or deactivate the LIPA functionality for a given H(e)NB or just for the selected UE. Moreover, MNO shall be able to perform any necessary measurement of the traffic parameters and of signaling performance, which will be related to LIPA. It shall also be allowed to the MNO to collect all information about fault management issues [32, 33].

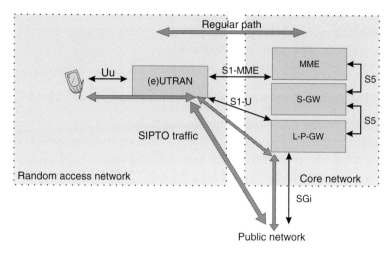

Figure 6.5 Selective IP traffic offloading (SIPTO) traffic flow.

6.3.3 Selective IP traffic offloading (SIPTO)

The concept of selective IP traffic offloading, known as SIPTO, has been introduced by 3GPP in Release 10 of its cellular network standards (3G and 4G only, GSM is not supported). In general, in the SIPTO case the portion of IP traffic both on H(e)NB access and on the cellular network can be offloaded to the local network. In other words, the whole selected traffic that will end somewhere inside the Internet can be offloaded from the core network managed by MNO, regardless whether the traffic originates in the home UE or in the UE located outside and connected to the (e)NodeB. SIPTO enables routing of that traffic through selection of an optimal path in the core network, or simply by bypassing it. MNO will benefit from shifting the traffic that is close to the core network edge – in such a situation the overloaded core or backhaul network would not be congested. One can imagine the situation where the best-effort traffic (e.g., file downloading or http data) can be offloaded, while online gaming or voice traffic will be managed by MNO.

As the breakout point in case of LIPA is always located in H(e)NB, in this situation the traffic will break out either at the H(e)NB level, or above the random access network (RAN) level, as it is illustrated in Figure 6.5. The basic idea is to select the S-GW and packet data network gateway (PDN-GW), which are closer either topologically or geographically to the radio network, and select them for data offloading. Similarly to the LIPA case, the simultaneous connections via SIPTO and via the core network have to be guaranteed; it is planned to keep the backward compatibility and allow pre-Release 10 devices to utilize the SIPTO functionality as well. Moreover, it is assumed that user interaction may not be necessary. Finally, roaming has to be offered to the user that uses SIPTO.

In order to maintain the subscription information the HSS has to be modified. As for LIPA, the HSS will indicate if an offload is allowed for the particular subscriber and APN, for both home and visited public land mobile network.

Interested readers are asked to search research literature, e.g., [10, 32–35], for more detailed information about LIPA and SIPTO solutions, including protocol stacks, messages exchange, and mobility management.

6.3.4 Seamless handover and traffic management: general considerations

One of the critical issues related to effective traffic offloading is the need of reliable and seamless handover between the regular cellular network and the WiFi access points, during which the IP address of the user will be preserved and no packet loss will be observed. Following [34], three mechanisms for IP-based data offloading can be classified, and are illustrated graphically in Figure 6.6.

- **Application-based switching**. In this solution, widely available in current terminals, all IP traffic can be redirected from e.g., the cellular network to WiFi from the application layer through switching between these two radio types; the main drawback of this solution is that the IP address is not preserved and this is the challenge for the application developer to keep the application active and operational during IP address change; please see Figure 6.6a.
- **WiFi mobility**. After introducing the dual stack concept for the IP networks, which provides joint management of IPv4 and IPv6 traffic, 3GPP has proposed in Release 8 the usage of the dual stack mobile IP (DSMIP) protocol in order to ease the handover between cellular and WiFi networks; in such an approach both server and the client (user equipment) have to be dual-stack capable and the concept of home agent has to be implemented; moreover, the local address (Care-of-Address) and Home Address (i.e. a permanent identifier) are created, which ensures that the exposed IP address is kept unchanged; the role of home agent is to bind the HoA with CoA. In such a case the whole traffic can be offloaded from the cellular to the WiFi network. Please see Figure 6.6b.
- **Starting from the 3GPP Release 10**. The concept of seamless switching of selected IP traffic together with simultaneous IP flow support has been proposed, which is based again on DSMIP. The goal behind that activity is to allow offloading of the selected traffic (such as video) while keeping the cellular/WiFi access. Here the concept is to allow registration of multiple CoAs to a single HoA, and also to bind different IP flows to different CoA or directly to HoA. Please refer to Figure 6.6c. It is also assumed that the ANDSF (access network discovery and selection function) mechanism, normally used for accessing the inter-system mobility policies, can be used for management of IP flow mobility. As the ANDSF relies on the direct communications between handset and OMA-DM (open mobile alliance – device management) server, the policies could be extended in such a way to indicate that for example audio traffic should always go through the cellular network while http can be routed via the WiFi network. More information about ANDSF will be provided later in this chapter.

Based on [11] two classes of the handover protocols can be used for seamless femto–WiFi traffic management.

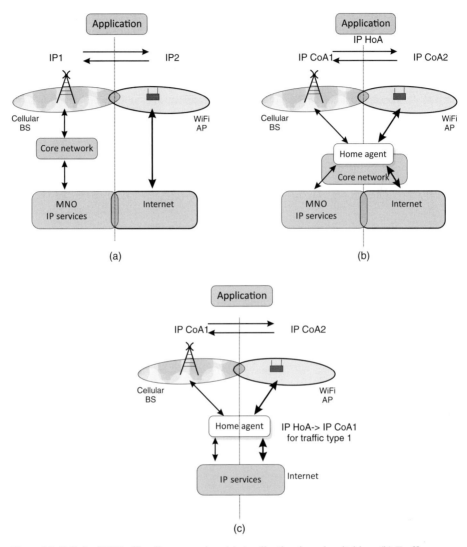

Figure 6.6 Cellular/WiFi offloading scenarios. (a) Application-based switching; (b) Traffic offloading to WiFi; (c) IP flow management (based on [5]).

- **First class**. Contains the protocols implemented at both user equipment and the network, known as host-based-mobility protocols; this class includes for example mobile internet protocol (MIP) or dual stack mobile IPv6 (DSMIPv6) [36, 37] protocols, where DSMIP is already approved for maintaining the data handover between 3G/4G networks and WiFiAP [10].
- **The second set**. Covers the protocols implemented only on the network side with reduced support from the user terminal; here one can include proxy MIP (PMIP) or GPRS tunneling protocol (GTP) [36, 10].

6.3.4.1 IP flow mobility

The idea of IP flow mobility, known as IFOM, is to allow the UE to establish and maintain the session with the same packed data network connection (PDN connection) simultaneously over the WiFi and 3GPP network. This concept assumes involvement of the user, thus IFOM is said to be more UE centric compared to the other concepts presented later in this chapter, and random access network (RAN) is not so highly involved in this case. A typical example is when the user has access to both networks (at the railway station or airport) and the voice call will be served by the cellular network, whereas the best-effort traffic will be offloaded to the WiFi network. Although the possibility to handover from 3GPP to non-3GPP networks has been introduced already in 3GPP Release 8, it is in Release 10 where the IFOM idea has been proposed. The main assumptions in IFOM are as follows [32]:

- to support data offload between 3GPP and WiFi networks in a given PDN connection
- to support handover between these two networks in a seamless manner
- configuration of IFOM can be realized either dynamically or statically.

6.3.4.2 Multi-access PDN connectivity (MAPCON)

The idea behind the IFOM is to enable the UE the selection of wireless network access (to 3GPP or WiFi, or other non-3GPP network) per IP flow basis. Contrarily, in multi-access packet-data-network connectivity, known as MAPCON, the concept is to provide the UE with one APN (equivalent to one PDN connection) for cellular access and one for non-3GPP access.

6.3.4.3 Non-seamless handover

The concept of non-seamless offload is also realized per IP flow basis. In that case the UE possess the opportunity to choose a non-3GPP access network; however, it is assumed that the data passing the non-3GPP network will not enter the 3GPP packet core network.

6.3.4.4 Multi-path TCP connection (MPTCP)

It is also worth mentioning that beside the multi-path communications over the CoAs, there is an alternative way for guaranteeing multi-path communication, mainly multi-path TCP connection [38]. In such a case both sides of that connection must support multi-path TCP interface creating the sub-TCP flows. Such a solution allows for seamless mobility for TCP connection.

The comparison between IFOM, MAPCON, and non-seamless handover is shown in Figure 6.7.

6.3.5 Access network discovery and selection function (ANDSF)

In order to enable offloading mechanisms among heterogeneous networks such as LTE and WiFi, 3GPP has focused on the issue of vertical handover optimization (which is required to seamlessly switch traffic from one network to the other) and has defined

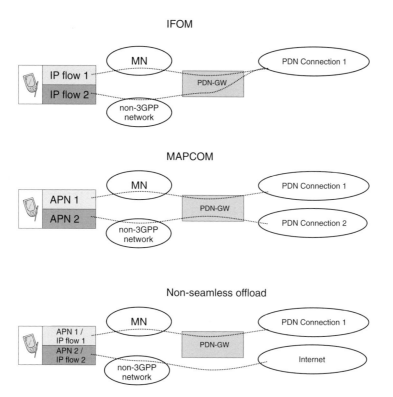

Figure 6.7 IFOM, MAPCOM, and non-seamless offload comparison (based on [39]).

the access network discovery and selection function (ANDSF) component. ANDSF has been defined by 3GPP in TS 22.278, TS23.402, and TS24.312 specifications. ANDSF has similarities with IEEE 802.21, which is the other well-known standard for performing seamless media-independent handover. The similarities between the two standards are related to the dissemination of candidate access network information to the MN, using a database/information server that stores the relevant data for each of the candidate networks that could be considered by the mobile node for switching. There are also some differences between the two solutions, related to the event and command services. ANDSF has also specific mechanisms for inter-system mobility and routing policies, whereas IEEE 802.21 does not include similar specifications. However, the basic mechanisms for vertical handover are common in both approaches. ANDSF may provide a list of access networks available in the vicinity of the MN, so there is no necessity for the MN to perform continuous background scanning. Additionally, ANDSF connection monitoring may be carried out to handover or perform targeted cell selection/re-selection measurement. Thus the MN saves energy power by performing much reduced background scanning, which costs a lot in terms of battery life. ANDSF selectively triggers the connection of the MN to a given threshold or the preferred available access network types based on policies. Once the MN receives the related policy ANDSF will apply the current conditions to match the policy. Policies can be customized for different access networks and may be independent MN types to enable seamless connectivity experience and network load balancing.

The messages sent to and from the ANDSF module in 3GPP network are represented by so-called ANDSF management objects being an XML document [40]. Typically the following messages can be exchanged between the UE and ANDSF entity:

- **UE location**. This message will be sent by the UE to ANDSF entity informing the latter about the location of the mobile user; this location can be based on geographical coordinates or information fetched from the cellular cell or WLAN areas.
- **Discovery information**. This management object will be delivered to the UE and contains the list of the available access networks that can be used by the mobile user; for example the ANDSF module, based on the UE location, can update the discovery information by providing to the user the set of closest WiFi access points in its vicinity.
- **Inter-system mobility policies**. Again, this object will be delivered by the ANDSF entity to the mobile user. ISMP object describes the policies defining the rules about which access networks shall be used by a certain UE, e.g., depending on its location, time of day, priorities, etc.
- **Inter-system routing policies**. This message is also generated by the ANDSF module and sent to the UE; only the MAPCOM, IFOM, or non-seamless offload capable users will be the destination of that message, since it contains information about the rules on which network shall be used on a per APN basis or per IP flow.

Beside the two above-mentioned types of policies, 3GPP in Release 11 identified also the access network discovery information policy, which indicates which is the preferable type of access technology (e.g., WiMAX, WLAN). Moreover, as the result of the analysis of the relationships between the standards provided by 3GPP and IEEE (mainly the concept of Hotspot 2.0 described in the next section), the WLAN selection policy as well as the preferred service policy list (PSPL) have been recently proposed by 3GPP. These sets of rules are used by the UE during identification of most preferred WiFi networks for making connection or roaming.

6.3.6 Hotspot 2.0

Areas covered by the WiFi signal are also counted to the small cells, regardless whether these are or are not managed by the MNO [39, 41]. Recently, the WiFi Alliance has published the new standards on so-called Hotspot 2.0, in which the new abilities have been provided to the WiFi device. In particular the Hotspot 2.0 compliant device would be able to discover existing public WiFi networks allowing for more efficient roaming among them. This standard is based on two other specifications, mainly: IEEE 802.11i [42] and IEEE 802.11u [43]. The application of the former specification guarantees the secure authentication and encryption for WiFi data. Extensible authentication protocol (EAP) in four separate schemes has been chosen for user authentication, i.e., EAP-SIM [44], EAP-AKA [45] and EAP-AKA' (AKA stands for authentication and key agreement) [46], EAP-TLS (transport layer security) [47], and its tunneled version EAP-TTLS (tunneled transport layer security) [48]. Such a differentiation results from the fact that various authentication methods are used by the UE and WiFi devices.

The second basis standard considered for Hotspot 2.0 devices, i.e., IEEE 802.11u, provides the query mechanisms that will allow the device to discover information about the possible roaming networks (both WiFi and cellular networks) and, at the same time, to learn the type of credentials that will be applied and used for the certain access network. Here the concept of subscription service provider (SSP) has been introduced, being the entity responsible for management of the user's subscriptions and credentials associated with them [41].The user can query the access point asking for the list of accessible SSPs regardless of the SSID of the access point. It has been guaranteed that the queries can be generated by the unauthenticated devices creating the possibility for faster and more efficient roaming (i.e., the roaming will be started only, for example, for such access points which will guarantee the accessibility of the selected SSP). For that purpose the access network query protocol (ANQP) has been proposed as the solution for discovering the network information prior to establishment of data connection.

It is worth summarizing the main features of the ANDSF and Hotspot 2.0 solutions, since these are complementary. As stated in [41], the ANDSF information will be available when:

- the device will only have access to the cellular network (the WiFi interface is not active)
- the device will have access to both the cellular and the WiFi network
- the device will be connected to WiFi only
- finally, when there will be no network, since the policies and discovery information can be prestored after these have been received during the prior connection.

Similarly, the Hotspot 2.0 information will be available when:

- the device will have access to both the cellular and the WiFi network
- the device will be connected to WiFi only.

When the device will only have connection to the cellular access network, the Hotspot 2.0 data will not be accessible. Moreover, in the case of no network connection usually no Hotspot 2.0 information will be available, since these are only received on the spot.

6.4 HetNets offloading: benefits analysis

The application of small cells addresses the problem of signal coverage inside buildings and reduces the traffic served by the macro-tier BSs by shifting it to small cells. WiFi technology is currently one of the most popular ways of achieving wireless internet access, not only in private premises but also in public places such as offices and stations. Thus means for intelligent traffic routing through the WiFi network are more and more frequently being considered by mobile operators. In such cases, various operational strategies have been analyzed taking into account the characteristics of traffic types, utilized protocols, and assumed business models. One key advantage of such solutions' use: the reduction of regular traffic in the mobile network thus decreasing the probability of congestion, the minimization of overall interference, and the powering down of unused APs when the whole traffic can be served by WiFi. The integration of WiFi and small cells may also result in benefits to both the end users and the operators in terms of services and applications. The operator may offer services to the users by using the WiFi network only, thus offering them the option to skip the cost of the data transmission. Furthermore there is the possibility for integrated services across the cellular and WiFi devices (e.g., monitoring and security). However, perhaps the utmost benefit when it comes to both operators and end users is simply that more spectrum is being made available to the communication service, allowing a higher data rate particularly when the services are being aggregated.

On the other hand, some important challenges can be listed, such as the need for overlapping coverage areas of the WiFiAP and small cell AP, of seamless handover or protocols translations, and of service inconsistencies between the two approaches such as frame error probability and SINR variability. The 3GPP is undertaking some ongoing standardization processes aiming at the definition of rules appropriate for data flow in IFW networks. It has also been stated that the congestion in the network can also result from the dense traffic in the backhaul or core networks, and application of IFW could be a solution here. One of the proposals is to offload the traffic by application of the local IP area (LIPA) concept, through which the whole connection is managed locally [49]. Finally, one can observe that simultaneous connections across both small cells and WiFi networks at the same time create additional degrees of freedom, among others new opportunities for flow segregation or flow mobility. Another pertinent observation is that a large degree of the strain/complexity on mobile networks is caused by the need to serve the users who are indoors with a very significant path loss to the base stations outside. Serving those users with the already-present indoor equipment, such as WiFiAPs or small cell APs if present, uses the propagation environment in our favor, as opposed to fighting against it, therefore greatly reduces the necessary transmission power. Moreover, if the indoor equipment is using the same frequency resource as the outdoor equipment, as is the case in femto cell deployments for example, the choice of the indoor coverage has the added benefit of naturally shadowing from the outside therefore providing an "automatic," efficient frequency reuse solution, albeit one that is difficult to manage on a local level. An important corollary here is the observation that a key to efficient mobile communications is to provide the given required data rate per area with the minimum possible sum transmission power in that area. Any transmission power above that minimum, as might be required, for example, to transmit through walls, is "polluting" the radio environment hence hindering the performance of frequency reuse (reducing its energy efficiency and capacity), through causing increased interference. Such observations do, however, have to be taken hand in hand with frequency planning requirements noting the benefit of conventional cellular topologies for such planned frequency usage.

The various costs and tradeoffs related to small cells deployments must also be considered; some examples are:

(i) fixed-term costs (e.g., cost of spectrum rights)
(ii) equipment costs (particularly, base stations)
(iii) cost of backhaul provision
(iv) installation costs
(v) operation and maintenance costs (notably, including the cost of energy).

Another key consideration is the charging for data: offloading to WiFi conventionally reduces the ability of the mobile network operator to charge for data, hence although having a positive effect in terms of reducing both the operation and maintenance cost (reduced energy consumption through power saving modes) and equipment and backhaul costs (number of necessary base stations and total amount of backhauled data), the revenue per customer will be reduced hence the end result might be a reduction in profit projection. However, should such charging be possible, e.g., through an integrated solution under control of the operator, then such offloading would, financially, be a no-brainer. Nevertheless, even in cases where charging is not possible, it might be inferred in some scenarios that the reduction in costs for the operator through such offloading (i.e., reduced energy and equipment expenditure) can outweigh the reduction in revenue, such that the net result of such offloading still nevertheless results in an increase in profit.

6.5 HetNets offloading assessment

In order to verify the opportunities created by WiFi offloading, a dedicated simulation model has been implemented, illustrating the behavior of the IFW network located in the residential or enterprise area. In particular a three-level office building (i.e., ground floor and two higher levels) is considered. The building contains $m = 5$ rooms of equal size on each side of every floor. The inside walls, of size $W_{inside} = 0.1$ m, and outside walls, of size $W_{outside} = 0.2$ m, have been assumed to be made of concrete with the corresponding attenuation of $A_{walls} = 10$ dB on every 10 cm of material. The attenuation of the floors has been calculated analogously.

The whole area is covered by the cellular network delivered by the MNO by means of the one LTE-compatible MBS located in the building vicinity. It is assumed that the coverage radius of that MBS does not exceed 0.5 km. Moreover, a flexible number of small cells (i.e., LTE-compatible FAPs and WiFiAPs) is deployed in the building premises allowing for efficient data offloading through small cells. The total number of FAPs and WiFiAPs have been also parameterized and equal to K_{FAPs} and $W_{WiFiAPs}$, respectively. Following the guidelines from the LTE standards it has been assumed that the finite number of N_{RBS} of resource blocks (RB) will be offered by the MBS depending on the selected bandwidth of the frequency band, i.e., $R_{RBS} \in \{6, 15, 25, 50, 75, 100\}$. For the sake of simplicity orthogonal resource allocation between the macro- and femto-tiers has been assumed. It has been realized in such a way that around 30% of the

MBS resource blocks have been shifted and assigned solely to the FAPs, thus each FAP contains $R_{FAPs} \in \{2, 5, 8, 18, 25, 33\}$ RBs. Finally, three non-overlapping channels (i.e., first, sixth, and eleventh, as in IEEE 802.11n standard) were available in every WiFiAP. Since the effective rate observed by the signed user in WiFi networks strongly depends on the medium access algorithms (i.e., based on CSMA/CA or RTS/CTS approach) the model used in [50, 51] has been applied, and is briefly presented later in this section.

The number of users (UE) located outside and inside the building has been parameterized and set to N_1 and N_2. The position of each user has been randomly generated in each iteration of the carried out Monte Carlo simulation.

6.5.1 WiFi interference model

Let us denote the probability that at least one transmission is present in the considered WiFi network as P_{tr}, and the probability that the transmission is successful as P_s, then the normalized throughput per one user can be calculated as:

$$S = \frac{P_s P_{tr} E[L]}{P_{tr} P_s T_s + (1 - P_s) P_{tr} T_c + (1 - P_{tr})\delta}, \tag{6.1}$$

where $E[L]$ and δ represents the average packet length and duration of the empty time slot, respectively. Hereafter, we assume that all users would transmit the same amount of data, thus $E[L] = L$. The values T_s and T_c denote the average time in which the channel is busy due to successful transmission or collision, accordingly, and – for the basic medium-access scheme – can be calculated as:

$$T_s = T_{DIFS} + T_{header} + L + T_{SIFS} + T_{ack} + 2\rho, \tag{6.2}$$

$$T_c = T_{DIFS} + T_{header} + L + T_{SIFS} + T_{ack}. \tag{6.3}$$

In the above formula T_{header} is the joint duration of the packet headers added in the PHY and MAC layer, and ρ is the assumed propagation delay. For the RTS/CTS scheme the above formulas have to be adapted accordingly:

$$T_s = T_{DIFS} + T_{header} + L + 3T_{SIFS} + T_{ack} + 4\rho + T_{RTS} + T_{CTS}. \tag{6.4}$$

$$T_c = T_{DIFS} + T_{RTS} + T_{SIFS} + T_{CTS}. \tag{6.5}$$

6.5.2 Path-loss models

One of the crucial parameters in the considered algorithms is the accurate value of the observable received SINR at the mobile terminals. Thus the path losses between the FAPs, WiFiAPs, and MBS on one side and every user on the second side have been calculated. In the link budget calculations the following formula has been used:

$$P_{rx} - SINR_{required} = P_{tx} + G_{tx} + G_{rx} + PL + C_{loss} - S(T_{ambient}), \tag{6.6}$$

where P_{rx}, P_{tx}, G_{tx}, and G_{rx} denote the received and transmit power, and the receive and transmit antenna gains, respectively. Furthermore, C_{loss} and $S(T_{ambient})$ are the cable loss (negative value) and the receiver sensitivity at given ambient temperature $T_{ambient}$, respectively. Finally, *PL* and $SINR_{required}$ represent the modeled path loss and the required SINR value required to receive correctly the transmitted signal of a given kind (i.e., cellular or WiFi). The exact values of the PL for FAPs have been calculated using the detailed models described in [2].

6.5.3 Proposed RRM and cost function

In order to verify the efficiency of selected traffic offloading via the WiFi network, it has been assumed that each user that has SINR higher than the minimum required value for guaranteeing user satisfaction can request one of two available types of services, generally called data and voice services. If the user is served by a cellular operator, then in the case where the user sends data it is allocated 3 RBs, while in the case of voice traffic, such a user will utilize 1 RBs. Additionally, the assignment of the user to the WiFi network leads to increase of occupancy of the WiFi channel, according to the interference model described in previous sections. It has also been considered that if the user has been assigned to the MBS, FAP, or WiFiAP the data will be delivered ideally to the receiver. The full-buffer approach has been implemented. Two radio resource management algorithms have been proposed that are based on the traffic type and on the user's location (i.e., inside or outside of the building). In general it has been stated that:

- voice traffic is prioritized over data traffic
- outside users can be only served by the MBS.

Then, the traffic of the inside users is scheduled in the following order – the algorithm attempts first to assign the voice traffic to the best available FAP. If that is not feasible, the algorithm then assigns the voice traffic to the MBS and if that also is not feasible, it finally directs the voice traffic to the WiFiAP, whereas the data traffic is scheduled in the first attempt to the WiFiAP, then to the FAP, and finally to the MBS. This algorithm is denoted hereafter as RRM-A. In the second approach (RRM-B) we change the order of resource allocation for the voice traffic generated by the inside users – all of the traffic is driven via the WiFi network.

6.5.4 Simulation results

In order to evaluate the possibilities of traffic offloading some extensive simulations have been made. All parameters have been set up according to the assumptions made in previous subsections. Moreover, 100 inside and outside users have been considered. During the experiments we have measured how much traffic can be offloaded from the macro cells to – generally speaking – small cells, i.e., femto cells and WiFiAPs. All simulation parameters have been set as described in the previous subsections. In Figures 6.8 and 6.9 the traffic distribution among the macro base stations, and femto and WiFi access points for fixed number of FAPs (i.e., equal to 1) and for an increasing

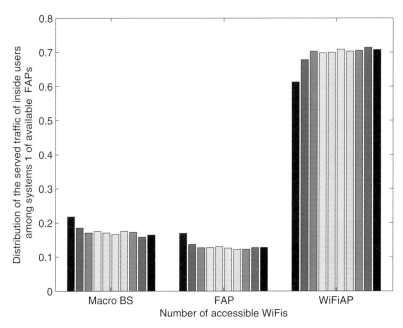

Figure 6.8 Traffic distribution among MBS, FAPs, and WiFis for a fixed number of FAPs (1), and increasing number of WiFiAPs; RRM-A.

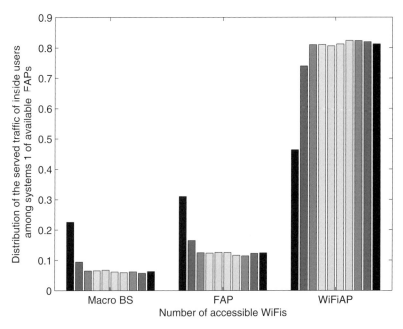

Figure 6.9 Traffic distribution among MBS, FAPs, and WiFis for fixed number of FAPs (1), and increasing number of WiFiAPs; RRM-B.

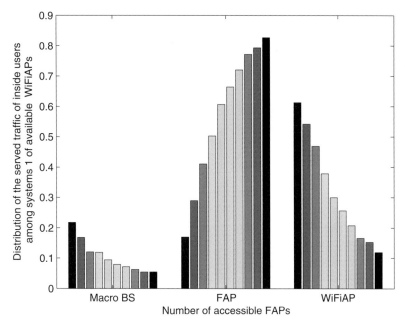

Figure 6.10 Traffic distribution among MBS, FAPs, and WiFis for fixed number of WiFiAPs (1), and increasing number of FAPs; RRM-A.

number of WiFiAPs is shown; accordingly, the first and second version of the radio resource management scheduler is applied. One can observe three groups of bars. In each set, the first bar corresponds to the case where there is only one WiFiAP inside the building, the second to the case where there are two FAPs inside the building, and so on. It can be noticed that the increase of number of WiFiAPs results in high traffic shift from FAPs, also the offload from the MBS to the FAP is observed. In the considered scenario, the optimal number of WiFiAPs is equal to three, since no further improvement is observed with the increase of WiFiAPs available inside the building. Of course, different results could be observed for different traffic and interference traffic models used for simulations. Analogously, in Figures 6.10 and 6.11 the traffic distribution among the MBS, FAPs, and WiFiAPs for fixed number of WiFiAPs (e.g., one) and increasing number of FAPs is illustrated; accordingly, a first and second RRM scheme was considered. Slightly different behavior is observed; with the increase of the number of FAPs, the percentage of traffic served by FAPs increases exponentially. The difference between the behavior of the two tested schemes (i.e., when the number of WiFiAPs and FAPs increases) is caused by the different modeling of data traffic. As in the FAPs case the spectrum occupancy depends on the number of free resource blocks, in case of WiFiAPs the upper bound of the number of served users depends on the maximum allowed interference level, which depends on the packet sizes L used for transmission.

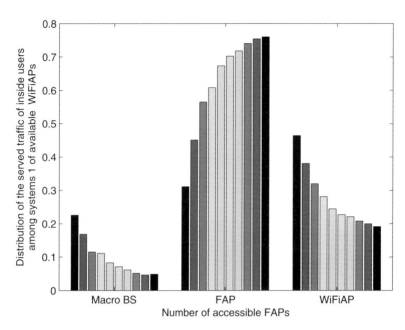

Figure 6.11 Traffic distribution among MBS, FAPs, and WiFis for fixed number of WiFiAPs (1), and increasing number of FAPs; RRM-B.

6.6 Conclusions

In this chapter, the analysis of various aspects of traffic offloading from the cellular networks to the small cells has been presented. Based on forecasts it is observed that future increases in data volume that will be served by mobile network operators will be very significant, leading to serious congestion problems. In this context, advanced techniques for traffic management have to be considered. Current solutions, such as LIPA, SIPTO, ANDSF, or Hotspot 2.0, can be seen as big, but still initial, steps toward higher integration between various wireless systems. In this light, the concept of WiFi data offloading together with reliable and seamless handover is of high importance. Presented simulation results in this chapter have proven that with the assumed quality of service a significant portion of the traffic can be shifted from the cellular to WiFi networks, reducing the load of the cellular network. The conclusions drawn from such observation can be utilized in various ways. For example, since part of the traffic will be served by WiFi networks, efficient cellular network planning tools can be applied, leading to better interference management in heterogeneous networks, higher energy efficiency, and – finally – for higher income or net profit.

Acknowledgments

This work has been supported by the ICT-ACROPOLIS Network of Excellence, www.ict-acropolis.eu, FP7 project number 257626. It has been also partially supported by the ICT-SOLDER project, www.ict-solder.eu, FP7 project number 619687, and by the Network of Excellence in Wireless Communications, ICT-NEWCOM, http://www.newcom-project.eu, (grant number 318306).

References

[1] Cisco (2011), "Cisco visual networking index: global mobile data traffic forecast update 2012–2017," www.slideshare.net/zahidtg/cisco-vni-2012.

[2] Cisco (2012), "Cisco visual networking index: global mobile data traffic forecast update 2012–2017," www.slideshare.net/zahidtg/cisco-vni-2012.

[3] Cisco (2013), "Cisco visual networking index: forecast and methodology, 20122017," www.cisco.com/en/US/solutions/collateral/ns341/ns525/ns537/ns705/ns827/white_paper_c11-481360.pdf.

[4] Fuxjager, P., Fischer, R., Gojmerac, I. and Reichl, P. (2010), "Radio resource allocation in urban femto-WiFi convergence scenarios," In 6th EURO-NF Conference on Next Generation Internet (NGI), pp. 1–8.

[5] Small Cell Forum (March 2013b), "Interference management in OFDMA femtocells," www.smallcellforum.org.

[6] Damnjanovic, A., Montojo, J., Wei, Y., et al. (2011), "A survey on 3GPP heterogeneous networks," *Wireless Communications, IEEE* 18(3), 10–21.

[7] Ghosh, A., Mangalvedhe, N., Ratasuk, R., et al. (2012), "Heterogeneous cellular networks: from theory to practice," *Communications Magazine, IEEE* 50(6), 54–64.

[8] Andrews, J. (2013), "Seven ways that HetNets are a cellular paradigm shift," *Communications Magazine, IEEE* 51(3), 136–144.

[9] SmallCellForum (2014a), www.smallcellforum.org.

[10] 3GPP (2013), "3GPP TS 23.402 V12.3.0 (2013–12); Technical Specification 3rd Generation Partnership Project; Technical Specification Group Services and System Aspects; Architecture enhancements for non-3GPP accesses (Release 12)."

[11] SmallCellForum (December 2013a), "Integrated femto–WiFi networks, version: 033.04.01," www.smallcellforum.org.

[12] Lopez-Perez, D., Valcarce, A., de la Roche, G. and Zhang, J. (2009), "OFDMA femtocells: a roadmap on interference avoidance," *Communications Magazine, IEEE* 47(9), 41–48.

[13] Chandrasekhar, V. and Andrews, J. (2009), "Spectrum allocation in tiered cellular networks," *Communications, IEEE Transactions* 57(10), 3059–3068.

[14] Abgrall, C., Strinati, E. and Belfiore, J. C. (2010), "Distributed power allocation for interference limited networks," *In* Personal Indoor and Mobile Radio Communications (PIMRC), 2010 IEEE 21st International Symposium, pp. 1342–1347.

[15] Garcia, L., Pedersen, K. and Mogensen, P. (2009), "Autonomous component carrier selection: interference management in local area environments for LTE-Advanced," *Communications Magazine, IEEE* 47(9), 110–116.

[16] Garcia, L., Costa, G. W. O., Cattoni, A., Pedersen, K. and Mogensen, P. (2010), "Self-organizing coalitions for conflict evaluation and resolution in femtocells," *In* Global Telecommunications Conference (GLOBECOM 2010), 2010 IEEE, pp. 1–6.

[17] Wu, Z., Huang, A., Zhou, H., Hua, C. and Qian, J. (2011), "Data fusion-based interference matrix generation for cellular system frequency planning," *International Journal of Communication Systems* 24, 1506–1519.

[18] Kliks, A., Nasreddine, J., Li, F., Zalonis, A., Dimitriou, N. and Ko, Y. (2012), "Interference management in heterogeneous wireless networks based on context information," *In* Wireless Communication Systems (ISWCS), 2012 International Symposium, pp. 251–255.

[19] Zalonis, A., Dimitriou, N., Polydoros, A., Nasreddine, J. and Mahonen, P. (2012), "Femtocell downlink power control based on radio environment maps," *In* Wireless Communications and Networking Conference (WCNC), 2012 IEEE, pp. 1224–1228.

[20] Zhao, Y., Le, B. and Reed, J. (2009), "Network support: the radio environment map", *In* B. Fette, ed., *Cognitive Radio Technology*, Butterworth-Heinemann.

[21] Bejerano, Y., Han, S. and Li, L. (2007), "Fairness and load balancing in wireless LANs using association control," *Networking, IEEE/ACM Transactions* 15(3), 560–573.

[22] Kim, H., de Veciana, G., Yang, X. and Venkatachalam, M. (2012), "Distributed alpha-optimal user association and cell load balancing in wireless networks," *Networking, IEEE/ACM Transactions* 20(1), 177–190.

[23] Siomina, I. and Yuan, D. (2012), "Load balancing in heterogeneous LTE: range optimization via cell offset and load-coupling characterization," *In* Communications (ICC), 2012 IEEE International Conference, pp. 1357–1361.

[24] Das, S., Viswanathan, H. and Rittenhouse, G. (2003), "Dynamic load balancing through coordinated scheduling in packet data systems," *In* INFOCOM 2003. Twenty-Second Annual Joint Conference of the IEEE Computer and Communications. IEEE Societies, Vol. 1, 786–796.

[25] Sang, A., Wang, X., Madihian, M. and Gitlin, R. (2008), "Coordinated load balancing, handoff/cell-site selection, and scheduling in multi-cell packet data systems," *Wireless Networks* 14(1), 103–120.

[26] Bejerano, Y. and Han, S. J. (2009), "Cell breathing techniques for load balancing in wireless LANs," *Mobile Computing, IEEE Transactions* 8(6), 735–749.

[27] Son, K., Chong, S. and de Veciana, G. (2007), "Dynamic association for load balancing and interference avoidance in multi-cell networks," *In* Modeling and Optimization in Mobile, Ad Hoc and Wireless Networks and Workshops, 2007. WiOpt 2007. 5th International Symposium, pp. 1–10.

[28] Ye, Q., Rong, B., Chen, Y., Al-Shalash, M., Caramanis, C. and Andrews, J. (2013), "User association for load balancing in heterogeneous cellular networks," *Wireless Communications, IEEE Transactions* 12(6), 2706–2716.

[29] Singh, S. and Andrews, J. (2014), "Joint resource partitioning and offloading in heterogeneous cellular networks," *Wireless Communications, IEEE Transactions* 13(2), 888–901.

[30] 3GPP (n.d.a), "3GPP TR 23.829 v10.0.0, Local IP access and selected IP traffic offload (LIPA-SIPTO) (Release 10)."

[31] 3GPP (n.d.b), "3GPP TS 23.261 v10.1.0, IP flow mobility and seamless wireless local area network (WLAN) offload (Release 10)."

[32] Sankaran, C. (2012), "Data offloading techniques in 3GPP Rel-10 networks: a tutorial," *Communications Magazine, IEEE* 50(6), 46–53.

[33] Samdanis, K., Taleb, T. and Schmid, S. (2012), "Traffic offload enhancements for eUTRAN," *Communications Surveys Tutorials, IEEE* 14(3), 884–896.

[34] QUALCOMM (March 2013), "3G/WiFi seamless offload."

[35] Gupta, R. and Rastogi, N. (2012), "LTE Advanced – LIPA and SIPTO," *White Paper by Aricent*, pp. 1–11.

[36] Kong, K.-S., Lee, W., Han, Y.-H., Shin, M.-K. and You, H. (2008), "Mobility management for all-IP mobile networks: mobile IPv6 vs. proxy mobile IPv6," *Wireless Communications, IEEE* 15(2), 36–45.

[37] Gondim, P. and Trineto, J. (2012), "DSMIP and PMIP for mobility management of heterogeneous access networks: evaluation of authentication delay," *In* Globecom Workshops (GC Wkshps), 2012 IEEE, pp. 308–313.

[38] IETF (2011), "RFC 6182, architectural guidelines for multipath TCP development."

[39] SmallCellForum (February 2014b), "Next generation hotspots (NGH)-based integrated small cell WiFi (ISW) networks, version: 089.03.01," www.smallcellforum.org.

[40] 3GPP(2011), "3GPP TS 24.312, access network discovery and selection function (ANDSF) management object (MO)."

[41] Orlandi, B. and Scahill, F. (2012), "WiFi roaming – building on ANDSF and Hotspot 2.0," White Paper Alcatel-Lucent and BT, pp. 1–44.

[42] IEEE (2011a), "IEEE standard for information technology telecommunications and information exchange between systems. Local and metropolitan area networks. Specific requirements. Part 11: Wireless LAN medium access control (MAC) and physical layer (PHY) specifications, Amendment 6: medium access control (MAC) security enhancements."

[43] IEEE (2011b), "IEEE standard for information technology telecommunications and information exchange between systems. Local and metropolitan area networks. Specific requirements. Part 11: Wireless LAN medium access control (MAC) and physical layer (PHY) specifications, Amendment 9: interworking with external networks."

[44] RFC (2006a), "4186, Extensible authentication protocol method for global system for mobile communications (GSM) subscriber identity modules (EAP-SIM)."

[45] RFC (2006b), "4187, Extensible authentication protocol method for 3rd Generation authentication and key agreement (EAP-AKA)."

[46] RFC (2009), "5448, Improved extensible authentication protocol method for 3rd Generation authentication and key agreement (EAP-AKA)."

[47] RFC (2008a), "5216, The EAP-TLS authentication protocol."

[48] RFC (2008b), "5281, Extensible authentication protocol tunneled transport layer security authenticated protocol version 0 (EAP-TTLSv0)."

[49] Khan, M., Khan, M. and Raahemifar, K. (2011), "Local IP access (LIPA) enabled 3G and 4G femtocell architectures," *In* Electrical and Computer Engineering (CCECE), 24th Canadian Conference, pp. 1049–1053.

[50] Wu, H., Peng, Y., Long, K., Cheng, S. and Ma, J. (2002), "Performance of reliable transport protocol over IEEE 802. 11 Wireless LAN: analysis and enhancement," pp. 599–607.

[51] Jung, B., Song, N. and Sung, D. (2013), "A network-assisted user-centric WiFi-offloading model for maximizing per-user throughput in a heterogeneous network," *Vehicular Technology, IEEE Transactions* (99), 1–1.

7 Required number of small cell access points in heterogeneous wireless networks

S. Alireza Banani, Andrew Eckford, and Raviraj Adve

How many small cell (SC) access points (APs) are required to guarantee a chosen quality of service in a heterogeneous network? In this chapter, we answer this question considering two different network models. The first is the downlink of a *finite-area* SC network where the locations of APs within the chosen area are uniformly distributed. A key step in obtaining the closed-form expressions is to generalize the well-accepted moment matching approximation for the linear combination of lognormal random variables. For the second model, we focus on a *two-layer* downlink heterogeneous network with frequency reuse-1 hexagonal macro cells (MCs), and SC APs that are placed at locations that do not meet a chosen quality of service from macro base stations (BSs). An important property of this model is that the SC AP locations are coupled with the MC coverage. Here, simple bounds for the average total interference within an MC makes the formulation possible for the percentage of MC area in outage, as well as the required average number of SCs (per MC) to overcome outage, assuming isolated SCs.

7.1 Introduction

Heterogeneous cellular networks (HCNs) are being considered as an efficient way to improve system capacity as well as effectively enhance network coverage [1, 2]. Comprising multiple layers of access points (APs), HCNs encompass a conventional macro cellular network (first layer) overlaid with a diverse set of small cells (SCs) (higher layers). Cell deployment is an important problem in heterogeneous networks, both in terms of the number and positioning of the SCs.

Traditional network models are either impractically simple (such as the Wyner model [3]) or excessively complex (e.g., the general case of random user locations in a hexagonal lattice network [4]) to accurately model SC networks. A useful mathematical model that accounts for the randomness in SC locations and irregularity in the cells uses spatial point processes, such as Poisson point process (PPP), to model the location of SCs in the network [5–10]. The independent placement of SCs from the MC layer, has the advantage of analytical tractability and leads to many useful SINR and/or rate expressions. However, even assuming that wireless providers would deploy SCs to support mobile broadband services, the dominant assumption remains that SCs are deployed randomly and independent of the MC layer [11]. Since the model does not incorporate dependencies between layers, it does not accurately capture the objective of SCs deployments in enhancing the coverage or capacity of the network and filling in coverage holes from the MC deployments [10, 12].

A significant problem is that, due to the lack of an accurate and tractable network model, there exist few design tools to aid network engineers; of specific interest here is the number of small cells that are required to guarantee a chosen quality of service in the network. Therefore, in this chapter, we concentrate on two different models for the location of SCs. As traditional cellular networks make way for heterogeneous networks, the consideration of such models is important for a better understanding of network challenges and limitations.

For the first model we consider the downlink of a *finite-area*, small cell network with *a fixed number of* SCs. The locations of SCs within the chosen area are distributed according to a binomial point process (BPP), i.e., SCs are distributed uniformly within the area of interest. The main motivation for considering a finite-area network as compared to an infinite network model is that, as recent work has shown [13], treating a finite-area network is an accurate model for cases with very high path-loss exponents (e.g,. path-loss exponent of $\alpha = 6$); for more realistic values such as the range of $\alpha = 2\text{--}4$, the infinite area assumption underestimates network performance significantly. As we will see, our results are consistent with this result. For such a finite-area network model, at first, we obtain an analytical expression that accurately approximates the worst-case user capacity as a function of number of SCs in the finite area. The worst-case capacity is chosen since it can be related directly to a coverage guarantee. Next, by inverting the sum capacity relation, we answer our title question: how many SC APs do we need to meet a coverage constraint? Here, we use a polynomial fit to tractably invert the relation between sum capacity and the number of APs.

For the second model, we focus on a two-layer downlink heterogeneous network where macro base stations (BSs) are located on a hexagonal grid. The aim is to enhance the capacity in the network, and to place SCs at those locations with poor rate coverage (defined as the probability that a particular user can achieve a target rate) from macro BSs (coverage holes). Under the assumption of isolated SCs[1], we provide analytical expressions for the average required number of SCs to cover the outage area.

The rest of this chapter is organized as follows. Section 7.2 is dedicated to the finite-area downlink system model with uniformly distributed SC APs, and Section 7.3 considers the dependent placement of SCs in a hexagonal grid network. Each section is supported with the simulation results and discussions. The notation used is conventional: matrices are represented using bold upper case and vectors using bold lower case letters; $(\cdot)^H$, and $(\cdot)^T$ denote the conjugate transpose, and transpose, respectively. A vector a $\sim$ $\mathcal{CN}(0, 1)$ comprises independent and identically distributed (i.i.d.) zero-mean complex Gaussian random variables, each with unit variance. $Q(x)$ represents the standard Q-function, the area under the tail of a standard Gaussian distribution. Finally, $\mathbb{P}\{\cdot\}$ denotes the probability of an event and $\mathbb{E}\{\cdot\}$ denotes expectation.

---

[1] An isolated SC does not receive interference from other SCs or the serving MC. This assumption is made to provide tractable analysis.

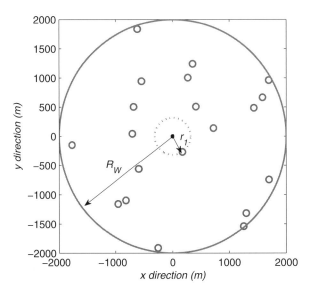

Figure 7.1 One realization of the location of APs based on a BPP in a circular region with radius $R_W = 2$ km and $N = 20$.

7.2 Finite-area network with uniformly distributed SC APs

7.2.1 APs geometry and user distance distribution

We consider the downlink of a finite-area reuse-1 network comprising N SC APs located in a circular area W with radius R_W centered at the origin. The location of APs follows a BPP, that is, we are dealing with N APs i.i.d. uniformly distributed in W. The AP locations partition the circular area into Voronoi cells. Users are distributed within the region W with each user connected to the closest AP and a single AP serving a single user. One realization of the location of APs, with $N = 20$ and $R_W = 2$ km, is depicted in Figure 7.1.

It is shown in [14] that the worst-case user capacity occurs at the center of W. As a result, we focus on the origin, i.e., the user of interest is located at the origin and communicates with its nearest AP. We therefore require the probability density function (PDF) of user distance to the nearest AP, denoted as r_1. Defining the region within a circle of radius r_1 as A_1 (denoted by the dashed circle in Figure 7.1), given that there are exactly N APs within area W, the conditional distribution of $N(A_1) = 0; A_1 \subseteq W$ is obtained as

$$\mathbb{P}\{\text{no BS in the area } A_1 | N \text{ BSs within } W\} = \mathbb{P}\{N(A_1) = 0 | N(W) = N\}$$
$$= \mathbb{P}\{N \text{ APs in } W \setminus A_1\} \quad (7.1)$$

where $W \setminus A_1$ denotes the complement of A_1, the region between the two circles with radii r_1 and R_W. Given that the N APs are i.i.d. uniformly distributed in W, the probability that a chosen AP is outside of A_i would be

$$\mathbb{P}\{\text{a chosen AP in } W \setminus A_1\} = \frac{\pi R_W^2 - \pi r_1^2}{\pi R_W^2} = 1 - \frac{r_1^2}{R_W^2}. \quad (7.2)$$

It follows that

$$\mathbb{P}\{N \text{ APs in } W \setminus A_1\} = \left(1 - \frac{r_1^2}{R_W^2}\right)^N. \quad (7.3)$$

The cumulative density function (CDF) and PDF of r_1 are

$$F_R(r_1) = \mathbb{P}\{R \leq r_1 | N \text{ BSs in } W\}$$

$$= 1 - \mathbb{P}\{N(A_1) = 0 | N(W) = N\} = 1 - \left(1 - \frac{r_1^2}{R_W^2}\right)^N, \quad (7.4)$$

$$f_{r_1}(r_1) = \frac{dF_R(r_1)}{dr_1} = \frac{2Nr_1}{R_W^2}\left(1 - \frac{r_1^2}{R_W^2}\right)^{N-1}; \ 0 \leq r_1 \leq R_W. \quad (7.5)$$

7.2.2 Signal-to-interference ratio

We consider an interference-limited network, i.e., we assume that the thermal noise is negligible as compared to the interference and can be ignored. This is justified in the current SC networks under consideration, which are typically interference limited [15]. Let h_1 denote the instantaneous channel from the nearest AP to the user located at the origin, and let $h_j; j = 2, \ldots, N$ denote the corresponding channels between the user and the remaining $N - 1$ interfering APs located outside the circle with radius r_1 (between the circles with radii r_1 and R_W). Similarly, let $r_j; j = 2, \ldots, N$ represent the distance from the j-th AP to the user at the origin, and let $PL(r_j)$ represent the path loss (in dB) over this distance. The instantaneous channels $h_j; j = 1, \ldots, N$ can be modeled as

$$h_j = \bar{h}_j / 10^{(PL(r_j) + L_j)/20}, \quad (7.6)$$

where $\bar{h}_j \sim CN(0, 1)$ represents the normalized complex channel gain, reflecting small-scale Rayleigh fading, from the j-th AP to the user, which is independent from $\bar{h}_i; i \neq j$; and where $L_j \sim CN(0, \sigma_L)$ represents lognormal fading, with standard deviation σ_L expressed in dB (the value of σ_L depends on the environment [16]). The path loss, in dB, can be expressed as $PL(r_j) = 10\alpha \log_{10} r_j$, where α is the path-loss exponent.

We assume that all APs transmit at the same power level. The instantaneous signal-to-interference ratio (SIR) of the user at a random distance r_1 from its nearest AP can be expressed as

$$\text{SIR}_{r_1} = \frac{\sigma_s^2 |h_1|^2}{I_r} = \frac{\sigma_s^2 |\bar{h}_1|^2 r_1^{-\alpha} z_1}{\sum_{j=2}^{N} \sigma_s^2 |\bar{h}_j|^2 r_j^{-\alpha} z_j}, \quad (7.7)$$

where σ_s^2 represents the power of the transmitted signals, I_r is the power of the interference from the $N - 1$ interfering APs, and $z_j = 10^{-L_j/10}, j = 1, \ldots, N$ with $z_j \sim LN(\mu_z = 0, \sigma_z = (0.1 \ln 10)\sigma_L)$, are independent lognormal RVs. Thus the instantaneous achieved SINR depends on r_1 (both via $PL(r_1)$ and I_r) as well as the instantaneous realizations of $\bar{h}_j$, and $L_j, j = 1, \ldots, N$. It is known that in an *infinite* area with *infinite* number of APs, the interference follows a shot-noise distribution [5, 17]. However, this is not true in the more realistic case of a *finite area with a finite number of APs*; this fact necessitates a new analysis technique. Here we present an accurate, if approximate, analysis.

It is worth noting that, like r_1, the distances $r_j; j = 2, \ldots, N$ are RVs, but with distance PDFs that differ from $f_{r_1}(r_1)$. For a given r_1, the distribution of the $N-1$ interfering APs in the area between circles with radii r_1 and R_W, denoted as B, is that of $N-1$ i.i.d. random points $(x_j, y_j); j = 2, \ldots, N$, uniformly distributed in B with common distribution $f_{x_j, y_j}(x_j, y_j) = 1/S(B) = 1/\pi(R_W^2 - r_1^2)$ expressed in Cartesian coordinates. With the change of variable $x_j = r_j \cos\theta_j$ and $y_j = r_j \sin\theta_j$, and then integrating over the resulting uniform distribution in $\theta_j, 0 \leq \theta_j \leq 2\pi$, the distance PDF of an individual AP location, for a given value of r_1, is given by

$$f_{r_j}(r_j) = \frac{2r_j}{(R_W^2 - r_1^2)}; \; r_1 \leq r_j \leq R_W; \; j = 2, \ldots, N. \tag{7.8}$$

7.2.3 Worst-case user capacity

In this section, we focus on worst-case capacity since it can be directly related to a coverage constraint. The interference is a weighted sum of lognormal random variables. The key step that allows for a tractable analysis is to use a lognormal approximation for the linear combinations of lognormal RVs such as $g = \sum_{i=1}^{n} a_i y_i$ where $y_i \sim LN(\mu_i, \sigma_i); \; i = 1, \ldots, n$ are independent lognormal RVs, and $a_i; i = 1, \ldots, n$ are also independent RVs, both among themselves and from $y_i; i = 1, \ldots, n$. The work in [18] presents several such approximations based on a generalization of the moment matching approximation (MMA) technique for the sum of lognormal RVs. By matching the first and second moments, a Gaussian approximation for the $\ln(g)$ has mean and standard deviation as $\mu = 2\ln(b_1) - 0.5\ln(b_2)$ and $\sigma = (-2\ln(b_1) + \ln(b_2))^{1/2}$, with

$$b_1 = \sum_{i=1}^{n} \mathbb{E}\{a_i\} \exp\left(\mu_{y_i} + \sigma_{y_i}^2/2\right), \tag{7.9}$$

$$b_2 = \sum_{i=1}^{n} \mathbb{E}\{a_i^2\} \exp\left(2\mu_{y_i} + 2\sigma_{y_i}^2\right)$$
$$+ \sum_{i=1}^{n} \sum_{i'=1; i' \neq i}^{n} \mathbb{E}\{a_i\}\mathbb{E}\{a_{i'}\} \exp(\mu_{y_i} + \mu_{y_{i'}} + (\sigma_{y_i}^2 + \sigma_{y_{i'}}^2)/2). \tag{7.10}$$

In the case of an interference-limited network, the achieved SIR is given by

$$\text{SIR}_{r_1} = \frac{\sigma_s^2 |\bar{h}_1|^2 r_1^{-\alpha} z_1}{\sum_{j=2}^{N} \sigma_s^2 |\bar{h}_j|^2 r_j^{-\alpha} z_j} = \frac{\omega_1 z_1}{\sum_{j=2}^{N} \omega_j z_j}, \tag{7.11}$$

where the denominator is a linear combination of $N-1$ lognormal RVs $z_j; j = 2, \ldots, N$ with coefficients $\omega_j; j = 2, \ldots, N$, which themselves are independent RVs. On the other hand, the numerator is a product of RVs; however, we still use (1.9)–(1.10) to approximate it as a lognormal RV given as $\text{SIR}_{\text{Num}} \sim LN(\mu_{Num}, \sigma_{Num})$, with

$$\mu_{\text{Num}} = 2\ln(\beta_1) - 0.5\ln(\beta_2) = \ln(\sigma_s^2/\sqrt{2}) + \ln r_1^{-\alpha}, \tag{7.12}$$

$$\sigma_{\text{Num}}^2 = -2\ln(\beta_1) + \ln(\beta_2) = \ln 2 + \sigma_z^2. \tag{7.13}$$

Note that we approximate the numerator as a lognormal RV despite it being a single RV with a complicated PDF. As we will see, this approximation is remarkably accurate.

Also, using (1.9)–(1.10), the denominator in (1.11) can be modeled as $\text{SIR}_{\text{Denom}} \sim LN(\mu_{\text{Denom}}, \sigma_{\text{Denom}})$, with $\mu_{\text{Denom}} = \mu_{I_r} = 2\ln(M_1) - 0.5\ln(M_2)$ and $\sigma_{\text{Denom}}^2 = \sigma_{I_r}^2 = -2\ln(M_1) + \ln(M_2)$, where

$$M_1 = \frac{2(N-1)\sigma_s^2 e^{\sigma_z^2/2}}{\alpha - 2}\left(\frac{r_1^{-\alpha+2} - R_W^{-\alpha+2}}{R_W^2 - r_1^2}\right), \tag{7.14}$$

$$M_2 = \frac{4(N-1)\sigma_s^4 e^{2\sigma_z^2}}{2\alpha - 2}\left(\frac{r_1^{-2\alpha+2} - R_W^{-2\alpha+2}}{R_W^2 - r_1^2}\right)$$

$$+ \frac{4(N-1)(N-2)\sigma_s^4 e^{\sigma_z^2}}{(\alpha - 2)^2}\left(\frac{r_1^{-\alpha+2} - R_W^{-\alpha+2}}{R_W^2 - r_1^2}\right)^2. \tag{7.15}$$

With numerator and denominator both modeled as lognormal RVs, the achieved SIR, conditioned on a given r_1, is also lognormal $\text{SIR}_{r_1} \sim LN(\mu_{\text{SIR}}, \sigma_{\text{SIR}})$ with $\mu_{\text{SIR}} = \mu_{\text{Num}} - \mu_{\text{Denom}}$ and $\sigma_{\text{SIR}}^2 = \sigma_{\text{Num}}^2 + \sigma_{\text{Denom}}^2$. Thus the worst-case achievable user capacity (in b/s/Hz) averaged over different realizations of AP locations, is obtained as

$$C_{\text{average}}^{\text{SIR}} = \int_0^{R_W} C_{\text{ergodic}|r_1}^{\text{SIR}} f_{r_1}(r_1) dr_1 = \int_0^{R_W} \mathbb{E}\{\log_2(1 + \text{SIR}_{r_1})|r_1\} f_{r_1}(r_1) dr_1, \tag{7.16}$$

where the ergodic capacity, for a given r_1, is the ensemble average over different realizations of channels in (7.6). In general, no closed-form expression can be found for the average capacity in (7.16) as a function of α. However, a tractable analysis is possible for specific values of α. Here, to allow for comparisons with results in the literature [5], we obtain an analytical expression for $\alpha = 4$ (this is not restrictive in the sense that, using similar approximations, closed-form expressions can also be obtained for other values of $\alpha > 2$). In cases of non-integer α it may be necessary to execute the integral numerically. Since this has to be done only once, this is not an onerous requirement.

For a given r_1, the ergodic capacity $C_{\text{ergodic}|r_1}^{\text{SIR}} = \mathbb{E}\{\log_2(1 + \text{SIR}_{r_1})|r_1\}$ is upper bounded by $\log_2(1 + \mathbb{E}\{\text{SIR}_{r_1}|r_1\})$. On the other hand, since the mean $\mathbb{E}\{\text{SIR}_{r_1}|r_1\} = e^{\mu_{\text{SIR}} + \sigma_{\text{SIR}}^2/2}$ is always greater than one, $\log_2(1 + \mathbb{E}\{\text{SIR}_{r_1}|r_1\})$ itself is lower bounded by $\log_2(\mathbb{E}\{\text{SIR}_{r_1}|r_1\})$. In general, there is no guarantee that $\log_2(\mathbb{E}\{\text{SIR}_{r_1}|r_1\})$ is lower than $C_{\text{ergodic}|r_1}^{\text{SIR}}$. However, for small values of μ_{SIR} and σ_{SIR}, $\log_2(\mathbb{E}\{\text{SIR}_{r_1}|r_1\}) \leq C_{\text{ergodic}|r_1}^{\text{SIR}}$ holds. Therefore an approximation for the $C_{\text{ergodic}|r_1}^{\text{SIR}}$, is

$$C_{\text{ergodic}|r_1}^{\text{SIR}} \approx \log_2(\mathbb{E}\{\text{SIR}_{r_1}|r_1\}) = \left(\mu_{\text{SIR}} + \sigma_{\text{SIR}}^2/2\right)/\ln 2. \tag{7.17}$$

With $\alpha = 4$, the mean and variance of the RV $\ln(\text{SIR})$ are obtained as

$$\mu_{\text{SIR}} \approx 2\ln R_W - 2\ln r_1 + \ln\frac{(N-2)^{1/2}}{(N-1)^{3/2}}$$
$$- \ln\sqrt{2} - \sigma_z^2/2 + \frac{1}{2}\ln\left(1 + \frac{2e^{\sigma_z^2}R_W^2}{3(N-2)r_1^2}\right), \quad (7.18)$$

and

$$\sigma_{\text{SIR}}^2 \approx \sigma_z^2 + \ln 2 + \ln\frac{N-2}{N-1} + \ln\left(1 + \frac{2e^{\sigma_z^2}R_W^2}{3(N-2)r_1^2}\right), \quad (7.19)$$

which gives the conditional ergodic capacity $C_{\text{ergodic}|r_1}^{\text{SIR}}$ as

$$C_{\text{ergodic}|r_1}^{\text{SIR}} \simeq \frac{1}{\ln 2}\left[2\ln\frac{R_W}{r_1} + \ln\frac{(N-2)^{1/2}}{(N-1)^{3/2}}\right.$$
$$\left. + \frac{1}{2}\ln\frac{N-2}{N-1} + \ln\left(1 + \frac{2e^{\sigma_z^2}R_W^2}{3(N-2)r_1^2}\right)\right]. \quad (7.20)$$

Also the average worst-case capacity is given by (for $N > 2$) [19]

$$C_{\text{ergodic}|r_1}^{\text{SIR}} \simeq \frac{1}{\ln 2}\left[\gamma + \frac{1}{2N} + \ln\frac{N(N-2)^{1/2}}{(N-1)^{3/2}} + \frac{1}{2}\ln\frac{N-2}{N-1} + \frac{1}{2}\ln\left(1 + \frac{2e^{\sigma_x^2}}{3(N-2)}\right)\right.$$
$$\left. + \left(\left(1 + \frac{2e^{\sigma_x^2}}{3(N-2)}\right)^N - 1\right)\left(\ln\left(\frac{1 + 1.5e^{-\sigma_x^2}(N-2)}{N-1}\right)\right) - \frac{1}{2(N-1)}\right]$$
$$(7.21)$$

where $\gamma = 0.578$ is the Euler–Mascheroni constant [20] and $\sigma_z^2 = (0.1 \times \ln 10)^2 \sigma_L^2$ for $\sigma_L^2 \leq 7.5$ dB.

Equation (7.21) is the key result. It provides a simple relationship between the average worst-case user capacity and the number of APs N, and the variance of the lognormal-fading σ_L^2 in an interference-limited network. Although (7.21) does not explicitly show its dependency on the radius R_W and the AP density λ, it can be rewritten as a function of R_W and λ by using the substitution $N = \pi R_W^2 \lambda$ in (7.21). It is worth noting that in the asymptotic case of infinite number of APs ($N \to \infty$), the average user capacity in (7.21) converges to

$$C_{\text{average}}^{\text{SIR}}|_{N\to\infty} \simeq \left(\gamma + e^{(2/3)e^{\sigma_z^2}}\ln\left(1.5e^{-\sigma_z^2}\right)\right)/\ln 2, \quad (7.22)$$

As we will see in the next section, simulations show that, under Rayleigh fading and no shadowing ($\sigma_L^2 \to 0$), the above asymptotic result approximates the corresponding result in [5] for an infinite network. The work in [5] analyzes the average capacity in the asymptotic case of $N \to \infty$ and $R_W \to \infty$ leaving the final result in the form of an integral that must be evaluated numerically. Here, we provide a closed-form approximation to a *lower bound* of the average user capacity. It is worth noting that the analysis approaches taken here and in [5] differ substantially.

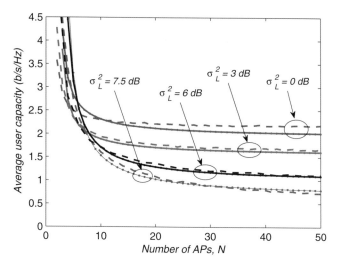

Figure 7.2 Average user capacity versus number of APs in a circular finite-area downlink network with $R_W = 10$ km. The dotted lines are the exact results from simulations; solid lines are the analytic results using (7.21).

As is well known, analyzing the original problem with all its complexities is intractable. The above result is based on approximating the SIR, conditioned on the connection distance, as a ratio of lognormal RVs, which is itself a lognormal RV. Averaging over the spatial distributions of the signal and interference source, requires multiple further approximations. While this averaging is done for the special case of $\alpha = 4$, the approach can be suitably modified for other values of the path-loss exponent. The final result is an explicit relation between the average worst-case user capacity and the number of APs in the chosen region. As we will show next, the approximations are accurate and help answer the original question – how many APs are required to guarantee a worst-case sum capacity.

7.2.4 Simulation results

In general, the worst-case average user capacity in (7.16) depends on various parameters of the network such as number of interfering APs, the radius of the finite-area R_W, and the path-loss exponent. In this section, we simulate the analytical formulations we provided in the previous sections for the example of $\alpha = 4$. Analytical formulations can also be given for different values of α, but the example provided here is sufficient to illustrate the approach. We consider a circular finite-area downlink with radius $R_W = 10$ km and the transmit signal power of each AP corresponds to $\sigma_s^2 = 20$ dBm. We note that the parameter values in the simulations are just for illustration purposes, and any other values only scale the results.

Figure 7.2 compares the approximate user capacity $C_{\text{average}}^{\text{SIR}}$ from the closed-form expressions in (1.21), with the exact results from Monte Carlo simulations with additive thermal noise with variance $\sigma_n^2 = -120$ dBm. The channels $h_j; j = 1, \ldots, N$ are generated according to (7.6) with the lognormal shadow fading for several values of σ_L^2. Although for different values of $N > 2$ and $\sigma_L^2 \leq 7.5$ dB, higher or lower capacities than the actual ones are predicted by the presented analytical expression, generally, there is a close match between the approximate and actual results (the approximate results are always within 10% of the actual capacity for $N > 10$).

In the special case of Rayleigh fading and no shadowing ($\sigma_L^2 \to 0$) and very large number of APs (where the system becomes a dense network), the work in [5] for the infinite area network provides a point of comparison and illustrates the effect of the approximations. Equation (7.22) predicts the average worst-case user capacity to be $C_{\text{average}}^{\text{SIR}}|_{N\to\infty} \approx 1.98$ b/s/Hz; this is based on using $\log_2(\mathbb{E}\{\text{SIR}_{r_1}|r_1\})$ as an approximation to $\mathbb{E}\{\log_2(1 + \text{SIR}_{r_1}|r_1)\}$ in (1.16). Based on $\log_2(1 + \mathbb{E}\{\text{SIR}_{r_1}|r_1\})$, numerical integration gives the average user capacity for $N \to \infty$ and $\sigma_L^2 \to 0$ as 2.25 b/s/Hz. This value decreases to 2.15 b/s/Hz for the average user capacity based on $\mathbb{E}\{\log_2(1 + \text{SIR}_{r_1}|r_1)\}$ which matches the actual capacity result given in [5]. This is expected since the true average user capacity uses $\mathbb{E}\{\log_2(1 + \text{SIR}_{r_1}|r_1)\}$ and is lower bounded by using $\log_2(\mathbb{E}\{\text{SIR}_{r_1}|r_1\})$ and upper bounded by using $\log_2(1 + \mathbb{E}\{\text{SIR}_{r_1}|r_1\})$ respectively. We note that it is only for $N \to \infty$ that the result from infinite area network matches the results obtained for the dense network. Since the average user capacity is a decreasing function in N (as illustrated in Figure 7.2), it is concluded that, for a finite number of APs, the infinite area network results underestimate the performance in a finite-area interference-limited network.

We began this chapter asking the question as to how many APs are needed to guarantee a required coverage in the network. To this end, we note that the presented capacity is the *worst-case* average user capacity given as a function of N. We can not simply invert the provided capacity associated with only the user at the center of W, in order to obtain N for a chosen target value of coverage in the network. Doing so is meaningless and ignores the contribution of capacity associated with other users in the network. Therefore the design has to be based on sum capacity across the N users being serviced in the network. The sum capacity is lower bounded by $NC_{\text{average}}^{\text{SIR}}$. As we saw in Figures 7.2 and 7.4, the per-user capacity converges and, hence, the sum capacity is $O(N)$; that is, the average sum capacity in a finite-area downlink network increases at least linearly with the number of APs.

Now, the required number of APs for a given target value of average sum capacity can be obtained by inverting the sum capacity lower bound of $NC_{\text{average}}^{\text{SIR}}$. Given the complexity of the expression we resort to a polynomial fit for the required number of APs, N that can be found from a non-linear least-squares curve-fitting in the form

$$N\left(C_{\text{sum}}^{\text{target}}\right) = \left\lfloor \sum_{i=0}^{P} a(i+1) \left(C_{\text{sum}}^{\text{target}}\right)^{P-i} \right\rfloor, \qquad (7.23)$$

where $\lfloor x \rfloor$ is the largest integer less than or equal to x and $\{a(i+1); i = 1, \ldots, P\}$ are coefficients found by curve fitting to the numerical values obtained by inverting the sum capacity lower bound for different values of N. As an example, for several values of σ_L^2 (the variance of the lognormal fading) Figure 7.3 illustrates the required number of APs given by (7.23) in an interference-limited network using a 6-th order polynomial fit

Table 7.1 The coefficients, $a(i)$, in the polynomial fit of (7.23) inverting the relationship between the sum rate and the number of APs in an interference-limited network for different values of σ_L^2.

Coefficients	$\sigma_L^2 = 0\,\text{dB}$	$\sigma_L^2 = 3\,\text{dB}$	$\sigma_L^2 = 6\,\text{dB}$	$\sigma_L^2 = 7.5\,\text{dB}$
$a(1)$	-3.74×10^{-17}	-1.8×10^{-11}	-3.28×10^{-10}	-5.62×10^{-9}
$a(2)$	2.55×10^{-9}	9.92×10^{-9}	1.076×10^{-7}	2.03×10^{-6}
$a(3)$	-6.87×10^{-7}	-2.14×10^{-6}	-1.36×10^{-5}	-2.79×10^{-4}
$a(4)$	-10^{-4}	2×10^{-4}	8.57×10^{-4}	0.0189
$a(5)$	-0.0065×10^{-11}	-0.0131	-0.0295	-0.6649
$a(6)$	0.7352	1.0072	1.6615	13.1036
$a(7)$	-4.5919	-5.7504	-14.1319	-90.3947

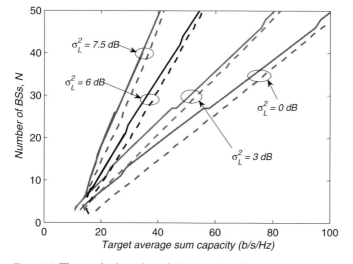

Figure 7.3 The required number of APs in an interference-limited network versus target value of sum capacity for different values of σ_L^2. The solid lines are the results from (7.23) and the dotted lines are from simulations.

(solid lines). The associated coefficients are given in Table 7.1. As is clear, the use of the lower bound only slightly overestimates the required number of APs. The approximately linear relationship is also clear.

An interesting result using Figure 7.3 is the impact of the fading environment on the required number of APs. For a target value of sum capacity, Figure 7.3 suggests a larger number of APs are required for more severe shadowing environments. As an example, for the target sum capacity of 50 b/s/Hz, the required number of APs is almost twice with $\sigma_L^2 = 6\,\text{dB}$ compared to the no-shadowing case $\sigma_L^2 = 0\,\text{dB}$.

7.2.5 Summary and conclusions

In this section we obtain the required number of APs to provide a required sum capacity in a finite (circular) area network. It is this focus on a finite area filled with a finite number of APs that distinguishes this work from that available in the literature. We use a BPP to model the location of APs. The formulation here accounts for the three important components of fading (path loss, lognormal shadowing, and Rayleigh fading), and can incorporate different network parameters. Closed-form approximations for the average worst-case user capacity can be obtained for a chosen path-loss exponent, α. In particular, for $\alpha = 4$, we derive accurate tractable analytical expressions for the worst-case average user capacity in an interference-limited network, where the worst-case capacity decreases monotonically with the number of APs.

The final result presented answers the question asked as to how many APs are required: Since the worst-case average user capacity provides a lower bound on the actual result, it can be used to invert the relationship between sum capacity and number of APs. Here we use a polynomial fit to obtain a closed-form expression that illustrates the dramatic impact the fading environment has on the required APs. Since the inversion is based on a lower bound on the sum capacity, the results are an upper bound. However, simulations show that the results are only a slight overshoot of the true values.

7.3 Dependent placement of SCs in a hexagonal grid network

In the previous section we analyzed a single layer of SC APs. This is generally the scenario considered in the literature in this area. In this section we focus on the *dependent placement* of SCs in a *two-layer* interference-limited network where macro BSs are located on a hexagonal grid. We consider a large cellular reuse-1 downlink HCN with macro BSs (first layer) located at the centers of hexagons in a hexagonal tessellation. The radius of the circumscribed circle of each of the hexagonal cells is denoted as r_{MC}. For such a symmetric model covering the entire plane, it is enough to concentrate on a single cell and interference from surrounding macro BSs has to be computed within that cell from multiple tiers of interferers. Since the first two tiers of interference have the most contributions in the overall received interference within a cell, here we concentrate on the first two tiers with a total of 18 interfering MCs: 6 in the first tier and 12 in the second tier.

We consider a fully loaded network with the total available bandwidth of W Hz and N subcarriers dividing the bandwidth equally to $W_0 = W/N$ Hz. It is assumed that no more than one subcarrier is assigned to a user. As a result, each user gets the fixed bandwidth of W_0. Also, we consider single-cell transmission with MC user association based on the strongest signal power (equivalent to maximum SIR) from nearby macro BSs. Let r_0 denote the distance of a user to the macro BS located at the center of an MC under consideration, and let r_k, $k = 1, \ldots, 18$ denote the corresponding distances between the user and the surrounding interfering macro BSs. We assume that all macro BSs transmit at the same power level of σ_M^2. Also, the SC AP location analysis is based on path loss and shadowing only.

The work in [19] provides tight and accurate lower and upper bounds on the average total interference power as a function of distance r_0 within a MC, which is given as

$$\hat{I}_{l/u}^{(\text{avg})}(r_0) \simeq \sigma_M^2 r_{\text{MC}}^{-\alpha} \exp\left(\sigma_z^2/2\right) \hat{\tilde{I}}_{l/u}^{(\text{avg})}(r_0/r_{\text{MC}}), \qquad (7.24)$$

Table 7.2 The coefficients, $a_{l/u}(i+1); i = 0, \ldots, P$, of the approximates of the lower and upper bounds on $\bar{I}^{avg}$ using third-order polynomial curve fitting for the two examples of path-loss exponents $\alpha = 3$ and $\alpha = 4$.

Coefficients	$a(1)$	$a(2)$	$a(3)$	$a(4)$
Upper bound; $\alpha = 4$	4.2482	−2.4301	0.7687	0.7469
Lower bound; $\alpha = 4$	1.1021	0.3650	0.1019	0.7784
Upper bound; $\alpha = 3$	2.4110	−0.8962	0.4137	1.5024
Lower bound; $\alpha = 3$	0.5138	0.7843	0.0109	1.5217

where $\hat{\bar{I}}_{l/u}^{(avg)}(r_0/r_{\text{MC}} = \bar{r}_0)$ is the polynomial fit for the lower and upper bound on the averaged total interference power normalized to $\sigma_M^2 r_{\text{MC}}^{-\alpha} \exp(\sigma_z^2/2)$, and $\sigma_z = (0.1 \ln 10)\sigma_L$. The subscript "$l/u$" in (7.24) refers to the lower or upper bound. The normalization makes the analysis independent from the macro BSs transmit power (σ_M^2), MC radius (r_{MC}), and the variance of lognormal shadowing (σ_L^2). The approximations $\hat{\bar{I}}_{l/u}^{(avg)}(\bar{r}_0)$ are found from a non-linear least-squares curve-fitting in the form of third-order (or fourth-order) polynomial functions as

$$\hat{\bar{I}}_l^{(avg)}(\bar{r}_0) = \sum_{i=0}^{P} a_l(i+1)\bar{r}_0^{P-i}, \quad (7.25)$$

$$\hat{\bar{I}}_u^{(avg)}(\bar{r}_0) = \sum_{i=0}^{P} a_u(i+1)\bar{r}_0^{P-i}, \quad (7.26)$$

where $\{a_{l/u}(i+1); i = 1, \ldots, P\}$ are coefficients obtained by curve fitting to the numerical values obtained from simulations. Table 7.2 gives the associated coefficients with the use of the third-order polynomial curve fitting, for the two examples of path-loss exponent $\alpha = 3$ and $\alpha = 4$.

From (7.24), a simple approximate formula for the instantaneous SIR is obtained by replacing the instantaneous total interference power in the denominator of SIR formula by the average total interference power, from which the following statistical SIR bounds are obtained as a function of distance r_0,

$$\text{SIR}_{l/u}(r_0) \simeq \underbrace{\left[\frac{(\max\{r_{\text{ref}}, r_0\}/r_{\text{MC}})^{-\alpha}}{\exp(\sigma_z^2/2) \sum_{i=0}^{P} a_{u/l}(i+1)(r_0/r_{\text{MC}})^{P-i}}\right]}_{\xi_{l/u}} z, \quad (7.27)$$

where z is a lognormal random variable with $z \sim LN(\mu_z = 0, \sigma_z = (0.1 \ln 10)\sigma_L)$ and r_{ref} is the close-in reference path-loss distance which is determined from measurements close to the transmitter. We note that, the SIR expression in (7.27) can be written as a function of the normalized distance $\bar{r}_0$ as well. Thus, for a given $\bar{r}_0$, the lower and upper bounds on SIR are modeled as lognormal random variables, $\text{SIR}_{l/u}(\bar{r}_0) \sim LN(\mu_{\text{SIR}_{l/u}}(\bar{r}_0), \sigma_{\text{SIR}_{l/u}})$ with $\sigma_{\text{SIR}_{l/u}} = \sigma_z$, and $\mu_{\text{SIR}_{l/u}}(\bar{r}_0) = \ln \xi_{l/u}(\bar{r}_0)$.

Even though the radii of the MCs are fixed within the hexagonal tessellation, users at the corners and the edges of the cell may receive the strongest signal from nearby macro

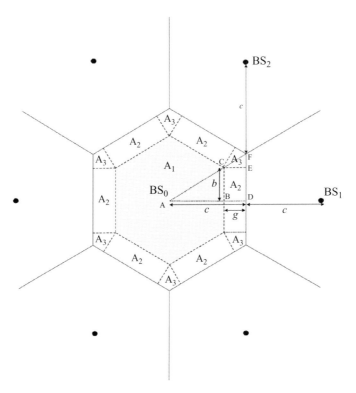

Figure 7.4 Separation of MC into regions A1, A2, and A3.

BSs at different realizations of lognormal shadowing. As a result, the user-association based on the strongest received SIR from nearby macro BSs motivates us to separate the area of each MC to the three regions as depicted in Figure 7.4. The users within region A_1 are associated with the macro BS located at the center of the cell, all the time. Users on region A_2 may receive a signal from either of the two nearby macro BSs, while the users on A_3 may communicate with any of the three nearby macro BSs.

In fact, the area constituting A_2 and A_3 can be interpreted as a guard region for A_1. In other words, it guarantees that users even at the edge of A_1 receive the strongest signal power from the macro BS in that cell with very high probability (essentially 1).

The width of the guard region can be controlled by the parameter g shown in Figure 7.4. For system formulation simplicity, we relate g to the radius of the inscribed circle of the cell, denoted as c, with $g = \gamma_g c = (\sqrt{3}/2)\gamma_g r_{MC}$ where $0 \leq \gamma_g \leq 1$ is now a flexible parameter chosen by the designer. Now, with $r_{A_1} = (1 - \gamma_g) r_{MC}$ it follows that

$$A_1 = (3\sqrt{3}/2)(1 - \gamma_g)^2 r_{MC}^2, \tag{7.28}$$

$$A_2 = 3\sqrt{3}\gamma_g(1 - \gamma_g) r_{MC}^2, \tag{7.29}$$

$$A_3 = (3\sqrt{3}/2)\gamma_g^2 r_{MC}^2. \tag{7.30}$$

7.3.1 Average number of required SCs

At first, we have to obtain the area in outage within an MC when there is no SC in use. With the fixed bandwidth of $W_0 = W/N$ available for each user, a macro BS can serve at most N users at a time. It follows that the per-user rate (in b/s) of $W_0 \log_2(1 + \text{SIR}_{\max}/\Gamma)$ is achieved at each point of the MC. This rate is equivalent to the per user capacity of $C = \log_2(1 + \text{SIR}_{\max}/\Gamma)$ in b/s/Hz. Here, Γ is called the *SNRgap* [4] and $\text{SIR}_{\max}$ is the maximum of SIR from nearby macro BSs. Therefore the coverage outage probability (COP) at each point of an MC, defined as the probability that a particular user can not achieve a target capacity C_0, is obtained as

$$\mathbb{P}\{C < C_0\} = \mathbb{P}\left\{\log_2\left(1 + \frac{\text{SIR}_{\max}}{\Gamma}\right) < C_0\right\} = \mathbb{P}\{\text{SIR}_{\max} < \Gamma(2^{C_0} - 1)\}. \quad (7.31)$$

A point is said to be in outage if the COP at that point is greater than a threshold η, i.e.,

$$\mathbb{P}\{C < C_0\} \geq \eta. \quad (7.32)$$

While each MC covers an area of AMC, each SC is assumed to provide service in an area of A_{SC}. As a crucial simplification of the geometry involved, we assume that the service area of an SC is also a hexagon – this allows us to place SCs within an MC in a hex-tile format. Furthermore, within each MC area, we assume a piecewise constant and uncorrelated shadowing model. This is akin to the popular block-fading model for small-scale fading. For convenience, the shadowing is assumed constant over the coverage area of a single SC, i.e., shadowing is considered to be constant within an A_{SC} and the shadowing realization is independent from one area of size A_{SC} to another.

With the assumption of isolated SCs, the average number of SCs that is required to cover the total MC area in outage is given as

$$\begin{aligned} N &= \left\lceil (A_1^{\text{outage}} + A_2^{\text{outage}} + A_3^{\text{outage}})/A_{\text{SC}} \right\rceil \\ &= \left\lceil (\delta^{(1)} A_1 + \delta^{(2)} A_2 + \delta^{(3)} A_3)/A_{\text{SC}} \right\rceil \end{aligned} \quad (7.33)$$

where A_i^{outage} is the average area of A_i in outage, $\delta^{(i)}$ is the associated fractional area in outage, A_{SC} is the area of an SC, and $\lceil y \rceil$ is the largest integer greater than or equal to y. Using the identities for $A_i; i = 1, 2, 3$ in (7.28)–(7.30), the average number of SCs can be rewritten as

$$N = \left\lceil \left(\delta^{(1)}(1 - \gamma_g)^2 + 2\delta^{(2)}\gamma_g(1 - \gamma_g) + \delta^{(3)}\gamma_g^2\right) \frac{r_{\text{MC}}^2}{r_{\text{SC}}^2} \right\rceil \quad (7.34)$$

in which a hexagonal cell layout for SCs with a circumscribed circle of radius r_{SC} is assumed.

We must now obtain the fractional area in outage associated with the regions $A_i; i = 1, 2, 3$, i.e., $\delta^{(1)}, \delta^{(2)}$, and $\delta^{(3)}$. Due to the symmetric structure of MCs, $\delta^{(i)}; i = 1, 2, 3$ can be calculated from 1/12th of the area of the MC (the triangular region ADF). Now, the triangle ABC and the rectangle BDEC in Figure 7.4 correspond to 1/12th of the area of A_1, and A_2, respectively. Similarly, the triangle CEF corresponds to 1/12th of the area of A_3.

For a point within ABC, the received SIR from BS_0, denoted as SIR_0, is the only SIR candidate that is accounted towards obtaining COP at that point. From (7.27) and (7.31), the lower/upper bound on COP at a point at normalized distance $\bar{r}_0$ from the center of an MC is obtained as

$$\mathbb{P}[C < C_0]_{l/u}^{ABC}(\bar{r}_0) = \mathbb{P}[SIR < \Gamma(2^{C_0} - 1)]_{l/u}$$
$$= \Phi\left(\frac{1}{\sigma_z}\left[\ln(\Gamma(2^{C_0} - 1)) - \ln\frac{\max\{\bar{r}_{\text{ref}}, \bar{r}_0\}^{-\alpha}}{\exp(\sigma_z^2/2)\sum_{i=0}^{P} a_{u/l}(i+1)\bar{r}_0^{P-i}}\right]\right) \quad (7.35)$$

where $\Phi(y) = 1 - Q(y)$ is the cumulative distribution function (CDF) of the standard normal distribution and $\bar{r}_{\text{ref}} = r_{\text{ref}}/r_{MC}$. Note that the COP does not depend either on the MC radius r_{MC} or the transmit power of the MCs.

A point in A_1 is in outage if the COP at that point is greater than threshold η, i.e., $\mathbb{P}[C < C_0]_{l/u}^{ABC}(\bar{r}_0) > \eta$. The COP in (7.35) is an increasing function of $\bar{r}_0$. Thus in order to obtain the area of an MC that is in outage, it is enough to obtain the value of $\bar{r}_0$ that gives $\mathbb{P}[C < C_0]_{l/u}^{ABC}(\bar{r}_0^{\text{opt}}) = \eta$. Any point out of this radius would be in outage. Since $\mathbb{P}[C < C_0]_{l/u}^{ABC}(\bar{r}_0)$ is not an easily invertible function in $\bar{r}_0$, the threshold value $\bar{r}_{0,l/u}^{\text{opt}}$ can be obtained numerically, by solving the following equation

$$\Phi\left(\frac{1}{\sigma_z}\left[\ln(\Gamma(2^{C_0} - 1)) - \ln\frac{\max\{\bar{r}_{\text{ref}}, \bar{r}_0\}^{-\alpha}}{\exp(\sigma_z^2/2)\sum_{i=0}^{P} a_{u/l}(i+1)\bar{r}_0^{P-i}}\right]\right) = \eta \quad (7.36)$$

It is worth noting that it is sufficient to obtain the optimal value of $\bar{r}_{0,l/u}^{\text{opt}}$ only once and restore the resultant values for any C_0 and σ_L in a look-up-table.

Now, if $r_{0,l/u}^{\text{opt}} = \bar{r}_{0,l/u}^{\text{opt}} r_{MC}$ is less that $\frac{\sqrt{3}}{2} r_{A_1}$, the circle $x_0^2 + y_0^2 = (r_{0,l/u}^{\text{opt}})^2$ is entirely inside the hexagon area A_1. As a result, $A_1^{\text{outage}} = A_1 - \pi(r_{0,l/u}^{\text{opt}})^2$. On the other hand, if $\frac{\sqrt{3}}{2} r_{A_1} \leq r_{0,l/u}^{\text{opt}} \leq r_{A_1}$, the circle $x_0^2 + y_0^2 = (r_{0,l/u}^{\text{opt}})^2$ is cut off by the edges of hexagonal region A_1. The area that is cut off by edges is given as $A_{\text{cut}} = 6(r_{0,l/u}^{\text{opt}})^2[\cos^{-1}\frac{\sqrt{3}r_{A_1}}{2r_{0,l/u}^{\text{opt}}} - \frac{1}{2}\sin(2\cos^{-1}\frac{\sqrt{3}r_{A_1}}{2r_{0,l/u}^{\text{opt}}})]$. Therefore the outage area is obtained from $A_1^{\text{outage}} = A_1 - A_1^{\text{coverage}} = A_1 - \pi(r_{0,l/u}^{\text{opt}})^2 + A_{\text{cut}}$, and the fraction area of A_1 in outage, $\delta_{l/u}^{(1)}$, is given by

$$\delta_{l/u}^{(1)} = 1 - \frac{2(\bar{r}_{0,l/u}^{\text{opt}})^2}{\sqrt{3}(1-\gamma_g)^2}\begin{cases}\frac{\pi}{3} & 0 \leq \frac{\bar{r}_{0,l/u}^{\text{opt}}}{(1-\gamma_g)} \leq \frac{\sqrt{3}}{2} \\ f_{l/u}(\bar{r}_0^{\text{opt}}) & \frac{\sqrt{3}}{2} \leq \frac{\bar{r}_{0,l/u}^{\text{opt}}}{(1-\gamma_g)} \leq 1 \\ 0 & \text{otherwise}\end{cases} \quad (7.37)$$

with $f_{l/u}(\bar{r}_0^{\text{opt}}) = \frac{\pi}{3} - 2[\cos^{-1}\frac{\sqrt{3}(1-\gamma_g)}{2\bar{r}_{0,l/u}^{\text{opt}}} - \frac{1}{2}\sin(2\cos^{-1}\frac{\sqrt{3}(1-\gamma_g)}{2\bar{r}_{0,l/u}^{\text{opt}}})]$.

The above results indicate that in the reuse-1 interference-limited networks, $\delta_{l/u}^{(1)}$ does not depend on the MC radius.

Now, for a point within BDEC, the maximum of the instantaneous received SIRs from BS_0 and BS_T is taken into account for obtaining the COP as

$$\mathbb{P}[C < C_0]_{l/u}^{\text{BDEC}} = \mathbb{P}\left[\max_{j=0,1} \text{SIR}_j < \Gamma(2^{C_0} - 1)\right]$$
$$= \mathbb{P}[\text{SIR}_0 < \Gamma(2^{C_0} - 1) \text{ and } \text{SIR}_1 < \Gamma(2^{C_0} - 1)]$$
$$= \prod_{j=0}^{1} \underbrace{\Phi\left(\frac{1}{\sigma_z}\left[\rho - \ln \frac{\bar{r}_j^{(-\alpha)}}{\exp\left(\sigma_z^2/2\right) \sum_{i=0}^{P} a_{u/l}(i+1)\bar{r}_j^{P-i}}\right]\right)}_{\Phi_j} \quad (7.38)$$

where $\rho = \ln[\Gamma(2^{C_0} - 1)]$. In (7.38), $\bar{r}_0$ and $\bar{r}_1$ are the associated normalized distances to BS_0 and BS_T, respectively. The points within BDEC nearly have equal distances to BS_0 and BS_T, so they nearly have equal COPs. Therefore we approximate $\delta_{l/u}^{(2)}$ with the average COP within BDEC. For obtaining the average COP within BDEC, it is more convenient to represent the distances r_0 and r_1 in Cartesian coordinates. With the origin set at the middle of the connecting line of BS_0 and BS_1, the average of $\mathbb{P}[C < C_0]_{l/u}^{\text{BDEC}}$ is obtained from the numerical integration in

$$\delta_{l/u}^{(2)} \approx \int_0^b \int_{-g}^0 \mathbb{P}[C < C_0]_{l/u}^{\text{BDEC}} f_{\text{BDEC}}(x,y) \mathrm{d}x \mathrm{d}y = B \int_0^{\frac{1-\gamma_g}{2}} \int_{-\frac{\sqrt{3}\gamma_g}{2}}^0 \Phi_0 \Phi_1 \mathrm{d}\bar{x} \mathrm{d}\bar{y} \quad (7.39)$$

where $f_{\text{BDEC}}(x,y)$ is the distribution of users within BDEC, that under the assumption of uniform user distribution, $f_{\text{BDEC}}(x,y) = 1/gq$ and $B = 4/(3\sqrt{3}\gamma_g(1-\gamma_g))$. From (7.39), $\delta_{l/u}^{(2)}$ is obtained irrespective of the MC radius.

Finally, for a point within the triangle CEF, the maximum of the received SIRs from BS_0, BS_1, and BS_2 is considered for COP calculation at each point. Therefore

$$\mathbb{P}[C < C_0]_{l/u}^{\text{CEF}} = \prod_{j=0}^{2} \underbrace{\Phi\left(\frac{1}{\sigma_z}\left[\rho - \ln \frac{\bar{r}_j^{(-\alpha)}}{\exp\left(\sigma_z^2/2\right) \sum_{i=0}^{P} a_{u/l}(i+1)\bar{r}_j^{P-i}}\right]\right)}_{\Phi_j} \quad (7.40)$$

As in the case for A_2, the average COP within CEF can approximate $\delta_{l/u}^{(3)}$ in A_3. Thus $\delta_{l/u}^{(3)}$ is approximated as

$$\delta_{l/u}^{(3)} \approx \frac{1}{A_{\text{CEF}}} \int_{-g}^{0} \int_{b}^{\frac{\sqrt{3}}{3}(x+g)+b} \Phi_0 \Phi_1 \Phi_2 \mathrm{d}y \mathrm{d}x$$
$$= \frac{8}{\sqrt{3}\gamma_g^2} \int_{\frac{\sqrt{3}\gamma_g}{2}}^{0} \int_{\frac{1-\gamma_g}{2}}^{\frac{\sqrt{3}}{3}(\bar{x}+\frac{\sqrt{3}\gamma_g}{2})+\frac{1-\gamma_g}{2}} \Phi_0 \Phi_1 \Phi_2 \mathrm{d}\bar{y} \mathrm{d}\bar{x}, \quad (7.41)$$

where $A_{\text{CEF}} = \sqrt{3}\gamma_g^2 r_{\text{MC}}^2/8$ is the area of triangle CEF. Since Φ_0, Φ_1, and Φ_2 do not depend on r_{MC}, the independency of $\delta_{l/u}^{(3)}$ with respect to the MC radius can be seen easily via (7.41).

By having $\delta_{l/u}^{(1)}, \delta_{l/u}^{(2)}, \delta_{l/u}^{(3)}$ at hand, the lower/upper bound on the percentage area of the MC in outage is obtained from

$$\delta_{l/u}^{MC} = \delta_{l/u}^{(1)}(1-\gamma_g)^2 + 2\delta_{l/u}^{(2)}\gamma_g(1-\gamma_g) + \delta_{l/u}^{(3)}\gamma_g^2 \tag{7.42}$$

From (7.37), (7.39), (7.41), and (7.42) we see that in the reuse-1 interference-limited networks, the percentage area in outage (without SCs in use) does not depend on the MC radius and BS transmit power, but does depend on the path-loss exponent α, variance of lognormal fading σ_L^2, capacity threshold C_0, and COP threshold η. It is worth emphasizing that the above claim is for the case with neglected noise.

With hexagonal layout for the SCs, the lower/upper bound on the average number of SCs $N_{l/u} = \lceil \delta_{l/u}^{MC} r_{MC}^2 / r_{SC}^2 \rceil$ is proportional to the square of the MS–SC radius ratio $(r_{MC}/r_{SC})^2$, via $\delta_{l/u}^{MC}$.

7.3.2 Simulations results

This section presents the results of simulations that validate the approximations used in the previous section and illustrate the results in terms of average required number of SCs. The details of the simulations are as follows. The transmit power of MCs and SCs are set to 43 dBm and 20 dBm, respectively. The additive noise power is –100 dBm, the standard deviation of lognormal shadowing is $\sigma_L = 4$ dB, the gap SNR is set to $\Gamma = 2$ dB, the COP threshold η is set to 0.5, and the parameters γ_g for adjusting the sizes of A_1, A_2, and A_3 is set as $\gamma_g = 0.2$. At $\gamma_g = 0.2$, the probability that the user at the areas A_1 and A_2 would pick the wrong serving BS (according to the strongest receive signal user association) is approximately 3%.

In the previous section, several approximations were used to derive the percentage of MC area in outage. Thus at first it is important to validate the approximations. Figure 7.5 plots the fractional area in outage using the approximate analytic results from previous section (the solid lines in the figure), and compares these results with the exact results obtained from Monte Carlo simulations (the dotted lines in the figure), for a PLE of $\alpha = 4$. The figure plots the fractional area in outage within each region of Figure 7.4 and the overall fraction in outage. The analytical results are the average of the lower and upper bounds, indicated by the subscript "avg" in the figure. As is clear from Figure 7.5, there is a very close match between the approximate analytic and simulated results for all cases with the differences approaching zero for moderate to high values of target rate C_0. In particular, for the example of $C_0 = 1$ b/s/Hz, the relative error in δ_{avg}^{MC} is 2.5%. Therefore a design based on the closed-form expressions provided in the previous section is valid. Figure 7.5b confirms that the fractional area in outage is independent of the macrocell radius (here $C_0 = 1$ b/s/Hz).

To illustrate the tightness of the bounds and the efficacy of the analysis in predicting the average number of required SCs, Figure 7.6 illustrates the lower and upper bound of the required average number of SCs for different values of capacity threshold C_0 in a reuse-1 network with $r_{MC} = 1$ km and SCs radius $r_{SC} = 150$ m for the examples of $\alpha = 3$ and 4 with $\eta = 0.5$. The exact results from simulations using the instantaneous interference in the SIR formula are also included in dashed lines. As is seen from the figure,

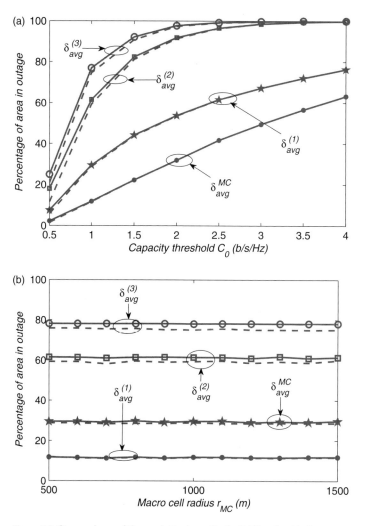

Figure 7.5 Comparison of the analytical results (solid lines) with the exact results obtained from Monte Carlo simulations (dotted lines) for different values of: (a) rate threshold C_0; (b) MC radius $r_{MC}(C_0 = 1\text{b/s/Hz})$.

the lower and upper bounds are in good match with the exact results. For $C_0 \gtrsim 1.5$ b/s/Hz, the lower and upper bound exactly lie on the exact results from simulations. Also, for smaller values of C_0, the average of the results from lower and upper bounds can be taken as the final required number of SCs since the average result matches exactly with the exact results. In the provided example, on average, each MC requires 13 SCs (3 SCs) for the target capacity of 1 b/s/Hz (0.5 b/s/Hz) to completely cover the area in outage if isolated SCs are used. The results can also be given for different values of network parameters, but the example provided here is sufficient to illustrate the approach. We note that the parameter values in the simulations are just for the illustration purposes, and any other values only scale the result.

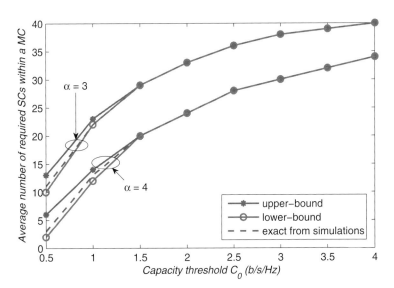

Figure 7.6 Lower and upper bounds on the required number of SCs for different values of capacity threshold C_0 in a reuse-1 network with $r_{MC} = 1$ km, and $r_{SC} = 150$ m. The exact results from simulations using the instantaneous interference in the SIR formula are also included for comparison in dashed lines.

It is worth noting that we do not consider resource allocation among SCs. As a result, without resource allocation used among SCs, there would be a residual outage area. The evaluation of the residual outage area is beyond the scope of this chapter and the interested readers are referred to [19]. Furthermore, the above results hold when the users are uniformly distributed within an MC. With non-uniform distribution of users within an MC, one has to account for hotspots as well as the low traffic density areas. Hotspots are identified as the areas with high traffic density where the number of active users (per an SC area) is greater than a threshold. Hotspots create localized strains on MC resources, so they have to be covered by SCs. On the other hand, there are regions with low traffic density where the number of active users (per an SC area) is below a threshold. Since there are few users within these regions, even though they may be in outage as a result of shadowing, it is not cost effective to be covered by SCs. A comprehensive algorithm has been presented in [21] where all the above factors have been accounted for in determining the required number of SCs within an MC.

7.3.3 Summary and conclusions

In this section, we considered the case where SCs are used for range extension and to fill coverage holes in a macro cell network. This leads to a *dependent* placement of SCs within MCs. Specifically, we obtain the average number of required isolated SCs to eliminate coverage holes in the downlink of a reuse-1 macro cell network. To provide a closed-form analysis, we developed accurate closed-form bounds on the average total interference within an MC. These bounds are particularly simple in the sense that they are expressed as functions of normalized distance to the center of the MC. The normalization makes the analysis free from MC radius and is also useful for quick system assessment and obtaining simple analytical expressions for SIR and ROP within MCs. We use these bounds to obtain, for a chosen target rate, the fractional area within an MC in outage (with no SCs in use) and the required average number of SCs (per MC) that can completely overcome outage with *isolated* SCs. The tools developed here allow for a quick assessment of the tradeoff between target rate, COP threshold, and required number of SCs.

References

[1] Chandrasekhar, V., Andrews, J. G., and Gatherer, A. (2008). "Femtocell networks: a survey," *Communications Magazine, IEEE* 46(9), 59–67.

[2] 4G Americas Report (2011). "4G mobile broadband evolution: 3GPP Release 10 and beyond," http://4gamericas.org.

[3] Wyner, A. D. (1994). "Shannon-theoretic approach to a Gaussian cellular multiple-access channel," *IEEE Trans. Information Theory* 40(6), 1713–1727.

[4] Goldsmith, A. (2005). *Wireless Communications*, Cambridge University Press.

[5] Andrews, J. G., Baccelli, F., and Ganti, R. K. (2011). "A tractable approach to coverage and rate in cellular networks," *IEEE Trans. Communications* 59(11), 3122–3134.

[6] Huang, K. and Andrews, J. G. (2013). "An analytical framework for multicell cooperation via stochastic geometry and large deviations," *IEEE Trans. Information Theory* 59(4), 2501–2516.

[7] Dhillon, H. S., Ganti, R. K., Baccelli, F., and Andrews, J. G. (2012). "Modeling and analysis of k-tier downlink heterogeneous cellular networks," *IEEE Journal Selected Areas in Communications* 30(3), 550–560.

[8] Dhillon, H. S., Kountouris, M., and Andrews, J. G. (2013). "Downlink MIMO HetNets: modeling, ordering results and performance analysis," *arXivpreprint arXiv:1301.5034*.

[9] Di Renzo, M., Guidotti, A., and Corazza, G. (2013). "Average rate of downlink heterogeneous cellular networks over generalized fading channels: a stochastic geometry approach," pp. 1–22.

[10] Haenggi, M. (2013). "A versatile dependent model for heterogeneous cellular networks," *arXiv preprint arXiv:1305.0947*.

[11] Andrews, J. G., Claussen, H., Dohler, M., Rangan, S., and Reed, M. C. (2012). "Femtocells: past, present, and future," *IEEE Journal Selected Areas in Communications* 30(3), 497–508.

[12] Wang, H., Zhou, X., and Reed, M. (2014). "Coverage and throughput analysis with a non-uniform small cell deployment," *IEEE Trans. Wireless Communications*.

[13] Vijayandran, L., Dharmawansa, P., Ekman, T., and Tellambura, C. (2012). "Analysis of aggregate interference and primary system performance in finite area cognitive radio networks," *IEEE Trans. Communications* 60(7), 1811–1822.

[14] Banani, S. A., Eckford, A. W., and Adve, R. (2015). "Analyzing the impact of access point density on the performance of finite area networks," submitted to *IEEE Trans. Communications*.

[15] Boudreau, G., Panicker, J., Guo, N., Chang, R., Wang, N., and Vrzic, S. (2009). "Interference coordination and cancellation for 4G networks," *IEEE Communications Magazine* 47(4), 74–81.

[16] Walke, B. (2001). *Mobile Radio Networks: Networking and Protocols*, Wiley Chichester.
[17] Haenggi, M. and Ganti, R. K. (2009). *Interference in Large Wireless Networks*, Now Publishers Inc.
[18] Pratesi, M., Santucci, F., and Graziosi, F. (2006). "Generalized moment matching for the linear combination of lognormal RVS: application to outage analysis in wireless systems," *IEEE Trans. Wireless Communications* 5(5), 1122–1132.
[19] Banani, S. A., Eckford, A. W., and Adve, R. (2014). "Analyzing dependent placements of small cells in a two-tier heterogeneous network with a rate coverage constraint," *IEEE Trans. Vehicular Technology* 1–13.
[20] Havil, J. (2003). "Gamma: exploring Eulers constant," *Australian Mathematical Society* 250.
[21] Banani, S. A., Eckford, A. W., and Adve, R. (2014). "Required number of small-cells in heterogenous networks with non-uniform traffic distribution," *IEEE Conf Information Sciences and Systems* pp. 1–5.

8 Small cell deployments: system scenarios, performance, and analysis

Mark C. Reed and He Wang

Mobile device data rates are increasing at effectively Moore's law [1] due to the fact that mobile devices are all integrated in silicon and thus are taking advantage of the reduction in geometry and increase in functionality and the number of transistors per die. The current cellular approach of using large outdoor base station towers to provide mobile broadband via wireless communications, however, does not scale efficiently to cope with the forecast 13-fold increase expected by 2017 [2]. Thus a need for a new approach is required. As we will discuss in this chapter, small cell deployments have the potential to provide a scalable solution to this demand where they are beginning to change the network topology into a so-called heterogeneous network (HetNet) containing a mix of different cell sizes and cell power levels, as shown in Figure 8.1. This will lead to a mix of macro cells, micro cells, pico cells, and femto cells. This deployment is not as uniform as outdoor macro cells with different cell sizes and a much more irregular deployment results.

As we will show in this chapter, small cell network (SCN) deployments provide, for the first time, a low-cost efficient scalable architecture to meet the expected demand. This technology was initially deployed in large scale when femto cells (residential small cells) were first deployed by a number of leading wireless operators (including Vodafone and AT&T). The solution was enabled by two key developments, namely:

- low-cost chip-sets (a so-called "system-on-chip"), which included the entire signal processing and most of the radio software stack, and
- high-speed internet access (greater than Mbps) was available, thus providing the so-called "backhaul" (the connectivity into the network) for these access points.

These developments mean that with volume the effective capital costs of small cells are negligible compared to the cost of deployment and operating costs.

Small cell deployments have a range of topologies and configurations that will depend on the location and the demand requirements. For example, downtown city areas will consist of a combination of small cells where they will exist:

1. on outdoor light poles to serve "hotspot" traffic needs around cafes and other places where users congregate
2. in city buildings to serve enterprise customers with high demands on throughput and reliability
3. in apartments or residential homes to serve private users' needs.

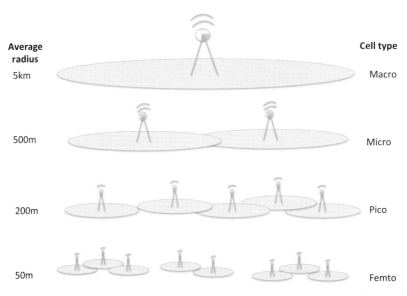

Figure 8.1 Example of a HetNet, showing access points and cell range for different cell types.

We will first discuss the different deployment scenarios and highlight some of the issues and challenges of each of these scenarios (residential, enterprise, outdoor hotspot). We also discuss the evolution of wireless networks from a homogeneous array of wireless base stations to a complete mix of access points, where each type of small cell has different power ranges, locations, backhaul connectivity, and costs associated with it.

As cellular networks are interference limited, the main challenge from the radio perspective is to minimize, avoid, or cancel interference in the network. Small cells introduce a new network topology where interference can be very significant and occurs in different ways to the conventional cellular networks [3, 4]. We exclusively consider the use of the same spectrum in each cell, thus interference between a user and a neighboring small cell/macro cell will occur. When the same spectrum is used in each cell this is called the reuse = 1 scenario and is commonly the case with both 3G Wideband Code-Division Multiple-Access (3G/WCDMA) and 4G Long Term Evolution (LTE) cellular network deployments.

The most costly part of any wireless network is the human resource needed to deploy, manage, and optimize the network. With such a massive increase in access points (ten to one hundred times as many), the approach to deployment and operations has to significantly change to avoid cost blowouts. Methods that were semi-manual involving technicians altering parameters or refining network design for a few thousand base stations don't scale when you have this many more access points. Coexistence of this varying array of wireless access is an important and challenging part of the move to HetNets. This means that tools and techniques are needed such that a high-power outdoor base station tower can efficiently coexist with a low-power residential small cell and that the entire system can be run to maximize benefits for all users in a fair and equitable way.

This chapter will highlight in some detail the open challenges in the radio access network and deploying small cells, where for each deployment scenario the key challenges are different and we will thus describe the important ones for each scenario. Other challenges that are not addressed in detail here include allocation of cell IDs to each small cell and determining the most cost effective and efficient way of connecting the small cells into the network. What sometimes occurs is that there is a tradeoff on small cell performance between the best location for radio access and the best location for backhaul connectivity.

In the final section of this chapter we will discuss analytical tools that are used to determine performance of these HetNets. We will describe the random spatial models and how we determine cell association, coverage probability, and single-user throughput. This model has been shown to map closely to real networks [5]; however, there are limitations and we describe these assumptions around this modeling approach. We perform a case study of six stages in small cell densification and show results from our analysis of data throughput as we add more and more small cells. What is interesting from these results is that the throughput grows in a linear way with the number of small cells.

8.1 System scenarios

8.1.1 Residential deployment

Residential deployments of small cells (also called femto cells) were the original deployment model of this new approach to wireless infrastructure [6–8]. In this scenario the small cell was placed in the home and connected to the network via a broadband internet connection. This was and still is a very disruptive deployment for mobile operators as it dramatically changed the deployment model of their infrastructure. Previously there was only a small number of towers and all the infrastructure was completely controlled and owned by the operator. With residential small cells the operator needed to use internet connectivity and potentially work with competitive internet service providers (ISPs) for network connectivity. Due to the sheer potential volume of customers new automated ways of provisioning and operational procedures were developed to keep deployment and operational costs low. This was very new compared to the semi-manual approach used in deploying and managing large macro cell sites.

Some of the features of residential small cells included the need to be "zero touch," this meant that there was no need for technicians to be involved with the deployment, i.e., there were fully automated systems for delivery, authentication, billing, and provisioning. When a high volume of deployments are forecast then the above features are essential to deliver cost-efficient automated procedures. Residential small cells were originally intended for connectivity to a few family members; however, this rapidly increased to support eight users and now up to 32 users. The residential small cell typically utilizes a closed subscriber group (CSG) configuration, meaning that it is not open to the public for connectivity; this prevents unwanted visitors connecting and thus is for private use only.

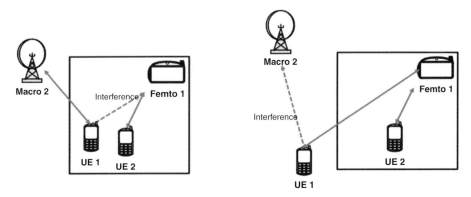

Figure 8.2 Interference scenarios between small cell and macro cells.

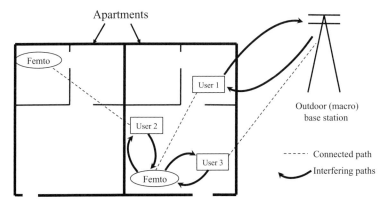

Figure 8.3 Interference scenarios between residential femto cell and users connected to an outdoor base station.

The CSG configuration creates some severe interference scenarios. In Figure 8.2 we show the two key interference scenarios. The left side figure shows how a macro connected user (UE1) that is visiting the home of a femto cell owner would interfere with the small cell, and likewise the small cell interferes with the mobile device. This scenario occurs due to the CSG constraint where even though the small cell signal is stronger, the user (UE1) cannot handover to the femto cell. The right side figure shows how a femto connected user (UE1) can interfere with a macro cell and likewise this macrocell can interfere with the user. This single user may not seem to create too much interference but if you have a scenario with a large number of femto connected users (with a whole area of deployed small cells) then this interference is additive and thus will be significant onto the macro cell.

Interference occurs in the uplink and downlink and can be between small cells as well as between small cells and macro cells as shown in Figure 8.3. Here User 2, who is visiting from a neighboring apartment and thus connected to their home femto, is interfering with the femto. Likewise the femto is interfering with User 2; this is due to

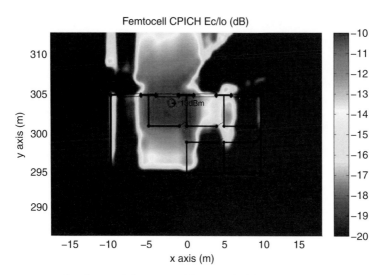

Figure 8.4 Signal strength heat map of femto cell with macro cell present.

the CSG constraint where User 2 is prevented from directly connecting to the femto that User 1 is connected to. Likewise a visitor, User 3, is connected to an outdoor macro cell, where he interferes with a femto in the downlink and likewise the femto interferes with User 3 in the uplink. Finally, User 1 is connected to the femto but interferes with the macro in the uplink (creating uplink noise rise) where the macro cell also interferes with User 3 in the downlink.

Each of these interference scenarios needs to be addressed and mechanisms to minimize the interference are important. Advanced multi-user receivers can also be used in the small cells [9] to minimize small cell user equipment (UE) transmission power, maximize uplink throughput, and also minimize uplink noise rise onto neighboring cells in the same band.

Figure 8.4 shows a residential home floor plan and the 3G/WCDMA pilot power coverage heat map in the downlink when the femtocell is located at the back of the home (as indicated by a black circle). The home is north (by 300 meters) of a macro cell. The femto cell coverage is limited due to the power of the neighboring macro cell, which penetrates the home. Figure 8.5 shows the pilot power coverage from the macro cell, which covers the remaining part of the home, however, not with the same level of pilot power as the femto; this configuration means that a user would be connected to the macro cell in a large area of the home and not take advantage of the low-cost wireless connectivity provided by the femto cell connection. This is a significant issue where the user would expect to be always connected to the femto cell while being in the residence.

As can be seen by the above example, with reuse = 1 wireless systems the indoor coverage performance is very dependent on placement of the small cell with respect to the distance from the dominating high-power macro cell. Improvements in this coverage can be achieved if the small cell power level can be adaptively set. Figure 8.6 shows the average percentage of indoor coverage for a residential small cell for different distances

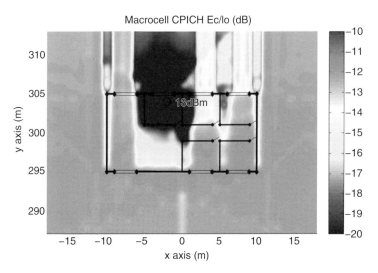

Figure 8.5 Signal strength heat map for macro cell with femto cell present.

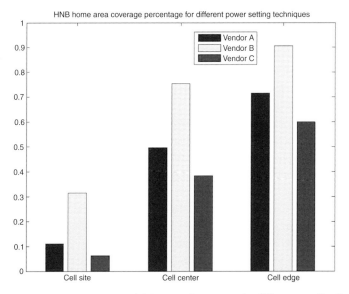

Figure 8.6 Average residential femto cell coverage for different small cell power-setting strategies.

from the macro cell (cell site = 100 m, cell center = 300 m, cell edge = 600 m). It can be clearly seen that for a co-channel 3G/WCDMA deployment coverage improves when the home is on the cell edge as it is no longer competing against the high-power macro cell. This is where residential femto cells have the most benefit as they can significantly improve the user's data throughput. The figure also shows this for three different vendors using different power-setting techniques. Vendor A's approach is a fixed power level, Vendor B is for an adaptive approach, and Vendor C is a hybrid approach that was used

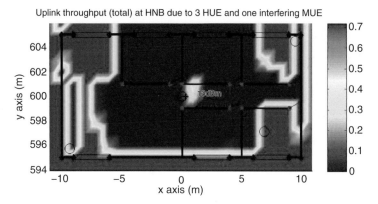

Figure 8.7 Average total throughput (kbps) at femto cell for three users and one macro interfering user. (Home is 600 m from macro station, no uplink attenuation, deadzone region is shown.)

in a deployed network. The adaptive approach uses signal strength measurements from the macro cell (via a mobile terminal receiver – or sniffer) to adjust the femto power to try to maximize indoor coverage; this approach would increase the femto cell power when it is close to the macro cell and lower the power when it is far from the macro cell. The hybrid approach used just two levels of power and switched between the low level when on the cell edge to the higher level when closer to the macro cell. What can be seen is that compared to fixed power levels the adaptive approach can improve indoor coverage by more than 15% on the cell edge and more than double coverage at the cell site. These are significant gains making adaptive small cell power setting an essential part of small cell deployments. What can also be seen with the Vendor C (hybrid) approach is that coverage is significantly reduced by using this scheme.

As highlighted above in Figure 8.3, 3G/WCDMA reuse = 1 wireless systems interference from macro-connected users onto the femto cell can severely limit data throughput due to the "noise rise." If enough power is sent from the macro user it can essentially overwhelm the femto cell with interference power. To avoid this the femto cell can attenuate the analog signal that is received [3]. This technique is called uplink attenuation and is widely used in deployed systems. The uplink attenuation reduces the femto cell front-end analog gains and thus attenuates the macro interfering user, minimizing the jamming. The femto cell connected users increase their power through closed-loop power control and can still connect. One consequence is that the femto-connected users now generate more power and thus more interference onto neighboring small cells and onto the macro cells.

In Figures 8.7 and 8.8 we highlight an example scenario of the effects of uplink attenuation on indoor coverage with the use of a heat map. We highlight the dead zones (areas of no coverage) created by a macro terminal that a femto cell user would experience in a residential home. Here the residence is 300 m from the macro cell and we highlight the deadzone area in the uplink. This deadzone area indicates that if a macro-connected user is in this area then all femto cell users' uplink would be in outage (i.e., jammed from the macro user). The clear desire is therefore to minimize the deadzone area.

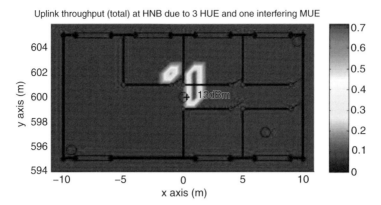

Figure 8.8 Average total throughput (kbps) at femto cell for three users and one macro interfering user. (Home is 600 m from macro station, 30 dB of uplink attenuation, deadzone region is shown.)

In Figure 8.7 there is no uplink attenuation so the deadzone is large and therefore if a macro cell is located in the majority of the home footprint the femto-connected users will be in outage (due to interference from the macro-connected user). In Figure 8.8 the 30 dB of uplink attenuation that the femto cell uses on the analog front end has a dramatic effect on the reduction in the deadzone area. Only if the macro-connected user is located very close to the small cell will it create uplink noise rise onto the femto, thus jamming the femto and not allowing femto-connected users to connect. The compromise for this solution is that each femto-connected user needs to transmit at 30 dB higher power to maintain their connection with the femto cell, thus draining their batteries quicker and potentially creating more uplink noise rise onto the macro cell.

In Figure 8.9 we show statistical results for different uplink attenuation strategies, which include no attenuation, 10 dB of attenuation, 30 dB of attenuation, and an adaptive uplink attenuation strategy. It is clear that the deadzone is more challenging at the cell edge, where the macro-connected user's power is the greatest. You can also see from this graph that the adaptive scheme and the high-attenuation schemes can provide nearly 90% coverage and provide gains of more than double the coverage over no uplink attenuation, which is very significant.

In summary, residential femto cells can be very effective at improving user experience; however, compensation for inter-cell interference from macro cells needs to be considered and taken into account. We have highlighted here some of the strategies that are used and the average improvements that can be achieved. Due to space limitations we try to highlight the trends and benefits of these interference avoidance techniques.

8.1.2 Enterprise deployment

Enterprise deployments differ from residential deployments in a number of important ways. Firstly, enterprise systems are more critical than residential, thus the reliability,

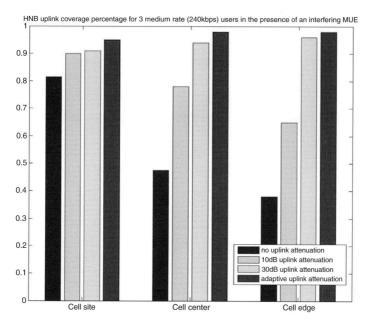

Figure 8.9 Average femto cell uplink coverage in the presence of an interfering macro user for different uplink attenuation strategies.

throughput, and overall performance of the system is expected to be greater to minimize outages and poor performance. Secondly, there are normally multiple small cells deployed on one floor of an office so the interference interaction is not just between an outdoor macro cell and the indoor small cell but also between indoor small cells (for a frequency re-use = 1 scenario with all small cells on the same frequency band this creates interference limited scenarios). Finally, one non-obvious challenge is the effect on performance due to the location of the small cells within the building. For example, small cells placed too close to the outer wall will generate cell boundaries both inside and outside the building, thus enabling users located outside to connect. These small cell-connected users have the potential to create significant (and sometimes direct line of sight) interference onto the macro system.

For this discussion we examine one floor of a building where the size is 60 m × 30 m, totally 1,800 square meters (approximately 19,375 square feet). The important things to consider about the configuration are: (1) there are multiple small cells on the floor, (2) the floor height may be above ground level and thus could suffer significant macro cell interference, (3) the location and power of the small cells needs to be determined, and (4) we want to minimize power leakage out of the building to minimize the opportunity for users outside the building to connect and thus create macro cell uplink interference (noise rise).

In Figure 8.10 we show a heat map for the pilot power of the small cells within an enterprise with five small cells. As can be seen here the power settings of the small cells

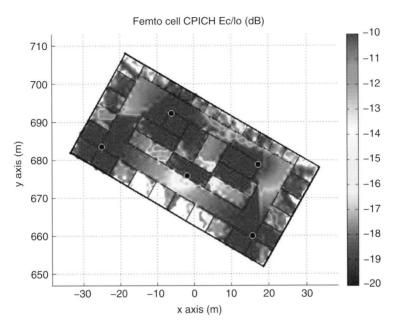

Figure 8.10 Small cell pilot power coverage for an enterprise 680 m from a macro cell tower.

cover most of the enterprise where the white areas are covered by the outdoor macro cells.

The objective in the enterprise is to provide maximum coverage without leaking power out of the building. Power leakage out of the building can enable users to connect outside the building thus creating interference between the small cell-connected user and the macro cell. Too little power from the small cell in the building creates coverage holes where users will be forced to connect to the macro cell system creating interference onto the indoor small cells (much like in Figure 8.2).

What is interesting to determine is the expected improvement in downlink throughput for an enterprise with the use of small cells. Figure 8.11 shows the average single-user throughput with and without small cells deployed for a varying number of active users in a 3G/WCDMA system, and we plot this for the building located at the cell edge (600 m from macro), cell center (300 m from macro), and cell site (100 m from macro). It can be seen that significant throughput improvements can be achieved for the enterprise building users. For buildings located on the cell edge, users can experience up to 50% improvement while users in a building at distance 100 m from a macro cell can experience throughput improvements of up to 25%.

In summary, enterprise small cells can be very beneficial to end users where, in a similar way as for residential femto cells, the benefits are maximized on the edge of the macro cell; however, substantial gains can also be achieved even at 100 m from the macro cell. With enterprise deployments it is important to be able to set the cell size to the boundary of the building. Any less and macro-connected users generate indoor interference, any more and small cell-connected users create outdoor interference.

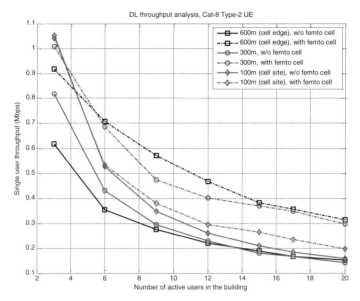

Figure 8.11 Enterprise downlink HSDPA single-user throughput over number of active users.

8.1.3 Outdoor hotspot deployment

Outdoor small cells is the most recent deployment model that is being developed and deployed by industry. In this scenario the small cells are placed on street light poles or the sides of buildings and enable the network to provide substantial capacity improvements. Unlike residential or enterprise solutions, outdoor small cells are expected to be deployed by the wireless operator. The challenges are substantial and include: securing the real-estate, providing metered power, obtaining backhaul (wired or wireless internet connectivity), and minimizing the cost of the deployment. The biggest difference with outdoor small cells includes (a) higher power levels, (b) support for mobility (handover functionality), (c) fully sealed weatherproof casings, and (d) site acquisition.

In Figure 8.12 we show a cell heat map for a three-sector macro cell with ten small cells placed within the coverage area. As can be seen, the users in this area would be allocated across all these different available cells. What is also noticeable and important is that the total area of the cell boundaries increases substantially meaning more users are in the handover region for longer, adding additional complexity to modem resources and also user-handover management. The next section of this chapter deals in more depth with this topic.

8.1.4 Heterogeneous wireless network

Above we have discussed residential, enterprise, and outdoor small cells. Our discussion of a heterogeneous network (HetNet) has been limited to small (femto, small cells) and macro cells. The next section uses a single wireless technology and combines different cell sizes where the cell topology leads to some unique design requirements. These

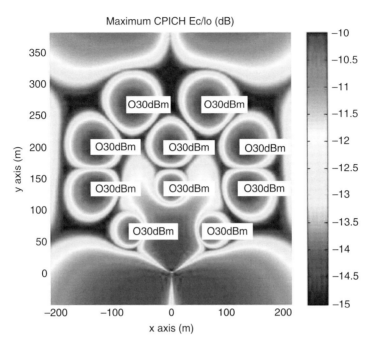

Figure 8.12 Heat map of pilot signal coverage for ten small cells combined with a three-sector macro cellular 3G/WCDMA system.

requirements vary for each wireless radio technology where we exclusively focus on 4G/OFDMA-based standards, where there is inter-cell interference but not intra-cell interference.

One of the other new challenges that HetNets bring is how best to connect a user when a mix of radio access technologies are available and what criteria should be used to determine this. The challenge is complex as there are different radio modems, different frequencies, different multiple access technologies, and different software protocol stacks for each radio technology. This approach to HetNets requires more sophisticated mechanisms of maximizing performance across radio technologies such as 3G, 4G, 5G, and even WiFi standards. This is one of the open challenges for HetNets where the design of robust, scalable optimization techniques is needed to provide the ultimately best connectivity to the end user.

8.2 Analytical model and performance analysis

Just like the discussion in Section 8.1, system-level simulations for both research and industry evaluations are extensively based on the grid-based cellular model, in which cellular base stations are placed on a two-dimensional grid (e.g., a square lattice or a more widely used hexagonal lattice according to the idealized cellular structure). Although analytical results can be obtained for particular user locations in the simplified grid model where only limited number of interferers are considered [10], analytical expressions of the received SINR and other useful performance metrics are not available for the more general cases.

The regular grid model only represents a highly idealized network setup. In practical deployments, cellular coverage may be affected by pragmatic considerations, such as the availability of certain site locations, substantially varying the base station locations and thus the cell boundaries. As more small cells are introduced in heterogeneous cellular networks, the increased complexity presents new challenges for system-level simulation. Furthermore, because of its analytical intractability, the grid-model-based simulation is unable to identify the key design parameters and their corresponding impacts on system performance [11].

8.2.1 Random spatial model

In HetNets, the overlaid small cells, deployed with little or no planning, are prone to be irregularly spaced and thus randomly scattered. In recent years, the extreme case of randomness has widely been regarded as a workable network model. Specifically, by randomly placing the base stations with no prior information involved, the base stations will be distributed as a two-dimensional Poisson point process (PPP). Mathematically, a two-dimensional homogeneous PPP can be defined by the requirement that the random variables (defined as the counts of the number of points inside each of a number of non-overlapping finite sub-regions of $\mathbb{R}^2$) should each have a Poisson distribution and should be independent of each other [12].

This PPP-based random spatial model is characterized by its Poisson distribution and independence assumption. Its accuracy has been proven even for the most planned scenario, i.e., the macro cell-only networks. In the studies conducted recently, it has been shown that compared with the practical network deployment, modeling the cellular network with base station (BS) locations drawn from a homogeneous PPP is as accurate as the traditional grid-based models to capture the main performance trends [11, 13]. Although this validation was based upon the comparison with one practical cellular deployment, there is still a belief that cellular deployment by operators (especially small cell deployment) becomes increasingly irregular and the random spatial model becomes even more suitable [14]. When the networks are densified, ideal base station locations are no longer available under the physical constraints of terrain, and high-rise buildings. In a new study [5], Blaszczyszyn et al. provided another important justification for the PPP-based model. They found that for any homogeneous BS distribution (including the one formed by grid models), as long as the lognormal shadowing variance is sufficiently high, any statistical measure based on the signal propagation is identical to an equivalent PPP-based model.

Surprisingly, the randomness introduced by the PPP-based model makes it tractable in terms of performance analysis, because of the closed-form results on interference and SINR distributions for certain signal attenuation laws [15, 16]. By accepting the PPP-based topology for cellular networks, important results on connectivity, coverage, and throughput have been successfully derived [17–22]. Due to the analytical tractability of this PPP-based random spatial model, deep insights on the important network parameters can be obtained.

8.2.2 The PPP-based model for small cell deployment

In this section, the PPP-based random spatial model is applied to a downlink K-tier cellular network with small cells deployed. The network assumes an orthogonal multiple access technique, like the orthogonal frequency-division multiple access (OFDMA) in Long Term Evolution (LTE). Based on the PPP-based random spatial model, the fundamental analysis results are provided in Section 8.2.3.

Here BSs are assumed to be located independently across different tiers, and the i-th tier BSs are spatially distributed as a two-dimensional homogeneous PPP Φ_i with density λ_i in the whole Euclidean plane, which have the same transmit power value $P_{\text{TX},i}$. The density λ_i is the expected number of i-th tier BSs per unit area. The method to simulate the i-th tier PPP on a region A will be: firstly draw a Poisson distributed number with mean $\lambda_i |A|$, say n, and then place n points uniformly at random in A.

An independent collection of mobile users is considered according to some independent homogeneous point process. Without any loss of generality, it is assumed that the randomly chosen mobile user under analysis is at the origin, which is otherwise called the *typical user* in the following discussion.

The standard power-loss propagation model is used with path-loss exponent $a > 2$ and path-loss constant L_0 at the reference distance $r_0 = 1$ m. (L_0 is regarded as a constant in this analysis since the shadowing effect is excluded here.) To include the random channel effects, such as fading and shadowing, it is assumed that the typical mobile user experiences independent Rayleigh fading from the serving and interfering BSs. The noise power is additive and constant with value σ^2.

8.2.2.1 Cell association model

All BSs are assumed to be open access and the mobile users are connected to the BS providing maximum long-term received power, which can be regarded as a widely used special case of the general cell association model [18]. Different from the cell association model employed in [19], the long-term received power from an i-th tier BS located at x is defined as $P_{\text{TX},i} L_0 \|x\|^{-\alpha}$. Consequently, the resultant cell association model can average out the fluctuation caused by Rayleigh fading and eliminate unnecessary handovers when the mobile users are located not far from cell boundaries. For the user of interest, we first need to determine the selected tier number and the BS location, κ and φ respectively, by determining the BS signal received with the maximum long-term power, i.e.,

$$[\kappa, \varphi] = \arg \max_{i \in \mathbb{K}, x \in \Phi_i} [P_{\text{TX},i} L_0 \|x\|^{-\alpha}], \tag{8.1}$$

in which $\mathbb{K}$ is defined as $\{1, 2, \ldots, K\}$. For simplicity, we use P_i to replace the product of $P_{\text{TX},i}$ and L_0, i.e., $P_i = P_{\text{TX},i} L_0$ for $i \in \mathbb{K}$.

By employing this cell association model, the cell regions for a cellular network with two overlaid small cell tiers are shown in Figure 8.13. The cell boundaries for macro

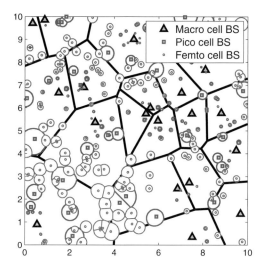

Figure 8.13 Cell association regions in a three-tier cellular network (macro cell, pico cell, and femto cell tiers), with density ratio $\lambda_3 = 4\lambda_2 = 20\lambda_1$. Transmit power levels are respectively $P_{TX,1} = 46$ dBm, $P_{TX,2} = 35$ dBm, and $P_{TX,3} = 20$ dBm.

cells, pico cells, and femto cells are marked by different shaded lines (thin gray for femto cells, thick gray for pico cells, and black for macro cells).

Some research results utilize the maximum instantaneous SIR or SINR for cell selection [19, 23, 24]; however, we employ the above cell association model as we believe it simulates practical approaches. In cellular networks, such as WCDMA and LTE [25], the mobile terminals perform cell access or cell reselection based on their received pilot channel power levels, and the measurement is performed over enough resource blocks in LTE (or time intervals in WCDMA) to average out the frequency-selective fading and fast fading, thus primarily taking large-scale propagation effects (such as path loss and shadowing effects) into account. Therefore associating mobile users to the BS with maximum instantaneous SIR or SINR (even if it is due to a peak value of a fast fluctuation) is not always a reasonable assumption compared with the model used here.

8.2.2.2 Tractable framework on coverage probability

Based upon the above-mentioned cell association model, all the interfering BSs will therefore have smaller values of long-term received power levels. Accordingly, the received downlink SINR expression is

$$\text{SINR} = \frac{P_\kappa h \|\varphi\|^{-\alpha}}{I + \sigma^2}, \tag{8.2}$$

in which h denotes the channel fading gain from the serving BS, and I is the cumulative interference from all BSs except the κ-th tier BS located at φ (the serving BS for the typical user), i.e.,

$$I = \sum_{i \in \mathbb{K}} I_i = \sum_{i \in \mathbb{K}} \sum_{x \in \Phi_i \setminus \{\varphi\}} P_i h_x \|x\|^{-\alpha}, \tag{8.3}$$

where h_x is the fading value for the interfering BS at the location of x, and I_i is defined as the interference component from the i-th tier. Due to the above-mentioned Rayleigh fading, the resultant channel power gains, h and $\{h_x : x \in \Phi_i\}$ follow the exponential distribution with the unitary mean value. In this study, no intra-cell interference is included since the orthogonal multiple access within a cell is assumed.

We use SINR$_i$ to denote the received SINR at the typical user served by the i-th tier for this uniform SCN deployment. Then the coverage probability at the typical user is $p_{c,i}(T) = \mathbb{P}[\text{SINR}_i > T]$ for the i-th tier, i.e., the probability of a target SINR T (or SINR threshold) achievable at the typical user. This coverage probability is also called the complementary cumulative distribution function (CCDF) of the received SINR.

Based upon the analytical framework provided by Jo et al. [18], the coverage probability at the i-th tier typical user can be expressed as

$$p_{c,i}(T) = \mathbb{P}[\text{SINR} > T \mid \text{typical user served by } i\text{-th tier}]$$
$$= 2\pi \lambda_{i,eq} \int_0^\infty x \exp\left(-\frac{T x^\alpha \sigma^2}{P_i}\right) \exp(-\pi x^2 \lambda_{i,eq}[1 + \rho(T, \alpha)]) dx, \quad (8.4)$$

where $\lambda_{i,eq}$ is defined as $\lambda_{i,eq} = \Sigma_j \lambda_i (P_j/P_i)^{2/\alpha}$, and the function $\rho(T, \alpha)$ is

$$\rho(T, \alpha) = T^{2/\alpha} \int_{T^{-2/\alpha}}^\infty \frac{1}{1 + u^{\alpha/2}} du. \quad (8.5)$$

It should be noted that $\exp(-T x^\alpha \sigma^2/P_i)$ and $\exp(-\pi x^2 \lambda_{i,eq} \rho(T, \alpha))$ respectively represent the impact of the noise and the interference.

8.2.2.3 Tractable framework on single-user throughput

The data throughput achievable at a single user is another important performance metric, especially when considering small cells and HetNets as a capacity solution, where this metric is determined by the cell load for the SCN deployment [26].

By assuming the mobile users are distributed according to an independent homogeneous PPP Φ_{MS} of density λ_{MS}, Singh et al. provided a method to approximate the distribution of the number of users sharing the resource with the typical user [21], which can be combined with the distribution of SINR$_i$ in (8.4) to estimate the distribution of single-user throughput.

Specifically, the number of other users sharing resource with the typical user served by the i-th tier is denoted by N_i, and the probability mass function (PMF) of N_i can be derived as

$$\mathbb{P}[N_i = n] \approx \frac{b^q}{n!} \cdot \frac{\Gamma(n+q+1)}{\Gamma(q)} \left(\frac{\lambda_{MS}}{\lambda_{i,eq}}\right)^n \left(b + \frac{\lambda_{MS}}{\lambda_{i,eq}}\right)^{-(n+q+1)}, \quad \text{for } n \in \mathbb{N}^+,$$
$$(8.6)$$

where $q = 3.61$, $b = 3.61$, and $\Gamma(x) = \int_0^\infty t^{x-1} e^{-t} dt$ is the standard gamma function, provided that the full buffer traffic model for every user and the Round-Robin scheduling with the most fairness are employed. It should be noted that the usable bandwidth is divided evenly among the in-cell users, which simulates the behavior of Round-Robin scheduling. Our method is similar to the one used in [16]. The throughput achievable at the typical user served by the i-th tier, denoted by $\mathcal{R}_i$, was derived in [21], i.e.,

$$\mathbb{P}[\mathcal{R}_i > \rho] = \sum_{n=0}^{\infty} \mathbb{P}[N_i = n] \cdot p_{c,i}\left(2^{(n+1)\rho/W} - 1\right), \qquad (8.7)$$

where the BS's bandwidth W is evenly allocated among all its associated users. It should be noted that the approximation is achieved by assuming the independence between the distribution of $SINR_i$ and N_i, where this assumption was proven to be accurate [21]. This therefore provides a technique to calculate the overall throughput of a multi-tier small cell network.

8.2.3 The impact of dense SCN deployment

Increasing the density of BSs has proven to be the only scalable method to provide the necessary throughput increase to meet the forecast demand in mobile broadband [27, 28]. When adding more small cells, mobile users are more likely to be offloaded to lightly loaded small cells, and each mobile user will compete with fewer users for a BS's wireless and backhaul resources, thus improving the end-user experience significantly [7].

When a cellular operator prepares to deploy an SCN, the first thing that needs to be determined is the benefits of the dense SCN deployment. Specifically, under a certain set of SCN parameters (e.g., the densities of macro and small cells, the transmit power settings, etc.), the throughput improvement achieved by the SCN offloading effect needs to be predicted for the business case and the network design.

Through the above-mentioned PPP-based analytical framework, the single-user downlink throughput performance for several scenarios can be obtained, in which different small cell densities are implemented (regarded as different phases of an SCN deployment). Here the small cell tier is assumed to be co-channel deployed in the regions originally with macro cell coverage. The transmit power settings for macro and small cell BSs are fixed as $P_{tx,1} = 46$ dBm and $P_{tx,2} = 30$ dBm respectively. The path-loss constant and exponent are assumed to be $L_0 = -34$ dB and $\alpha = 4$, respectively, and the thermal noise power is $\sigma^2 = -104$ dBm. The single-user throughputs demonstrated in our results are the achievable rates over the available 10 MHz bandwidth, that is, $W = 10$ MHz.

The SCN offloading effect can be directly observed from the changed cellular association regions with and without SCN deployment in Figure 8.14. By keeping the macro cell density fixed at $\lambda_1 = 1$ km^{-2} and increasing the small cell density from 0 (no SCN deployed) to 20 km^{-2} (dense SCN deployment), the area of macro-served regions has been continuously reduced. The figure also shows that the association area of the small cell highly depends on its distances to neighboring macro cell BSs, provided that the constant power setting assumption is used.

In Figure 8.15 we show the actual single-user throughput increase that can be achieved by the introduction of small cells. We show the throughput for λ_2 is 0, 1, 5, 10, 15, and 20 km^{-2} corresponding to the six scenarios in Figure 8.14, where both the analytical and simulation results are shown. We show the improvement for macro users, small cell

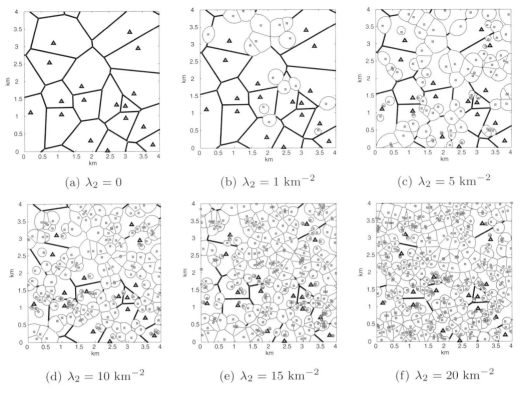

Figure 8.14 Illustration of cellular association regions with and without SCN deployed, PPP-based model. Transmit power levels are $P_{TX,1} = 46$ dB and $P_{TX,2} = 30$ dB for macro and small cells respectively. The macro cell density is kept as $\lambda_1 = 1$ km^{-2}, and the small cell density λ_2 is 0, 1, 5, 10, 15, and 20 km^{-2} respectively.

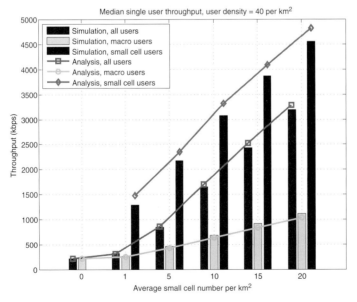

Figure 8.15 Throughput median values achievable at a randomly chosen user, over six scenarios with different small cell densities, i.e., $\lambda_2 = 0, 1, 5, 10, 15,$ and 20 km^{-2}.

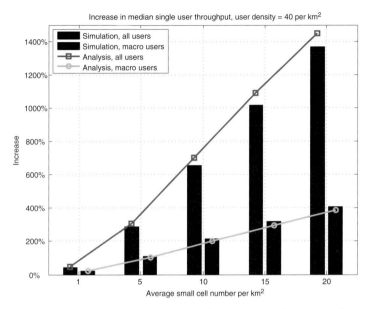

Figure 8.16 The increase of throughput median value over the macro-only scenario, for five scenarios with different small cell densities, i.e., $\lambda_2 = 1, 5, 10, 15,$ and 20 km^{-2}.

users, and overall benefit. Note that the macro users benefit due to users being offloaded onto the small cells. What can be observed is that the achievable throughput is nearly linear with the number of small cells, and this is very encouraging as it shows that such use of small cells is truly a scalable solution to address forecast demand.

In Figure 8.16 we show the **relative** improvement with the introduction of small cells. What we can see is that the gains are linear and that we obtain roughly a 1300% gain in throughput achievable at a randomly chosen user. What is also important to note is that this gain doesn't seem to flatten out, even with dense SCN deployment, and this is also encouraging as the benefits of small cells will keep providing benefits to users.

It should be noted that the throughput improvement comes from two major sources: Firstly, by offloading part of the users to the newly deployed SCN, fewer users will be left to share the macro cell resources, which makes the macro users' performance improved. This effect can be clearly observed from both Figures 8.15 and 8.16, where the median macro user throughput with $\lambda_2 = 20$ km^{-2} has been increased by more than 400% over the macro-only scenario. Secondly, some mobile terminals originally served by macro cells will be offloaded to the less congested small cells and thus benefit as the throughput of the small cells is much better than the macro cells, as demonstrated in Figure 8.15.

8.3 Summary

In the first section we have overviewed the three key system scenarios for small cell deployments, namely (a) residential, (b) enterprise, and (c) outdoor hotspots. With the residential scenario we discussed interference that can occur from the terminals onto infrastructure and also from the infrastructure onto user terminals. We specifically studied worst-case interference scenarios where femto cell-connected users are outdoors and macro cell-connected users are indoors. We showed signal-strength heat maps to highlight how indoor coverage can be significantly reduced by high-power macro cells nearby. We showed strategies to partially overcome this effect through adaptive femto cell power management. We also highlighted how macro cell-connected users in a home can cause radio jamming onto a femto cell and thus significantly compromise its performance, and we demonstrated with results how this can be overcome through the use of uplink attenuation techniques in the femto cell, significantly improving uplink indoor coverage and reducing the deadzone. We considered enterprise deployments and discussed the different objectives and requirements of this scenario. We discussed how multiple small cells create more challenges to minimize interference and the overall best improvements (compared to macro-only) are achieved when the enterprise building is on the cell edge. We highlight the need to achieve good indoor coverage without providing significant leakage of pilot signal power out of the building. Finally, we introduced outdoor deployments but left the details to the second section of this chapter.

In the second section we discussed the analytical model, which is used to characterize the new small cell network (or HetNet) topology. We discussed the random spatial model assumptions, the modeling technique, and how performance can be determined. This analytical tool can be used for a whole range of scenarios to analyze coverage probability and throughput. We considered dense small cell networks and defined six scenarios with different densities of small cells, which represent various stages of network enhancement that is expected over the coming years as small cell densification of the wireless infrastructure occurs. Using these six scenarios we determined the median single-user throughput that can be achieved. We found that the overall increase in throughput grows at a linear rate with the number of small cells. For example, with averagely 20 small cells per macro cell (in a 1 km^2 area) we could achieve a median single-user throughput of just over 3 Mbps, this corresponds to nearly a 14 times increase over a macro-only scenario. These are remarkable improvements and highlight that small cells can provide both a scalable and efficient way to meet the future demands of mobile broadband.

References

[1] S. Cherry, "Edholm's law of bandwidth," *IEEE Spectrum*, 41(7), 58–60, July 2004.
[2] "Cisco visual networking index: Global mobile data traffic forecast update, 2012–2017," *Cisco Systems, Inc. White Paper*, Feb. 2013.
[3] M. Yavuz, F. Meshkati, S. Nanda, *et al.* "Interference management and performance analysis of UMTS/HSPA+ femtocells," *IEEE Commun. Mag.*, 47(9), 102–109, Sept. 2009.
[4] Femto. Forum, "Interference management in UMTS femtocells," *Femto Forum White Paper*, Dec. 2008.
[5] B. Baszczyszyn, M. K. Karray, and H. P. Keeler, "Using Poisson processes to model lattice cellular networks," *Proc. 32nd IEEE Int'l Conf. on Computer Communications (INFOCOM13)*, Turin, Italy, 2013, pp. 773–781.

[6] S. R. Saunders, S. Carlaw, A. Giustina, R. R. Bhat, V. S. Rao, and R. Siegberg, *Femtocells: Opportunities and Challenges for Business and Technology*. New York, NY: John Wiley & Sons Ltd., 2009.

[7] J. G. Andrews, H. Claussen, M. Dohler, S. Rangan, and M. C. Reed, "Femtocells: Past, present, and future," *IEEE J. Select. Areas Commun.*, 30(3), 497–508, Apr. 2012.

[8] V. Chandrasekhar, J. G. Andrews, and A. Gatherer, "Femtocell networks: A survey," *IEEE Commun. Mag.*, 46(9), 59–67, Sept. 2008.

[9] M. Zhao, M. Ruan, R. Chan, *et al.*, "Hardware realization of UMTS femtocell modem with uplink interference cancellation," in *Communications and Information Systems Conference (MilCIS), 2012 Military*, 2012, pp. 1–5.

[10] A. Goldsmith, *Wireless Communications*. New York, NY: Cambridge University Press, 2005.

[11] J. G. Andrews, F. Baccelli, and R. K. Ganti, "A tractable approach to coverage and rate in cellular networks," *IEEE Trans. Commun.*, 59(11), 3122–3134, Nov. 2011.

[12] D. Stoyan, W. S. Kendall, and J. Mecke, *Stochastic Geometry and its Applications*, 2nd edn. New York, NY: John Wiley & Sons Ltd., 1995.

[13] R. K. Ganti, F. Baccelli, and J. G. Andrews, "A new way of computing rate in cellular networks," *Proc. IEEE International Conf. on Commun. (ICC'11)*, Kyoto, Japan, June 2011, pp. 1–5.

[14] C. S. Chen, V. M. Nguyen, and L. Thomas, "On small cell network deployment: A comparative study of random and grid topologies," *Proc. IEEE 76th Vehic. Tech. Conf. (VTC'12-Fall)*, Québec City, Canada, Sept. 2012, pp. 1–5.

[15] M. Haenggi and R. K. Ganti, *Interference in Large Wireless Networks*. Hanover, MA: Now Publishers Inc., 2009.

[16] A. Busson, "An overview of results on ad hoc network performances using spatial model," *Proc. IEEE Symp. on Computers and Commun. (ISCC'09)*, Sousse, Tunisia, July 2009, pp. 36–41.

[17] P. Madhusudhanan, J. G. Restrepo, Y. Liu, T. X. Brown, and K. R. Baker, "Multi-tier network performance analysis using a shotgun cellular system," *Proc. IEEE Global Commun. Conf. (GLOBECOM'11)*, Houston, USA, Dec. 2011, pp. 1–6.

[18] H.-S. Jo, Y. J. Sang, P. Xia, and J. G. Andrews, "Heterogeneous cellular networks with flexible cell association: A comprehensive downlink SINR analysis," *IEEE Trans. Wireless Commun.*, 11(10), 2012, 3484–3495.

[19] H. S. Dhillon, R. K. Ganti, F. Baccelli, and J. G. Andrews, "Modeling and analysis of K-tier downlink heterogeneous cellular networks," *IEEE J. Select. Areas Commun.*, 30(3), Apr. 2012, 550–560,

[20] C. C. Wang, T. Q. S. Quek, and M. Kountouris, "Throughput optimization, spectrum allocation, and access control in two-tier femtocell networks," *IEEE J. Select. Areas Commun.*, 30(3), 561–574, Apr. 2012.

[21] S. Singh, H. S. Dhillon, and J. G. Andrews. "Offloading in heterogeneous networks: modeling, analysis and design insights". http://arxiv.org/abs/1208.1977.

[22] Y. Zhong and W. Zhang. "Multi-channel hybrid access femtocells: A stochastic geometric analysis". http://arxiv.org/abs/1108.1257.

[23] H. S. Dhillon, R. K. Ganti, and J. G. Andrews, "A tractable framework for coverage and outage in heterogeneous cellular networks," *Proc. Information Theory and Applications Workshop (ITA 11)*, San Diego, USA, Feb. 2011, pp. 1–6.

[24] H. S. Dhillon, R. K. Ganti, F. Baccelli, and J. G. Andrews, "Coverage and ergodic rate in K-tier downlink heterogeneous cellular networks," *Proc. 49th Annual Allerton Conference*

on *Communication, Control, and Computing (Allerton'11)*, Monticello, USA, Sept. 2011, pp. 1–6.
[25] H. Holma and A. Toskala, *LTE for UMTS: OFDMA and SC-FDMA Based Radio Access*. New York, NY: John Wiley & Sons Ltd., 2009.
[26] J. G. Andrews, "Seven ways that HetNets are a cellular paradigm shift," *IEEE Commun. Mag.*, 51(3), 136–144, 2013.
[27] S. Landstrom, A. Furuskar, K. Johansson, L. Falconetti, and F. Kronestedt, "Heterogeneous networks: increasing cellular capacity," *Ericsson Review*, no. 1, pp. 1–6, 2011.
[28] M.-S. Alouini and A. J. Goldsmith, "Area spectral efficiency of cellular mobile radio systems," *IEEE Trans. Veh. Technol.*, 48(4), 1047–1066, 1999.

9 Temporary cognitive small cell networks for rapid and emergency deployments

Akram Al-Hourani, Sithamparanathan Kandeepan, and Senthuran Arunthavanathan

This chapter introduces the concept of temporary cognitive small cell networks (TCSCN) as a supplement infrastructure to LTE-Advanced macro networks, and examines how the cognitive capabilities can enable the rapid and temporary nature of such deployments. Temporary networks are suitable for disaster-recovery scenarios where the nominal macro network is severely affected or completely paralyzed. In addition to that, such temporary networks can address the sudden increase in wireless traffic in certain geographic areas due to public events. The approach in realizing the cognitive capabilities is achieved by exploiting the latest LTE-Advanced HetNet features, as well as by presenting novel techniques for intelligently mitigating interference between the macro network base stations and the introduced temporary infrastructure. Simulation results are presented in order to show the enhancement of the wireless service when such temporary networks are deployed together with the proposed cognitive capabilities. At the end of this chapter an overview will be provided about open research directions that are fundamental for further possible realization of temporary cognitive small cell networks.

9.1 Introduction

The recent developments in broadband wireless networks have added an unprecedented level of reliability and bandwidth efficiency to the cellular communications, leading to the emergence of new user applications that would not be possible without such enhancements. In fact, commercial wireless networks are nowadays an irreplaceable gear in modern economies and societies, not only used for voice and video communications but also as a carrier for many mission-critical businesses and applications, such as wireless video monitoring, transportation signals, logistics tracking, and automatic power meter reading.

With this increasing dependency on telecommunication networks, the total failure of the economic system and public services would be massive if such networks are disrupted by means of a natural disaster such as a flood, earthquake, or tsunami, or by man-made attack. Among critical users of telecommunication networks are public safety agencies that are progressively depending on broadband telecommunication, making the need for providing a reliable broadband wireless service one of the top priorities for both commercial and public safety network operators.

Temporary small cells are seen as one of the viable solutions assisting network operators in tackling a sudden disruption of the wireless service, or a sudden increase in service demand generated by public events. These events include the potential move of the crowd, generating a heavy demand on cellular network capacity. From this perspective, temporary small cells can provide an economical solution in many cases where it is not feasible to provide a permanent redundant network capacity for such events.

The *temporary* nature of these small cells will allow a short-term supplementary service that can assist the macro network in areas either with disturbed coverage or areas suffering an overwhelming traffic demand, while the small coverage nature of such cells can substantially decrease the realization expenses, due to the reduction in implementation cost when using low-power radio frequency (RF) transmitters.

Several challenges arise from the *rapid* and *temporary* nature of the proposed networks. These challenges are due to the lack of network layout design and planning. Since most of the configuration parameters should be rapidly self-learned on the spot by the small cell access point including: (i) intercell interference mitigation with both the macro network and other small cell eNBs, (ii) optimum power level control, and (iii) spectrum sensing and white space discovery for efficient opportunistic spectrum access, these self-learning requirements can only be realized using cognitive radios and cognitive techniques that have the capability of spectrum analysis and interference mitigation.

9.2 The concept of temporary cognitive small cell networks

In this chapter, we distinguish between the temporary cognitive small cell network denoted as (TCSCN), and the temporary cognitive access point denoted as (TCAP). TCAP is the main enabler for the TCSCN network, acting as a power-limited base station and a capacity-limited call-switch. Accordingly, each TCAP can form an independent cell or *cluster* that is capable of routing intra-cluster traffic locally, without the dependency on the macro network EPC (in LTE the core network is called *Evolved Packet Core* or EPC). The *temporary* aspect of the TCSCN implicitly mandates a rapid deployment capability, aiming to introduce service gains swiftly for a short term and for a small geographic area between (30 m to 1,000 m cell-radius). These service gains can be listed as follows:

- indoor coverage extension
- cell-edge coverage extension
- serving high-bandwidth locally terminated traffic
- relieving congested hotspots
- supporting disaster-recovery networks.

The self-learning capabilities of the TCSCN should include the self-organization, self-configuration, and self-management, with minimal intervention from the macro network. These self-learning capabilities are required because of the unpredictable deployment locations of TCAPs, since TCAP deployments are based on sudden local needs as listed previously. Self-learning features need advanced processing and intelligence capabilities to be embedded in the TCAP. These capabilities can be facilitated using cognitive radios (CR) [1] that continuously monitor and assess the geo-localized radio spectrum in order to construct what is called the *radio environment map* (REM), with the objective of mitigating interference on/from the macro network, as will be elaborated on later in Section 9.6, while the concept of REM will be briefly elaborated in Section 9.6.3.

The remainder of this section will further illustrate the concept and the architecture of the TCAP and its types, while the rest of the chapter is organized as the following: in Section 9.3 an overview of the network model is provided, illustrating where the TCSCN fits and in which scenarios the TCSCN could be enabled. In Section 9.6, an overview of cognitive radios is provided, elaborating on how the CRs can construct the radio environment map by periodic spectrum sensing. Section 9.4 elaborates on the high-level procedure steps of enabling the TCSCN and integrating it with the parent network, focusing on the adoption of interference mitigation techniques, proposed in Section 9.5 and then simulated in Section 9.7, showing the expected enhancements in the network coverage. Finally Section 9.8 provides the conclusion remarks.

9.2.1 Access points variants

The rapid nature of deploying TCAP demands easy-to-deploy devices, which implies considerable limitations imposed on the size, weight, transmission power, backhauling options, and operating time. In the light of the previous constraints, the possible realization variants of TCAPs that are proposed in this chapter are the following:

- Hand-held or back-mounted TCAP units with limited transmission power and battery lifetime. This type can greatly facilitate the operational work of public-safety personnel in tactical field operations.
- Wheeled-trailer TCAP for larger coverage areas: this type is enabled by auto-deployment capabilities and power support units, as depicted in Figure 9.1. Such capabilities include auto-tracking backhaul antenna, generator set, and solar panels [2]. This type can be used for supplementing both terrestrial networks and aerial networks (where base stations are airborne).

Temporary access points can operate in two different modes according to their location with respect to users and with respect to the parent networks; these modes are listed below:

- **primary mode**, where TCAPs can perform traffic switching and routing as explained earlier
- **secondary mode**, where TCAPs act as decode-and-forward (DF) relay nodes [3], mainly for coverage extension of either a primary mode TCAP or a macro eNB.

DF relay nodes are not standardized under LTE Release 10; however, a general description is provided in 3GPP document [4], under the name "Type-2" Relay Nodes. Figure 9.2 depicts the general concept of the primary and the secondary mode TCAPs.

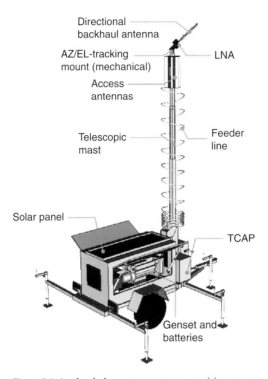

Figure 9.1 A wheeled-type temporary cognitive access point.

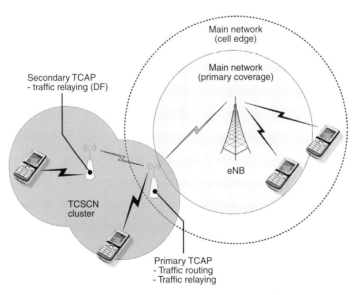

Figure 9.2 Primary and secondary types of temporary cognitive access points.

9.2.2 Access point architecture

TCAPs are designed to handle light local traffic efficiently by performing local packet routing and prioritization. In order to enable these complex tasks, TCAPs need to be fitted with a miniature version of an EPC [5] capable of handling low to medium data throughput rates. On the other hand, a TCAP shall be capable of routing external traffic from/toward its attached users, via a backhaul link. The most feasible realization of backhaul links is by using wireless connections. Wired solutions would not be practical and might impose considerable delays in TCAPs deployment, which contradicts the rapidity objective of TCSCN. Wireless backhauling options can be classified into the following categories according to the embraced technology and the communication method:

- **Inband backhauling**, where both the access link (i.e., between TCAP and UE) and the backhaul link (i.e., between TCAP and MeNB) are operating within the same spectrum band. In this method, interference between the two links must be avoided, otherwise the received signal in the backhaul will be entirely interfered. This interference is due to the large power difference between the access link transmission-power and the backhaul link received-power, which can easily exceed 100 dB. From this perspective, both the access and the backhaul links cannot operate simultaneously, and certain transmission arrangements shall be embraced. In fact, 3GPP has tackled this issue in LTE Release 10 through the standardization of the relay nodes (Type 1: inband half-duplex). Readers are referred to 3GPP standard [6] for further details
- **Outband backhauling**, where the access link and the backhaul link are operating in two different spectrum bands, either *non-contiguous* bands, or *contiguous, with sufficient guard band in between*. Outband backhauling could be realized using variant technologies, depending on the TCAP purpose and its target deployment period. These technologies are summarized below:
    (i) **Point-to-point (PtP) microwave link with line-of-sight (LoS) conditions** can usually achieve very high bandwidth. However, having a clear LoS is a questionable issue, since TCAP deployment can take place in random locations that might not favor a backhaul LoS.
    (ii) **Point-to-multi-point (PtMP) non-line-of-sight microwave links (NLoS)** might be considered as one of the promising solutions, as some recent studies suggest [7]; however, these NLoS links require precise pointing, from the TCAP side, that can be achieved using an automated azimuth elevation pointing mechanism.
    (iii) **Outband LTE back hauling** is the most feasible solution for hand-held or back-held TCAPs, since it can overcome the pointing issue. This method is having no or very minor impact on 3GPP LTE standardization as per 3GPP technical report [4].

A high-level block diagram of TCAP is illustrated in Figure 9.3 showing the distinct logical functional entities of (i) the access entity and (ii) the backhaul entity. The link between cognitive radio engine and the self-interference coordination engine is highly important when operating using the inband backhauling method. The mini EPC is

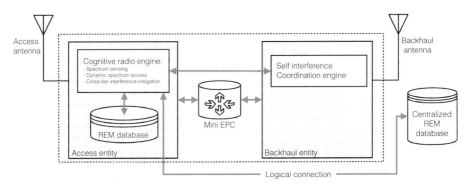

Figure 9.3 Temporary cognitive access point block diagram.

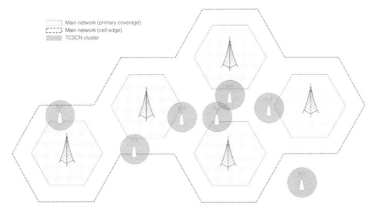

Figure 9.4 Temporary cognitive small cells supplementing a terrestrial main network.

responsible for routing intra-cell traffic, which will also allow the TCAP to function independently from the parent network in case of the total lack of macro network coverage or the backhauling link.

9.3 Network model

This section presents the adopted network model in which the TCSCN can be deployed. There are two distinct types of networks proposed here where TCSCN can substantially supplement the primary coverage. These network types are listed below.

- The first network type is based on conventional macro eNBs and is called here the "terrestrial main network," designated as TMN, providing the coverage for the intended geographical area (refer to Figure 9.4). Usually such macro eNBs are roof-top or tower mounted in a certain geometrical distribution, ideally approximated to a hexagon grid of one eNB at the center of the cell/hexagon. Some modern studies suggest

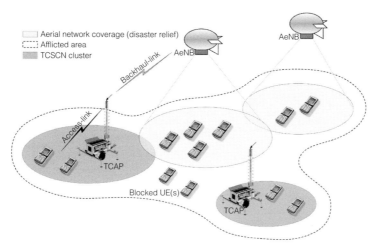

Figure 9.5 Temporary cognitive small cells supplementing an aerial rapid network.

representing multi-tier networks using stochastic geometry, such representation can lead to tractable estimations and formulas for SINR levels [8] and tier-association probability [9]. In this network type, TCSCN fills coverage gaps in poor serviced areas that might exist in both indoor and outdoor locations.

- The second type of network, is the "aerial rapid network" (ARN) (refer to Figure 9.5), where the primary coverage is provided by eNBs lifted onboard low-altitude platform aerostats such as balloons and quadcopters, or any other quasi-stationary aerial platform. These types of networks are mainly planned to act as a robust recovery for TMN networks in case of a natural disaster afflicting a certain geographic area. A good example of such aerial rapid network development efforts is the ongoing European Union-funded project "ABSOLUTE" [10] (aerial base stations with opportunistic links for unexpected and temporary events) focusing on low altitude platforms (LAP) that are considered as one of the solutions for the rapid emergency networks. In ABSOLUTE, two tightly interconnected network segments are proposed: an air segment and a terrestrial segment. The former consists of an innovative helium balloon-kite structure carrying LTE aerial base stations, while the ground segment consists of portable land rapid deployment units providing satellite backhauling connectivity (not shown in the figure). It is important to note that TCSCN is essential in complementing such aerial coverage.

The coexistence of both TMN and ARN network types can certainly happen. For example, in the wake of a disaster the ARN will be gradually implemented during which the TCSCN can play the vital role of helping the interconnection of users in dark spot areas. Figure 9.6 depicts the concept of the coexistence of both (A) the TMN network and (B) the ARN network, in addition to the underlying (C) TCSCN. In this chapter, both TMN and ARN networks will be referred to as the macro network.

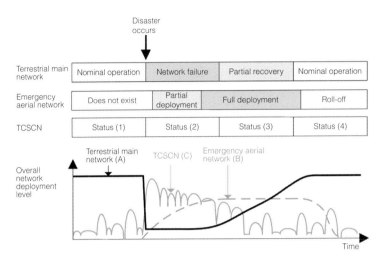

Figure 9.6 The operational states of the TCSCN.

Based on the above-mentioned network types, TCSCN has three main deployment scenarios:

- deployment of TCSCN as a supplementary network for standalone ARN (refer to Figure 9.5)
- deployment of TCSCN as a supplementary network for standalone TMN (as in Figure 9.4)
- deployment of TCSCN as a supplementary network for coexisting ARN and TMN
- partial standalone deployment of TCSCN in poorly serviced or unserviced areas.

9.4 Deployment process

The decision to initiate (enable) a certain TCSCN cluster is taken by the TCAP and depends on several triggers. One of them is the automatic service threshold trigger, which is initiated due to the lack of proper service level for a certain predefined time, where service availability can be defined as the following:

$$\text{Service Availability} = \begin{cases} 1 & \text{SINR}_{UE} > \text{SINR}_{th} \\ 0 & \text{SINR}_{UE} \leq \text{SINR}_{th} \end{cases} \quad (9.1)$$

where SINR_{th} is given by:

$$\text{SINR}_{th} = \text{SINR}_{min} + \rho \quad (9.2)$$

and SINR_{min} is the required SINR for decoding the minimum modulation and coding scheme (MSC) in LTE, while ρ is a tuning parameter that depends on the nominal service level for attaining the target application over the LTE network. For example: a certain (kbps) throughput to attain a minimum quality of video streaming. Other triggers that can initiate the TCSCN cluster are listed below:

- specific local demand for high bandwidth group communication (automatic trigger), e.g., streaming video locally to the firefighters' team leader from the dispatched team members
- user-initiated TCSCN cluster (manual trigger)
- search and rescue operation that expects victims to be outside the parent network service cell radius (manual trigger).

Figure 9.6 shows the various operational stages of a public-safety LTE network when affected by a natural disaster, and represents the generic case when TCSCN exists as a complementary network for both terrestrial and aerial coverage. TCSCN will have the below operational status accordingly:

- **status (1)** TCSCN clusters could be initiated anytime according to the previously mentioned *triggers*
- **status (2)** TCSCN clusters will fill the considerable gaps left by ARN and TMN networks and are essential at this point
- **status (3)** TCSCN clusters will still fill the coverage gaps left by ARN network and the gradually recovering TMN network
- **status (4)** TCSCN will go back to normal operation; similar to status (1).

If a UE inside the affected areas detects an established TCSCN cluster and opts to join this cluster, it should synchronize with the primary synchronization signal (PSS) of the corresponding TCAP, after which the cell attachment continues through the normal random access procedure.

On the other hand, terminating (disabling) the TCSCN cluster could be triggered by several conditions as listed below:

- parent coverage is restored and is above the $SINR_{th}$ level for a certain predefined period of time (automatic trigger)
- the TCAP is not capable of maintaining the TCSCN due to battery drain (automatic trigger)
- user decision (manual trigger)
- parent network decision (ARN or TMN), e.g., severe interference caused by a TCAP (automatic trigger).

9.5 Cognitive interference mitigation techniques

Whether the TCSCN is intended to be used under an ARN or under a TMN, some special interference mitigation techniques shall be taken into consideration for such deployments, in order to avoid the disruption of the macro network coverage and service. This might happen when both the main and the TCSCN networks are operating on the

Table 9.1 Spectrum access methods.

Scheme	Implementation techniques	Access types
Inband	LTE interference mitigation methods (frequency domain, time domain).	Semi-static
	Spectrum sensing for the macro LTE network.	Dynamic
Outband	Dedicated component carrier (only in LTE-Advanced).	Static
	Spectrum sensing for other systems (e.g., TV white spaces).	Dynamic

same spectrum band (inband scheme). In this scheme, two different approaches can be implemented. (i) The first approach is achieved by cross-tier partial or full coordination; the key here is that both the TCSCN and the main network are aware of the existence of each other. (ii) The second approach is when the macro network is not aware of the existence of the TCSCN, and hence the latter shall perform opportunistic spectrum access by constructing a radio environment map (REM) based on spectrum sensing as it will be illustrated in Section 9.6.

On the other hand, the situation becomes much simpler when the TCSCN is operating on a separate component carrier (outband scheme), that is because both cross-tier interference and self-interference can be avoided. However, an outband scheme can only be achieved by either (i) dedicating a specific component carrier for the TCSCN network by the telecom operator and making sure that it is vacant at all times. This scheme is only applicable in LTE-Advanced Release 10 that allows up to five component carriers to be utilized by a single system. The second scheme (ii) is achieved by dynamically accessing *other systems*' spectrum bands in a cognitive manner, i.e., by performing continuous spectrum sensing and measurements. Here, *other systems* might refer to any wireless communication system other than LTE; a famous example is when using the white spaces in the underutilized TV band. In Table 9.1, the possible different spectrum-access scenarios are illustrated.

The rest of this chapter illustrates the interference mechanisms that are considered as strong candidates for implementation in TCSCN, bearing in mind that the target is to mitigate cross-tier interference, i.e., the interference between the macro network and TCSCN, assuming that the co-tier interference is minimal due to the clustering nature of TCSCN network. This chapter is not targeting the self-interference caused by the backhauling on LTE, since this particular issue has been addressed and standardized by 3GPP [6].

The new inter-cell interference coordination techniques presented in Release 10, overtook the implicit inefficiency in previous releases, especially when dealing with new elements introduced as part of the heterogeneous network (HetNet) such as relay nodes, femto cells, and remote radio heads. These techniques, as seen from the time-spectral resources space, can be categorized into two distinct groups: a *time domain group* and a *frequency domain group*. The former concentrates on controlling the transmission of the subframes and symbols of the overlaid eNBs (such as femto cells and pico cells), while the latter focuses mainly on fractional frequency reuse (FFR) methods [11]. The commonality between all techniques is that they lie inside the

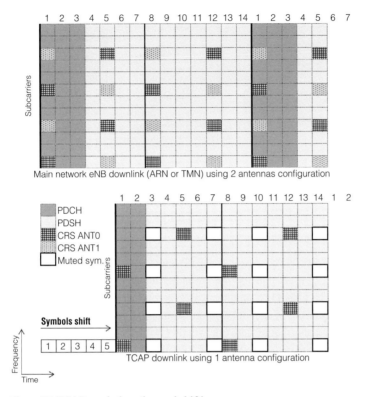

Figure 9.7 TCAP symbol muting and shifting.

constraint triangle of *bandwidth efficiency, dynamicity,* and *implementation complexity*. For instance, one technique might yield a high spectrum efficiency but at the same time will require a massive exchange of signaling information between the different network nodes.

Considering the rapid and temporary nature of TCSCN, two selected semi-static solutions are explained below for cross-tier interference mitigation: the time-domain solution and the frequency-domain solution.

9.5.1 Time-domain solution

Time-domain interference mitigation requires reliable signaling between the parent network and the deployed TCAPs. One of the practical interference mitigation approaches is using OFDM *symbol muting* with *shift* [12]. In this method subframes of the TCAP are shifted forward by five symbols with respect to the parent eNB subframes (refer to Figure 9.7). Shifting will mitigate the interference caused by the MeNB's PDCCH on TCAP's PDCCH, while muting will mitigate the interference by TCAPs on MeNB's cell-specific reference signal (CRS) that is quite vital for UEs to perform radio-link monitoring measurements [13], in particular, to carry out the RS–SINR calculations (received signal–signal to interference and noise ratio). Low values of RS–SINR will

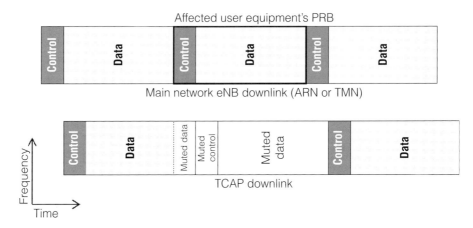

Figure 9.8 TCAP muted subframe.

result in the UE falsely considering the radio link as unusable and terminating the connection with the eNB [12]; another reason to protect the reference signals is that they play an essential role in channel estimation and OFDM demodulation within the UE.

Furthermore, the macro network whether a TMN or ARN can report to a certain TCAP if any of the cell-edge UEs has a severe interference caused by its downlink. In this case the TCAP will configure the interfering subframes as *almost blank subframes* (ABSFs) [14] in order to further mitigate the interference on both PDCCH and PDSCH (see Figure 9.8).

9.5.2 Frequency-domain solution

Various frequency-domain techniques, such as dynamic power control, dynamic fractional frequency reuse, and formation of groups [11], are implementable in OFDM systems. However, for the specific TCSCN application we are discussing the static fractional frequency reuse (FFR) since it requires less cross-tier coordination, i.e., less coordination between the macro network and the TCSCN, and it can even be used in the case of the total absence of cross-tier signaling, in contrast to the dynamic FFR methods (DFFR). DFFR could be smoothly used in the current LTE commercial deployments, since eNBs have an excellent mutual interface X2 allowing ICIC/eICIC information exchange, a privilege that might not exist in the case of the TCAP backhaul link.

Soft FFR, on the other hand, is a static type FFR that has high spectrum efficiency. Its essence is to divide the available spectrum into N sub-bands, where N is the number of cells in a frequency reuse cluster. Figure 9.9 illustrates the proposed reuse pattern, where the available subcarriers are divided into three groups (designated as G1, G2, and G3). The MeNB's cell center can utilize all three groups, while at the cell edge one of the subcarrier groups is limited in terms of power, that is in order to allow TCAPs to operate on this *reduced* carrier. For example, looking at the same figure, in the left

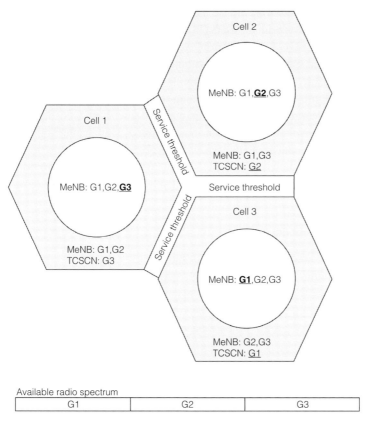

Figure 9.9 Fractional frequency reuse for TCSCN. The underlined frequency band is shared between the macro network and the TCSCN.

cell, MeNB will transmit a full power for G1 and G2 and a slightly reduced power for G3. Accordingly, TCAPs will be able to use G3 at the cell edge. However, if a TCSCN cluster needs to initiate within the central region of the cell then the corresponding TCAP needs to utilize the time domain interference mitigation techniques. On the other hand, in outer cell areas where the service threshold condition is reached (i.e., $SINR_{UE}f_j$ $\leq SINR_{th}$), the TCAP will have more freedom in accessing the spectrum, since the cross-tier interference probability is much lower.

9.6 Spectrum sensing and radio environment learning

Performing cross-tier coordinated interference mitigation might not always be possible, especially in cases such as the aftermath of a natural disaster, or when coordination between the main network and TCSCN is not pre-planned or pre-configured. Accordingly, TCAPs will resort to their spectrum-sensing capabilities [15, 16] in order to learn the surrounding radio environment, which will allow the utilization of the main network resources in an opportunistic manner. In this scheme the macro network eNBs (MeNB) and the associated MUEs are always assigned higher spectrum access priority and therefore are considered as primary users (PU), while TCSCN and its users are considered as secondary users (SU) of the spectrum. According to this consideration, TCAPs will be able to access the spectral resources without interfering with the PU system.

This scheme can only be achieved if the TCAPs are periodically performing spectrum sensing. There are many spectrum-sensing techniques available, such as energy based sensing, cyclo-stationary based sensing, matched filter based sensing, waveform based sensing, cooperative sensing, and distributed sensing. This chapter mainly focuses on the energy detection model due to its ease of implementation, and because it is considered the least complex form in terms of computational processing. Therefore this methodology is ideal for TCAPs where the power is limited.

9.6.1 Energy-detector model

In order to evaluate the energy-detector model, we start with the form of the received signal in a random wireless channel, i.e., a channel subjected to small-scale fading and shadowing models. The form is shown in the following equation:

$$r(t) = \frac{1}{\sqrt{L(d)}} h(t)s(t) + w(t) \tag{9.3}$$

where $r(t)$ is the received signal, $s(t)$ is the transmitted signal, $h(t)$ represents the random variations in wireless channel gain, $w(t)$ is the additive white Gaussian noise (AWGN), and $L(d)$ is the path loss between the transmitting node and the receiving node at a distance d. The total received energy of the wireless signal is the input of the test statistic ϵ for the energy-detector model. The test statistic used by the energy detector to perform the decision of the presence of the PU signal is represented in the time domain as in the following:

$$\epsilon = \int_0^t |r(t)|^2 dt \tag{9.4}$$

where ϵ is the energy of the received signal. The test statistic ϵ or the energy of the received signal is then represented in a discrete domain as shown below:

$$\epsilon = \sum_{n=0}^{N_s-1} |r(n)|^2 \tag{9.5}$$

where N_s is the number of samples per estimate. If T is the signal duration and F_s is the sampling frequency, therefore $N_s = T \times F_s$ is known as the time-bandwidth product. The mean signal-to-noise ratio of the received signal in terms of power ratio is defined as $\rho = P_s/P_n = 1/N_s(E/N_0)$, where P_s is the mean power of the signal term $h(t)s(t)/L(d)$, P_n is the power of the AWGN term $w(t)$ in (9.3), E is the mean energy, and N_0 is the noise power spectral density.

9.6.1.1 Detection process of the energy detector

The detection process is dependent on the energy of the received signal or the test statistic, and is based upon two hypotheses as shown in the following [17]:

$$\begin{aligned} H_0 &: r(t) = w(t) \\ H_1 &: r(t) = \frac{1}{\sqrt{L(d)}} h(t) s(t) + w(t) \end{aligned} \tag{9.6}$$

As one can observe, the hypothesis H_0 defines the case when the received signal is purely the AWGN term $w(t)$ in (9.3) only, and characterizes the absence of the PU signal. The second case defined by the hypothesis H_1 characterizes the presence of the PU signal with the addition of the noise floor. The decision criterion on whether the PU signal is present or not is based on the level of the detected energy. So that only if the detected energy is above a certain threshold λ, does the system consider the PU as present. It is important to note here that finding the optimal detection threshold λ for a given detection criterion is a challenging task, especially with fading channels.

9.6.1.2 Theoretical detection performance

The overall detection performance of the energy detector is quantified by the two probabilities: *probability of detection P_D and probability of false alarm P_F*. The probability of detection P_d is defined as:

$$P_D = \mathbf{P}(\epsilon > \lambda | H_1) \tag{9.7}$$

and the miss-detection probability is given by $P_M = 1 - P_D$. The probability of false alarm P_F is defined as:

$$P_F = \mathbf{P}(\epsilon > \lambda | H_0) \tag{9.8}$$

We assume here that the TCSCN utilizes channels that are subjected to Rician fading. Thus we present theoretical closed-form expressions for Rician fading along with simulated results. The Rician channel follows a Rician distribution with the instantaneous energy per noise spectral density ratio, and is well documented in the literature [1, 18].

The energy-detection performance varies when the received signal component has a time-varying envelope. The mean probability of detection $\bar{P}_D$ for fading channels is obtained by averaging P_D in (9.7) over the distribution of the instantaneous energy per noise spectral density ratio denoted as γ. Mathematically, this averaging is performed as per the following:

$$\bar{P}_D = \int_0^\infty P_D(\gamma) f_\Gamma(\gamma) d\gamma \tag{9.9}$$

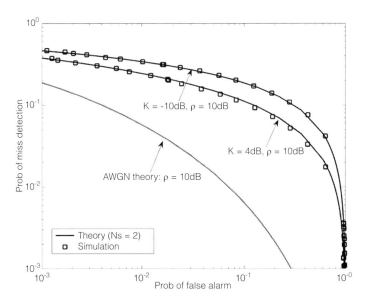

Figure 9.10 Complementary ROC curves for the Rician-envelope fading channel, with $N_s = 2$, comparing theory vs. simulations.

The above integral has been solved for the Rician channel in existing literature [19]. The mean detection probability of the Rician channel is given for the special case of $N_s = 2$ as in existing literature [1, 18, 19],

$$\bar{P}_D = Q\left(\sqrt{\frac{2K\bar{\gamma}}{1+K+\bar{\gamma}}}, \sqrt{\frac{\lambda(1+K)}{1+K+\bar{\gamma}}}\right) \quad (9.10)$$

where Q represents the Q-function. The false-alarm probability in the Rician channel is expressed as in the following:

$$P_F = \Gamma\left(\frac{N_s}{2}, \frac{\lambda}{2}\right) \quad (9.11)$$

where $\Gamma(.,.)$ is the normalized incomplete upper gamma function.

9.6.1.3 Simulation results of the spectrum sending

The detection performance of the energy detector is analyzed for the utilized Rician channel by means of plots termed as receiver operating characteristic (ROC) curves, which are plots of the detection probability P_D vs. the false-alarm probability P_F. The detection performance can alternatively be analyzed by the complementary-ROC (CROC) curve, which is a plot of the miss-detection probability P_M vs. the false-alarm probability P_F. Theoretical results are provided for the Rician channel with limitations for the closed-form expression. The Rician-envelope fading channel provides theoretical results from the closed-form expression up to values of $N_s = 2$ only. Figure 9.10 displays the detection performance in a Rician channel subject to lognormal shadowing where the limitation of the time-bandwidth product $N_s = 2$ is applied as in the theoretical literature.

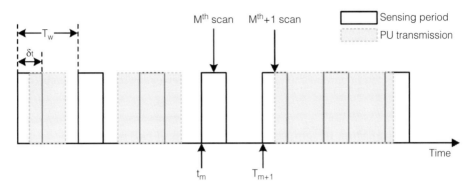

Figure 9.11 Generic periodic sensing model.

Therefore one observes that the simulated and theoretical models tally together for the time-bandwidth product of $N_s = 2$. This denotes the fact that the simulated model holds true for the theoretical expression, thus leading to further expansion of the simulation for the generic case for values of $N_s > 2$.

9.6.2 Periodic sensing of the radio environment

The radio environment map (REM) is constantly updated by means of periodic cognitive sensing [20] performed by the TCAP. The function of the periodic sensing allows the TCFCN to detect any incoming PU transmissions and thus preventing any interference with the PU, by muting the SU transmissions. The periodic sensing is required for updating of the REM, especially in stages when the PU system is recovering from a natural disaster.

Figure 9.11 displays a generic model for periodic sensing. The boxes represent the sensing durations of the CR engine in the TCAP, which is done in a periodic manner T_w. The sensing durations of the TCAP to detect ongoing or incoming PU system transmissions are denoted as δ_t. During δ_t, the TCAP performs energy-based sensing to learn about the environment and build the REM or update the REM periodically. The light-shaded boxes are referred to as the PU system transmissions.

9.6.3 Building the radio environment map

The REM is an integrated database based on geo-localized measurements that characterizes the radio environment for CRs. The database is either static or dynamic thus requiring spatio-temporal processing and interpolation. The REM can be altered to specific needs such as limited static data or dynamic data of large volumes. The REM performs the overall task of characterizing and integrating the radio environment data, which consists of the following: geographical information, available services and networks, regulation and policy, activity profile of radio devices, and experience of the CR system. As shown in Figure 9.3, a TCAP consists of a sensing module that is part of a cognitive engine (CE) and also consists of an REM database. The CE first obtains situation awareness by querying the REM for any existing data in the centralized network REM database, and then determines the utility function that best fits its current situation [21, 22]. The dual communications between the local REM database and the CE from one side, and the centralized network REM database and CE from the other side, provide an improved method of optimization and accurately updated knowledge of the radio environment.

9.7 Simulation of temporary cognitive small cell networks

The goal of the presented simulation is to illustrate the need for TCSCN, in order to encourage further research and studies concerning this domain, and is meant to reproduce the common LTE coverage difficulties in indoor environments. The simulation model is partially derived from preliminary guidelines discussed in 3GPP for indoor and home femto cell simulation [23], while the results compare the spatial-probability performance of three coverage scenarios.

- The first scenario represents the performance of a conventional standalone network comprising an eNB placed at 250 m to the north-east side of the subject building. The building has no TCSCN coverage.
- The second scenario shows the results of introducing a TCSCN cluster inside the building, with two TCAPs. The first is working in the primary mode, while the second is working in the secondary mode. The TCSCN in this scenario does not perform any cross-tier interference coordination.
- The third scenario demonstrates the effect of using the cross-tier interference mitigation as suggested in Section 9.5, so the indoor UE has the freedom to attach to either the macro network directly or to the TCSCN inside the building.

The building is assumed to have dimensions of 50 m × 50 m with 25 apartments inside it, of 10 m × 10 m each [24]; UEs are assumed to be placed randomly inside the building with uniform distribution. The target building is located within the parent network cell, which has an antenna radiation pattern as follows:

$$\text{Gain}(\theta) = \text{Gain}(\theta_o) - \min\left\{12\left(\frac{\theta}{\theta_{3\text{dB}}}\right)^2, \text{FBR}\right\} \quad (9.12)$$

where $\theta_{3\text{dB}} = 70°$ is the half power beamwidth, and $\text{FBR} = 20$ dB is the front-to-back gain ratio. The antennas of the primary and the secondary TCAPs are assumed to have a 0 dBi gain (on the access side), which is the most probable future realization for handheld TCAP units. The propagation model is approximated to WINNER II indoor model, type A1 (room to room) [25], assuming thick walls of 12 dB penetration loss each. The rest of the simulation parameters are similar to [26].

We define the service outage probability of a receiver (r) for certain scenario (s) by:

$$P_{r,s}(\text{Outage}) = \mathbf{P}\{\text{SINR}_{r,s} < \text{SINR}_{\text{th}}\} \quad (9.13)$$

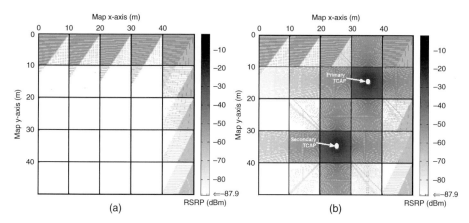

Figure 9.12 Simulation results for indoor coverage: (a) without TCSCN, (b) with TCSCN (cross-tier interference coordination enabled).

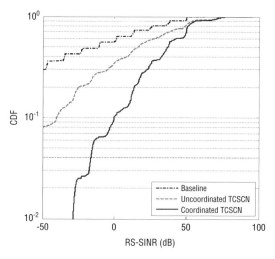

Figure 9.13 Spatial RS-SINR CDF comparison for the three simulation scenarios.

$SINR_{th}$ depends on the target application minimum throughput and on user requirements. In this simulation the value of $SINR_{th} = 10$ dB is used [24], which is the LTE commonly accepted value for a minimum-level service.

The simulation results are depicted in Figure 9.12 in terms of the reference signal received power (RSRP), showing a clear enhancement in the coverage when introducing TCSCN in Figure 9.12b, compared to the coverage serviced from a standalone macro network in Figure 9.12a.

The quantitative results are shown in Figure 9.13, which depicts the cumulative distribution function (CDF) of the signal to interference-plus-noise ratio (SINR) of the three simulation scenarios explained earlier, calculated over the entire simulation area. The CDF is showing that at a service threshold (SINR = 10 dB), the spatial blockage percentages are 63.7%, 40.9%, and 13.2% for scenarios (i), (ii), and (iii) respectively. Accordingly, the overall blockage rate reduction enhancement in this particular example is 50.5% when using the proposed interference mitigation schemes.

9.8 Conclusion

This chapter provided a general overview of LTE-based temporary cognitive small cell networks and highlighted the most important use case scenarios. The scenarios include temporary events, poorly covered areas, and supplementing disaster-recovery networks. In addition, this chapter showed how TCSCN could intelligently construct a radio environment map by periodically sensing the radio frequency spectrum. Two interference mitigation techniques were suggested for the semi-static spectrum access method, and simulated at the end of the chapter showing a significant enactment in network service availability. The concept of temporary cognitive small cell networks is still in its early stages. However, it is also growing rapidly, especially for such applications as public safety and disaster recovery.

Acknowledgement

The research related to this chapter was partially funded by the ABSOLUTE project from the European Commission's Seventh Framework Programme (FP7-2011-8) under the Grant Agreement FP7-ICT-318632 [10].

References

[1] K. Sithamparanathan and A. Giorgetti, *Cognitive Radio Techniques: Spectrum Sensing, Interference Mitigation, and Localization*. Artech House, 2012.

[2] A. Al-Hourani and S. Kandeepan, "Cognitive relay nodes for airborne LTE emergency networks," *7th International Conference on Signal Processing and Communication Systems, IEEE*, 2013.

[3] M. Imawara, H. Takashashi, and S. Nagata, "Relay technology in LTE-Advanced," *NTT Docomo Technical Journal*, 12(2), 2010.

[4] 3GPP, "Technical Specification Group Radio Access Network, further advancements for E-UTRA physical layer aspects (Release 9)," 3rd Generation Partnership Project (3GPP), TR 35.814, 2010. www.3gpp.org/ftp/Specs/html-info/36814.htm.

[5] Fraunhofer. (2013) OpenEPC project by Fraunhofer FOKUS. www.openepc.net.

[6] 3GPP, "Technical Specification Group Radio Access Network, physical layer for relaying operation (Release 11)," 3rd Generation Partnership Project (3GPP), TS 35.216, 2012. www.3gpp.org/ftp/Specs/html-info/36216.htm.

[7] M. Coldrey, J.-E. Berg, L. Manholm, C. Larsson, and J. Hansryd, "Non-line-of-sight small cell backhauling using microwave technology," *Communications Magazine, IEEE*, 51(9), 78–84, 2013.

[8] H. Dhillon, R. Ganti, F. Baccelli, and J. Andrews, "Modeling and analysis of k-tier downlink heterogeneous cellular networks," *IEEE Journal on Selected Areas in Communications*, 30(3), 550–560, 2012.

[9] H. ElSawy, E. Hossain, and D. I. Kim, "HetNets with cognitive small cells: user offloading and distributed channel access techniques," *Communications Magazine, IEEE*, 51(6), 28–36, 2013.

[10] "EU-FP7 ICT IP Project ABSOLUTE," 2013. www.absolute-project.eu/reports/publications.

[11] N. Saquib, E. Hossain, and D. I. Kim, "Fractional frequency reuse for interference management in LTE-Advanced Hetnets," *Wireless Communications, IEEE*, 20(2), 113–122, 2013.

[12] 3GPP, "Comparison of time-domain eICIC solutions," 3rd Generation Partnership Project (3GPP), Discussion and Decision R1-104661, 2010. www.3gpp.org/ftp/tsg_ran/wg1_rl1/TSGR1_62/Docs.

[13] E. Dahlman, S. Parkvall, and J. Skold, Chapter 10. *4G LTE/LTE-Advanced for Mobile Broadband*. Academic Press, 2011.

[14] D. Lopez-Perez, I. Guvenc, G. De la Roche, M. Kountouris, T. Quek, and J. Zhang, "Enhanced intercell interference coordination challenges in heterogeneous networks," *Wireless Communications, IEEE*, 18(3), 22–30, 2011.

[15] S. Kandeepan, R. Piesiewicz, T. C. Aysal, A. R. Biswas, and I. Chlamtac, "Spectrum sensing for cognitive radios with transmission statistics: considering linear frequency sweeping," *EURASIP Journal on Wireless Communications and Networking*, 2010, p. 6.

[16] S. Kandeepan, A. Giorgetti, and M. Chiani, "Distributed 'ring-around' sequential spectrum sensing for cognitive radio networks," In *IEEE International Conference on Communications (ICC)*. IEEE, 2011, pp. 1–6.

[17] S. Arunthavanathan, S. Kandeepan, and R. J. Evans, "Spectrum sensing and detection of incumbent-UEs in secondary-LTE based aerial-terrestrial networks for disaster recovery," *1st International IEEE Workshop on Emerging Technologies and Trends for Public Safety Communications (ETPSC), CAMAD Conference*, 2013.

[18] F. F. Digham, M. Alouini, and M. K. Simon, "On the energy detection of unknown signals over fading channels," *IEEE Transactions on Communications*, 55(1), 21–24, 2007.

[19] S. Atapattu, C. Tellambura, and H. Jiang, "Performance of an energy detector over channels with both multipath fading and shadowing," *IEEE Transactions on Wireless Communications*, 9(12), 3662–3670, 2010.

[20] S. Kandeepan, A. Giorgetti, and M. Chiani, "Periodic spectrum sensing performance and requirements for legacy users with temporal and noise statistics in cognitive radios," In *GLOBECOM Workshops, IEEE*. IEEE, 2009, pp. 1–6.

[21] Y. Zhao, L. Morales, J. Gaeddert, K. K. Bae, J.-S. Um, and J. H. Reed, "Applying radio environment maps to cognitive wireless regional area networks," In *DySPAN 2007. 2nd IEEE International Symposium on New Frontiers in Dynamic Spectrum Access Networks*. IEEE, 2007, pp. 115–118.

[22] Z. Wei, Q. Zhang, Z. Feng, W. Li, and T. A. Gulliver, "On the construction of radio environment maps for cognitive radio networks," *arXiv preprint arXiv:1302.6646*, 2013.

[23] 3GPP, "Simulation assumptions and parameters for FDD HeNB RF requirements," 3rd Generation Partnership Project (3GPP), RAN WG4 (Radio) Meeting No. 51R4-092042, 2009. www.3gpp.org/ftp/tsg_ran/wg4_radio/TSGR4J51/Documents.

[24] 3GPP, "Relay radio transmission and reception," 3rd Generation Partnership Project (3GPP), TR 36.826, 2013. www.3gpp.org/DynaReport/36826.htm.

[25] P. Kyosti and others, "WINNER II Channel Models," Tech. Rep. D1.1.2, 2007. www.ist-winner.org/WINNER2-Deliverables/D1.1.2v1.1.pdf.

[26] A. Al-Hourani and S. Kandeepan, "Temporary cognitive femtocell network for public safety LTE," *1st International IEEE Workshop on Emerging Technologies and Trends for Public Safety Communications (ETPSC), CAMAD Conference*, 2013.

10 Long-term evolution (LTE) and LTE-Advanced activities in small cell networks

Qi Jiang, Jinsong Wu, Lu Zhang, and Shengjie Zhao

10.1 Introduction

The general definition of a small cell is the low-powered radio access node operating in licensed and unlicensed spectrum with the smaller coverage of ten meters to one or two kilometers, compared to a mobile macro cell with a range of a few tens of kilometers. With the introduction of this new concept, the heterogeneous network (HetNet) constructed with different layers of small cells and large cells can deliver the increased bandwidths, reduced latencies, and higher uplink (UL) and downlink (DL) throughput to end users. Since 2009, the standard evolution of the small cell related topics has been studied in 3GPP (The 3rd Generation Partnership Project) LTE (long-term evolution) and LTE-Advanced. The following sections in this chapter will introduce the standardization progress of LTE and LTE-Advanced in small cells.

10.1.1 Definition of small cells in 3GPP LTE-Advanced

In 3GPP LTE and LTE-Advanced, small cells can generally be characterized as either relay nodes, or pico cells (also referred to as hotzone cells), controlled by a pico eNodeB, or femto cells, controlled by a Home evolved NodeB (HeNB). The common features among the relays, pico cells, and femto cells are low transmission power node and independent eNB functionality, while the typical different features can be summarized as follows:

1. Relay node [1, 2]. A relay node (RN) is a network node connected wirelessly to a source eNodeB, called the donor eNodeB. According to the different implementation types of the relay node into wireless network, the roles of the relay node played are also different.
2. Pico cell. A pico cell usually controls multiple small cells, which are planned by:
   a. The 3rd Generation Partnership Project (3GPP), which unites six telecommunications standard development organizations (ARIB, ATIS, CCSA, ETSI, TTA, and TTC), known as organizational partners, and provides their members with a stable environment to produce the highly successful reports and specifications that define 3GPP technologies.
   b. The evolved Node B could be abbreviated as eNodeB or eNB by the network operator in a similar way as the macro cells [3]. The pico cell is usually open to all users (open subscriber group (OSG))[4].

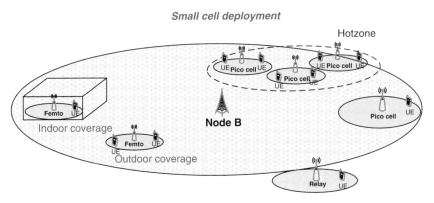

Figure 10.1 Illustration of small cell deployment.

3. Femto cell [5]. The transmission power of the femto cell is usually even lower compared to that of relay node and pico cells. A femto cell is typically designed to cover a house or apartment and is accessible only to a limited group of users, which is named the closed subscriber group (CSG) cell. Recently a new emerging neighborhood femto cell network to deploy outdoor femto cells as an open access model in malls and neighborhoods has also been designed and applied.

10.1.2 Deployment scenarios for small cell in 3GPP LTE-Advanced

From the perspective of the 3GPP LTE-Advanced, there is no restriction about the deployed spectrum band for the small cells. The small cells can be deployed on the same spectrum band as the macro cells or a dedicated spectrum band. For the latter scenario, it is preferred to assign a higher frequency to deliver high data rates within a limited area without causing excessive interference. Figure 10.1 demonstrates a deployment of the small cells in a current cellular network. According to the characteristics for the different types of small cells, the following features are preferred.

1. Relay node. Due to the wireless backhaul connection with the macro cell, the relay node is best deployed where a fixed-wired connection is unavailable or not necessary all the time. One standard scenario for relay is in temporary coverage filling or coverage extension of the macro cell coverage, such as in wireless communication enablement in a disaster area or rural area. Another standard scenario is enhanced hotspot coverage, such as at a stadium when a game is being played or an office building during peak hours. According to the wireless connection to the network, more flexible deployment can be obtained via relays compared to other small cells from the perspective of the operator. However, due to single-hop or multi-hop relay connection between the user equipment (UE) and the core network, the latency brought by the involved relays should also be considered.
2. Pico cell. Without the differentiation of the transmission power levels, the pico cell can be treated as a macro cell as well. Generally, the use cases of pico cells are

extensive for a variety of scenarios. A pico cell can improve coverage extension and capacity in areas traditional macro cells cannot easily provide, such as in buildings, offices, shopping malls, outdoor stadiums, and rural areas. A traditional wired backhaul interface used between macro cells is also applied between macro cell and pico cell. Both the Ethernet cabling and the fiber can be used for the backhaul connection, which is based on the requirements of the operators. The pico cell also may have the advantages of low-cost deployment, simplified radio frequency (RF) units, and low operating expense (OPEX).

3. Femto cell. Unlike relays and pico cells, the initial objective scenarios of femto cells are for home or small business use, especially for indoor cases. In 3GPP terminology, a Home Node B (HNB) can be treated as a 3G femto cell and a Home eNode B (HeNB) can be treated as an LTE femtocell. Unlike the deployment manner of relays and pico cells, which are controlled by the operators, the deployment of femto cells is likely to be owner controlled. The CSG can be configured by the owner to restrict the accessed users. After the completion of payment of the internet connection to femto cell traffic route or the authorization fee on the assigned carrier, the owner can operate its own "small network" according to its own requirement without any control from the operator, theoretically. Since there is no network control for the placement and carrier selection of the femto cells, the interference between the femto cell and the wider network will be a problem.

10.2   Relay eNodeB in LTE-Advanced

As a key new feature of LTE-Advanced, relay node has been introduced in Release 10 of the LTE specifications [6]. The different types of relays have been studied widely, based on their specific characteristics. In this section, the relevant standard evolutions of the relays will be introduced.

10.2.1   Relay definition

During the period of UMTS (universal mobile telecommunications system) and Release 8 of LTE [7, 8], the initially defined relays were in the form of repeaters in legacy radio interface technologies, which can be treated as simple devices to just amplify and forward the transmission signals from macro base stations to UEs without baseband processing, backhaul network installations, or subscription fees for access to the fixed public networks. Although the repeater-only relays are useful for coverage improvement and easily implemented, they have some significant drawbacks, such as the potential amplified interference and the uncontrolled operation separated from the operation and maintenance (O&M) functionality.

Currently, there is a kind of developed relay node (RN) under the full control of the radio access network present for LTE wireless networks. The controllable RN is a network node wirelessly connected to and controlled by a source eNodeB (donor eNodeB), while with similar monitoring and scheduling capabilities as an eNodeB to

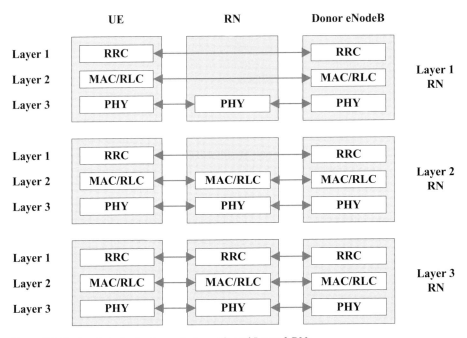

Figure 10.2 Protocol stacks for Layer 1, Layer 2, and Layer 3 RNs.

the connected UE. In contrast to a repeater, an RN processes the received signal before forwarding it. Depending on the process level, the RN can be classified into Layer 1, Layer 2, and Layer 3 relays, which range from an enhanced repeater to a fully fledged eNodeB with a wireless backhaul connection.

1. For Layer 1 RN: pure radio frequency (RF) processing or RF processing with some extra baseband processing.
2. For Layer 2 RN: medium access control (MAC) functions such as scheduling are enabled. On the other hand, some Layer 3 functions, such as radio resource control (RRC), can also be located within the eNodeB rather than the RN.
3. For Layer 3 RN: it has its own PCI and can support the RRC functionalities, such as mobility between RNs, and all Layer 1 and Layer 2 functions are supported within such an RN. This RN can be classified as Type 1/1a/1b RN, described as follows.

Figure 10.2 shows a simplified diagram of the protocol stacks for Layer 1, Layer 2, and Layer 3 RNs [9, 10].

In order to keep aligned our understanding of the 3GPP, the following terminology is used as follows.

1. Donor eNodeB/cell: the source eNodeB/cell from which the RN receives its signal.
2. Relay cell: the coverage area of the RN.
3. Backhaul link: the link between the donor eNodeB and the RN.
4. Access link: the link between the RN and a UE.
5. Direct link: the link between the donor eNodeB and a UE.

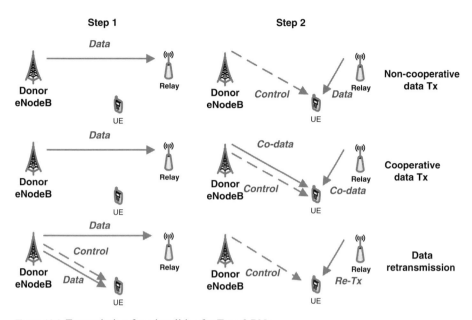

Figure 10.3 Transmission functionalities for Type 2 RNs.

6. Inband/outband: an inband RN uses the same carrier frequency for the backhaul link as for the access link; otherwise, the RN is named outband RN.
7. Half/full duplex: a half-duplex RN cannot receive on the backhaul link at the same time as transmitting on the access link, and vice versa, whereas a full-duplex RN has sufficient antenna isolation to be able to operate without this restriction. This distinction applies to inband RNs only, since outband RNs are always full-duplex.

From the perspective of 3GPP standardization, the key point of the RN definition is whether it can be identified by the UE and the backward compatibility to legacy UEs. In the relevant 3GPP LTE-Advanced discussions, two main types of RNs have been proposed, Type 1 (1/1a/1b) and Type 2 RNs [6].

1. Type 1, Type 1a, and Type 1b RNs. All these types of RNs can be classified as Layer 3 RNs. They can be differentiated from the normal eNodeB with their own independent physical cell identifiers (PCIs) and perform like an eNodeB. Type 1 RNs are inband half-duplex RNs, while Type 1a RNs are outband. Type 1b RNs are inband full-duplex RNs with sufficient isolation between the received and transmitted signals.
2. Type 2 RNs. Unlike Type 1 RNs, Type 2 RNs do not have their independent PCIs and control channels, and can be classified as Layer 2 RNs. Type 2 RNs are usually used as complementary or enhanced nodes for data transmission, where the donor coverage is not good enough. Figure 10.3 shows an example of these transmission functionalities for Type 2 RNs.

Although two kinds of RNs have been proposed in 3GPP, LTE-Advanced only supports Type 1 and Type 1a RNs in Release 10 of the specifications. No specific support has been provided for Type 1b RNs, although such RNs may be able to be deployed via implementation-dependent means. In the following, our discussion will be focused on the specifications on Type 1 and Type 1a RNs.

10.2.2 Backhaul and access resource allocation

According to the definition of both Type 1 and Type 1a RNs, they need to receive and transmit signals from and to the donor eNodeB via the backhaul link, while transmitting and receiving signals to and from the UE on the access link. Some coordination or separation is needed to avoid self-interference between transmission and receiving. Both frequency domain and the time domain separation have been studied in 3GPP LTE-Advanced [11].

1. Frequency domain separation. A typical frequency domain separation is the design of Type 1a RNs where two independent carriers are applied on the backhaul and access link, respectively. If the separation of these two carriers is large enough, there will be no self-interference generated. For this kind of separation, no extra changes will occur when the RNs are implemented into the macro cell dominating cellular network. The key issue is whether there are extra significant benefits to allocate two independent carriers on RNs, since an independent carrier is always a luxury resource for the operator.
2. Time domain separation. The key progress during 3GPP LTE-Advanced study for relay is the time domain separation between the backhaul and the access link. The time domain multiplexing (TDM) mechanism is applied for Type 1 RNs. For the timeline flow, part of the time window will be allocated to the backhaul link and the rest will be allocated to the access link. Since the Type 1 RN is a half-duplex RN, the UE associated on the RN will lose the connection to the eNodeB due to lack of reference signals (RS), synchronization signal, broadcasting channel, and control channels such as physical downlink control channel (PDCCH). Further, in order to keep accurate network measurement for the legacy UEs (Rel-8/9 UE), some specific indication about the separation of the backhaul and access link is also required.

In LTE-Advanced, the concept of the multi-media broadcast single frequency network (MBSFN) subframe is browsed here to realize the notification of the separation to legacy UEs. During the operation of the RNs, radio resource control (RRC) signaling will be used to indicate to the UE that certain subframes are assigned to MBSFN transmission. In these subframes, the UE will only receive the control signals and RS in the first two OFDM (orthogonal frequency division multiplexing) symbols, and ignore the remaining part of the subframe for both data reception and measurement purposes. Due to the existence of the control channel for legacy UE in the beginning of both the access and backhaul link subframe, a transition gap for Tx(transmit) – Rx(receive) turnover will be implemented.

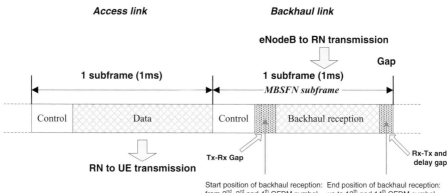

Figure 10.4 MBSFN subframe configuration for Type 1 RNs.

According to the 3GPP RAN1 (radio access network layer 1) specification [11], the starting symbol of the RN's backhaul downlink reception windows is configured by RRC (radio resource control) signaling as the start of the second, third, or fourth OFDM symbol. The largest gap at the end of the backhaul reception is one whole OFDM symbol. Based on these requirements, the available OFDM symbol for backhaul link reception will range from 10 (from 4th to 13th) to 13 (from 2nd to 14th without gaps) in the normal cyclic prefix (CP) case. Figure 10.4 shows the MBSFN subframe configuration for Type 1 RNs.

10.2.3 Relay architecture

Currently, the Type 1 and Type 1a relay included in Rel-10 can support the full eNodeB functionality, including termination of the radio protocols and the S1/X2 interfaces. The corresponding supported architecture for RNs in 3GPP LTE-Advanced can be found in Figure 10.5, which is referred by 3GPP specification TS 36.300 [12]. A new network interface called Un between RN and donor eNodeB is defined [12]. From the architecture, it can be found that the donor eNodeB provides S1 and X2 proxy functionality between the RN and other eNodeBs, which includes mobility management entities (MMEs) and serving gateways (S-GWs).

Figures 10.6 and 10.7 provide the user plane and control stack protocol for RN, which is referred by [12]:

1. For the user plane, the donor eNodeB provides the S-GW and packet data network (PDN) gateway (P-GW) functionality for the RN, which includes the management of the evolved packet system (EPS) bearers.
2. For the control plane, there is one S1 interface relation between the RN and the donor eNodeB, and one S1 interface between the donor eNodeB and each MME. The donor eNodeB processes and forwards all S1 messages between the RN and the MMEs for all UE dedicated procedures.

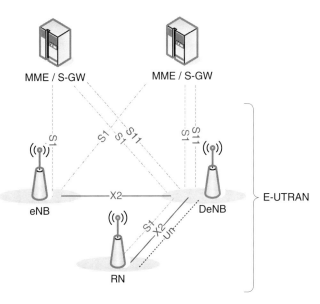

Figure 10.5 Overall E-UTRAN architecture supporting RNs.

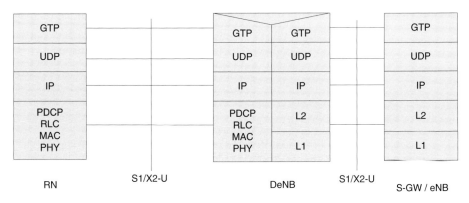

Figure 10.6 RN user plane protocol stack.

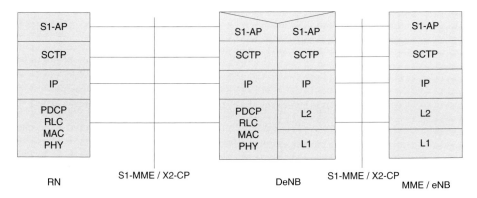

Figure 10.7 RN control plane protocol stack.

According to the stack protocols of the RN, the RN can be treated as a UE when performing the initialization and the backhaul connection with the donor eNodeB. During the procedure of the initialization, the configuration parameters will be transmitted to the RN, including a list of donor eNodeBs to which it is allowed to attach, from an RN O&M server. After that, the RN will detach from the network as a UE and perform like an eNodeB to receive the signals from the real mobile users.

10.2.4 Backhaul link design

Compared to the legacy UE connection to eNodeB, the appearance and the corresponding two links sustained by RN provide some specific designs of the backhaul link on the current cellular network. In this section, these specific designs will be provided below.

10.2.4.1 Backhaul reference signals
Both Rel-8/9 defined cell-specific reference signals (CRS) and Rel-10 defined demodulation reference signals (DM-RS), which can be reused on the backhaul link [11]. Since the tail gap is configured for Rx-Tx transition and propagation delay (see Figure 10.4), the DM-RS located in the second slots of one subframe cannot be used (antenna port 7 to 10, located in the last two OFDM symbols of one subframe). Besides, a further restriction is that the backhaul link in Release 10 is limited to a maximum of four spatial layers, and DM-RS antenna ports 11 to 14 have never been used for backhaul link transmission.

10.2.4.2 Backhaul control channels in downlink
Due to the one or two OFDM symbols PDCCH transmission to UE on the MBSFN subframe, as well as the timing synchronization between the access and backhaul link, the RN cannot receive the PDCCH from the donor eNodeB. In order to resolve this, a new backhaul link control channel named as relay PDCCH (R-PDCCH) is provided. It is located in the legacy PDSCH region and multiplexed with the data channel in a frequency domain multiplexing (FDM) manner. The example of multiplexing among PDCCH, R-PDCCH, and PDSCH is shown in Figure 10.8.

10.2.4.3 Backhaul control channel for uplink
The physical uplink control channel (PUCCH) transmission by RN will be generally in the same way as the transmission between UE and donor eNodeB. There are only two slight differences. The first one is that the resource used for backhaul PUCCH will be reserved for high-layer signaling. Consequently, in contrast to the PDCCH, there is no relationship between the resources used for R-PDCCH and PUCCH transmission.

10.2.4.4 Backhaul data channels
The data transmission for both the downlink and uplink on the backhaul link applies the same physical channels defined (PDSCH and PUSCH) for the access link. For the multiplexing of the PDSCH and R-PDCCH for backhaul link, three alternatives are

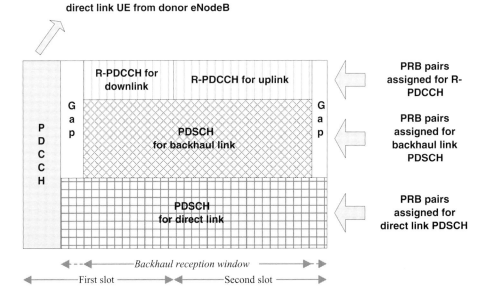

Figure 10.8 Backhaul control channel design for RN.

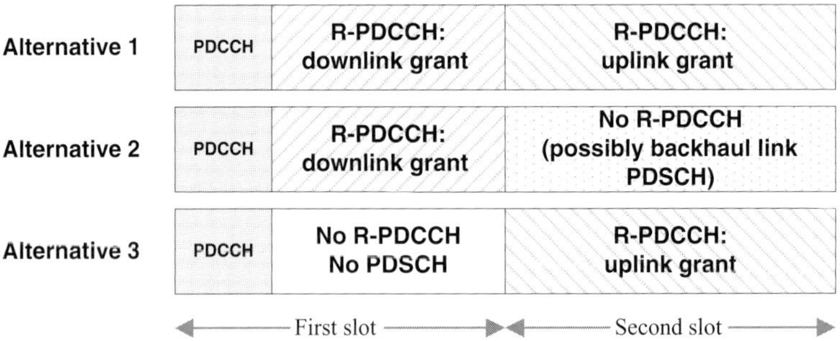

Figure 10.9 Multiplexing of R-PDCCH and PDSCH.

applied as follows (shown in Figure 10.9, for simplicity, the gap between the Tx-Rx turnover has not been included):

1. Alternative 1: a downlink grant in the first slot and an uplink grant in the second slot.
2. Alternative 2: only one downlink grant in the first slot.
3. Alternative 3: only one uplink grant in the second slot.

10.2.4.5 Backhaul data channels

For outband RN, there is no special restriction on the scheduling or timeline requirement between the donor eNodeB and RN. For inband RN, due to the additional link for backhaul link transmission, the MBSFN subframe for downlink transmission, as well as the available transmission resources for the corresponding feedback to or from RN with a proper timeline, some specific subframe configuration for backhaul link transmission should be designed.

For FDD operation, the subframe configured for backhaul link transmission follows a periodicity of 8 ms (microseconds), which keeps aligned with the UE to eNodeB transmission. An 8-bit bitmap is therefore sufficient to configure the downlink backhaul subframes. The timing association between the uplink grant and uplink transmission, as well as timing association between the backhaul link PDSCH transmission and corresponding ACK/NACK on uplink, maintain the same as that using legacy principles.

For TDD operation, the design of the backhaul and access link subframe configurations is more complicated than that for FDD (frequency-division duplexing) mode. Among seven different types of uplink–downlink configuration, Configuration 0 and 5 cannot support the backhaul transmission due to quite limited downlink or uplink subframes within one radio frame (two DL subframes for Config. 0 and one UL subframe for Config. 5). Different kinds of configurations specified for backhaul links, as well as the timeline association criterion, have been specified in [6].

10.3 Pico eNodeB in LTE-Advanced

The standardization discussion for pico eNodeB has been studied in Rel-10 and Rel-11 versions of inter-cell interference coordination [11]. The main focus of LTE-Advanced for pico eNodeB is on the support of heterogeneous network deployment where the macro cell and pico cells share the same frequency. The pico cell can improve the user experience. However, as a double-edge sword, the overlaying deployment also results in co-channel interference between the macro cell and pico cells. This section will focus on the enhanced solution and corresponding specification effort to resolve this issue restricted in the macro–pico scenario. The macro–femto scenario will be discussed in the next section.

10.3.1 Inter-cell interference for LTE-Advanced

In the homogeneous macro-cellular network, the UE is basically served by the strongest cell BS with the strongest receiving power. Due to the restriction of the transmission power, as well as less antenna gain compared with that in a macro cell, the difference of the receiving power from macro cell and pico cell is almost 25 dBm in the 3GPP specification if the same path loss is assumed. According to this, the number of UEs served by the pico cell will be quite limited, which would result in quite limited gains of the pico cell deployment. In order to fully explore the cell-splitting gain, the serving eNodeB can intentionally "bias" the handover offset values of some UEs in RRC_JCONNECTED mode to transfer them to the pico cells, which is known as cell range expansion (CRE)[13, 14].

If an UE is transferred in this way, the desired signal from the pico eNodeB would be even lower compared to the interference received from the macro eNodeB. Figure 10.10 shows the heterogeneous network interference scenarios in downlink and uplink. Due to

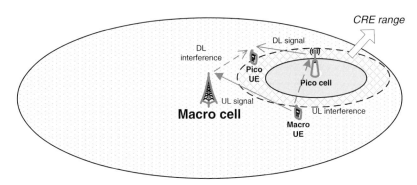

Figure 10.10 Interference scenario for heterogeneous network.

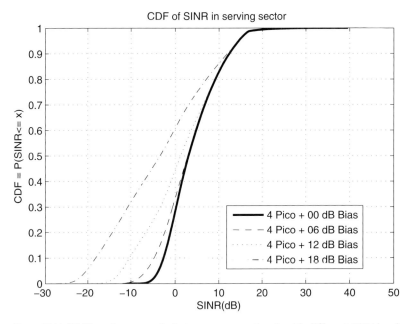

Figure 10.11 SINR performance for heterogeneous network with different CRE levels.

CRE implementation, the major interference to border pico UE in downlink is generated from the macro cell BSs while the major interference to pico cell BS in uplink is generated from macro UEs.

Figure 10.11 shows the SINR (signal to interference-plus-noise ratio) performance of all UEs in a heterogeneous scenario referred by [15]. The figure shows how SINR performance degrades with the bias increase.

10.3.2 Almost blank subframe (ABS)

The interference mitigation and coordination applied in Release 8 and 9 for the downlink data channel are mainly based on fractional frequency reuse and frequency hopping.

Correspondingly, the interference mitigation in the uplink data channel is mainly based on open-loop or closed-loop power control. The corresponding X2 signaling, such as relative narrowband Tx power (RNTP), UL high interference indication (HII), and UL interference overload indication (OI), are defined to support these interference mitigation mechanisms [16]. Actually, for PDCCH, there is no specific inter-cell interference coordination mechanism proposed in Release 8 and 9, and the enhancement of the robustness is mainly based on the increase of duplication transmissions, such as the increase of the aggregation levels [17].

However, all these interference mitigation methods proposed for homogeneous networks cannot accommodate the interference scenario for heterogeneous network deployment. In order to improve the performance of both the control and data channels, the time-domain multiplexing such as enhanced inter-cell interference coordination method (eICIC) has been included into the specification of the Release 10 version [16]. The general mechanism of this eICIC method is to mute the transmission of data and corresponding control channel on some dedicated downlink subframes of one layer of cells (such as macro cells) to reduce the interference to the other layer (such as pico cells) [18]. These muted subframes are defined as almost blank subframes (ABSs) [18].

In the specification description, the ABSs are defined as subframes with reduced downlink transmission power and/or activity. Ideally, ABS can be totally blank with only remained CRS when configured by a macro cell in a macro–pico scenario, in order to remove as much as possible interference towards pico UEs. In practice, since the ABS configuration would also cause resource losses in the macro cell, the transmission and the exact content of the ABSs would be an implementation choice, which takes the total system gains from interference reduction versus the loss of transmission resources from being unable to transmit PDCCH/PDSCH in the ABSs. As one example, the UL scheduling indication (UL grant) for PUSCH transmission in PDCCH, as well as the corresponding HARQ ACK/NACK, can also be transmitted on ABSs, although the other DL assignment has been muted.

Another key characteristic for ABSs is that the backward-compatible signals should also remain in ABS for Release-8 or 9 UE camping on. That means that the cell-specific reference signal (CRS), primary synchronization signal (PSS), secondary synchronization signal (SSS), paging channel, broadcast channel (PBCH and SIB-x), should be transmitted in ABSs if necessary. However, even if these backward-compatible signals remain in the ABSs, the effect of interference mitigation toward the data and control channel of the pico UE on corresponding subframes is still significant due to the reduced activities on other data and control channels on the macro cell. Figure 10.12 shows an example of the ABS configuration between macro cell and pico cell for FDD mode. The corresponding X2 enhancement and measurement enhancement will be introduced in the next section.

10.3.3 X2 interference coordination for ABS

In order to accommodate the different traffic loads and UE deployment distributions in heterogeneous networks, the ABS configuration between the macro cell and pico cells

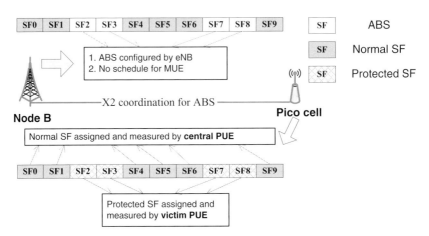

Figure 10.12 Example of ABS configuration between macro cell and pico cell.

can be adjusted with a high degree of flexibility. That means that both the percentage and the location of the ABSs among the entire subframes can be updated to explore the maximum gain introduced by the deployment of the pico cells. For this purpose, the time-domain inter-cell coordination signaling via the X2 interface is defined in Release 10, where the pattern of ABSs configured by one eNodeB can be indicated to its neighbors.

A new information element called ABS information is defined to provide the information about which subframes of the sending eNodeB are configured as ABS and which subset of ABSs are recommended for measurement configurations toward the UE. Two bitmaps are used to represent these two pieces of information. The first bitmap, ABS pattern info, indicates the set of ABS subframes to assist the receiving eNodeB with its scheduling operations. Each bit of an ABS bitmap informs the receiving eNodeB of the sending cell's intention regarding its transmission activity. This information can be used by the receiving eNodeB to arrange its scheduling operations to avoid transmitting in non-ABS subframes to victim UEs (UEs assumed to be in high-interference areas such as the cell edge). For the convenience of discussions in this chapter, we introduce two non-standard terms (not defined in LTE specifications), protected subframe and normal subframe. The receiving eNodeB in the pico cell can schedule its subframe, called protected subframe, occupying the same time-frequency slot as the ABS (note that the ABS is the subframe transmitted by the macro cell) for UEs to reduce interference. The rest of the non-protected subframes from the pico cell are called normal subframes. The second bitmap, measurement subset, is a subset of the first subset and is expected to be less variable and intended to recommend to the receiving eNodeB a suitable set of subframes that can be used for steady measurements without the interference from the sending eNodeB. One use case for this second bitmap is that although the ABS pattern info may be adapted periodically per hundreds of milliseconds, the second pattern may not be changed. Accordingly, the UE measurement, especially for cell reselection and handover, can also be based on this second pattern in a steady way.

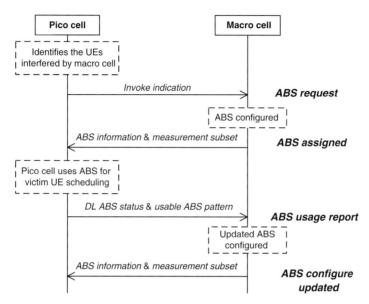

Figure 10.13 Typical ABS coordination procedure.

In order to be aligned with the HARQ time round trip times (RTTs), the ABS bitmaps have the following periodicities:

1. 40 ms for FDD
2. 20 ms for TDD configurations 1–5
3. 70 ms for TDD configuration 0
4. 60 ms for TDD configuration 6.

Further, in order to implement the ABS configuration in a proper two-direction coordination manner instead of one-direction notification, several extra indication elements (IEs) are designed:

1. Invoke indication: to request the ABS configuration from the sender to the receiver cells.
2. DL ABS status: to aid the eNodeB designating ABS to evaluate the need for modification of the ABS pattern. It contains the percentage of used ABS resources (may be a subset of the assigned ABSs indicated in ABS pattern info).
3. Usable ABS pattern info: a bitmap with each bit to indicate "ABS" that has been designated as protected from inter-cell interference by the eNodeB-1 (such as the macro cell), and available to serve this purpose for DL scheduling in the eNodeB-2 (such as pico cells).

A typical message exchange over the X2 interface for ABS coordination between a macro cell and a pico cell is illustrated in Figure 10.13.

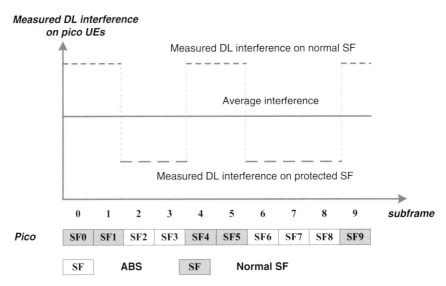

Figure 10.14 Typical interference fluctuation for ABS configuration.

10.3.4 Restricted UE measurement

Due to the ABS configuration on the macro cell, the measured interference on normal subframes would be different from that on protected subframes (as shown in Figure 10.14). According to the definition of the UE implementation in Release 8 or 9, there is no differentiation among subframes from measurement perspective, which means that the UE will treat all the subframes the same way and just report the measurement results on the assigned subframes. It can be an averaged result of all the subframes or an instant result according to the eNodeB assignment. In order to overcome this fluctuation among the protected subframes and the normal subframes, a restricted measurement dedicated to some subframes instead of all subframes has been proposed, which includes the measurements of radio link monitoring (RLM), radio resource management (RRM), and channel state information (CSI):

1. RLM measurement [19]: the RLM procedure is used for a UE to monitor the downlink radio link quality of the serving cell for the purpose of determining whether the synchronization has been lost and radio link failure (RLF) has occurred. In order to avoid unnecessary loss of synchronization reported by a UE, the RLM measurements should be made in the protected subframes from pico cells only when ABS from a macro cell is configured.
2. RRM measurement [19]: for RRM-related measurements such as intra-frequency cell reselection, the UE needs to perform both RSRQ (reference signal received quality) and RSRP (reference signal received power) measurements. Since the RSRQ results are derived by RSSI (received signal strength indicator), which depends on whether the measurement is made in ABSs only or in all subframes, the RSRQ measurement should be restricted to a subset of the subframes. For RSRP results,

since it is focused on the receiving power of the CRS, which is not relevant to the types of subframes, according to the functionality of the RRM measurement, as well as the ABS configuration, some new patterns defined in Release 10 are designed.
  i. Pattern 1: applies to RRM measurement on the serving cell, which is the same pattern as applied for RLM measurement.
  ii. Pattern 2: applies to RRM measurement on certain intra-frequency neighbor cells with specifically indicated physical cell identities (PCIs).
  iii. CSI measurement [20]: CSI measurements, especially the CQI reports, provide the serving eNodeB with information related to the SINR, which is useful for scheduling and selection of an appropriate modulation and coding scheme (MCS) applied on each UE. When ABSs are configured in an interfering cell, the SINR measured for the UEs served in the interfered cell, especially UE located on the cell border, will fluctuate dramatically from subframe to subframe. In order to exploit the potential performance gain of ABSs for scheduling and link adaptation, the CSI measurements made by these UEs should be performed on restricted subsets of the subframes.

In order to get an accurate CSI report, two independent sets of subframes can be configured per UE. It can be a combination of a normal subframe and a protected subframe from pico cells, or ABSs configured by two different neighbor macro cell eNodeBs. In the latter example, the two signaled subsets of subframes could consist of one subset that is common to both the interfering eNodeBs and the eNodeB that contains only the protected subframes configured by one of the interfering pico cell eNodeBs.

Currently, there are two kinds of the CSI reports defined in 3GPP LTE, periodic CSI reporting and aperiodic CSI reporting. For the periodic reporting, two sets of reporting periodicities and offsets are configured for the UE; each set is associated with one group of subframes. For the aperiodic case, the UE reports CSI for whichever set of subframes contains the "CQI reference resource" corresponding to the aperiodic CSI trigger.

10.3.5 Further eICIC (FeICIC) for LTE-Advanced

As we mentioned in Section 10.3.1, one of the key features for Release 10 eICIC is the introduction of CRE to expand the coverage of the pico cell and explore the corresponding gain of the frequency reuse by ABS configuration. With the implementation of the ABS, the control and data channel performance can be improved. However, due to consideration of the backward compatibility, some downlink physical channels of the pico cells may still suffer serious interference coming from macro cells, especially for the CRE implemented scenario. We list some potential inter-cell interference as follows.

1. Broadcast channel collision: there are two ways to deliver the broadcast information to UE in 3GPP LTE systems. One way is by the master information block (MIB) based on PBCH, which is transmitted on the central 6 physical resource block of the entire bandwidth on subframe 0. Another way is by secondary information block (SIB) based on the scheduling of PDSCH by common search space (CSS) configured in PDCCH on some dedicated subframes. As an example, SIB-1, which contains

parameters required to determine whether a cell is suitable for cell selection, as well as the information about the time-domain scheduling of the other SIBs, would be transmitted on PDSCH in the first subframe during each 80 ms slot. Since no matter whether or not the ABS is configured, the above-mentioned broadcast information will remain in the dedicated position and subframes, which would result in serious interference during the reception of the corresponding broadcast channels of the pico cells.

2. CRS interference on PDSCH: similar to the PBCH and SIB information, the CRS remains for the transmission in PDSCH unless it is located in a MBSFN subframe. Thus the interference of the macro cells CRS would impact the PDSCH transmission of the pico cell even if it is located in a protected subframe from a pico cell corresponding to ABS from a macro cell.

3. CRS collision: based on current resource mapping of the CRS, the possible position of the CRS can only be three alternatives from frequency domain perspective if two or above two antenna ports are configured. Then, if more than three pico cells are located in one macro cell and generate interference to each other, the CRS collision between pico cells needs to be considered. This problem is usually related to the dense deployment of the pico cells.

For the above-mentioned problems, LTE Release 11 FeICIC (Further eICIC) has identified several aspects of enhancements on both the UE and eNodeB sides.

1. Network assistance for broadcast channel acquisition and CRS interference cancellation: the serving cell will deliver the CRS configuration of the neighbor cell(s) to help the UE to cancel the interference generated by CRS of the neighboring cells. The network assistance information can include the number of CRS ports and subframes containing CRS in the data region. Further, for SIB-1 acquisition, the interfered cell can use dedicated signaling to provide SIB-1 for UEs in the CRE zone, such as the protected subframes from pico cells instead of the fixed subframes assigned for SIB-1. For the enhancement of the MIB, the specification only enhances the performance requirement of the MIB reception on the UE side.

2. CRS interference suppression/cancellation for collision: in the case of colliding CRS scenario, for the purpose of RLM/RRM and CSI feedback corresponding to one of the configured subsets, 3GPP specifications recommend the UE to perform suppress or cancel the CRS interference from the cell(s) that are included in the cell list. Further, some other mechanisms, such as the proposal of reduced power ABS, instead of the zero power ABS defined in Release 10, have also been discussed in Release 11 FeICIC.

10.4 Home eNodeB in LTE-Advanced

Basically, a home eNodeB (HeNB) is a small cell applied for home installation and indoor coverage. To avoid including some vendor-dependent functionalities, such as closed subscriber group (CSG) restriction on UE access, the standardization efforts of

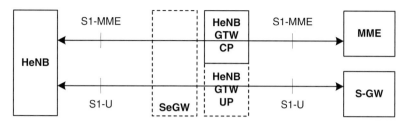

Figure 10.15 HeNB logical architecture.

the HeNB have been mainly on the architecture between the HeNB and the core network, and the interference management for HeNB.

10.4.1 Home eNodeB architecture

10.4.1.1 Architecture overview

The architecture of the HeNB is to design a flexible and scalable network with respect to the number of the HeNBs. According to the number of the deployed HeNBs, a new unit called an HeNB gateway (GTW) is optionally deployed to manage the HeNBs from the perspective of the evolved packet core (EPC). The capacity of the HeNB GTW can be up to several tens of thousands of HeNBs. It can be deployed with either control plane (CP) only or with both control plane and user plane (UP). The logical architecture is illustrated in Figure 10.15. When an HeNB GTW is deployed, it serves as a concentrator for the CP S1-MME interface. Each HeNB therefore has only one stream control transmission protocol (SCTP) association toward the HeNB GTW, and each mobility management entity (MME) likewise has only one SCTP association toward the HeNB GTW. The S1-U interface may also optionally be terminated in the HeNB GTW UP. In this case the UP connection for one E-UTRAN radio access bearer (E-RAB) of one UE consists of two GPRS tunneling protocol (GTP) tunnels instead of one. The HeNB GTW is assumed to be transparent for the S1 interface, i.e., the HeNB treats the HeNB GTW exactly as if it were an MME and conversely the MME treats the HeNB GTW as a regular eNodeB.

Supported mode

Principally, the HeNB can support three kinds of access modes:

1. Closed access mode: the HeNB is associated with one or more specific CSGs identified by CSG identifiers (CSG IDs). Only UEs that have owned the corresponding CSG IDs included in their CSG subscription list are allowed accesses. This is the most common mode for HeNBs.
2. Hybrid access mode: as for the closed access mode, the HeNB belongs to a particular CSG. It provides service to all UEs but gives preferential treatment to UEs that include the corresponding CSG ID in their CSG subscription list; such UEs are called "CSG members."
3. Open access mode: the HeNB behaves as a regular eNodeB, named OSG mode.

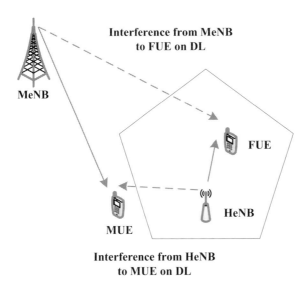

Figure 10.16 Macro–femto downlink interference scenarios.

10.4.1.2 Mobility

The HeNB can support the mobility from HeNB to an eNodeB, from eNodeB to HeNB, normal mobility from HeNB to HeNB via S1 handover, and the optimized mobility from HeNB to HeNB via X2 handover.

10.4.2 Interference management

Interference management is a key issue for heterogeneous network deployments of macro cells and small cells in LTE. The problem becomes even more serious for femto cells operating in closed access mode on the same carrier frequency as the macro cells. Unlike the macro–pico cell scenario, the UE cannot freely handover to or measure the RSRP/RSRQ of the femto cell if it is configured with CSG. In addition, due to the absence of X2 coordination with the macro cell, the interference coordination of the femto cells from the network perspective cannot be managed easily. In the following subsections, several interference scenarios and the corresponding solutions to resolve the interference have been identified.

10.4.2.1 DL interference from macro to femto

UE camped on the femto cell (known as femto UEs or FUEs) are more susceptible to interference from the macro cell when they are located closer to the macro cell, since the transmission power of the macro cell is much bigger than that of the femto cell and the received interference power will be higher. FUEs are also more susceptible when located far away from the serving HeNB, especially if they are outside the houses or apartments in which the femto cells are deployed. Figure 10.16 shows an example of this kind of interference scenario. Potential interference mitigation schemes for this scenario may include:

1. Control channel protection: to arrange the PDCCH, synchronization signals (PSS/SSS), and physical broadcast channel (PBCH) in an orthogonal way in time-domain and/or frequency domain, such as via applying subframe boundary offset n femto cells relative to the macro cells. However, this method may not be appropriate for TDD mode.
2. Data channel protection: to avoid the scheduling of the PRBs with high interference from the macro cell. The HeNB can confirm the frequency partition information of the macro cell via several ways, such as by configuration or surveillance of the macro cell transmission.

10.4.2.2 DL interference from femto to macro

Downlink transmissions from the macro cell will suffer serious interference from the transmissions from a femto cell if the macro UE is close to the femto, such as close to the window of a house or apartment with deployed femto cells. This can cause a "deadzone" surrounding an HeNB. Such a deadzone is larger for HeNBs near the edge of the macro cell, where the signal received from the macro cell is weakest, or for MUEs located indoors in the coverage of a CSG HeNB. Figure 10.16 shows an example of this kind of interference scenario. Potential interference mitigation approaches for this scenario may include:

1. Enabling hybrid or open access if possible: in the case of hybrid access, the power settings of the HeNB could be adapted differently to the closed access case, taking into account the total system performance (macro plus hybrid cell) and the resources consumed via "visiting" non-CSG UEs.
2. Downlink power setting: the HeNB can limit the maximum downlink power (or power per PRB) according to its environment, such as set the power to achieve a tradeoff between coverage and interference, based on the estimated path loss between the HeNB and the victim MUEs. This could be done through detecting uplink transmissions at the HeNB, or by means of measurement reports from victim MUEs to the serving MeNB if it is possible to signal this information to the HeNB.
3. Time-domain coordination making use of almost blank subframes (ABS): ABSs contain only certain essential transmissions, leading to a reduction in interference to victim UEs. Typically, an aggressor HeNB would set up a pattern of ABSs resulting in reduced interference to victim MUEs. In Release 10, there is no X2 interface between HeNBs and macro eNodeBs, and therefore ABS patterns at an HeNB have to be configured either by operation and maintenance (O&M) or by the HeNB itself.

10.4.2.3 UL interference from macro to femto

The uplink transmission of the FUE will suffer interference from the uplink transmission of the MUE nearby. Due to the large difference in transmission power between the macro cell and femto, this kind of interference is more serious when MUE and HeNB are both situated indoors. Figure 10.17 shows an example of this kind of interference scenario.

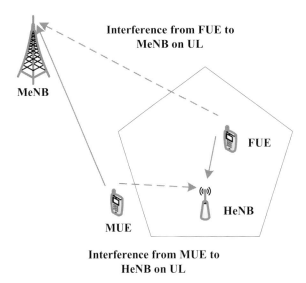

Figure 10.17 Macro–femto uplink interference scenarios.

Potential interference mitigation approaches for this scenario may include:

1. Uplink power control: the HeNB can control the power of its FUEs to overcome the interference from neighboring MUE.
2. Uplink control channel protection (PUCCH): to keep the orthogonal resource allocation between the HeNB's PUCCH and macro cell's PUCCH.

10.4.2.4 UL interference from femto to macro

The uplink transmission of the MUE will also suffer interference from the uplink transmission of the FUE nearby. Figure 10.17 shows an example of this kind of interference scenario. Potential interference-mitigation approaches for this scenario are similar to those for the above-mentioned scenario for UL interference from macro to femto.

10.4.2.5 Interference among femtos

Interference among femto cells is also a problem to be considered in both downlink and uplink transmissions. The similar approaches, such as control channel protection, ABS, and UL/DL power control can also be applied here.

10.4.3 Network listening mode

Due to the restriction on UE for CSG, there is no information about the measurement results from MUE for the femto cell. A new network listening mechanism to support the measurement of the surrounding interference on the femto cell has been designed. Based on this mechanism, some of the above-mentioned interference mitigation schemes can be realized [21, 22]. HeNBs are therefore commonly designed to have a network listen mode (NLM) of operation, which involves making measurements and decoding system information from neighboring eNodeBs. This can be performed during the initial system setup procedure and updated periodically thereafter. Several aspects can be included into the NLM operation:

1. Uplink interference power: to infer the presence of nearby MUEs.
2. Determination of cell IDs and CSG status/ID: by decoding the SI of the neighbor cells.
3. Co-channel RSRP and RS transmission power measurement: to estimate the path loss to neighbor cells, which can be used for uplink and downlink power control at the HeNB.
4. Reference signal received quality (RSRQ) measurement: to be used together with RSRP to determine the reliability of coverage of a macro cell, to help determine a suitable power setting for a femto cell operating in hybrid access mode. In addition, TDD femto cells may obtain time synchronization from neighboring macro cells.
5. Uplink interference power: to infer the presence of neighbor MUEs; determination of cell IDs and CSG status/ID: by decoding the SI of the neighbor cells; co-channel RSRP and RS transmission power measurement: to estimate the path loss to neighbor cells, which can be used for uplink and downlink power control at the HeNB.

10.5 Small cell enhancement in Release 12

In the above four sections, the standardization evolutions for small cell related topics have been discussed according to different kinds of definitions. Actually, during the manuscript preparation of this chapter, the new round of 3GPP standardization discussions for Release 12 features has already been launched in all RANs. A key study topic in Release 12 version is small cell enhancement. In this section, we will summarize the current proposed technical issues and corresponding potential solutions for the small cell enhancement identified scenarios. Although not all of these solutions have been included in the final specifications, it will provide the guidance for the following standard work item phase.

10.5.1 Scenarios for small cell enhancement

Unlike the previously discussed small cell deployment, the scenarios discussed in small cell enhancement (SCE) are even more complicated: in the context of SCE, both the scenario where the macro cells and small cells are deployed on the same carrier (co-channel case), and the scenario where the macro cell and small cells are deployed on different carriers (non co-channel case). In addition, the small cell standalone case is also included. For the implementation aspects, the first one is the backhaul requirement. Both the ideal backhaul and non-ideal backhaul, which is highly relevant to the backhaul coordination capacity and latency, have been considered. The second one is

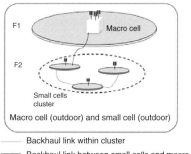
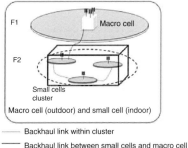
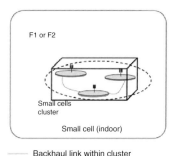

Figure 10.18 Four typical deployments for SCE.

a group of small cells. More dense small cell deployment has been considered compared to Release 10 or 11 small cell related discussion. Figure 10.18 has shown the four typical deployment scenarios discussed in Release 12 SCE study item, which is referred to in [23].

10.5.2 Spectrum efficiency improvement

One key aspect discussed in SCE is to improve the spectrum efficiency in the physical layer including the control channel, data channel, and reference signals. Several major schemes have been identified as the potential direction for further efforts.

10.5.2.1 Higher order modulation

Currently the highest modulation order in LTE and LTE-Advanced system is 64 QAM. For some SCE scenarios, the UEs may experience extremely high SINRs, which make the 256 QAM modulation for downlink transmission possible. Through implementation of the 256 QAM, more efficient spectrum usage could be achieved.

10.5.2.2 Overhead reduction for downlink UE-specific reference signal

According to the deployment of the SCE, low frequency-selective and low time-selective fading provide the possibility for overhead reduction of downlink UE-specific reference signal. One major concept is to reduce the occupied resource elements (REs) of the DM-RS by half in the frequency domain, which means only support port 7 and 8 instead of all. By doing this, the saved REs can be used for data and control channel transmission to improve spectrum efficiency.

10.5.2.3 Overhead reduction for uplink UE-specific reference signal

Similar to the objective of the downlink reference signal reduction, the uplink reference signals can also be reduced in the following two aspects:

1. reduction of number of the reference signal symbols per subframe
2. reduction of number of subcarriers carrying the reference signal.

10.5.2.4 Control channel enhancement

Currently, the control channel capacity becomes a bottleneck when a large number of small size packets arrive simultaneously. Because of the low UE mobility, leading to relatively time-invariant channel conditions for frequency-selective scheduling and link adaptation, it may be less necessary to schedule UE in a separate downlink control indication (DCI) message for each packet. Multi-subframe and cross-subframe scheduling is proposed to enable a new operation with a single DCI message to allocate resources in multiple subframes.

10.5.2.5 Enhanced power control/adaptation

Small cell downlink power control refers to the adaptation of small cell transmission power, including both control channel power and data channel power. The small cell can optimize its transmission power of the downlink control or data channel to achieve the interference mitigation to the neighboring cells. Both the frequency-domain power optimization on different resource block and the time-domain power optimization on different time-slot (subframes) can be applied.

10.5.3 Operation efficiency improvement

Due to multi-carrier deployment, as well as the dense distribution of the large number of small cells, the operation efficiency should also be improved to enhance the performance from the network perspective.

10.5.3.1 Small cell on/off

In order to operate the whole network in a more environmentally friendly manner to accommodate the potential traffic fluctuation, as well as to mitigate the potential interference originated by dense small cell deployment, the small cell on/off operation has been proposed as a key alternative technology for Release 12 SCE. According to the time-scale difference during operation, the operation of the small cell on/off can be classified in semi-static manner and dynamic manner. For the semi-static way, the criteria used for the on/off decision may be traffic load increase/decrease, UE arrival/departure (i.e., UE–cell association), and packet call arrival/completion. For the dynamic way, the small cells may be turned on/off at subframe level, following criteria such as packet arrival/completion and the need for interference coordination/avoidance in subframe time scales. In other words, at the moment of a packet arrival, the small cells can be turned on immediately and transmit the packet to a UE, and it can be turned off at the moment of completion of the packet. The small cell can be turned on/off immediately based on the need for interference coordination/avoidance. Due to the introduction of small cell on/off, some corresponding mechanisms need to be considered further.

10.5.3.2 Mechanism for small cell wake-up

The criterion for small cell "off" may be a straightforward operation of no UE served or quite limited service delivered. However, the mechanism of small cell "on" can have several options of implementation. According to the Release 12 discussion in SCE, there are two major mechanisms for small cell wake-up operation. One is based on the detection of feedback from UE via the small cell's downlink signals. Both the legacy reference signals, such as PSS/SSS/CRS with reduced density, and the new discovery signals, can be used for UE for discovery. Based on this option, the "off" cell will mute all the control and data but keep the discovery signals transmission. For downlink-based wake-up manner, it can be operated in a network control manner since the UE location information can be acquired by the network. Some centralized node can be used to optimize the on/off operation. Another option is based on the detection of uplink signals transmitted by UE, such as PRACH or sounding RS (SRS). For this option, the "off" cell can mute all downlink transmission and keep on "overhearing" the uplink transmission by surrounding UEs. It is more like a cell autonomous operation for uplink based wake-up operation.

10.5.3.3 Dual connectivity

Another key technology proposed in Release 12 SCE is the concept of dual connectivity [15]. Due to the imbalance of the UL/DL between macro and small cells, and mobility robustness problem under dense deployment, the terminology of "dual connectivity" is proposed. Different from the legacy UE association on one network point (such as eNodeB), the dual-connected UE will assume to acquire the radio resources provided by at least two different network points connected with non-ideal backhaul. Furthermore, each eNB involved in dual connectivity for a UE may assume different roles. Those roles do not necessarily depend on the eNodeB's power class and can vary among UEs. According to the different separation of functionalities for user plane and control plane, different alternatives of architecture have been summarized in [24].

10.5.4 Inter-cell synchronization improvement

It is well known that TDD systems require inter-cell synchronization on the same frequency or different frequencies within the same band; in addition, for both TDD and

FDD, it would be essential to consider the synchronization mechanisms between cells to bring benefit to the existing features (such as (F)eICIC, CoMP (coordinated multiple-points processing) as well as carrier aggregation) and some potential techniques for small cell enhancements (e.g., efficient small cell on/off and discovery). For small cell enhancement scenario, because of potential non-ideal backhaul, the legacy inter-cell synchronization mechanism, such as GNSS or synchronization over backhaul network, may not be available. Currently, two kinds of enhancement have been proposed for inter-cell synchronization improvement in Release 12 SCE.

10.5.4.1 Network listening

The target cell monitors the network listening RS (e.g., CRS, CSI-RS, and PRS) of the source cell directly to maintain synchronization with the source cell. When the target cell monitors the source cell, the target cell mutes its own transmission at least when the target cell and the source cell are in the same frequency.

10.5.4.2 UE-assisted synchronization

The synchronization between the source cell and the target cell can be achieved by some information provided by or obtained from UEs. The indication of stratum level is also needed when UE-assisted synchronization is applied. It may include the following steps:

1. UEs provide information to the network or transmit UL signals for network to detect
2. eNodeBs exchange information and the target cell adjusts its timing
3. possible measurements on the UE and eNB sides including the propagation delays between the UE and both eNodeBs.

10.6 Summary

The introduction of small cells represents a significant new step in the LTE radio access networks as it is evolved for LTE-Advanced. The incorporation of small cells into an LTE network can offer attractive cost advantages and more spectrum efficiency for network operators compared to a homogeneous deployment of macro eNodeBs. It can be foreseen that relevant research efforts into small cells and standardization evolutions would significantly improve current wireless communication networks.

References

[1] Loa, K., Wu, C., Sheu, S., Yuan, Y., Chion, M., Huo, D., and Xu, L. (August 2010). "IMT-advanced relay standards," *IEEE Communications Magazine* 48(8), 40–48.

[2] Hoymann, C., Chen, W., Montojo, J., Golitschek, A., Koutsimanis, C., and Shen, X. (February 2012). "Relaying operation in 3GPP LTE: challenges and solutions," *IEEE Communications Magazine* 50(2), 156–162.

[3] Okino, K., Nakayama, T., Yamazaki, C., Sato, H., and Kusano, K. (June 2011). "Pico cell range expansion with interference mitigation toward LTE-Advanced heterogeneous

networks," *Proc. IEEE International Conference on Communication (ICC) Communications Workshops.*

[4] Mukherjee, S. (June 2011). "UE coverage in LTE macro network with mixed CSG and open access femto overlay," *Proc. IEEE International Conference on Communication (ICC) Communications Workshops.*

[5] Andrews, J. G., Claussen, H., Dohler, M., Rangan, S., and Reed, M. C. (Apr. 2012). "Femtocells: past, present, and future," *IEEE J. Select. Areas Commun.* 30(3), 497–508.

[6] 3GPP1 (Sept. 2012). "Evolved universal terrestrial radio access (E-UTRA); Physical layer for relaying operation (Release 11)," 3GPP Technical Specification 36.216 V11.0.0, www.3gpp.org

[7] 3GPP2 (Jan. 2013). "UTRA repeater radio transmission and reception (Release 11)," 3GPP Technical Specification 25.106 V11.1.0, www.3gpp.org.

[8] 3GPP3 (Mar. 2013). "FDD repeater radio transmission and reception (Release 11)," 3 GPP Technical Specification 36.106 V11.2.0, www.3gpp.org.

[9] 3GPP4 (June 2008). "R1-082024 A discussion on some technology components for LTE-Advanced."

[10] 3GPP5 (June 2008). "R1-082397 Discussion on the various types of relays," Panasonic, 3GPP TSG RAN1 WG Meeting #54, Warsaw, Poland, www.3gpp.org.

[11] 3GPP6 (Mar. 2010). "Evolved universal terrestrial radio access (E-UTRA); further advancements for E-UTRA physical layer aspects (Release 9)," 3GPP Technical Report 36.814 V9.0.0, www.3gpp.org.

[12] 3GPP7 (Sept. 2013). "Evolved universal terrestrial radio access (E-UTRA) and evolved universal terrestrial radio access network (E-UTRAN); Overall description; Stage 2 (Release 11)," 3GPP Technical Specification 36.300 V11.7.0," www.3gpp.org.

[13] 3GPP9 (2009). "Study item: enhanced inter-cell interference control (ICIC) for non-carrier aggregation (CA) based deployments of heterogeneous networks for LTE," 3GPP Features and Study Items (Rel-10), www.3gpp.org/ftp/Specs/html-info/FeatureListFrameSet.htm.

[14] 3GPP8 (2010). "Study item: further enhanced non CA-based ICIC for LTE," 3 GPP Features and Study Items (Rel-11), www.3gpp.org/ftp/Specs/html-info/FeatureListFrameSet.htm.

[15] 3GPP10 (Apr. 2010). "R1-101874 Co-channel control channel performance for HetNet," Alcatel-Lucent, Alcatel-Lucent Shanghai, 3GPP TSG RAN WG1 Meeting #60bis, Beijing, P. R. China, www.3gpp.org.

[16] 3GPP11 (Sept. 2013). "Evolved universal terrestrial radio access network (E-UTRAN); X2 application protocol (X2AP) (release 11)."

[17] Qin, M., Liu, L., Lan, C., and Takeda, K. (June 2013). "Search space design in enhanced physical downlink control channel for LTE-Advanced," *Proc. IEEE Vehicular Technology Conference (VTC Spring).*

[18] Damnjanovic, A., Montojo, J., Wei, Y. *et al.* (June 2011). "A survey on 3GPP heterogeneous networks," *IEEE Wireless Commun.* 18(3), 10–21.

[19] 3GPP12 (Mar. 2013). "Evolved universal terrestrial radio access (E-UTRA); requirements for support of radio resource management Release 11," 3GPP Technical Specification 36.133 V12.3.0, www.3gpp.org.

[20] 3GPP13 (Sept. 2013). "Evolved universal terrestrial radio access (E-UTRA); radio resource control (RRC); protocol specification," 3GPP Technical Specification 36.331 V12.1.0, www.3gpp.org.

[21] 3GPP14 (Sept. 2012). "Evolved universal terrestrial radio access (E-UTRA); FDD Home eNode B (HeNB) radio frequency (RF) requirements analysis," 3GPP Technical Report 36.921 V11.0.0, www.3gpp.org.

[22] 3GPP15 (Sept. 2012). "Evolved universal terrestrial radio access (E-UTRA); TDD Home eNode B (HeNB) radio frequency (RF) requirements analysis," 3GPP Technical Report 36.922 V11.0.0, www.3gpp.org.

[23] 3GPP16 (Sept. 2013). "Small cell enhancements for E-UTRA and E-UTRAN: physical layer aspects (Release 12)," 3GPP Technical Report 36.872 V12.0.0, www.3gpp.org.

[24] 3GPP17 (June 2013). "Study on small cell enhancements for E-UTRA and E-UTRAN: higher layer aspects (Release 12)."

11 Game theory and learning techniques for self-organization in small cell networks

Prabodini Semasinghe, Kun Zhu, Ekram Hossain, and Alagan Anpalagan

11.1 Small cell networks

The tremendous increase of bandwidth-craving mobile applications (e.g., video streaming, video chatting, and online gaming) has posed enormous challenges to the design of future wireless networks. Deploying small cells (e.g., pico, micro, and femto) has been shown to be an efficient and cost-effective solution to support this constantly rising demand since the smaller cell size can provide higher link quality and more efficient spatial reuse [1]. Small cells could also deliver some other benefits such as offloading the macro network traffic, providing service to coverage holes and regions with poor signal reception (e.g., macro cell edges). Following this trend, the evolving 5G networks [2] are expected to be composed of hundreds of interconnected heterogeneous small cells.

Figure 11.1 gives an illustration of a heterogeneous network (HetNet) where a macro cell is underlaid with different types of small cells. Different from the cautiously planned traditional network, the architecture of a HetNet is more random and unpredictable due to the increased density of small cells and their impromptu way of deployment. In this case, the manual intervention and centralized control used in traditional network management will be highly inefficient, time consuming, and expensive, and therefore will be not applicable for dense heterogeneous small cell networks. Instead, **self-organization** has been proposed as an essential feature for future small cell networks [3, 4].

The motivations for enabling self-organization in small cell networks are explained below.

- Numerous network devices with different characteristics are expected to be interconnected in future wireless networks. Also, these devices are expected to have "plug and play" capability. Therefore the initial pre-operational configuration has to be done with minimum expertise involvement.
- With the emergence of small cells, the spatio-temporal dynamics of the networks has become more unpredictable than legacy systems due to the unplanned nature of small cell deployment. Therefore intelligent adaptation of the network nodes is necessary. That is, the self-organizing small cells need to learn from the environment and adapt with the network dynamics to achieve the desired performance.

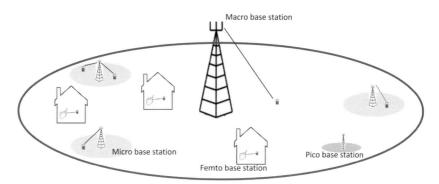

Figure 11.1 A heterogeneous network.

- Improper or uncoordinated power and spectrum allocation paradigms can lead the small cells to cause severe inter-tier and intra-tier interference. Therefore resource allocation is a key issue for interference management in heterogeneous small cell networks. Centralized control will be highly inefficient and time consuming for a dense network due to the high computational power and the huge amount of information exchange required. Instead, small cell base stations (SBSs) should be capable of taking individual decisions on resource allocation with local interactions.
- Self-organization of the network will also prevent possible human mistakes in configuration and network management, which can drastically degrade the performance of the network and can result in extensively long recovery times. Also, enabling self-organization could reduce a considerable amount of operational and capital expenditure (OPEX/CAPEX).

The Small Cell Forum, which is an organization that supports and promotes the wide-scale adoption of small cell technologies, claims that small cells are the first commercial example of a self-organizing network in practice [5].

There are ongoing projects that develop the self-organizing paradigms for small cell networks involving both academia and industry. BeFemto (**B**roadband **e**volved **Femto** network) is one such project, which focuses on developing femto cell technologies for LTE-A systems [6]. They also plan to provide guidelines for standardization of the next generation femto cell technologies. SOCRATES (**S**elf-**O**ptimization and self-**C**onfigu**RAT**ion in wirel**E**ss network**S**) also targets developing self-organizing paradigms for small cell networks in 3GPP LTE interface [7, 8]. The SOCRATES project was partnered by several leading telecommunication companies in Europe including Nokia Siemens Networks (in Poland and Germany), Vodafone (United Kingdom), and Ericsson AB (Sweden). The End-to-End Efficiency (E^3) project [9, 10] works on integrating the heterogeneous network infrastructures into a scalable and efficient cognitive framework with self-organizing capabilities. In addition to that they also focus on research, regulation, and standardization perspectives of cognitive radio networks.

11.2 Self-organization

The concept of self-organization is not new and can be widely observed in many natural systems and phenomena (e.g., collective behaviors of ants and social insects, flocks of cranes, generation of laser light, and planetary systems). Extensive efforts have been taken by researchers to model the self-organizing behavior of natural systems mathematically and these models can be borrowed and adapted to develop self-organizing algorithms for artificial systems [11]. First, it is essential to understand the basic properties, requirements, and design concepts of a self-organizing system.

As self-organization is a concept being used in many different fields, the term has been defined in many different ways based on the context. A globally accepted precise and concise definition of self-organizing networks (SONs) has not yet been presented. However, in the area of wireless communication, the standardization of technical specifications for self-organizing LTE and LTE-A networks has been initiated by the 3rd Generation Partnership Project (3GPP) in Release 8 and Release 9 [12, 13] and the Next Generation Mobile Networks (NGMN) Alliance [14, 15]. In this section, we will illustrate the concept of self-organization and its basic cornerstones in the framework of cellular networks.

The basis of a self-organizing system is its *autonomous and intelligent adaptivity*, i.e., the ability to respond to external environmental changes. Much literature in the context of wireless networks also suggests that a self-organizing network should be capable of *learning* from environmental dynamics and adapt to them accordingly [16–18]. Specifically, for small cell networks, detecting the environmental dynamics can be done based on local interactions with other nodes and/or through spectrum sensing. In [19], the authors explain that the adaptive behavior of each member of a self-organizing set should also lead the whole system to form a global pattern, which is denoted as the *emergent behavior*.

Based on the above notions, the basic cornerstones of a self-organizing small cell network are identified as follows:

- autonomous and intelligent adaptivity
- ability to learn from the environment
- emergent behavior.

In addition to the aforementioned properties, researchers also discuss *distributed control* where each node in the network has to take individual decisions on its own behavior. Distributed control is a desirable feature for self-organizing small cell networks. In 3GPP Release 11, the specifications divide self-organizing networks into three categories as given below.

- Centralized SON: self-organizing algorithms are executed in the operation, administration, and management (OAM) system.
- Distributed SON: self-organizing algorithms are executed at the network node level.
- Hybrid SON: algorithms are executed at both the OAM and network node levels.

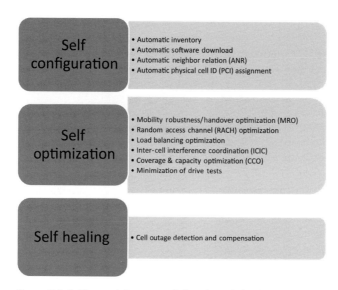

Figure 11.2 Self-organizing network functions defined by 3GPP.

Distributed resource allocation is essential for the provision of distributed control in self-organizing networks. Several distributed resource allocation algorithms for small cell networks have been proposed in the recent literature, which will be discussed later in this chapter.

11.2.1 Self-organizing functionalities

In general, the self-organizing process of a small cell network can be split into three phases, i.e., pre-operational phase, operational phase, and failure recovery phase. These three phases commonly correspond to *self-configuration, self-optimization*, and *self-healing*, which are also referred to as *Self-X* functionalities [3, 20, 21].

During the standardization process for LTE SON, 3GPP has defined a set of use cases and associated functions in Releases 9, 10, and 11 [22–24], which are described in Figure 11.2. The Next Generation Mobile Networks (NGMN) Alliance also highlights several operational use cases for the introduction of SON features for mobile networks. The NGMN divides the SON-related use cases into four categories i.e., planning, deployment, monitoring, and maintenance [25]. However, most of the steps in planning are not covered by SON functions, therefore we only list the use cases of the latter three categories in Figure 11.3, which are similar to the 3GPP SON user cases. The network parameters such as neighborhood list and handover settings are considered as radio parameters while IP addresses and QoS requirements are considered as transport parameters.

A brief overview of the operation and associated functions of each phase of self-organization is given below.

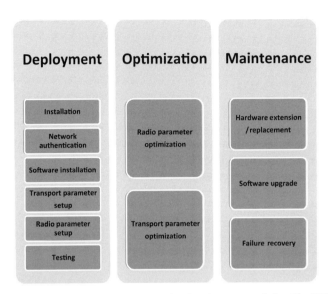

Figure 11.3 Self-organizing network-related use cases defined by NGMN.

- *Self-configuration.* Self-configuration is performed in the pre-operational process during which the small cell base stations (SBSs) connect to the network and execute their initialization algorithms automatically while providing plug-and-play capabilities to the network nodes. This functionality is composed of basic set-up of the base station and the initialization of network parameter settings.

  Specifically, an SBS is expected to automatically configure its IP address once it is connected to the network. This can be done by using the dynamic host configuration protocol (DHCP). Then the SBS can communicate with the OAM center and small cell access gateway for authentication. This procedure is called automatic inventory. Once the SBS is connected to the core network it can download and install the required software (i.e., automatic software download). The SBSs are also expected to set transport parameters such as transport layer QoS setting and radio parameters (e.g., neighborhood list and handover settings). The assignment of a PCI (physical cell ID) is also done in the self-configuration phase.

  As small cells are usually deployed in the coverage area of macro cells, frequency reuse scheme plays a major role in interference control and frequency selection is important, which needs to be decided at the self-configuration phase. An SBS should identify its allowable frequency band before entering into the operational phase. One option is to use universal frequency reuse (UFR) with cross-tier interference constraints and another option is to split the existing bandwidth for each tier. In the latter case there will be no cross-tier interference; however, the spectral efficiency can be less than that in UFR [26]. A detailed description of frequency selection will be given in Section 11.3. Enabling self-configuration process lessens or avoids involvement of manual expertise during the installation phase.

- *Self-optimization.* The main task of self-optimization is to automatically adjust certain parameter settings to adapt with the network dynamics for the optimal performance. In order to perform self-optimization, the network nodes need to measure certain network parameters (e.g., number of users, traffic patterns, and traffic load) and collect the information about the network conditions (e.g., channel gains). Then this information can be used to optimize network performance.

    In recent literature, many approaches have been proposed to realize self-optimization in small cell networks. Some of the prominent game theory-based approaches will be discussed in the latter parts of this chapter.

    Resource allocation-based inter-cell interference coordination is one of the mostly targeted issues in self-organizing networks. Resource allocation settings (e.g., channel allocation and power allocation) and scheduling are significant in inter-cell interference coordination. Different criteria can be used for performance optimization depending on the objectives. Several commonly used optimization objectives are as follows: thoughput/data rate/SINR maximization, coverage maximization, load balancing, and power minimization.

    Note that multiple objectives can also be merged together by defining a proper payoff function [27, 28].

    These objectives can be further categorized as system-centric objectives and user-centric objectives. System-centric objectives focus on optimizing the total network performance rather than individual performance. This type of approaches generally relies on a considerable amount of information exchange among network nodes and a centralized controller is usually required. An example is the maximization of the total network throughput with the constraint of a maximum transmit power. In comparison, user-centric objectives focus on individual performance at each node (e.g., maximizing the individual rate) rather than the overall performance. These types of objectives are common for self-organizing networks since they are more likely to rely on local interactions among the nodes.

- *Self-healing.* Self-healing enables the network to have the ability to detect, diagnose, compensate, and recover from failures and abnormal status. Accordingly, the self-healing process is mainly composed of three functions [29]: fault detection, fault diagnosis, and fault recovery.

    Traditional healing approaches may not be feasible due to the existence of a large number of heterogeneous base stations and their random nature of deployment. Instead, methods for self-healing would be required. Firstly, the problems should be detected from performance measurement (e.g., abrupt performance degradation) or event-driven report. In this case, periodic monitoring should be performed. Then a diagnosis process can be performed to determine the cause of the failure (e.g., software or hardware) according to which the corresponding compensation and recovery schemes can be performed. In the case of software faults, the base station may try several actions such as reloading of a backup of software, activation of a fallback software load, and downloading a software unit and reconfiguration. In the case of hardware faults, the base station may use redundant resources [30].

11.2.2 Characteristics of self-organizing algorithms

Comprehending the significant and necessary features of a self-organizing algorithm is important and essential for the design of self-organizing small cell networks. In this section, we summarize the important characteristics of self-organizing algorithms as follows.

1. *Stability.* In the context of self-organizing networks stability is defined as [3]: "An algorithm or adaptation mechanism that is able to consistently traverse a finite number of states within an acceptable finite time." That is, a self-organizing algorithm should be able to converge within acceptable iterations. Note that for game theory-based algorithms, there could exist more than one equilibrium point. Certain conditions and initial points may be required for the algorithm to converge to the desired equilibrium point. Also, the delay in information exchange may result in delayed convergence or oscillations around an equilibrium point [28].
2. *Robustness.* This is the ability of an algorithm to reach back to a stable state within a bounded duration of time in case of an unexpected change in the system or environment, which makes the system deviate from a stable state. Small cell base stations may be more vulnerable to failures than cautiously planned macro base stations. Self-organizing algorithms should be capable of bringing the system back to an equilibrium state. In this regard, robustness can also be viewed as a part of self-healing functionality.
3. *Scalability.* The complexity of self-organizing algorithms should not increase in an unbounded manner with the increase of network size. The scalability poses certain complexity requirements on the algorithms. Specifically, less complex algorithms, which occupy less computation resource (e.g., CPU and memory), could make the network more scalable. Also, the amount of information exchange should not increase unbounded with increase in the number of network nodes. Learning through local interactions can prevent the system from extensive information exchange.
4. *Agility.* The network should respond to the environmental changes within a reasonable duration of time. Agility depends on backhaul constraints of the nodes as the information has to be exchanged prior to decision making. It also depends on the computational power of the network nodes. Global information exchange can make the system respond too sluggishly, while responding to temporary changes may also result in oscillations between states. Therefore perfect agility is considered as one of the most difficult conditions to be fulfilled for a self-organizing network.

11.3 Issues and challenges in self-organizing small cell networks

Enabling self-organization for small cell networks poses a number of issues and challenges, which should be fully understood. In this section, we identify the main design issues and challenges for self-organizing small cell networks.

1. *Interference mitigation.* Due to the scarcity of the available bandwidth allocated for wireless networks, small cells have to share the same transmission bandwidth with the existing macro network, which results in both cross-tier and co-tier interference. With the increasing density of the small cell networks, interference mitigation, which is essential for self-organizing small cell networks, becomes more challenging.
2. *Resource management.* Guaranteeing the efficient coexistence of a large number of small cells with traditional macro cells from the perspective of resource allocation is a fundamental issue [31]. Self-organizing algorithms should be capable of performing resource allocation to achieve optimal performance. Note that resource allocation objectives may vary depending on the requirement. For example, cross-tier and intra-tier interferences can be mitigated through proper power and sub-channel allocation. In addition to that most other use cases categorized under self-optimization phase (e.g., load balancing, coverage and capacity optimization, and handover optimization) can also be achieved by using appropriate resource allocation. It is desirable for SON entities to take independent decisions on resource allocation without any centralized control. Therefore developing distributed or semi-distributed resource-management techniques for self-organizing small cell networks is one of the key issues.
3. *Access control.* A mobile user in a multi-tier network is capable of connecting to either macro base station or a small cell base station provided that the user is in the coverage area of both cells. This decision can be taken by the users based on the receive power of the pilot signal. On the other hand, the base stations can also decide how many users and which users should be accepted to be served in order to meet their own requirements (e.g., maximizing the total capacity and load balancing). Decisions on access control are expected to be taken distributively in SONs.
4. *Learning and reasoning.* Devising suitable learning techniques for self-organizing small cell networks is one of the major challenges. Self-organizing entities are expected to collect network information during the learning process. The learning technique should be strong enough to develop a sufficient knowledge base that can be used by the self-organizing entities to exploit the available resources efficiently. This also involves issues such as deciding the information collection rate and achieving a balance between the exploration and exploitation tradeoff. Reasoning refers to the decision process to achieve optimal or desired network performance according to the knowledge base obtained during the learning process.
5. *Computation cost.* The SBSs may not have as high processing power as that of traditional macro base stations. In this case, complex algorithms that require high computation power may not be suitable for small cells. Designing low-complexity self-organizing algorithms for small cells is a major challenge.
6. *Imperfect information.* With certain self-organizing algorithms, the SBSs are expected to exchange information with nearby nodes (i.e., local interactions). However, this information can be distorted due to the noisy backhaul and can be delayed due to the time taken in processing and transmission. In addition to local interactions, many algorithms also rely on channel state information (CSI), while CSI can also be distorted or temporally unavailable due to the fading experienced by feedback

channels. Also, if the status of each channel is estimated by spectrum sensing, the sensing result can be inaccurate.

This imperfect information could affect the self-organizing algorithms from two aspects. First, the performance of the algorithms could degrade due to the use of inaccurate information. Second, the stability of the algorithms may not be guaranteed due to the delayed information [28, 32]. Therefore dealing with imperfect information also poses a significant challenge to the design of self-organizing algorithms for small cell networks. The consideration of imperfect information and quantification of its effect can be found in several works such as [32–34].

7. *Limited backhaul.* Unlike macro base stations, which have a separate backhaul, SBSs such as femto base stations connect to the core network via an IP-based backhaul such as DSL. The same backhaul link may also be used for inter-cell coordination and periodic information exchange required by self-organizing algorithms. The limited capacity of backhaul and the possible latency and errors introduced are considerable issues in the context of self-organizing small cell networks. Also note that the backhaul can be hybrid (e.g., coexistence of both wired and wireless backhaul) with different constraints [35]. These backhaul limitations and constraints should be taken into account when developing self-organizing algorithms for small cell networks. Security is also a significant issue since the backhaul may not be owned by the same operator.

11.4 Game theory for self-organizing small cell networks

11.4.1 Fundamentals of game theory

Game theory provides a rich set of mathematical tools for modeling and analyzing interactive decision-making problems in which the interests of agents (i.e., players) may conflict with each other. It is a well developed area in applied mathematics and has been used primarily in economics to model competitions in markets.

In recent years, game theory has also been widely adopted to solve many problems in the area of wireless communications [36, 37]. A number of works have explored the applications of game theory for the analysis and optimization of various issues in wireless systems, in most cases to solve resource-allocation problems in a competitive environment.

A non-technical definition of a game is given as follow. A game is a process in which the agents select certain strategies from their own strategy sets and obtain payoffs according to the strategies of all agents. The choice of a strategy can be made both simultaneously and non-simultaneously. In addition, an agent may make decisions multiple times according to the game rule. A game consists of a set of players, a set of strategies available to those players, and a specification of payoffs for each combination of strategies.

1. *Set of players* $\mathcal{N}$. The set of decision makers involved in the game. The players are assumed to be rational or bounded rational depending on the type of the game.

2. *Set of strategies* $(S_i)_{i \in \mathcal{N}}$. Strategies are the options that a player can select depending on the state of the game. Here S_i denotes the set of strategies of player $i \in \mathcal{N}$. A player's strategy could contain a single action, multiple actions, or probability distribution over multiple actions. As common in game theory, S_{-i} denotes the strategies of all players other than i. The state of a game depends on the strategies taken by all the players (i.e., $[s_i, s_{-i}]$). Note that different players could have different strategy sets.
3. *Payoff* π_i. The payoff represents the preference of each player under the current strategy profile. The payoff could be modeled as a cost function $c_i(s_i, s_{-i})$, a utility function $u_i(s_i, s_{-i})$, or a combination of both (e.g., in the form of equation (11.1)), where the cost function represents the cost of performing certain strategies (e.g., transmit power), which need to be minimized, and the utility function represents the gain (e.g., profit of service providers), which needs to be maximized.

$$\pi_i(s_i, s_{-i}) = u(s_i, s_{-i}) = c(s_i, s_{-i}) \tag{11.1}$$

It is straightforward to see that a player's payoff depends not only on her own strategy but also on the strategies of all other players.

11.4.2 Motivations of using game theory for self-organizing networks

The motivations of using game theory for self-organizing small cell networks are summarized below.

- The heterogeneous network nodes in small cell networks can be deployed by different operators/users. The performance of one network could be easily affected by the behavior of other networks. In this case, modeling of interactive behavior would be required. Different from optimization models in which the mutual impact among different entities during the decision-making process cannot be accurately taken into account, game theory models provide a mathematical framework to analyze the competitive or cooperative interactions among the players in a multi-player system.
- Different network nodes could have different QoS requirements and can be self-interested. Each node takes individual decisions (e.g., on resource allocation and scheduling) to meet her own requirements rather than optimizing the system-wide performance. In this case, these nodes may have conflicting interests. Such self-interested behavior can be easily modeled by using game theory (e.g., by formulating a non-cooperative game). The "self-interest" of the nodes can be modeled in terms of performance metrics such as capacity, delay, throughput, interference, and signal to interference-plus-noise ratio (SINR), in the payoff.
- The basic keystones of a self-organizing network as defined in Section 11.2 are *ability to learn from environment, autonomous adaptivity and emergent behavior capability*. In the context of game theory, the players could adapt their decisions to obtain a better payoff (i.e., learning and adaptation). Also, after several adaptation iterations, the game could reach the equilibrium (emergent behavior). The above mentioned properties of a self-organizing network can be attained by devising self-organizing algorithms based on game theory.

- Centralized algorithms could be highly inefficient for dense heterogeneous wireless networks due to the complexity of the algorithms and the amount of information exchange. Accordingly, distributed control is a desirable feature for self-organizing small cell networks as explained in Section 11.2. Game theory provides a natural tool to develop distributed self-organizing algorithms as it allows local interactions and individual decision making. Local interactions will reduce the amount of information exchange among the nodes and as a result the network becomes more scalable and more capable of operating with limited backhaul conditions.

11.4.3 Types of games

Different game models (e.g., non-cooperative/cooperative, static/dynamic) have been used to address self-organizing problems in small cell networks, the choice of which depends on the characteristics of the network, the applications, and also the objectives.

Different game-theory models may differ considerably in structure from many aspects, e.g., number of players, number of strategies, and payoffs. The number of players may vary in different games. If a game has only one player, the game becomes an optimization problem. We call it a two-person game, or multiple-person game, if the game has two or more players, respectively. In different games, the number of strategies for players can be either finite (e.g., in a rock–scissors–paper game) or infinite (e.g., in a pricing game). The analysis of a finite strategy game and an infinite strategy game are different. The summation of payoffs of all players may also differ in different models. In general, this summation can be zero, a non-zero constant number, or any arbitrary value. The game process is an important aspect in the game structure. The players in a game may take actions simultaneously, in a certain order, or in a repeated fashion, according to which the game can be referred to as a *static game*, a *dynamic game*, and a *repeated game*, respectively. In addition, the assumptions of players' rationality are different. Most of the game theory models assume perfect rationality of players, while some models consider that the players are with limited rationality (i.e., bounded rationality). According to the above analysis, game models can be divided into the following categories.

11.4.3.1 Non-cooperative vs. cooperative games

Non-cooperative games are the most popular games. In non-cooperative games, the players are commonly considered to be rational and self-interested who have fully or partially conflicting interests. Each player selects the strategy to optimize her own payoff function. For non-cooperative games, the most commonly used solution concept is Nash equilibrium (NE), the definition of which is given as follows.

DEFINITION 11.1 *Nash equilibrium*: let $s_i \in \mathcal{S}_i$ and $s_{-i} \in \mathcal{S}_{-i}$. Then the NE strategy profile (s_i^*, s_{-i}^*) is defined as:

$$\pi_i(s_i^*, s_{-i}^*) \geq \pi_i(s_i, s_{-i}^*) \tag{11.2}$$

for all $s_i \in \mathcal{S}_i$ and for all $i \in \mathcal{N}$.

When the game reaches an NE, none of the players can improve their payoff by changing strategy unilaterally. There are also other solution concepts such as correlated equilibrium, which can be considered as a generalized version of NE [38], evolutionary equilibrium, and dominant-strategy equilibrium. We also discuss some of the other solution concepts that have been applied in the context of self-organizing small cell networks later in this chapter.

Recently, cooperation among network nodes for improving both individual and system-wide performance has attracted much attention. The players can make agreements and cooperate. A cooperative game provides analytical tools to model and analyze the cooperative behavior of rational players who may form coalitions. In this case, the members of each coalition cooperate to maximize the coalition payoff and the competition is among coalitions instead of among individual players.

11.4.3.2 Static vs. dynamic games

A static game is one in which a single decision (time irrelevant but may contain multiple actions) is made by each player, and each player has no knowledge of the decisions made by other players before making their own decision. Decisions are made simultaneously (or their order is irrelevant). A game is dynamic if the order in which the decisions are made is important or the strategy itself is time dependent. For dynamic games, the dynamics can be abstracted from different aspects, which lead to different types of dynamic games listed as follows.

(i) *Dynamic nature in games' play order*. The dynamic nature in games' play (decision) order leads to the development of multi-stage game (e.g., Stackelberg game). In this case, the decisions are made asynchronously and the games' play order is important. The players who move later can observe the decisions of the players who move first and then make the decisions accordingly. Note that if multiple players exist in one stage, the competition within this stage is usually formulated as a stage game.

(ii) *Dynamic nature in time dependency*. The dynamic nature in the time dependency leads to the development of a differential game and evolutionary game. For a differential game, the strategy of a player is time dependent (i.e., function of time t). That is, the player seeks a best response strategy considering the entire time horizon. For an evolutionary game, the players adapt their strategies according to the time-varying system state.

11.4.3.3 Games with special structures

Note that in general the existence and uniqueness of equilibrium as well as the convergence of best response dynamics to the equilibrium cannot be guaranteed. However, based on the characteristics of the game formulation, some special structures of games can be identified with which the games show remarkable properties in terms of the existence and convergence of pure strategy NE. Two of those special structure games, which are useful in deriving self-organizing solutions, are discussed below.

1. *Supermodular games*: These games are characterized as the games with strategic complementarities. The "increment" of strategy of one player will be unprofitable to

other players. Therefore the best response of other players would also be an increment of their strategies. The technical definition is given as follows.

DEFINITION 11.2 *Supermodular game*: let $\mathcal{G} = (\mathcal{N}, (\mathcal{S}_i)_{i \in \mathcal{N}}, (\pi_i)_{i \in \mathcal{N}})$, where $\mathcal{N}$ is the player set, $(\mathcal{S}_i)_{i \in \mathcal{N}}$ is the strategy set, which is a subset of Euclidean space, and (π_i) is the payoff of the ith player. The game $\mathcal{G}$ is said to be supermodular if the following conditions are satisfied [39]:

i. $(\mathcal{S}_i)_{i \in \mathcal{N}}$ is a compact subset of $\mathbb{R}$.
ii. $(\pi_i)_{i \in \mathcal{N}}$ is continuous.
iii. s_i and s_{-i} show increasing differences which are equivalent to the condition $\frac{\partial^2 \pi_i(s_i, s_{-i})}{\partial s_{ih} \partial s_{ik}} \geq 0$ for all $k \neq h$.

The following properties can be observed in supermodular games [40, 41].

i. Best responses are monotonically increasing.
ii. Pure strategy NE exists.
iii. NE can be attained using greedy best response algorithms.
iv. If the NE is unique, it is also globally stable.

2. *Potential games*: A game is categorized as a potential game if the motivation of all players to change their strategy can be expressed using a single global function (i.e., the potential function). In such games, obtaining the NE is equivalent to the maximization of the potential function.

DEFINITION 11.3 *Exact potential game*: let $\mathcal{G} = (\mathcal{N}, (\mathcal{S}_i)_{i \in \mathcal{N}}, (\pi_i)_{i \in \mathcal{N}})$ be a non-zero sum non-cooperative game where $\mathcal{N}$ is the player set, $(\mathcal{S}_i)_{i \in \mathcal{N}}$ is the strategy set, and (π_i) is the payoff of the ith player. The game $\mathcal{G}$ is an exact potential game if there exists an exact potential function $\Phi : \mathcal{S} \to \mathbb{R}$ for all $i \in \mathcal{N}$ such that

$$\Phi(s'_i, \mathbf{s}_{-i}) - \Phi(s''_i, \mathbf{s}_{-i}) = \pi(s'_i, \mathbf{s}_{-i}) - \pi(s''_i, \mathbf{s}_{-i}), \qquad (11.3)$$

where $\mathbf{s}_{-i} \in \mathcal{S}_{-i}$ and $s'_i, s''_i \in \mathcal{S}_i$.

In other words, the change in individual payoff gained by any player by unilaterally deviating to another strategy is the same as the difference in the corresponding values of the potential function. In ordinal potential games the signs of the differences are similar.

DEFINITION 11.4 *Ordinal potential game*: let $\mathcal{G} = (\mathcal{N}, (\mathcal{S}_i)_{i \in \mathcal{N}}, (\pi_i)_{i \in \mathcal{N}})$ be a non-zero sum non-cooperative game where $\mathcal{N}$ is the player set, $(\mathcal{S}_i)_{i \in \mathcal{N}}$ is the strategy set and (π_i) is the payoff of the ith player. The game $\mathcal{G}$ is an exact potential game if there exists an exact potential function $\Phi : \mathcal{S} \to \mathbb{R}$ for all $i \in \mathcal{N}$ such that

$$\text{sgn}[\Phi(s'_i, \mathbf{s}_{-i}) - \Phi(s''_i, \mathbf{s}_{-i})] = \text{sgn}[\pi(s'_i, \mathbf{s}_{-i}) - \pi(s''_i, \mathbf{s}_{-i})], \qquad (11.4)$$

where $\mathbf{s}_{-i} \in \mathcal{S}_{-i}, s'_i, s''_i \in \mathcal{S}_i$, and sgn denotes the sign function.

Note that the above definitions are only valid for static potential games. Potential games have the following remarkable properties [36, 42].

i. Every finite exact or ordinal potential game has at least one pure strategy NE.
ii. Both best response dynamics and better response dynamics converge to the pure NE.
iii. The NE is unique if
   - S is compact and convex
   - Φ is continuously differentiable on the interior of S
   - Φ is strictly concave on S.

11.4.4 Price of anarchy and price of stability

Game theoretic approaches may not guarantee the optimal performance, i.e., an equilibrium solution of the game may not be the optimal solution for the problem. This inefficiency of the game theoretic solutions may occur due to the selfish behavior of the players. To measure the inefficiency of equilibrium solutions of a game, two popular concepts, i.e., the price of anarchy (PoA) and the price of stability (PoS) [43] can be defined as follows.

11.4.4.1 Price of anarchy

Price of anarchy is defined as the ratio between the payoffs at the worst equilibrium (i.e., the equilibrium point that gives the least payoff) and the optimal centralized solution to the problem. PoA can vary for different payoff functions.

11.4.4.2 Price of stability

Price of anarchy can be significantly small for games with multiple equilibria even if only one equilibrium point is inefficient. Hence price of stability is defined as the ratio between payoff received at the best equilibrium and the optimal (best possible) payoff.

Note that PoA and PoS are both equal for games with unique equilibrium.

11.4.5 Design of payoff functions

Game theory was initially proposed and developed for economics and social sciences. Therefore properly fitting those game models in the context of communication engineering is challenging. Specifically, defining the payoff functions based on the network performance metrics (e.g., achievable data rate, delay, and transmit power), modeling the network dynamics (e.g., randomness of the wireless channel, randomness of the user locations and base station deployment, and mobility of the users), meeting the requirements defined by the standards and realization of the SON characteristics have to be considered within the scope of the game. Among all these, defining a proper payoff function is one of the key challenges. The payoff function quantifies the perceived preference or the satisfaction level of a player. In the context of self-organizing small cells, the user-satisfaction level may depend on one or multiple performance metrics given as follows:

- individual performance (e.g., rate, SINR, and delay)
- global network performance
- interference level caused to other network nodes
- power/energy consumption
- user fairness.

As self-organizing small cell technologies are still in their infancy, there is no well-defined framework for designing the payoff functions. To this end, we will introduce some general approaches and guidelines on how payoff functions can be designed for various applications and objectives in the context of self-organizing small cell networks.

A payoff function $\pi(x)$ is expected to satisfy the following criteria.

1. The non-stationary property: $\frac{d\pi(x)}{dx} > 0$, which states that the payoff increases with the preference or satisfaction.
2. The risk-aversion property: $\frac{d^2\pi(x)}{dx^2} < 0$, which states that the payoff function is concave. In other words, the marginal payoff of satisfaction decreases with increasing level of satisfaction.

Depending on the objective, behavior, and rationality of the network nodes, different payoff functions are defined in the wireless communications literature. The payoff/utility functions that can be applied in the context of small cell networks are discussed below.

11.4.5.1 Payoff functions for power consumption

Power/energy conservation is crucial in small cell networks as they might be operated in an energy-limited environment (e.g., power supplied by a battery). Work [44] defines a simple energy-aware payoff function as follows:

$$\pi_i(e) = \frac{E_{tot}}{e_i}, \qquad (11.5)$$

where E_{tot} is the total energy available for each player and e_i is the energy required by player i for transmission. Players would try to achieve a higher payoff by reducing the transmission power.

11.4.5.2 Payoff functions for individual performance

Instead of direct power minimization as shown in equation (11.5), it is more appropriate for self-organizing algorithms to perform power control in such a way that the desired performance can be satisfied. The following logarithmic payoff function with individually perceived SINR as the input parameter can capture the self-interest of network nodes and is used for power control in [45, 46]:

$$\pi_i(s_i, s_{-i}) = \log(\gamma_i(s_i, s_{-i})), \qquad (11.6)$$

where γ_i is the SINR of the ith player. Such a logarithmic payoff function and its extensions are the most popular payoff functions used in the context of resource allocation due to their simplicity and mathematical tractability [47]. For example, such a form of payoff can be used for subcarrier allocation (in OFDMA networks) and joint power-subcarrier allocation as well.

Another widely used payoff function is the Shannon capacity or the maximum achievable rate that can be considered as an extended version of logarithmic function of SINR as shown below:

$$\pi_i(s_i, \mathbf{s}_{-i}) = \ln(1 + \gamma_i(s_i, \mathbf{s}_{-i})). \tag{11.7}$$

11.4.5.3 Fairness utility function

One of the desired objectives of resource allocation is to provide fairness among users instead of obtaining the optimum performance. The most widely used payoff function, which guarantees fairness, is given below:

$$u(x) = \begin{cases} \dfrac{x^a}{a}, & \text{if } a < 0, \\ \log x, & \text{if } a = 0, \end{cases} \tag{11.8}$$

where $a \leq 0$. By twice differentiation of (11.8) with respect to x we obtain

$$\frac{du(x)}{dx} = \begin{cases} x^{a-1}, & \text{if } a \neq 0, \\ \dfrac{1}{x}, & \text{if } a = 0, \end{cases} \tag{11.9}$$

and

$$\frac{d^2 u(x)}{dx^2} = \begin{cases} (a-1)x^{a-2}, & \text{if } a \neq 0, \\ \dfrac{-1}{x^2}, & \text{if } a = 0. \end{cases} \tag{11.10}$$

It can be observed that the function given in equation (11.8) has both non-stationary and risk-aversion properties for all $x > 0$ since $\frac{du(x)}{dx} > 0$ and $\frac{d^2 u(x)}{dx^2} < 0$.

11.4.5.4 System payoff functions

In self-organizing enabled small cell networks, a group of densely deployed small cells could form a cluster and cooperate with each other to enhance the performance of the cluster [48]. In addition to that, cooperative games can also be formulated to design self-organizing algorithms for small cells. Accordingly, cooperative payoff functions, which reflect the overall network/cluster performance, are required.

The simplest and most intuitive cooperative payoff function would be the sum capacity/rate of the cluster/network as shown below:

$$\pi_i(\mathbf{s}) = \sum_{j \in \mathcal{N}_i} C_j(\mathbf{s}), \tag{11.11}$$

where $\mathcal{N}$ is the set of players in the ith cluster which cooperates with each other and C_j is the capacity of the jth player.

11.4.5.5 Multi-dimensional payoff function

The payoff function can be designed considering multiple performance metrics. In such cases, these multiple metrics could appear in the payoff function (in most cases in a product form). One typical example is given as follows:

$$\pi_i = \pi_i^{rate} \pi_i^{delay}. \tag{11.12}$$

11.4.5.6 Payoff function with cost

For a strategy adopted by a player, there could be a cost associated with it (e.g., cost of using bandwidth, power consumption) or it may affect the performance of other players (e.g., cause interference). This issue can be modeled by introducing certain cost functions into the payoff function. In particular, the payoff function (some may refer to this as *net utility*) can be defined to reflect both the satisfaction of the player (modeled by utility function) and the cost (e.g., price per unit resource) as follows:

$$\pi_i(s_i, \mathbf{s}_{-i}) = u_i(s_i, \mathbf{s}_{-i}) - mx, \tag{11.13}$$

where $u_i(s_i, \mathbf{s}_{-i})$ is the utility based on the user satisfaction and m is the price paid for each resource x.

Work [49] uses a net utility function with logarithmic payoff as given below:

$$\pi_i(s_i, \mathbf{s}_{-i}) = a_i \log(1 + \gamma_i(s_i, \mathbf{s}_{-i})) - b_i m \gamma_i(s_i, \mathbf{s}_{-i}), \tag{11.14}$$

where γ_i is the SINR of the ith user, a_i and b_i are weighting parameters and m is the cost for the received SINR. The gain of maximizing γ_i could be neutralized by the cost associated with the received SINR.

The following form of payoff function (equation 11.15) is used in [28] to limit the interference caused to the macro users by the downlink transmission of small cells:

$$\pi_i(s_i, \mathbf{s}_{-i}) = w_1(\pi(\gamma(s_i, \mathbf{s}_{-i}))) - w_2(I_m - T), \tag{11.15}$$

where w_1 and w_2 are biasing factors, which can be determined based on which network (i.e., macro or small cell network) should be given priority in resource allocation. I_m is the interference caused to the nearest macro user and T is the macro user interference threshold. When the interference caused to the nearest macro user (I_m) exceeds a certain threshold, small cell base stations are demotivated to allocate resources to their user even if it increases the individual payoff. At the same time, such payoff function encourages the SBSs to use resources (i.e., transmit power and OFDMA subcarrier) as long as it does not exceed the interference threshold of the macro users.

Guaranteeing the existence of equilibrium is one of the essential features of any game formulation. It is straightforward that the existence of equilibrium, convergence, and stability of the equilibrium is highly related to the payoff function and the structure of the game. Therefore special payoff function can also be designed to fit the game model into special structures (e.g., super-modular, potential). Polynomial time computability is another important feature of a payoff function. Besides, when it comes to self-organizing small cell networks, the ability to compute with local information or with reduced information exchange is also highly desirable.

11.5 Game theory-based resource management for self-organizing small cells

Resource management aims for efficient usage of scarce resources (e.g., power and spectrum) as well as for interference management when it comes to the underlying small cell networks. Besides, some other issues such as load balancing and coverage optimization can also be eventually modeled as resource allocation problems. In general, the resource allocation in orthogonal frequency-division multiple access (OFDMA)-based small cell networks can be categorized into three classes: subcarrier allocation, power allocation, and joint subcarrier-power allocation.

Game theory-based resource management is one of the most addressed issues in the context of self-organizing small cell networks. Different types of games are used to address the above issues depending on the objective and the network characteristics. In the following, formulations of the selected game models for devising self-organizing distributed resource management algorithms are discussed.

11.5.1 Non-cooperative game-based decentralized power allocation

The power-allocation problem of a self-organizing small cell network can be modeled by a non-cooperative game in which the players are the small cell nodes (e.g., SBSs or users). The strategy of a player is the allocation of transmit power. The strategy selection of a player will impact the payoff of other players. Specifically, the transmission power selection of a player creates a positive or negative impact on the payoff of other players due to the possible increase or decrease of interference. The payoff function of the players can be chosen appropriately according to the design objective.

11.5.1.1 Non-cooperative game-based downlink power control

In [39], a non-cooperative game is used to model the downlink transmission power allocation problem among the SBSs. A system with one central macro cell and several underlaid closed-access small cells is considered. It is also assumed that the distances between an SBS and its associated users are almost the same, hence all users served by this SBS have equal rate and the rate of a user of small cell $i(R_i)$ is given by

$$R_i = \frac{1}{N_i} \log \left(1 + \frac{h_{i,i} P_i}{I + \sum_{j \neq i} h_{i,j} P_j} \right), \quad (11.16)$$

where N_i is the number of users associated with base station i, $h_{i,j}$ is the average channel gain from small cell base station i to users in small cell j, I is the noise power plus the interference from the macro base station, and P_i is the transmit power of small cell base station i. The small cell base stations are the players of the game, each of which is self-interested and tries to increase its own capacity.

The strategy set for base station i is defined as $S_i = [0, P_{max}]$, where P_{max} is the maximum allowable transmit power for an SBS. The payoff function is then defined considering three factors:

- achievable average rate of the small cell base station
- fairness of the system
- transmit power.

The payoff increases with the increasing average rate, while it decreases with the increasing transmit power due to the increased interference caused to neighboring cells. Also, fairness should also be considered among the base stations. Therefore the payoff function is defined as follows:

$$\pi(P_i) = N_i \log(R_i) - \beta P_i, \tag{11.17}$$

where β is a positive constant.

This game is shown to be a supermodular game and accordingly the NE can be achieved by using best response dynamics (see Section 11.4.3). Based on best response dynamics, the following power control algorithm is derived. Small cell base stations update their transmit power periodically to best response to the current strategy profile of other base stations. Eventually the algorithm converges to the NE.

Algorithm 11.1 Non-cooperative game-based downlink power allocation algorithm

---

1: Initialize
2: **repeat**
3: Measure noise and interference from other SBSs
4: Calulate the payoff by substituting in equation (11.17)
5: Find P_i that maximizes equation (11.17)
6: Update P_i
7: Wait until the next update time
8: **until** SBS turns off.

The performance of the above algorithm is evaluated numerically in [39], which proves the capability of the algorithm to be implemented in a real environment while providing fairness to the small cell users. However, the algorithm shows slightly degraded performance than the centralized system, which delivers optimal performance. This is due to the selfish decentralized behavior of the users.

11.5.1.2 Non-cooperative game-based uplink power control

Uplink power allocation in small cell networks can also be modeled as a non-cooperative game [50, 51]. Specifically, a non-cooperative game-based distributed uplink power control algorithm is proposed in [51]. The power control is performed distributively based on SINR adaptation while mitigating the interference caused to the macro base station. A single macro cell and a set of N underlaid small cells are considered. Each base station serves only one user at a time with a guaranteed SINR requirement.

In order to protect the macro base station from interference due to the uplink transmission of the small cell users, the macro user is also considered as a player in the game. In this case, the player set consists of the macro user and the small cell users denoted by, $\mathcal{N} = \{0, 1, \ldots, N\}$, where index 0 denotes the macro user and the indices $1, 2, \ldots, N$ denote small cell users. The strategy of each player i is its transmit power denoted by p_i. The payoff function for the macro user is given by

$$\pi_0(p_0, \boldsymbol{p}_{-0}) = -(\gamma_0 - \Gamma_0)^2, \tag{11.18}$$

where Γ_0 is the target SINR and γ_0 is the received SINR of the macro user. The received SINR of any user is given by

$$\gamma_i = \frac{p_i h_{i,i}}{\sigma^2 + \sum_{j \neq i} p_j h_{i,j}}, \quad (11.19)$$

where σ^2 is the noise power and $h_{i,j}$ is the channel gain between users i and j.

Each small cell user also tries to maximize her own individual SINR while meeting the minimum SINR requirement, Γ_i. The payoff of a small cell user is given by

$$\pi_i(p_i, \boldsymbol{p}_{-0}) = R(\gamma_i, \Gamma_i) + b_i \frac{C(p_i, \boldsymbol{p}_{-0})}{\sigma^2 + \sum_{j \neq i} p_j h_{i,j}}, \quad (11.20)$$

where b_i is a weighting factor. The reward function, $R(\gamma_i, \Gamma_i)$ and the penalty function $C(p_i, \boldsymbol{p}_{-0})$ are defined as follows:

$$R(\gamma_i, \Gamma_i) = 1 - \exp(-a_i(\gamma_i - \Gamma_i)), \quad (11.21)$$

where a_i is a constant and

$$C(p_i, \boldsymbol{p}_{-0}) = -p_i h_{0,i}. \quad (11.22)$$

The reward increases with γ_i until the threshold Γ_i is met. Once the received SINR exceeds the minimum requirement, the reward decreases with γ_i, which discourages the small cell users to increase their power. By equation (11.22), the small cell users are given a penalty with the increase of transmit power. The penalty is scaled by interference and noise ($\sigma^2 + \sum_{j \neq i} p_j h_{i,j}$) in equation (11.33) to ensure that small cells experiencing higher interference are less penalized. Note that the payoff function is a monotonically increasing concave function of γ_i for fixed p_i. Also, for fixed γ_i, the payoff is a monotonically decreasing concave function of p_i.

The existence of Nash equilibrium for the above uplink power control game can be proven by employing the following theorem from [52–54].

THEOREM 11.1 *A Nash equilibrium exists in game* $\mathcal{G} = (\mathcal{N}, (\mathcal{S}_i)_{i \in \mathcal{N}}, \pi_i(.)_{i \in \mathcal{N}})$ *if, for all* $i = 0, 1, \ldots, N$,

1. $(\mathcal{S}_i)_{i \in \mathcal{N}}$ *is a non-empty, convex, and compact subset of some Euclidean space* $\mathbb{R}^{N+1}$.
2. $\pi_i(\boldsymbol{s})$ *is continuous in* $\boldsymbol{s}$ *and quasi-concave in* p_i.

The uplink transmit power at the NE (denoted by $\boldsymbol{p}^*$) is given by following two equations [51]:

$$p_0^* = \min\left(\frac{I_0(\boldsymbol{p}_{-0}^*)}{g_{0,0}} \Gamma_0, p_{max}\right), \text{ when } i = 0, \quad (11.23)$$

$$p_i^* = \min\left(\frac{I_i(\boldsymbol{p}_{-i}^*)}{g_{i,i}} \left[\Gamma_i + \frac{1}{a_i} \ln\left(\frac{a_i g_{i,i}}{b_i g_{0,i}}\right)\right]^+, p_{max}\right), \text{ when } i \neq 0, \quad (11.24)$$

where $[x]^+ = \max(x, 0)$ and $I_i(\boldsymbol{p}_{-i}^*) = \sigma^2 + \sum_{j \neq i} p_j^* h_{i,j}$.

In order to devise a distributed power control algorithm, which converges to the NE, [51] uses the *standard interference* function defined in [55].

DEFINITION 11.5 *Standard interference function*: $f(p)$ is a standard interference function if the following conditions are satisfied for all $p > 0$:

1. Positivity, $f(p) > 0$.
2. Monotonicity, if $p' > p$ then $f(p') > f(p)$.
3. Scalability, for all $\alpha > 1$, $\alpha f(p) > f(\alpha p)$.

Yates [55] showed that an iterative power control algorithm, which calculates the power at next iteration k + 1 according to the rule $p^{k+1} = f(\mathbf{P})$, converges to a unique fixed point if $f(\mathbf{P})$ is a standard interference function.

The received SINR is $\gamma_i = \frac{p_i}{I_i}$, according to which the equations (11.23) and (11.24) can be modified to form a distributed iterative power control algorithm. The individual uplink transmit power is updated as follows:

$$p_0^{k+1} = \min\left(\frac{p_0^k}{\gamma_0^k}\Gamma_0, p_{max}\right), \text{ when } i = 0, \tag{11.25}$$

$$p_i^{k+1} = \min\left(\frac{p_i^k}{\gamma_0^k}\left[\Gamma_i + \frac{1}{a_i}\ln\left(\frac{a_i g_{i,i}}{b_i g_{0,i}}\right)\right]^+, p_{max}\right), \text{ when } i \neq 0. \tag{11.26}$$

Both equations (11.25) and (11.26) are standard interference functions. Therefore, the power control algorithm converges to a unique fixed point, which is the NE defined in (11.23) and (11.24).

More importantly, the algorithm can be executed distributively with minimal network overhead, and therefore would be suitable for resource allocation in a self-organizing small cell network.

11.5.2 Non-cooperative game-based subcarrier allocation

In addition to power allocation, subcarrier allocation (we consider OFDMA networks) is also an essential part for self-organizing resource allocation. The distributed subcarrier allocation problem can also be modeled as a non-cooperative game in a similar way as that for power control. In this case, the strategy set should represent the selection of available subcarriers for each node of the network. In the following, we give a descriptive example for uplink-distributed subcarrier allocation in OFDMA-based small cell networks based on non-cooperative games.

Work [56] proposes a decentralized method for small cells in a two-tier network to individually select the most appropriate subset of resource blocks in order to mitigate both cross-tier and co-tier interference. In the model, the macro network consists of 19 macro cell sites, each of which has three hexagonal sectors. Small cells are deployed inside the macro cell according to the 5 × 5 grid model specified in the 3GPP simulation scenario given for urban deployment in [57]. Twenty-five apartments are arranged according to a 5 × 5 grid and each of these apartments would have a small cell (femto cells in this case) with a probability of p_d. U_M number of macro users are randomly and uniformly located in each sector and U_S number of small cell users are randomly and uniformly located in each apartment. It is also assumed that the total bandwidth W is divided into K resource blocks.

The resource block (RB) allocation is modeled as a non-cooperative game $\mathcal{G} = (\mathcal{N}, (\mathcal{S}_i)_{i \in \mathcal{N}}, \pi_i(.)_{i \in \mathcal{N}})$, where N is the set of small cell users. The strategy S of each player i is the selection of a subset of RBs. Two payoff functions π_1 and π_2 are considered (given in equations (11.27) and (11.28)), where π_1 takes into account only the co-tier interference between SBSs while π_2 considers both co-tier and cross-tier interferences. Let each small cell user select H number of resource blocks for transmission. S and M are the total number of small cell users and macro users, respectively. The selected set of resource blocks by small cell user i is given by $\mathcal{R}_i = k_i^1, k_i^2, \ldots, k_i^H$. For all $i, j \in \mathcal{N}$ and $x, y \in 1, 2, \ldots, H$, $\delta_{k_i^x, k_j^y}$ is an interference indicator function, where $\delta_{k_i^x, k_j^y} = 1$ if RBs k_i^x and k_j^y are the same and $\delta_{k_i^x, k_j^y} = 0$, otherwise. The two payoff functions are given below.

$$\pi_i^1(s_i, s_{-i}) = \sum_{x=1}^{H} \sum_{y=1}^{H} \left(-\sum_{j=1, j \neq i}^{S} g_j^{b_i} p_j^{k_j^y} \delta_{k_i^x, k_j^y} - \sum_{j=1, j \neq i}^{S} g_j^{b_j} p_i^{k_i^y} \delta_{k_j^x, k_i^y} \right), \quad (11.27)$$

and

$$\pi_i^2(s_i, s_{-i}) = \sum_{x=1}^{H} \left[\sum_{y=1}^{H} \left(-\sum_{j=1, j \neq i}^{S} g_j^{b_i} p_j^{k_j^y} \delta_{k_i^x, k_j^y} - \sum_{j=1, j \neq i}^{S} g_j^{b_j} p_i^{k_i^y} \delta_{k_j^x, k_i^y} \right) \right.$$
$$\left. + \sum_{z=1}^{L} \left(-\sum_{m=1, m \neq i}^{M} g_m^{b_i} p_m^{k_m^z} \delta_{k_i^x, k_m^z} - \sum_{m=1, m \neq i}^{M} g_j^{b_j} p_i^{k_i^z} \delta_{k_j^x, k_i^z} \right) \right], \quad (11.28)$$

where $g_j^{b_i}$ is the channel gain between the small cell j and the SBS, b_i, $g_j^{b_m}$ denotes the channel gain between the small cell j and the MBS b_m, and $p_j^{k_j^y}$ is the transmit power of user j on its selected RB k_j^y.

The two terms in equation (11.27) represent the co-tier interference caused at the base station i by other base stations and the co-tier interference caused to other SBSs by SBS i, respectively. The additional two terms in equation (11.28) measure the cross-tier interference.

The existence and the uniqueness of the NE are proved by showing that the above game is an exact potential game. The corresponding potential functions and the proof can be found in [56]. In potential games, the best response dynamics always converges to a pure strategy NE. The best response strategy is given by

$$s_i^{t+1} = \arg\max_{s' \in \mathcal{S}} \pi_i \left(s', s_{-i}^t \right). \quad (11.29)$$

The small cell users are assumed to be able to sense the spectrum in order to select the best set of RBs as S_i^{t+1}.

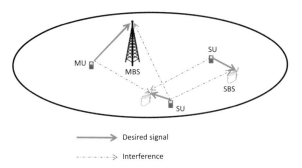

Figure 11.4 Uplink interference scenario in a two-tier network.

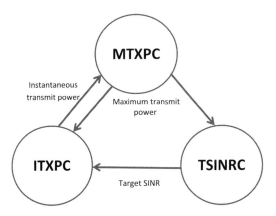

Figure 11.5 Complementary TRi-control loop.

Giupponi in [58] also formulates the downlink resource (joint subcarrier and power) allocation problem as a potential game. The payoff function is designed to model both co-tier and cross-tier interferences and a distributed resource allocation algorithm is proposed.

11.5.3 Complementary TRi-control loop for uplink interference mitigation

Small cell users may cause considerable amount of interference to the macro base station in uplink and vice versa (as shown in Figure 11.4). [59] proposes a self-organizing uplink interference management architecture called complementary TRi-control loop (CTRL), which is composed of three control loops as shown in Figure 11.5, which are explained below.

- *MTXPC – maximum transmit power control loop.* This determines the maximum possible transmit power for small cells in order to provide the uplink protection for the macro network. The decision is taken based on the feedback uplink load margin information of the macro base station.

- *TSINRC – target SINR control loop.* The duty of target SINR control loop is to determine the required uplink SINR for each SBS with minimum possible information exchange among them. A non-cooperative game is formulated and the decisions are taken considering the maximum transmit power decided at the MTXPC loop.
- *ITXPC – instantaneous transmit power control loop.* The actual transmit power allocation is done at this loop. The transmit power is allocated to achieve the target SINR calculated at the TSINRC loop with the constraint on maximum transmit power.

The maximum transmit power control loop is modeled as a Q estimation problem in adaptive control theory. The MTXPC loop calculates the maximum transmit power for each SBS user in a self-organizing manner. The complete model can be found in [59].

The target SINR control loop is formulated as a non-cooperative game in which NE is obtained as the solution. The players are the set of small cell base stations. The strategy set (S) is composed of the set of possible transmit powers on each resource block. The transmit power vector of each user i is given by $p_i = (p_{i,1}, p_{i,2}, \ldots, p_i, K)$, where $p_{i,k}$ is the transmit power of user i on subcarrier k. $p_{i,k}$ must be less than the maximum transmit power $P_{i,k}$ obtained at the MTXPC loop. The vector p is composed of the transmit powers of all the users in all SBSs. Let $\kappa_3 \mathcal{K}$ and $\mathcal{N}_f$ denote the set of subcarriers and the set of users connected to SBS i, respectively. $b_{i,k}$ is the normalized time period that the user i transmits on resource block k. The payoff function of each player f is given by

$$\pi_f(\boldsymbol{p}, \boldsymbol{b}) = \sum_{i \in \mathcal{N}_f} \sum_{k \in \mathcal{K}} b_{i,k} W \log_2 \left(1 + \frac{\gamma_{i,k}}{c}\right) - \sum_{i \in \mathcal{N}_f} \sum_{k \in \mathcal{K}} b_{i,k} \mu_{i,k} p_{i,k}, \quad (11.30)$$

where W is the size of a resource block, $\omega = -\ln(5BER)/1.6$ is a constant to achieve a given bit error rate (BER), $\mu_{i,k}$ is the price paid for the interference caused and $\gamma_{i,k}$ is the target SINR of user i on subcarrier k.

There exists an NE (given in equation 11.31) for the above game if $\mu_{i,k}$ is large enough.

$$\gamma_{i,k}^* = \max \left(\left[\frac{W h_i^k}{(\ln 2) I_{i,k}(\boldsymbol{p}_{-i}) \mu_i^k} - \omega \right]^+, \frac{h_{i,k} P_{i,j}}{I_{i,k}(\boldsymbol{p}_{-i})} \right), \quad (11.31)$$

where $I_{i,k}$ is the interference (including macro cell interference) plus the thermal noise at user i on resource block k and h_i^k is the channel gain from user i to its own base station on subcarrier k. The proof of the existence of an NE is based on the fact that the payoff function is continuous and quasi-concave and the strategy set is a non-empty, compact, and convex subset in the Euclidean space. The complete proof can be read from [59].

Based on the interactions among the above-proposed three control loops, the spatial reuse of spectrum within small cells is enabled without degrading the performance of the macro tier. The operation of CTRL does not require any changes in the resource management of the macro network and converges distributively to a stable solution.

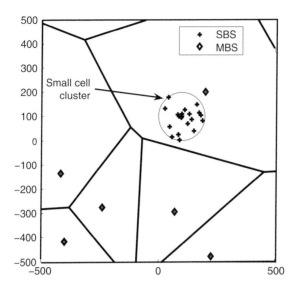

Figure 11.6 Small cell cluster underlaid with a macro network.

11.5.4 An evolutionary game approach for self-organization with reduced information exchange

As we have seen in the previous examples, traditional game theory (e.g., Nash equilibrium problem) relies on the rational decisions of the players. The players are expected to choose their strategies rationally as the best responses to the strategies of other players. The rationality implies complete information and strong computation capability of each player to calculate the best response to other players' strategies. This assumption may be too strong for densely deployed nodes in a self-organizing small cell network. As a solution, the distributed resource allocation problem can be formulated as an evolutionary game [28, 60, 61].

Evolutionary game theory (EGT) was originally developed to analyze the evolution of populations of biological species with bounded rationality. Instead of selecting the strategy that gives the best response to the other users' strategies, in EGT, each player selects a strategy by replication and can adapt its selection for a better payoff (i.e., evolution). Accordingly, EGT focuses on the dynamics of the strategy adaptation in the population. A population is the set of players involved in the game. The behavior of the population can be described by the number of its members choosing each pure strategy. The success of a strategy is reflected by the proportion of members in the population using it. In the following, we provide an example of formulating the downlink subcarrier selection and transmit power allocation of a small cell network as an evolutionary game [28].

The downlink transmission of an OFDMA-based two-tier cellular network composed of macro cells and an underlying self-organizing small cell cluster is considered here (as shown in Figure 11.6). The spatial distributions of the small cell base stations and macro base stations follow two independent point processes in $\mathbb{R}^2$ with densities λ_f and

λ_m, respectively. Each macro user is attached to the nearest macro base station and each small cell user is located at a distance r_f from its serving base station. Each SBS serves only one user at a time and selects one subcarrier to serve that user. The macrocell and the small cells share the same set of orthogonal subcarriers denoted by $\mathcal{K} = \{1, 2, \ldots, K\}$. They are also capable of selecting a transmit power level from a finite set of values, which is denoted by $\mathcal{L} = \{1, 2, \ldots, L\}$. Each SBS should select a suitable subcarrier–power combination, which is referred to as the "transmission alignment" of that SBS. The set of transmission configurations (i.e., strategy set) is denoted by $\mathcal{S}$. For each subcarrier k, there is a maximum aggregate interference threshold that can be caused by the entire small cell cluster to the macro users, which is denoted by $T^{(k)}$.

The small cell base stations form the player set of the game, denoted by $\mathcal{N}$. The strategies $\mathcal{S}$ available for each player are the set of transmission alignments. In the context of an evolutionary game, the set of players also constitutes the population. Denote by n_s the number of SBSs selecting pure strategy $s \in \mathcal{S}$. Then the frequency of strategy s used in the population is given by

$$x_s = \frac{n_s}{N}, \tag{11.32}$$

where the frequency x_s is also referred to as the population share of pure strategy s. The population shares of all strategies add to 1. The payoff is a function of the utility of an SBS when certain transmission alignment is used and the interference caused to the nearest macro user and is given by

$$\pi_s = \pi_l^{(k)} = w_1\big(\mathcal{U}\big(\mathrm{SINR}_l^{(k)}\big)\big) - w_2\big(I_m^{(k)} - T^{(k)}\big), \tag{11.33}$$

where $\mathrm{SINR}_l^{(k)}$ is the received SINR of a small cell user served by subcarrier k and power level l, w_1 and w_2 are biasing factors, and $I_m^{(k)}$ is the aggregate interference created by the small cell cluster on subcarrier k at the nearest macro user.

Specifically, in [28], two utility functions are considered, which are given as follows:

$$\mathcal{U}_1\big(\mathrm{SINR}_l^{(k)}\big) = \mathbf{E}\big[\mathrm{SINR}_l^{(k)}\big], \tag{11.34}$$

and

$$\mathcal{U}_2\big(\mathrm{SINR}_l^{(k)}\big) = \mathbf{E}\big[\ln(1 + \mathrm{SINR}_l^{(k)})\big]. \tag{11.35}$$

Based on two utility functions, two payoff functions can be defined ($\pi_s^{(1)}, \pi_s^{(2)}$) and hence two games are formulated.

$$\mathcal{G}^1 = \big(\mathcal{N}, \mathcal{S}, \pi_s^{(1)}\big)\big), \tag{11.36}$$

and

$$\mathcal{G}^2 = \big(\mathcal{N}, \mathcal{S}, \pi_s^{(2)}\big)\big). \tag{11.37}$$

The evolutionary equilibrium (EE) is the solution concept for both $\mathcal{G}^1$ and $\mathcal{G}^2$. In the context of the evolutionary game for transmission alignment selection, each SBS will adapt its strategy according to its received payoff. This is referred to as the evolution of the game during which the strategy adaptation of SBSs will change the population share, and therefore the population state will evolve over time. The strategy adaptation process and the corresponding population state evolution can be modeled and analyzed by *replicator dynamics* [62], which is a set of ordinary differential equations defined as follows:

$$\dot{x}_s(t) = x_s(t)(\pi_s(t) - \bar{\pi}(t)), \qquad (11.38)$$

for all $s \in \mathcal{S}$, with initial population state $x(0) = x_0 \in \mathbb{X}$, where $\mathbb{X}$ is the state space which contains all possible population distributions. Here π_s is the payoff of each SBS choosing transmission alignment s and $\bar{\pi}$ is the average payoff of the entire population. The equilibrium point of the game can be obtained by solving the replicator dynamics. Evolutionary equilibrium is the point where the replicator dynamics is equal to zero. In other words, when the system is p_f at equilibrium, the fractions of the population choosing each strategy remain constant.

As the game (either $\mathcal{G}^1$ or $\mathcal{G}^2$) is repeated, each SBS observes its own payoff and compares it with the average payoff of the system. Then, if its payoffs are less than the average, in the next period, the SBS randomly selects another strategy. The proposed distributed resource allocation algorithm is given in Algorithm 11.2.

Algorithm 11.2 Evolutionary game-based distributed resource allocation

---

1: **Initialize**: the SBSs choose a transmission alignment randomly and set $i = 1$.
2: **repeat**
3: **Exploitation**: each SBS transmits on the selected transmission configuration and observes the received utility. The utility and the transmission alignment information are then sent to the central controller.
4: **Learning**: a central controller calculates the average payoff of the population and the population state and broadcasts it to all SBSs.
5: **Update**: each SBS compares its own payoff with the average payoff of the population. If the payoff is less than the average, the SBS randomly selects another subcarrier for transmission.
6: $i = i + 1$
7: **until** $i \geq \text{Max}_i$ (maximum number of iterations that the algorithm can execute)

The stability of the equilibrium point can be analyzed by using a stochastic geometry approach. The spatial distributions of macro base stations and small cell base stations are approximated by Poisson point processes (PPP). Then the expected SINR for any population distribution can be derived in terms of the population shares (see [28] for the stochastic geometry-based derivation). The expressions obtained for the above-mentioned utility functions (when path-loss exponent equals 4) are as given below:

$$\mathbf{E}\left[\text{SINR}_l^{(k)}\right] = \frac{8p_l}{A^2\left(\lambda_m\sqrt{p_m} + \lambda_f^{(k)}\mathbf{E}[\sqrt{p_f}]\right)^2}, \qquad (11.39)$$

and

$$\mathbf{E}[r_l^{(k)}] = \int_{t=0}^{\infty} \exp\left(\frac{-A}{2\sqrt{p_l}}(\lambda_m\sqrt{p_m} + \lambda_f^{(k)}\mathbf{E}[\sqrt{p_f}])\sqrt{e^t - 1}\right) dt, \quad (11.40)$$

where p_l is the transmit power of level l, p_m is the transmit power of MBSs, $P_f \in \{p_1, p_2, \ldots, p_L\}$ denotes the transmit power of SBSs, $A = \pi^2 r_s^2$, and $\lambda_f^{(k)}$ is the density of SBSs transmitting on subcarrier k, which is also assumed to be uniformly randomly distributed.

The probability mass function of the transmit power of any interferer (i.e., pf in (11.39) and (11.40)) can be directly obtained from the proportions of the population selecting each strategy. For transmission alignment corresponding to subcarrier k and power level l, the PMF (which can be used to find $\mathbf{E}[\sqrt{p_f}]$) of the transmit power of a generic interferer is given as follows:

$$\Pr(p_f = p_j) = \begin{cases} \dfrac{n_j^{(k)}}{\sum_{t=1}^{L} n_t^{(k)} - 1}, & \text{if } j \neq l, \\ \dfrac{n_j^{(k)} - 1}{\sum_{t=1}^{L} n_t^{(k)} - 1}, & \text{if } j = l, \end{cases}$$

or equivalently,

$$\Pr(p_f = p_j) = \begin{cases} \dfrac{x_j^{(k)}}{\sum_{t=1}^{L} x_t^{(k)} - \dfrac{1}{N}}, & \text{if } j \neq l, \\ \dfrac{x_j^{(k)} - \dfrac{1}{N}}{\sum_{l=t}^{L} x_t^{(k)} - \dfrac{1}{N}}, & \text{if } j = l, \end{cases}$$

where $n_j^{(k)}$ is the number of players selecting subcarrier k and power level j and $x_j^{(k)} = \frac{n_j^{(k)}}{N}$. For a network with two orthogonal subcarriers and one transmit power level, the interior evolutionary equilibrium in game $\mathcal{G}^1$ can be shown to be asymptotically stable [28].

Simulations show that $\mathcal{G}^2$ converges faster than $\mathcal{G}^1$. The impact of delay in information exchange is also analyzed numerically, which shows that the system converges to the equilibrium under small delays. However, when the delay is larger than a certain bifurcation point, the system will diverge. Also there is no guarantee that the system will converge to the same equilibrium point as the delay-free system. The key features of the above-discussed evolutionary game-based algorithm are as follows: simplicity, reduced information exchange than other non-cooperative game-based algorithms, and fairness.

In [28], the performance of the above algorithm is compared with the optimal performance obtained by a centralized resource allocation, which acts as a benchmark. A gap exists between the maximum payoff and the payoff obtained by the EGT-based algorithm. Also the gap increases with the number of base stations in both $\mathcal{G}^1$ and $\mathcal{G}^2$.

Up to this end, we have studied some basic examples of using different game models in order to devise self-organizing algorithms for small cells. Other customized algorithms, which satisfy various constraints, can be built on top of these basic examples. For example, in [63] a self-organizing interference management paradigm is proposed while taking into account the constraints due to the presence of heterogeneous backhauls. The problem is formulated as a non-cooperative game and a fully distributed learning algorithm is devised based on reinforcement learning (RL), which converges to an equilibrium solution.

Cognitive radio (CR) enabled small cells [64, 65], which can sense the spectrum, are also proposed as a solution for interference mitigation. CR-enabled SBSs can opportunistically allocate both licensed and unlicensed frequency bands to the users in order to avoid interference. [66] uses a correlated equilibrium-based approach [67] to mitigate co-tier interference among cognitive femto access points for the downlink OFDMA LTE networks. Correlated equilibrium is preferred for a self-organizing network raher than NE as it allows devising decentralized and adaptive algorithms. In [66], the spectrum allocation competition among cognitive base stations is formulated as a non-cooperative game. The spectrum allocation is done using two payoff functions, i.e., global payoff and local payoff. The global function provides fairness among players considering the total network performances while the local payoff function is based on individual performance measures.

A summary of the game models discussed in this chapter is given in Table 11.1.

11.6 Learning techniques for self-organizing small cell networks

In the previous sections of this chapter, we have observed that in order to achieve the equilibrium, small cells have to be aware of the environment (e.g., by sensing or local interaction) and react accordingly by adjusting their resource allocation policies. Also, due to the dynamics of the wireless environment, most of the system parameters need to be adjusted as well. Such adaptation for obtaining better performance and/or for reaching the equilibrium can be viewed as a learning process that is crucial for self-organization in small cells. Accordingly, different learning techniques can be applied. Specifically, distributed learning techniques such as RL and Q-learning have recently gained significant attention from the research community. Also, in the past few years there has been a growing interest in applying machine learning techniques for wireless networks.

In this section, we will introduce the basics of some commonly used learning techniques and provide examples on how those techniques can be applied for self-organizing

Table 11.1 Game models for self-organizing small cell networks.

Objective	Game type	Player set	Strategy	Solution
Power allocation [39]	Non-cooperative, supermodular	SBSs	Downlink transmit power	NE
Power allocation [50]	Non-cooperative	SUs	Uplink transmit power	NE
Power allocation [51]	Non-cooperative	MUs & SUs	Uplink transmit power	NE
Resource block allocation [56]	Non-cooperative, potential	SUs	OFDMA RBs for uplink	NE
Joint power–subchannel allocation [58]	Non-cooperative, potential	SBSs	A composite of downlink transmit power and subcarrier	NE
Power allocation [59]	Non-cooperative	SBSs	Uplink transmit power	NE
Joint power–subcarrier allocation [28, 61, 68]	Evolutionary	SBSs	Downlink power–subcarrier combination	EE
Forming a group for interference draining [69]	Cooperative coalition formation	SBSs	Coalitions	RC
Joint power–subcarrier allocation [70]	Non-cooperative	SBSs	A composite of downlink transmit power and subcarrier (mixed strategy)	LE
Forming a cluster for joint beamforming [34]	Cooperative coalition formation	SBSs	Coalitions	RC
Joint power–spectrum allocation [38]	Non-cooperative	SBSs	Downlink power level and frequency band	ε-CCE
Coverage optimization [48]	Non-cooperative	SBSs	Downlink transmit power	NE
Resource block (RB) allocation in cognitive SBSs [66]	Non-cooperative	SBSs	Downlink RBs	Correlated equilibrium

small cell networks. Specifically, we focus on the adaptations of different learning techniques to learn the equilibria in game-based self-organizing systems.

11.6.1 Reinforcement learning

Reinforcement learning (RL) [71] can be simply explained as mapping situations into actions in such a way that the cumulative reward is maximized. A learning agent has to *explore* the environment and *exploit* what it has already explored to get a better payoff. The basic elements of RL are explained below.

- *Policy* (Φ). A policy is a mapping of the set of states given by $\mathcal{S}$ into actions ($\mathcal{A}$). Generally policies are stochastic.
- *Reward function* (u). Similar to the payoff/utility function in game theory, the reward function reflects an agent's preference of that state. Simply, the reward function maps the state-action pair into a numerical value. The agent's objective is to maximize its total reward in the long run.
- *Model* of the environment. The model reflects the behavior of the environment.

Reinforcement learning has been widely used in the field of cognitive radio networks [72]. Here we introduce an example (based on [70] and [73]) of applying RL to learn and reach equilibrium in a self-organizing small cell network.

Let us consider the downlink transmission of a macro base station and a set of N underlaid small cell base stations (denoted by $\mathcal{N}$) each of which transmits to one user at a time. $\mathcal{K} = \{1, \ldots, K\}$ is the set of orthogonal subcarriers shared by both tiers. At each time slot, the macro base station serves one macro user over each subcarrier and also each SBS selects one subcarrier to transmit. Each SBS is capable of selecting a transmission power level from a finite set of power levels. The combination of the power level and the subcarrier is termed as a *transmission alignment*. The problem is to select suitable transmission alignments for the downlink transmission of SBSs while protecting the macro users from interference.

The above problem can be modeled as a mixed-strategy non-cooperative game. SBSs form the player set $\mathcal{N}$ and the available transmission alignments form the action set $\mathcal{A}_i$. The mixed strategy (s_i) vector of user i is given by

$$s_i = (\alpha_{i,a_1}, \alpha_{i,a_2}, \ldots, \alpha_{i,a_L}) \in \Delta(\mathcal{A}_i), \tag{11.41}$$

where L is the total number of transmission alignments and α_{ial} is the long-term probability of the SBS taking action α_l.

Two games are formulated ($\mathcal{G}^1$ and $\mathcal{G}^2$) based on two payoff functions as follows:

$$\pi_i^1(a_i(n), \boldsymbol{a}_{-i}(n)) = \sum_{k \in \mathcal{K}} \log_2\left(1 + \gamma_i^k(n)\right)_{\mathbb{K}\{\gamma_0^k(n) > \Gamma_0^k\}}, \tag{11.42}$$

$$\pi_i^2(a_i(n), \boldsymbol{a}_{-i}(n)) = \sum_{i \in \mathcal{N}} \sum_{k \in \mathcal{K}} \log_2\left(1 + \gamma_i^k(n)\right)_{\mathbb{K}\{\gamma_0^k(n) > \Gamma_0^k\}}, \tag{11.43}$$

where γ_i^k is the SINR at the user served by SBS i on subcarrier k, γ_0^k is the SINR at macro user receiving on subcarrier k, Γ_0^k is the SINR threshold at macro user on subcarrier k, and n denotes the time step.

The long-term average value of the payoff (for both games) is given by

$$\bar{\pi}_i(s_i, s_{-i}) = \sum_{a \in \mathcal{A}} \pi_i(a_i, \boldsymbol{a}_{-i}) \Pi_{j=1}^N \alpha_{j,a_j}. \tag{11.44}$$

The solutions obtained for the above games are in the notion of logit equilibrium [74]. Before defining the logit equilibrium, it is necessary to understand the concept of smoothed best response (SBR).

DEFINITION 11.6 *Smoothed best response*: the smoothed best response of player i with parameter m_i is given by

$$\beta_i^{m_i}(s_{-i}) = (\beta_{i,i}^{m_i}(s_{-i}), \ldots, \beta_{i,L}^{m_i}(s_{-i})), \quad (11.45)$$

where

$$\beta_{i,l}^{m_i}(s_{-i}) = \frac{exp(m_i \bar{u}_i(e_l^{(L)}, s_{-i}))}{\sum_{t=1}^{L} exp(m_i \bar{u}_i(e_l^{(L)}, s_{-i}))}, \quad (11.46)$$

in which the vector $e_l^{(L)} = (e_{l,1}^{(L)}, e_{l,2}^{(L)}, \ldots, e_{l,L}^{(L)}) \in \mathbb{R}^L$ denotes the sth vector of the canonical base spanning the space of real vectors of dimension S, (i.e., $e_{l,t}^{(L)} = 0$ for $t \in \{1, 2, \ldots, L\} \setminus \{l\}$ and $e_{l,l}^{(L)} = 0$). Note that SBR is equivalent to best response when $m_i \to \infty$. For finite $m_i > 0$, SBR assigns high probabilities to the actions associated with high average payoffs.

Using the above definition, the logit equilibrium is defined as follows:

DEFINITION 11.7 *Logit equilibrium*: a strategy profile $s^* = (s_1, s_2, \ldots, s_N) \in \Delta(A_1) \times \cdots \times \Delta(A_N)$ is logit equilbrium with parameters $m_{i(\forall i \in \mathcal{N})} > 0$ of the $\mathcal{G}^1$ or $\mathcal{G}^2$ if

$$s_i^* = \beta_i^{m_i}(s_{-i}^*), \quad (11.47)$$

where $\beta_i^{m_i}$ is the smoothed best response of player i with parameter m_i.

Note that s^* is an e-equilibrium with $\epsilon = \max_{i \in \mathcal{N}}(\frac{1}{m_i} \ln(L))$. Also, it can be observed that $\epsilon \to 0$ for large m_i, which means ϵ-equilibrium reaches a pure strategy NE when m_i is large enough.

As both the above games are finite games, the existence of the LE can be proved following Theorem 1 in [75]. Due to the fact that $\mathcal{G}^2$ is a potential game, it can be proved that the convergence of the SBR dynamics is guaranteed for $\mathcal{G}^2$ if each SBS possesses the complete information of the strategies of other SBSs [76]. For self-organizing small cell networks, [70] proposes an RL-based technique in order to learn and reach the equilibrium. Each SBS makes an estimation (given by equation (11.48)) on their own instantaneous payoff $((a_k(n), a_{-k}(n)))$ based on user feedback as follows:

$$\tilde{\pi}_i(n) = \pi(a_i(n), \boldsymbol{a}_{-i}(n)) + \epsilon_{i,a_i(n)}(n), \quad (11.48)$$

where $\epsilon_{i,a_i(n)}(n)$ represents the error of the estimation due to thermal noise and it is also assumed that $E[\epsilon_{i,a_i(n)}(n)] = 0, \forall i$.

Each SBS should estimate the expected utility it achieves with each of its actions in order to build the SBR. Two coupled RL processes are proposed in [70] to achieve the LE. The first RL process allows SBSs to build an estimate of the vector of average payoffs $\bar{\pi}_i(., \pi_{-i}(n))$ using observations $\tilde{\pi}_i(n)$, where $\bar{\pi}_i(., \pi_{-i}(n)) = (\bar{\pi}_i(e_1^{(L)}, \pi_{-i}(n)), \ldots, \bar{\pi}_i(e_L^{(L)}, \pi_{-i}(n)))$. The first process is given by equation (11.49). The second RL process (given in equation (11.50)) uses the vector of estimated average payoffs at time n to update the transmission probability vector $s_i(n)$, where w_i^1 and w_i^2 are learning parameters.

$$\hat{\pi}_{i,l}(n) = \hat{\pi}_{i,l}(n-1) + w_i^1(n)\mathbb{I}_{\{a_i(n)=l\}}(\tilde{\pi}_i(n) - \hat{\pi}_{i,l}(n-1)). \tag{11.49}$$

$$s_{i,l}(n) = s_{i,l}(n-1) + w_k^2(n)\big(\tilde{\beta}_{i,l}^{(m_i)}(\hat{\mathbf{u}}_i(n)) - s_{i,l}(n-1)\big). \tag{11.50}$$

The parameters should satisfy the following conditions:

$$\lim_{T\to\infty} \sum_{t=1}^{T} w_i^1(t) = +\infty,$$

$$\lim_{T\to\infty} \sum_{t=1}^{T} \big(w_i^1(t)\big)^2 < +\infty,$$

$$\lim_{T\to\infty} \sum_{t=1}^{T} w_i^2(t) = +\infty,$$

$$\lim_{T\to\infty} \sum_{t=1}^{T} \big(w_i^2(t)\big)^2 < +\infty,$$

$$\lim_{T\to\infty} \frac{w_i^1(t)}{w_i^2(t)} = 0,$$

and

$$\forall i \in \mathcal{N},\ w_i^2 = w^2, \tag{11.51}$$

or

$$\forall i \in \mathcal{N}/\{N\} \lim_{T\to\infty} \frac{w_i^2(t)}{w_i^2(t+1)} = 0. \tag{11.52}$$

[70] proves that the convergence point (if there is one) of the above given RL algorithm is an LE. In addition to that it is also proved that the convergence of $\mathcal{G}^2$ is always guaranteed as it is a potential game and all players share identical interests.

11.6.2 Q-learning

Q-learning proposed by Watkins in [77] is also a form of reinforcement learning technique that can be used to find an optimal decision policy for any given finite Markov decision process (MDP) problem without knowledge of the transition probabilities. It has also been shown that the Q-learning algorithm converges to the optimal policy for the systems with centralized control [71, 77]. Q-learning has recently been applied in the field of cognitive radio and wireless communications. For example, [78] investigates the problem of network selection in a heterogeneous network and [65] applies Q-learning based learning technique for channel selection in multi-user cognitive radios. A Q-learning based distributed resource allocation algorithm is devised in [79] in order to reduce interference in a network where small cells coexist with the macro network.

In the following, we provide an example of the use of Q-learning for downlink resource allocation in a two-tier small cell network [60].

Small cell base stations (denoted by the set $\mathcal{N}$) form the player set and universal frequency reuse with K subcarriers is considered. It is also assumed that there is only one macro user receiving on each subcarrier at a given time. Each of these macro users has a minimum SINR requirement. Similar to the system model of the example given in Section 11.6.1, each SBS is able to select its transmission power from a finite set of values. Therefore the set of subcarrier-power level combinations (i.e., transmission alignments) defines a possible set of actions for each SBS.

The states for each player i at time t are defined as follows.

$$s_i(t) = \left(s_i^{(1)}(t), s_i^{(2)}(t), \ldots, s_i^{(K)}(t)\right), \tag{11.53}$$

where $s_i^{(K)}(t)$ takes the value 0 if the SBS i violates the QoS constraint for macro user on subcarrier k and $s_i^{(K)}(t) = 1$, otherwise. The action and utility vectors of each SBS at time t are given by

$$a_i(t) = \left(a_i^{(1)}(t), a_i^{(2)}(t), \ldots, a_i^{(K)}(t)\right), \tag{11.54}$$

and

$$u_i(t) = \left(u_i^{(1)}(t), u_i^{(2)}(t), \ldots, u_i^{(K)}(t)\right), \tag{11.55}$$

where $a_i^{(1)}(t) \in \{0, 1\}$.

Each SBS i observes its current state $s_i(t)$ and takes an action $a_i(t)$ based on the decision policy $\Phi = s \rightarrow a$. Our objective is to find an optimal decision policy Φ^*.

For each SBS, a Q-function maintains the knowledge of other players based on which the decisions can be taken individually without interacting with other players.

The expected discounted reward over a finite horizon is given by

$$V^{\Phi}(s) = E\{\gamma^t \times r(s_t, \Phi^*(s_t)) | s_0 = s\}, \tag{11.56}$$

where $0 \leq \gamma \leq 1$ is the discount factor at time t and r is the reward. The above equation can be re-written as

$$V^{\Phi}(s) = R(s, \Phi^*(s)) + \gamma \sum_{s' \in S} p_{s,s'}(\Phi(s)) V^{\Phi}(s'), \tag{11.57}$$

where $R(s, \Phi^*(s))$ is the mean value of the reward $r(s, \Phi(s))$, $p_{s,s'}$ is the transition probability from state s to s' and $0 \leq \gamma \leq$ is the discount factor. The optimal policy Φ^* gives the optimal discounted reward $V^*(s)$. Hence,

$$V^*(s) = V^{\Phi(s)^*} = \max_{\forall a}(R(s, a)) + \gamma \sum_{s' \in S} p_{s,s'}(a) V^*(s')). \tag{11.58}$$

A Q value is maintained to learn the expected discounted reward. For an agent who takes action a when it is at state s and then follows the policy Φ the expected discounted utility (which is the Q value) is given by

$$Q^*(s, a) = R(s, a) + \gamma \sum_{s' \in S} P_{s,s'} V^{\Phi}(s'). \tag{11.59}$$

Each agent keeps trying all action-state combinations with a non-zero probability. The Q-learning algorithm utilizes the obtained reward at each step to update the Q values according to the following equation:

$$Q_t((s),(a)) = (1-\alpha)Q_{t-1}((s),(a)) + \alpha[r_t((s),(a)) + \gamma \max_{b \neq a} Q_{t-1}(v,b)], \quad (11.60)$$

where α is the learning rate.

Simulation results show that this technique reaches the convergence after few iterations for the above-mentioned system model. However, the evolutionary game-based learning algorithm in Section 11.5.4 shows faster convergence than the Q-learning based algorithm. The faster convergence is achieved at the expense of more information exchange among base stations.

11.6.3 Regret-matching learning

Regret-matching [80] is a learning technique that can converge to a correlated equilibrium (CE) in finite games. The notion of CE is based on having a correlating mechanism for the players. This correlating mechanism provides a probability distribution over the set of actions of each player, which provides an assignment *recommendation* for each action. Such assignment recommendation is said to be in CE if none of the players would benefit by deviating from the recommendation. For a more formal definition of CE, we first define the game $\mathcal{G} = \{\mathcal{N}, (\mathcal{S}_i)_{i \in \mathcal{N}}, (\mathcal{P}_i)_{i \in \mathcal{N}}\}$ where $\mathcal{N}$ is the set of players, $\mathcal{S}_i$ is the action set of the ith player and $\mathcal{P}_i$ gives the set of payoffs that can be obtained by the ith player. $\mathcal{S} = \Pi_{i \in \mathcal{N}}(\mathcal{S}_i)$ is the set of N-tuples of the strategies. Let s denote any element in $\mathcal{S}$, s^i denotes an element of $\mathcal{S}_i$ and $\pi_i \in \mathcal{P}$ is the payoff of the player i.

DEFINITION 11.8 *Correlated Equilibrium*: a probability distribution ψ over $\mathcal{S}$ gives a correlated equilibrium for the game $\mathcal{G}$ if $\forall i \in \mathcal{N}, \forall k \in \mathcal{S}_i$, and $\forall s_{-i} \in \mathcal{S}_{-i}$,

$$\sum_{s_{-i} \in \mathcal{S}_{-i}} \Psi(s)(\pi^i(k, s_{-i}) - \pi^i(s, s_{-i})) \leq 0, \quad (11.61)$$

where $\mathcal{S}_{-i}$ is the set of actions played by the opponents of player i. Every Nash equilibrium is also a correlated equilibrium, which corresponds to the case where the recommendations are not correlated at all.

In the regret-matching algorithm, a player would take decisions in order to minimize the regret. The regret of a player's playing action s' is defined as the difference between the average payoff that the player would have achieved if she played the action s' all the time and the average current payoff. Regret of player i playing action s at the nth step is defined as follows.

$$r_i^{(s')}(n) = \frac{1}{n-1} \sum_{t=1}^{n-1} (\pi_i(s', s_{-i}(t)) - \pi_i(s(t), s_{-i}(t))), \quad (11.62)$$

where $s(t)$ denotes the action played by the corresponding player (k in the above equation) at time step t. The steps of the regret-matching algorithm are given in Algorithm 11.3.

Algorithm 11.3 Regret-matching algorithm

1: For $t = 1, 2, 3, \ldots$
2: Calculate the regret for each user using equation (11.62).
3: Obtain the regret vector for each player, i.e., $\mathbf{R}_i(n) = (\forall s \in \mathcal{S} : r_i^{(s)}(n))$.
4: Obtain the probability distribution $\mathbf{\Psi}_i(n)$ by normalizing $\mathbf{R}_i(n)$.
5: The action played at time step n is chosen according to the probability distribution $\mathbf{\Psi}_i(n)$.

It is also known that regret-matching learning can converge toward pure strategy NE points of exact potential games [81]. However, the regret-based learning algorithm assumes that each player can calculate the expected payoff that it would have achieved by playing any action other than the current action. Therefore a considerable amount of information exchange might be needed for this algorithm to be implemented.

In the context of self-organizing small cells, the aforementioned regret-based learning algorithm can be modified to implement a fully decentralized algorithm, which is only based on the SINR feedback of the users to the base station [38]. This modified algorithm converges to an ϵ-coarse correlated equilibrium.

11.6.4 Learning by cooperation

The performance of learning mechanisms can be significantly improved by enabling cooperation among the learners [82, 83]. Small cell base stations that cooperate with neighboring base stations in order to speed up and improve their learning process are called *docitive base stations*. Cooperation is generally done via the backhaul. Network nodes are expected to select other nodes, which operate under similar conditions, to learn from. The similarities between two base stations are captured by a *gradient*, which is defined based on the network architecture.

Several different cases of docition can be identified based on the degree of docition [84, 6].

- Startup docition. When a small cell base station connects to the network for the first time, it can learn the policies from other SBSs with similar gradients by exchanging Q tables.
- IQ-driven docition. SBSs with similar gradients share their policies periodically.
- Performance-driven docition. Base stations share their policies with less expert nodes, based on their ability to meet a pre-defined QoS target.

In addition to the above-discussed techniques, there are a number of other learning techniques that are potentially applicable for self-organizing small cell networks. For instance, the logit learning algorithm and its variants (i.e., max-logit algorithm and binary logit algorithm) converge to NE in potential games [85]. Learning automata [86], which is a branch of adaptive control theory, is also another potential technique to implement distributed learning in self-organizing small cell networks. Stochastic learning automata-based channel selection algorithm has been proposed for opportunistic spectrum access in cognitive radio networks in [87]. This stochastic learning automata-based algorithm can converge to a pure strategy NE point for any exact potential game. Since development of distributed learning techniques for small cell networks has attracted significant attention from the research community recently, new learning techniques are still emerging.

11.7 Conclusion

In this chapter, we have discussed game theory approaches and learning techniques for self-organization in small cell networks. First, we have given an overview of self-organizing networks including the motivations of enabling self-organizing functionalities in densely deployed small cells. Then a brief introduction to game theory has been given and the motivations of using game theory in self-organizing small cell networks have been discussed. Also, some widely used game models have been explained. Then a few examples have been given to explain how game theory can be used to solve the problem of self-organization in small cell networks. This chapter has also discussed learning techniques that can be used in small cell networks.

Some of the future research issues in designing self-organizing small cell networks are outlined below.

1. *Incomplete information games.* In order to address the issue of incomplete information, most of the existing algorithms use RL-based techniques as we discussed in this chapter. However, models to address partial information can also be developed using Bayesian games. By using the Bayesian theorem, a belief on the parameters of other players can be constructed. The solution concept obtained in such games is the Bayesian–Nash equilibrium.
2. *Multi radio access technology (Multi-RAT).* In future networks, different radio access technologies (e.g., WiFi, small cells) are expected to be integrated in order to provide seamless service to the user. Access control between different technologies should be done in a self-organizing way to achieve the optimal performance.
3. *Signaling overhead-optimal performance tradeoff.* There is always a tradeoff between the signaling overhead and optimal performance of a network. A network may deliver optimal performance with complete information but the signaling cost for implementing such algorithms would be higher. On the other hand, the algorithms that rely on less information or incomplete information may deliver slightly degraded performance. Addressing this issue and quantifying the tradeoff is significant in order to achieve near-optimal or optimal performance in self-organizing networks.
4. Context-awareness. Context awareness, which is a powerful feature in many intelligent systems, is recently proposed to be applied for enhancing self-organizing features in small cell networks. The idea is to utilize the context information, i.e., information from the users' environment, behavior, and social media, to enhance the

provision of services and applications. The algorithms should be devised considering the efficient exploitation of context-aware information taken from different sources. The reliability of the different information sources would also be an important issue.

References

[1] V. Chandrasekhar, "Femtocell networks: a survey," *IEEE Communications Magazine*, pp. 59–67, September 2008.

[2] P. Demestichas and A. Georgakopoulos, "5G on the horizon: key challenges for the radio-access network," *IEEE Vehicular Technology Magazine*, pp. 47–53, September 2013.

[3] O. Aliu, A. Imran, M. Imran, and B. Evans, "A survey of self organisation in future cellular networks," *IEEE Communications Surveys and Tutorials*, vol. 15, pp. 336–361, 2012.

[4] H. Claussen, L. T. Ho, and L. G. Samuel, "An overview of the femtocell concept," *Bell Labs Technical Journal*, vol. 13, pp. 221–245, 2008.

[5] N. Johnson, "Small cell forum and LTE small cells," *The Small Cell Forum, Tech. Rep.*, 2011. www.smallcellforum.org/smallcellforu_resources/pdfs7/

[6] M. Bennis and L. Giupponi, "Interference management in self-organized femtocell networks: the BeFEMTO approach," *2nd International Conference on Wireless Communication, Vehicular Technology, Information Theory and Aerospace Electronic Systems Technology (Wireless VITAE)*, Chennai, 2011, pp. 5–10.

[7] J. V. den Berg and R. Litjens, "SOCRATES: Self-Optimisation and self-ConfiguRATion in wirelEss networkS," *Fourth ERCIM Workshop on eMobility*, 2008.

[8] T. Jansen, M. Amirijoo, U. Turke, *et al.* "Embedding multiple self-organisation functionalities in future radio access networks," *IEEE 69th Vehicular Technology Conference*. IEEE, 2009, pp. 1–5.

[9] EU FP7 project. End-to-End Efficiency (E3). https://ict-e3.eu.

[10] E. Bogenfeld and I. Gaspard, "Self-X in radio access networks," *E3 White Paper v1. 0*, vol. 22, 2008.

[11] E. Bonabeau, M. Dorigo, and G. Theraulaz, *Swarm Intelligence:From Natural to Artificial Systems*. New York, Oxford University Press 1999, vol. 4.

[12] 3GPP TS 32.500, "Telecommunication management; self-organizing networks (SON); concepts and requirements."

[13] 3GPP TR 36.902, "Evolved universal terrestrial radio access network (E-UTRAN); self-configuring and self-optimizing network (son) use cases and solutions."

[14] NGMN Alliance, "Next generation mobile networks beyond HSPA and EVDO," *White PaperDec*, vol. 5, 2006.

[15] NGMN Alliance, "Next generation mobile networks recommendation on SON and O&M requirements," *Req. Spec. v1*, vol. 23, 2008.

[16] S. Haykin, "Cognitive dynamic systems," *IEEE International Conference on Acoustics, Speech and Signal Processing, 2007.*, vol. 4, 2007, pp. IV–1369–IV–1372.

[17] Haykin, "Cognitive radio: brain-empowered wireless communications," *IEEE Journal on Selected Areas in Communications*, vol. 23, pp. 201–220, 2005.

[18] A. Spilling, A. Nix, M. Beach, and T. Harrold, "Self-organisation in future mobile communications," *Electronics and Communication Engineering Journal*, vol. 12, pp. 133–147, 2000.

[19] C. Prehofer and C. Bettstetter, "Self-organization in communication networks: principles and design paradigms," *IEEE Communications Magazine*, vol. 43, no. 7, pp. 78–85, 2005.

[20] S. Hämäläinen, H. Sanneck, C. Sartori *et al.*, *LTE Self-Organising Networks (SON): Network Management Automation for Operational Efficiency*. John Wiley & Sons, 2012.

[21] I. Viering, M. Dottling, and A. Lobinger, "A mathematical perspective of self-optimizing wireless networks," *IEEE International Conference on Communications*, 2009, pp. 1–6.

[22] H. Hu, J. Zhang, and X. Zheng, "Self-configuration and self-optimization for LTE networks," *IEEE Communications Magazine*, pp. 94–100, February 2010.

[23] H. Sanneck, Y. Bouwen, and E. Troch, "Context based configuration management of plug and play LTE base stations," *IEEE Network Operations and Management Symposium – NOMS 2010*, pp. 946–949, 2010.

[24] "Self-optimizing networks: the benefits of SON in LTE," 4G Americas, 2013. www.4gamericas.org.

[25] NGMN Alliance, "Use cases related to self organising network: Overall description," 2007.

[26] Y. Choi, C. Kim, and S. Bahk, "Flexible design of frequency reuse factor in OFDMA cellular networks," *IEEE International Conference on Communications*, vol. 4, 2006, pp. 1784–1788.

[27] O. Ileri, S.-C. Mau, and N. B. Mandayam, "Pricing for enabling forwarding in self-configuring ad hoc networks," *Journal on Selected Areas in Communications*, vol. 23, pp. 151–162, 2005.

[28] P. Semasinghe, E. Hossain, and K. Zhu, "An evolutionary game approach for distributed resource allocation for self-organizing small cells," *IEEE Transactions of Mobile Computing*, 2013.

[29] M. Dirani and Z. Altman, "Self-organizing networks in next generation radio access networks: application to fractional power control," *Computer Networks*, vol. 55, no. 2, pp. 431–438, 2011.

[30] 3GPP TS 32.541, "Evolved universal terrestrial radio access network (E-UTRAN); self-organizing networks (son); self-healing concepts and requirements (Release 11)."

[31] A. Attar, V. Krishnamurthy, and O. N. Gharehshiran, "Interference management using cognitive base-stations for UMTS LTE," *IEEE Communications Magazine*, vol. 49, pp. 152–159, 2011.

[32] K. Zhu, E. Hossain, and D. Niyato, "Pricing, spectrum sharing, and service selection in two-tier small cell networks: a hierarchical dynamic game approach," *IEEE Transactions on Mobile Computing*, 2013.

[33] X. Huang and B. Beferull-Lozano, "Non-cooperative power allocation game with imperfect sensing information for cognitive radio," *IEEE International Conference on Communications (ICC)*. IEEE, 2012, pp. 1666–1671.

[34] S. Guruacharya, D. Niyato, M. Bennis, and D. Kim, "Dynamic coalition formation for network MIMO in small cell networks," *IEEE Transactions on Wireless Communications*, 2013.

[35] R. Madan, A. Sampath, N. Bhushan, A. Khandekar, J. Borran, and T. Ji, "Impact of coordination delay on distributed scheduling in LTE-A femtocell networks," in *IEEE Global Telecommunications Conference (GLOBECOM 2010)*, 2010, pp. 1–5.

[36] Z. Han, D. Niyato, W. Saad, T. Basar, and A. Hjorungnes, *Game Theory in Wireless and Communication Networks*. Cambridge University Press, 2012.

[37] K. Akkarajitsakul, E. Hossain, D. Niyato, and D. I. Kim, "Game theoretic approaches for multiple access in wireless networks: a survey," *IEEE Communications Surveys and Tutorials*, vol. 13, pp. 372–395, 2011.

[38] M. Bennis, S. M. Perlaza, and M. Debbah, "Learning coarse correlated equilibria in two-tier wireless networks," *IEEE International Conference on Communications (ICC)*, 2012, pp. 1592–1596.

[39] E. J. Hong, S. Y. Yun, and D.-H. Cho, "Decentralized power control scheme in femtocell networks: a game theoretic approach," *IEEE 20th International Symposium on Personal, Indoor and Mobile Radio Communications*, 2009, pp. 415–419.

[40] D. M. Topkis, *Supermodularity and Complementarity*. Princeton University Press, 1998.

[41] E. Altman and Z. Altman, "S-modular games and power control in wireless networks," *IEEE Transactions on Automatic Control*, vol. 48, pp. 839–842, 2003.

[42] A. B. MacKenzie and L. A. DaSilva, "Game theory for wireless engineers," *Synthesis Lectures on Communications*, vol. 1, pp. 1–86, 2006.

[43] N. Nisan, *Algorithmic Game Theory*. Cambridge University Press, 2007.

[44] M. K. H. Yeung and Y.-K. Kwok, "A game theoretic approach to power aware wireless data access," *IEEE Transactions on Mobile Computing*, vol. 5, pp. 1057–1073, 2006.

[45] J. Huang, R. A. Berry, and M. L. Honig, "Distributed interference compensation for wireless networks," *IEEE Journal on Selected Areas in Communications*, vol. 24, pp. 1074–1084, 2006.

[46] H. Li, Y. Gai, Z. He, K. Niu, and W. Wu, "Optimal power control game algorithm for cognitive radio networks with multiple interference temperature limits," *Vehicular Technology Conference, Spring 2008*. IEEE, 2008, pp. 1554–1558.

[47] P. Reichl, B. Tuffin, and R. Schatz, "Logarithmic laws in service quality perception: where microeconomics meets psychophysics and quality of experience," *Telecommunication Systems*, pp. 1–14, 2011.

[48] L. Huang, Y. Zhou, X. Han, Y. Wang, M. Qian, and J. Shi, "Distributed coverage optimization for small cell clusters using game theory," *IEEE Wireless Communications and Networking Conference (WCNC)*. IEEE, 2013, pp. 2289–2293.

[49] S. Subramani, T. Basar, S. Armour, D. Kaleshi, and Z. Fan, "Noncooperative equilibrium solutions for spectrum access in distributed cognitive radio networks," *3rd IEEE Symposium on New Frontiers in Dynamic Spectrum Access Net-works, DySPAN 2008*. IEEE, 2008, pp. 1–5.

[50] W. Ma, H. Zhang, W. Zheng, and X. Wen, "Differentiated-pricing based power allocation in dense femtocell networks," *15th International Symposium on Wireless Personal Multimedia Communications (WPMC)*. IEEE, 2012, pp. 599–603.

[51] V. Chandrasekhar, J. G. Andrews, T. Muharemovic, Z. Shen, and A. Gatherer, "Power control in two-tier femtocell networks," *IEEE Transactions on Wireless Communications*, vol. 8, pp. 4316–4328, 2009.

[52] I. L. Glicksberg, "A further generalization of the kakutani fixed point theorem, with application to nash equilibrium points," *Proceedings of the American Mathematical Society*, vol. 3, pp. 170–174, 1952.

[53] J. B. Rosen, "Existence and uniqueness of equilibrium points for concave n-person games," *Econometrica: Journal of the Econometric Society*, pp. 520–534, 1965.

[54] G. Debreu, "A social equilibrium existence theorem," *Proceedings of the National Academy of Sciences of the United States of America*, vol. 38, p. 886, 1952.

[55] R. D. Yates, "A framework for uplink power control in cellular radio systems," *IEEE Journal on Selected Areas in Communications*, vol. 13, pp. 1341–1347, 1995.

[56] I. W. Mustika, K. Yamamoto, H. Murata, and S. Yoshida, "Potential game approach for self-organized interference management in closed access femtocell networks," *IEEE 73rd Vehicular Technology Conference (VTC Spring)*. IEEE, 2011, pp. 1–5.

[57] 3GPP TR 36.814 v9.0.0, "Evolved universal terrestrial radio access (E-UTRA); further advancement of E-UTRA physical layer aspects (Release 9)," March 2010.

[58] L. Giupponi and C. Ibars, "Distributed interference control in OFDMA-based femtocells," *IEEE 21st International Symposium on Personal Indoor and Mobile Radio Communications (PIMRC)*. IEEE, 2010, pp. 1201–1206.

[59] J.-H. Yun and K. G. Shin, "Adaptive interference management of OFDMA femtocells for co-channel deployment," *IEEE Journal on Selected Areas in Communications*, vol. 29, pp. 1225–1241, 2011.

[60] M. Bennis, S. Guruacharya, and D. Niyato, "Distributed learning strategies for interference mitigation in femtocell networks," *IEEE Global Telecommunications Conference (GLOBECOM 2011)*. IEEE, 2011, pp. 1–5.

[61] P. Semasinghe, K. Zhu, and E. Hossain, "Distributed resource allocation for self-organizing small cell networks: an evolutionary game approach," *IEEE GLOBECOM Workshops*, 2013.

[62] J. W. Weibull, *Evolutionary Game Theory*. MIT Press, 1997.

[63] S. Samarakoon, M. Bennis, W. Saad, and M. Latva-aho, "Backhaul-aware interference management in the uplink of wireless small cell networks," *IEEE Transactions of Wireless Communications*, 2013.

[64] G. Gur, S. Bayhan, and F. Alagoz, "Cognitive femtocell networks: an overlay architecture for localized dynamic spectrum access [dynamic spectrum management]," *IEEE Wireless Communications*, vol. 17, pp. 62–70, 2010.

[65] H. Li, "Multi-agent Q-learning of channel selection in multi-user cognitive radio systems: a two by two case," *IEEE International Conference on Systems, Man and Cybernetics, SMC*. IEEE, 2009, pp. 1893–1898.

[66] J. W. Huang and V. Krishnamurthy, "Cognitive base stations in LTE/3GPP femtocells: a correlated equilibrium game-theoretic approach," *IEEE Transactions on Communications*, vol. 59, pp. 3485–3493, 2011.

[67] R. J. Aumann, "Correlated equilibrium as an expression of Bayesian rationality," *Econometrica: Journal of the Econometric Society*, pp. 1–18, 1987.

[68] M. Nazir, M. Bennis, K. Ghaboosi, A. B. MacKenzie, and M. Latva-aho, "Learning based mechanisms for interference mitigation in self-organized femtocell networks," *Forty-Fourth Asilomar Conference on Signals, Systems and Computers*, 2010, pp. 1886–1890.

[69] F. Pantisano, M. Bennis, W. Saad, M. Latva-aho, and R. Verdone, "Enabling macrocell–femtocell coexistence through interference draining," *Wireless Communications and Networking Conference Workshops*, 2012, pp. 81–86.

[70] M. Bennis, S. M. Perlaza, P. Blasco, Z. Han, and H. V. Poor, "Self-organization in small cell networks: a reinforcement learning approach," *IEEE Transactions on Wireless Communications*, vol. 12, pp. 3202–3212, 2013.

[71] R. S. Sutton and A. G. Barto, *Reinforcement Learning: An Introduction*. Cambridge University Press, 1998, vol. 1, no. 1.

[72] M. Bkassiny, Y. Li, and S. Jayaweera, "A survey on machine-learning techniques in cognitive radios," *IEEE Communications Surveys Tutorials*, vol. 15, pp. 1136–1159, 2013.

[73] M. Bennis and S. M. Perlaza, "Decentralized cross-tier interference mitigation in cognitive femtocell networks," *IEEE International Conference on Communications (ICC)*. IEEE, 2011, pp. 1–5.

[74] D. L. McFadden, "Quantal choice analaysis: a survey," *Annals of Economic and Social Measurement*, vol. 5, no. 4. NBER, 1976, pp. 363–390.

[75] R. D. McKelvey and T. R. Palfrey, "Quantal response equilibria for normal form games," *Games and Economic Behavior*, vol. 10, pp. 6–38, 1995.

[76] J. Hofbauer and W. H. Sandholm, "On the global convergence of stochastic fictitious play," *Econometrica*, vol. 70, pp. 2265–2294, 2002.

[77] C. J. C. H. Watkins, "Learning from delayed rewards." PhD dissertation, University of Cambridge, 1989.

[78] D. Niyato and E. Hossain, "Dynamics of network selection in heterogeneous wireless networks: an evolutionary game approach," *IEEE Transactions on Vehicular Technology*, vol. 58, pp. 2008–2017, 2009.

[79] M. Bennis and D. Niyato, "A Q-learning based approach to interference avoidance in self-organized femtocell networks," *IEEE GLOBECOM Workshops*. IEEE, 2010, pp. 706–710.

[80] S. Hart and A. Mas-Colell, "A simple adaptive procedure leading to correlated equilibrium," *Econometrica*, vol. 68, pp. 1127–1150, 2000.

[81] N. Nie and C. Comaniciu, "Adaptive channel allocation spectrum etiquette for cognitive radio networks," *Mobile Networks and Applications*, vol. 11, pp. 779–797, 2006.

[82] M. N. Ahmadabadi and M. Asadpour, "Expertness based cooperative Q-learning," *IEEE Transactions on Systems, Man, and Cybernetics, Part B: Cybernetics*, vol. 32, pp. 66–76, 2002.

[83] M. Tan, "Multi-agent reinforcement learning: Independent vs. cooperative agents," *Proceedings of the Tenth International Conference on Machine Learning*, vol. 337. Amherst, MA, 1993.

[84] A. G. Serrano, L. Giupponi, and M. Dohler, "Befemto's self-organized and docitive femto-cells," *Future Network and Mobile Summit*. IEEE, 2010, pp. 1–8.

[85] Y. Song, S. H. Wong, and K.-W. Lee, "Optimal gateway selection in multi-domain wireless networks: a potential game perspective," *Proceedings of the 17th Annual International Conference on Mobile Computing and Networking*. ACM, 2011, pp. 325–336.

[86] K. S. Narendra and M. Thathachar, "Learning automata: a survey," *IEEE Transactions on Systems, Man and Cybernetics*, pp. 323–334, 1974.

[87] Y. Xu, J. Wang, Q. Wu, A. Anpalagan, and Y.-D. Yao, "Opportunistic spectrum access in unknown dynamic environment: a game-theoretic stochastic learning solution," *IEEE Transactions of Wireless Communications*, pp. 1380–1391, 2012.

12 Energy efficient strategies with BS sleep mode in green small cell networks

Hong Zhang and Jun Cai

The traditional mobile cellular networks are often designed so that a base station (BS) is always under uninterrupted working condition without considering the dynamic nature of user traffic, which results in an inefficient usage of energy. How to improve the system energy efficiency in order to achieve green networking is our major concern in this chapter. Beginning with a comprehensive review of the related works in literature, we introduce a self-organized BS virtual small networking (VSN) protocol so as to adaptively manage BSs' working states based on heterogeneity of user traffic changing in space and time. Motivated by the fact that low-traffic areas can apply a more aggressive BS-off strategy than hotspots, the proposed method is targeted at dividing BSs into groups with some similarity measurements so that the BS-off strategy can be performed more efficiently. Numerical results show that our proposals can save energy consumption on the entire cellular network to a great extent.

12.1 Introduction

As demand increases for more energy-efficient technologies in wireless networks, to tackle critical issues such as boosting cost on power consumption and excessive greenhouse gas emissions, the concept of green networking has drawn great attention in recent years. In fact, during the last decades, people have witnessed that the carbon footprint of the telecommunications industry has been exponentially growing due to the explosive rise of service requirements and subscribers' demands. The concern on reducing power consumption comes from both environmental and economical reasons. With respect to the environment, the information and communications technology (ICT) industry is responsible for approximately 2% of current global electricity demands, with 6% yearly growth in ICT-related carbon dioxide emission (CO_2-e) forecast till 2020 [1]. With respect to economics, the power consumption for operating a typical base station (BS), which needs to be connected to the electrical grid, may cost approximately $3,000/year, while off-grid BSs, generally running on diesel power generators in remote areas, may cost ten times more [2]. As more than 120,000 new BSs are deployed annually [3], there is still no end in sight for the development of mobile communications with a large amount of new subscribers and a constant desire for upgrading user equipment from 2G to 3G, and then to 4G. The continuing growth in power consumption and carbon footprint of operating cellular networks has led to an emerging trend of addressing power efficiency between the service providers and standard regulators. The term "green" is widely mentioned in current research and the concept of "green networking" can be defined with various strategies and goals in different research fields. In wireless networks, the concepts of "green networking" [4] and "green radio" [5] have been described from the environmental, economical, and regulatory points of view. In this chapter, for general telecommunication networks, we define *green networking* as "the implementation of energy-aware network technologies, protocols, and products, optimizing resources usage to reduce energy consumption, providing better service with guaranteed quality of service (QoS), and establishing next-generation economical and ecological telecommunication networks."

12.1.1 BS sleep mode techniques

A typical cellular network mainly consists of a core network taking care of switching, and the deployed BSs providing radio frequency interface and services to mobile users (MUs). From the power consumption point of view, the BSs and the attributed backhaul networks consume approximately 60 billion kWh per year, which corresponds to CO_2-e of 40 million tons per year or approximately the annual gas emissions of 8 million cars [6]. It is reported that the power consumption of BSs is responsible for 60 to 80% of the total power demand of the entire network [7]. At present, all BSs are working in the "always-active state," regardless of the associated traffic level for each individual BS. In other words, a BS remains consuming power as usual even when there is no traffic load in its coverage. Moreover, traditional BS deployment is designed to satisfy peak traffic requirements. In fact, the worldwide average peak utilization rates of the cellular networks are merely at 65% in 2011. Since the BSs are often underutilized in normal operation, this leaves much room for saving network power consumption. This has motivated extensive research on switching BSs between the operational (active) mode and the non-operational (sleep) mode[1] so as to achieve power saving with respect to the fluctuations of traffic in space and time [8, 9].

Many power-saving efforts have been made toward switching off some underutilized BSs. For instance, the authors in [9] investigated the feasibility of reducing the number of active BSs, called BS-off strategies, by considering the day–night behavior of mobile users. Some typical cellular network configurations, such as crossroads, Manhattan, and hexagonal cell layout, are considered. The authors concluded that switching off some underutilized BSs is possible to achieve large power saving. In [10], the authors developed dynamic BS power management for cellular networks and derived the power-saving ratio. The concept of cell zooming, which adjusts the cell size dynamically based on traffic variations was presented in [11]. The cell size can be reduced when the BS detects less traffic in its coverage, while the adjacent cells enlarge their cell size to cover the gap correspondingly. In practice, cell zooming can be implemented by adjusting the physical parameters, such as BS transmit power, or by cooperation and

---

[1] We use terms operational (active, power on) mode and non-operational (sleep/dormant, power off) mode to indicate BS working modes interchangeably in this chapter.

relay among BSs. The authors in [12] demonstrated the insufficiency of cell zooming in some cases and discussed the feasibility of deploying smaller but more cells to increase energy efficiency. The smaller cell size has the advantages of consuming less power and being easily managed. Another similar proposal on dynamically adjusting cell size in a multi-layer cellular architecture, called "cell breathe," was introduced in [13]. Also, a decentralized BS sleeping algorithm was proposed for the long-term evolution (LTE) system, where the BS will be turned off if it has the smallest utility value based on the provided data rate for service and the maximum operational power [14]. In [15], the fundamental tradeoffs in green cellular networks were analyzed in terms of energy, spectrum, deployment efficiency, bandwidth, and delay, etc.

12.1.2 Deployment of small cell networks

The heterogeneous network, commonly known as a HetNet, is a promising technique for next-generation wireless networks in response to the explosive growth on the request of faster wireless data services, more coverage, as well as less cost on infrastructure construction and energy consumption. HetNets consist of multi-tier cellular networks including macro cells, micro cells, pico cells, femto cells, and relay base stations. Such multi-tier network cooperation encourages the deployment of small-size cells, called small cell networks (SCNs), overlaid with the existing macro cell networks [16].

Currently, the evolution of wireless technology has made an isolated system (with single type of BS) reach capacity limit determined by information theory [17]. Thus further enhancement can only be achieved by developing advanced wireless topologies, one of which is cooperative SCNs with self-organizing features. The main benefits of this solution are summarized as follows.

- The deployment of SCNs can eliminate dead spots in the conventional macro cell layout and improve network capacity in hotspots. In reality, there are always some areas with dense population, which may even require different types of service such as public, private, and business services. There may also exist areas with sparse population where users may locate in coverage holes. Moreover, the wireless network traffic in some areas may be quite dynamic and unpredictable. Under all these scenarios, the cooperative SCN is an ideal solution to extend the coverage of the hotspots and uncovered areas with low implementation cost, fast adaptation to the wireless traffic fluctuation, and ability to provide various services.
- The deployment of small cells is more cost effective compared to the traditional macro cell layout. The distributive placement of small cells can be based on a rough knowledge of coverage and traffic density, while the placement of macro BSs generally needs careful network-wide planning and centralized operation. In addition, the SCNs can implement self-organizing network (SON) techniques [18], so that a smart network configuration in a cognitive approach with self-learning, intelligent decisions, and dynamic resource management can be adopted. The cooperation among small cells can also maximize the overall system capacity, coverage, and spectrum efficiency. Moreover, SON enables the large deployment of small cells with economic viability

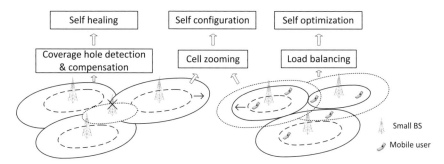

Figure 12.1 The main features of self-organizing small cell networks.

by greatly reducing the costly human intervention in operation, administration, and maintenance [19, 20].
- The deployment of SCNs can make the next-generation wireless network more energy efficient, and contribute to global energy conservation and CO_2-e reduction. A small cell can be powered on or off more adaptively without worrying about the large cooling system as equipped in macro BSs. In addition, the small cells are likely to be turned off more often if there is no associated traffic due to their relatively small coverage. Moreover, from a network point of view, the SCNs can operate cooperatively and adaptively by matching the wireless traffic variation, so as to selectively switch off some underutilized BSs during a certain period with the cells that remain on taking care of the coverage holes [9]. The number of switched-off small cells can be controllable based on the monitored network traffic levels. Such a BS power on/off strategy in SCNs can significantly reduce the energy consumption of the entire network [11].

The deployment of self-organizing networks is also reflected in the current 3rd Generation Partnership Project (3GPP) standard (e.g., 3GPP TS 32.521). As shown in Figure 12.1, the SCN aims to enable the network to optimize, reconfigure, and recover itself, so as to reduce energy expenditure and improve network performance and flexibility. For self-healing, the cells can detect coverage holes, diagnose defects, and make compensation by adapting the cell coverage. For self-configuration, the cells can manage the transmit power, access control, IP address, and connectivity configuration. For self-optimization, the cells can perform load balancing, inter-cell interference coordination, coverage and capacity optimization. In [16], the authors indicated that cell size reduction is the simplest and most effective way to increase system capacity. They also discussed the challenges of deploying SCNs including self-organization, interference management, mobility, and security. In [19], the authors discussed offloading and distributed channel access techniques for the deployment of SCNs with coexistence of macrocells, in which small cells serve as offloading spots in the radio access network to offload users and their associated traffic from congested macro cells. It showed that small cells could overcome the deployment challenges and efficiently coexist in a multi-tier cellular network with cognition capabilities (e.g., spectrum sensing). In [21], the authors indicated that self-organizing is effectively the only viable way to achieve optimal performance in future wireless networks in a cost-effective manner. Also, the main objectives of self-organizing networks towards SCN perspective can be summarized as coverage expansion, capacity optimization, energy efficiency, and QoS enhancement.

The motivation of our work comes from two aspects. First, the current green networking strategy pays little attention to self-organizing SCNs. In fact, the small cells' configuration can be more adaptable and manageable with respect to network traffic distribution. Second, most existing works considering BS sleep mode adopted a unified BS-off strategy over the entire cellular network, and missed the consideration of the potential heterogeneity of traffic distribution among different zones. In fact, zones with low traffic can apply more aggressive BS-off strategy than the hotspots, so that further improvement on power saving can be achieved. We proposed to form BS virtual small networks (VSNs), which can be considered as partitions of the entire cellular network. Each VSN can be self-organized distributively based on the cooperation of small cells, powering off the underutilized small cells by monitoring the current traffic distribution, and guaranteeing the required QoS at the same time.

The remainder of this chapter is organized as follows. Section 12.2 describes the general system model of a wireless virtual small network (VSN). In Section 12.3, the methodologies of determining BS realistic traffic level and estimating mean network traffic are investigated. Also, we propose the self-organized BS VSN firefly (BSVF) algorithm, as well as the BS-off matching scheme and BS on/off weight. Simulation results in terms of power consumption, CO_2-e, convergence rate on VSN average traffic and the number of VSNs are shown in Section 12.4. Finally, Section 12.5 concludes the chapter.

12.2 System model and problem formulation

We consider an infrastructure-based cellular network in an urban environment consisting of multiple micro BSs. Let $\mathcal{N}$ denote a grand coalition of homogeneous micro hexagonal cells serving this urban area. Each cell with radius $\mathcal{R}$ is covered by one micro BS located at the center and all BSs are equipped with omnidirectional antennas. The Euclidean distances, r_{ij}, of any two neighboring BSs i and j are the same.

DEFINITION 12.1 *Neighboring BS.* BS i is the neighboring BS of BS j if BS i and BS j have adjacent geographical covering regions. In hexagonal cells layout of cellular networks, we have $r_{ij} = \sqrt{3}\mathcal{R}$. Define n_j as the number of neighboring BSs of BS j. Since we consider identical micro-hexagonal cells layout, we have $n_j^{max} = 6$. In the cellular network, each BS and its neighboring BS have opportunities to be grouped into a cell set $\mathcal{C}$, each of which is called a virtual small network (VSN). At the initial stage, the total number of VSNs is assumed to be K and any VSN in the network is randomly formed. The detail on determining K is provided in Section 11.3.4. Each VSN, $\mathcal{C}_k$, consists of $c_k, k = \{1, 2, \ldots, K\}$, neighboring BSs. Thus, we have $\mathcal{N} = \cup_{k \in \mathcal{K}} \mathcal{C}_k$.

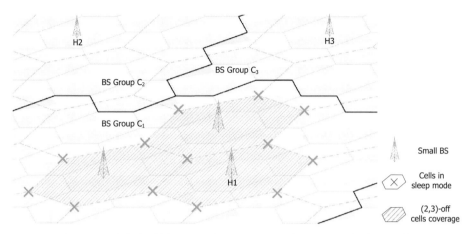

Figure 12.2 System model of cellular networks. (It consists of BS groups C_1, C_2, and C_3 with BS headers $\mathcal{H}_1$, $\mathcal{H}_2$, and $\mathcal{H}_3$, respectively. A (2, 3)-off strategy is adopted in C_1, in which the power consumption can be reduced by switching off two BSs in every three BSs).

When the VSNs are set up, the BS located at the center of each VSN is initialized to be the VSN controller, $\mathcal{H}_k$, whose task is to collect data from member BSs within the same VSN and coordinate collaboration among all member BSs. To reflect the practical urban environment, each BS i may experience different traffic densities changing with space and time. For example, peak traffic hours usually appear in office districts in the daytime or residential regions at night. For any VSN, define Ω as a set of all available BS-off strategies. For explanation purpose, we adopt the widely used (v, u)-off strategies [9][12] as an example. An (v, u)-off strategy means switching v out of u active BSs to sleep mode, where $v < u$ and both u and v are positive integers.

The system model is shown in Figure 12.2 and illustrated with an example of (2, 3)-off strategy. In this chapter, we further introduce the following three requirements on implementation.

R1. The radio coverage and service of BSs in sleep mode can be maintained by the remaining active BSs, i.e., the QoS of MUs has to be guaranteed in dormant cells. The power increment due to the coverage enlargement is omitted, which follows the similar discussions on micro cell layout as shown in [22].

R2. The entire cellular network is considered in a distributed manner, in which direct information exchange is not allowed between BS i and BS j if they belong to different VSNs. It means the entire network condition is unknown for an individual BS. However, the neighboring BSs can exchange information such as traffic level, BS power on/off states, and VSN association status.

R3. The traffic distribution has the same profiles in daily measurement. Hence, the network power consumption is measured in a daily interval and the time is divided into fixed-length intervals as time slots.

12.2.1 BS power consumption model for micro cells

The power consumption of a micro BS consists of static and dynamic power consumptions, denoted by $\mathcal{P}_{sta}$ and $\mathcal{P}_{dyn}$, respectively. $\mathcal{P}_{sta}$ mainly consists of power cost by power amplifier (PA), power supply, air conditioning, etc. $\mathcal{P}_{dyn}$ indicates the dynamic power consumption, which represents the power cost for data transmission. Obviously, $\mathcal{P}_{dyn}$ depends only on the traffic load of a BS. We also take into consideration the power cost in the transmitter transient period, when the BS working states transform between the active and sleep modes, denoted as $\mathcal{P}_{tc}$, as defined in LTE 3GPP Release 11 [23]. It is essential to prevent frequent transition between different BS working modes. In summary, the power consumption model of BS i can be formulated as

$$\mathcal{P}_i = \mathcal{P}_{sta} + \mathcal{P}_{dyn} + \mathcal{P}_{tc} \qquad (12.1)$$
$$= (1+\rho)\mathcal{P}_{sta} + \mathcal{P}_{dyn} \qquad (12.2)$$

where $\mathcal{P}_{tc}$ is a constant and is derived as a ratio of BS static power consumption, i.e., $\mathcal{P}_{tc} = \rho \mathcal{P}_{sta}$, where ρ denotes the power ratio of transmitter transient cost, and $\rho = 0$ if the BS working mode is unchanged.

12.2.2 Traffic distribution model for MUs

We introduce a new traffic distribution model for the homogeneous cellular network in order to better match the reality by revising the one in [24]. The revision is based on the fact that most BSs do not experience maximum traffic load during a day, especially those BSs located outside the hotzones. Moreover, the micro cell deployment implies that the number of MUs served by each BS is small, which leads to more significant random traffic variations among cells. The reformatted traffic model in a cell is

$$f(t) = \frac{\mu}{2^v}[1 + \sin(\pi t/12 + \eta)]^v + \xi(t) \qquad (12.3)$$

where $f(t)$ is the normalized traffic fluctuating with time t. μ is a uniformly distributed random variable in the interval [0,1], which controls the peak traffic rate in a micro cell, $v = 1, 3,$ or 5 determines the abruptness of the traffic model. A larger v means that the traffic curve has a steeper slope, while the average traffic load is lower. η is a uniformly distributed random variable in the interval $[\frac{3\pi}{4}, \frac{7\pi}{4}]$, which determines the distribution of traffic with different peak hours among BSs. $\xi(t)$ is a Poisson distributed random process, which is used to model the random fluctuation of the total traffic. Note that the modified traffic model can restrict the peak hours in a rational period instead of a whole day.

12.2.3 Problem formulation

Our objective is to minimize total power consumption of the entire cellular network by designing the VSN formation protocol with diversified (v, u)-off strategy. The optimization procedure is considered within a certain period of time $\mathcal{T}$, e.g., a daily based measurement. Also, let T denote the total number of time slots. The optimization problem can be formulated as

$$\min \mathcal{P}_{nw} = \int_0^T \left(\sum_{i=1}^N \mathcal{P}_i \right) dt$$

$$\approx \sum_{T=0}^T \sum_{k=1}^K \sum_{j=1}^{c_k} \mathcal{P}_{j,k}^{(\Omega_k)}(t) \cdot w_{j,k}(t) \tag{12.4}$$

$$\text{s.t.} \quad \sum_{k=1}^K c_k = N \tag{12.5}$$

$$\mathcal{P}_{j,k}^{(\Omega_k)}(t) \geq 0, \forall j \in \mathcal{N}, \forall k \in \mathcal{K} \tag{12.6}$$

$$\sum_{k=1}^K w_{j,k}(t) = 1, \forall j \in \mathcal{N} \tag{12.7}$$

$$\sum_{i=1}^{c_k} w_{i,k}(t) \leq n_j^{max}, \text{ when } w_{j,k}(t) = 1 \tag{12.8}$$

$$v < u \leq c_k, \forall k \in \mathcal{K}, (v, u) \in \Omega_k \tag{12.9}$$

$$\sum_{j=1}^{u^{(\Omega_k)}} f_{j,k}(t) \leq (u-v)^{(\Omega_k)} f_{max}, \forall k \in \mathcal{K} \tag{12.10}$$

where $\mathcal{P}_i$ denotes the power consumed by BS I, $\mathcal{P}_{j,k}^{(\Omega_k)}$ represents the power consumption of BS j in VSN $\mathcal{C}_k$ under the BS-off strategy Ω_k. The binary value $w_{j,k} \in (0,1)$ denotes the association status between BS j and VSN $\mathcal{C}_k$. If $w_{j,k} = 1$, it means BS j is associated to VSN $\mathcal{C}_k$. Otherwise, $w_{j,k} = 0$. Constraint (12.5) satisfies the total number of BSs in all VSNs and that in grand coalition N are the same. Constraint (12.6) limits the power consumption of BS to be non-negative values. Constraint (12.7) means that a BS is associated with only one VSN at any time t. If BS j is a member of VSN $\mathcal{C}_k$, constraint (12.8) limits the number of BS j's neighboring BSs, which are associated with VSN $\mathcal{C}_k$, cannot exceed n_j^{max}. Since different BS-off strategy is applied in each VSN, and there are v in u BSs to be powered off, constraint (12.9) means that the number of BSs in VSN $\mathcal{C}_k$ has to be larger than u given Ω_k. Constraint (12.10) assures the remaining active BSs can guarantee the service of sleep BSs under certain BS-off strategy Ω_k, while the aggregated traffic loads of both active BSs and the covered sleep BSs cannot exceed overall predefined maximum traffic load f_{max} in (12.10), $f_{j,k}(t)$ denotes the instant traffic load in BS j of VSN $\mathcal{C}_k$ at time t.

12.3 Methodologies

In this chapter, we will introduce a solution to problem (12.4) in a distributed manner, which is more feasible for practical implementations. The basic idea is to adaptively divide BSs into cell groups based on monitored traffic level.

DEFINITION 12.2 *Traffic level:* the traffic level is a term to describe the traffic condition in wireless networks based on the predefined measurements and provided level of service. We claim that the traffic condition can be translated to three levels with linguistic terms: low, moderate, and high traffic levels.

We will first describe the framework of the proposed solution and then provide details in parameter estimation, VSN formation, BS-off strategy selection, and BS on/off weight in the following sections.

12.3.1 Reception, analysis, and control (RAC) process

The proposed solution framework consists of three procedures, called *reception*, *analysis*, and *control*.

- In the *reception* stage, each BS collects information for VSNs formation, such as traffic load, VSN association status, and routing table from all member BSs. The gathered information is further calculated to obtain realistic traffic levels in the corresponding BS surrounding area. The details on calculating the realistic traffic level in a cell is shown in Section 12.3.2.
- In the *analysis* stage, based on collected realistic traffic levels, each BS estimates the mean network traffic by using hidden a Markov model (HMM) with details introduced in Section 12.3.3.
- In the *control* stage, following the motivation to operate separate BS-off strategies in each VSN, the entire network is partitioned into several VSNs. The VSN arrangement and selected BS-off strategy are updated with respect to varying traffic levels by adopting the proposed BSVF algorithm, and all VSNs run the best BS-off strategy so as to switch off the underutilized BSs to reduce power consumption.

Figure 12.3 shows the flow chart of RAC process. Each VSN chooses the optimal BS-off strategy in one time slot. After that, with observed traffic variation, new VSNs will form and each VSN reselects its corresponding BS-off strategies. By doing this, the power consumption of the entire network can be reduced significantly.

12.3.2 Realistic traffic level determination

In the traditional works considering cellular networks, the traffic level is measured merely by the number of MUs associated with each BS [24, 25]. Different from that, in this chapter the realistic traffic level of a cell is defined as a time-varying parameter by taking into consideration the number of MUs, the average distance between MUs and the BS, and the timeline. In reality, the association between a BS and MUs may become unstable when most MUs locate at the edge of a cell due to handover effects. Moreover, there is significant difference when comparing the network traffic in daytime and at night. Thus if a cell has a high traffic level, it has to satisfy three conditions: (i) the cell has a high traffic load, (ii) most MUs are near the BS, and (iii) the timeline lies in busy hours of a day. Since it is hard to directly determine the accurate bounds of the factors with respect to realistic traffic level of a cell, the calculation of the realistic traffic level is addressed by a fuzzy inference system (FIS) as a multiple input and one output problem [26, 27].

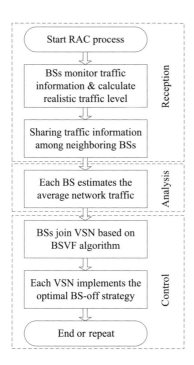

Figure 12.3 The strategy of distributed management of small cell networks, i.e., *reception, analysis,* and *control* processes.

In this FIS, the number of MUs, the average distance between MUs and the BS, and the timeline are considered as fuzzy values [28]. Thus the BS can make real-time decisions upon this soft computing method even with incomplete information due to, for example, measurement errors.

For analysis purposes, we take a single BS into consideration and all other BSs follow the same procedure. Let z_i be the input vector representing the information collected by the BS in the *reception* stage with $z = (z_1, z_2, z_3)^\Phi$, where z_1 denotes the number of MUs, z_2 denotes the average distance between MUs and the BS, z_3 denotes the timeline, and Φ is a linguistic label set. We initialize $\Phi(z_i)$, $i = \{1, 2, 3\}$, with terms $\Phi(z_1) = \{light(QL), medium(QM), heavy(QH)\}$, $\Phi(z_2) = \{near(DN), middle(DM), far(DF)\}$ and $\Phi(z_3) = \{idle(TI), moderate(TM), busy(TB)\}$. Let l^Φ be the output vector, i.e., the realistic traffic level, and $\Phi(l) = \{very\ low(FVL), low(FL), medium(FM), high(FH), very\ high(FVH)\}$. We first determine the degree of inputs to which they belong to each of the fuzzy sets $\mathcal{S}_i^\Phi$, and define it as

$$\mathcal{S}_i^\Phi = \{(z_i, \psi_{\mathcal{S}_i^\Phi}(z_i)) | z_i \in \mathcal{D}_i\}. \qquad (12.11)$$

In Equation (12.11), the fuzzy set $\mathcal{S}_i^\Phi$, on a universe of discourse $\mathcal{D}_i$ is characterized by a set of ordered pairs, which are represented by a generic element z_i and its degree of membership function $\psi_{\mathcal{S}}(z_i)$ that takes values in the interval [0,1] with triangular and trapezoidal curves. Figure 12.4 defines how each point in the input space is mapped to a membership value (or degree of membership) between 0 and 1. For z_1 and z_2, since

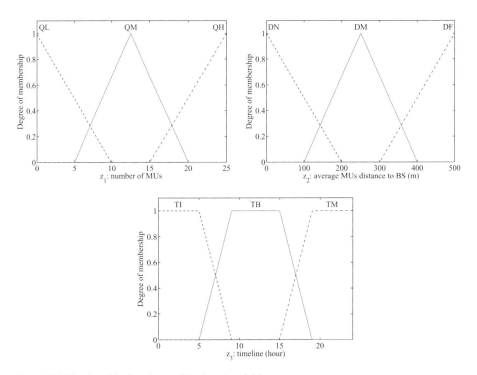

Figure 12.4 Membership functions of the input variables.

there are precise values for the corresponding linguistic label, medium and middle, the triangular membership functions are applied. For the parameter timeline, the trapezoidal membership function is adopted because it is unlikely to define an exact timeline in a busy period. Since it is possible that some inputs have more than one corresponding value for the degree of membership function, the *max* operator is adopted to decide the membership degree. The intersection of the degrees of membership functions within two fuzzy sets (e.g., S_1 and S_2) can be determined by [29],

$$\psi_{S_1 \cap S_2}(z) = \max(\psi_{S_1}(z), \psi_{S_2}(z)). \qquad (12.12)$$

After the inputs are fuzzified, the *if-then* rules are required to map them to the cases of various term combinations. For instance, one of the *if-then* rules can be: if x_1 is $S_1^{\Phi(z_1)}$, x_2 is $S_2^{\Phi(z_2)}$ and x_3 is $S_3^{\Phi(z_3)}$, then l is Φ_l. In total, we have $3^3 = 27$ fuzzy rules for consideration. The fuzzy outputs with *if-then* rules are listed in Table 12.1.

The inputs to the defuzzification process are an aggregated fuzzy set. The centroid defuzzification method is adopted to determine the center of gravity, which is given by

$$l = \frac{\sum_{i=1}^{I} z_i \psi_{S_i^{\Phi}}(z_i)}{\sum_{i=1}^{I} \psi_{S_i^{\Phi}}(z_i)}. \qquad (12.13)$$

We then obtain a crisp output value l, which represents the realistic traffic level. Based on FIS, we can regulate mapping rules in terms of linguistic labels rather than numbers, and obtain the realistic traffic levels in cells by combining the key factors: time, user

Table 12.1 Fuzzy rules to evaluate realistic traffic level in terms of the level of timeline: idle, moderate, and busy.

	DN	DM	DF
Timeline is idle			
QL	FL	FVL	FVL
QM	FM	FL	FVL
QH	FH	FM	FL
Timeline is moderate			
QL	FM	FL	FVL
QM	FH	FM	FL
QH	FVH	FH	FM
Timeline is busy			
QL	FH	FM	FL
QM	FVH	FH	FM
QH	FVH	FVH	FH

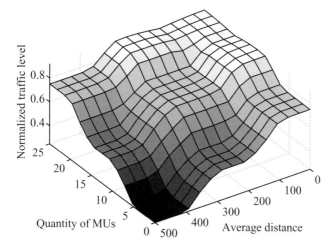

Figure 12.5 The mapping surface on traffic level in a cell upon FIS.

number, and density. In Figure 12.5, the defuzzified output of normalized realistic traffic level is presented by using Equation (12.13). Note that even if the traffic load in a cell is high, it may direct to different realistic traffic levels with respect to the other two factors.

12.3.3 Mean network traffic estimation

The mean network traffic is defined as a time-varying parameter, denoting the average traffic level of the entire network in a time slot. The mean network traffic is a necessary condition for VSN formation such that the traffic level in any VSN can match the mean network traffic so as to select an optimal BS-off strategy network wide. However, the mean network traffic is unknown to individual BSs in the distributed manner. The estimation can be viewed as a procedure that an individual BS needs to know the timely mean network traffic (i.e., the hidden states to the BS) under the knowledge of realistic traffic levels from the BS itself and all neighboring BSs (i.e., the observable states). Such viewing motivates the application of a hidden Markov model (HMM) to estimate mean network traffic.

We introduce the following variables for modeling HMM. Let $t = \{0, 1, \ldots, T\}$ denote discrete time slots. Random variable $\mathcal{X} = (x_0, x_1, \ldots, x_T)$ denotes the hidden states representing the mean network traffic. $\mathcal{X}$ is stable during the estimation process. Random variable $\mathcal{Y}^{(q)} = (y_{q_0}, y_{q_1}, \ldots, y_{q_T})$ denotes a sequence of observed average traffic levels by a BS in observable state q_t. In other words, $y_{qt} \in \mathcal{Y}^{(q)}$ indicates the average traffic level that a BS can calculate based on the realistic traffic levels of the neighboring BSs. The state transient matrix and the confusion matrix are denoted by $\mathcal{A} = (a_{nm}) = \mathcal{P}r(x_{n_t}|x_{m_{t-1}})$, and $\mathcal{B} = (b_{nm}) = \mathcal{P}r(y_n|x_m)$, respectively, and both $\mathcal{A}$ and $\mathcal{B}$ do not vary in time. Let the initial state probability be π_n. Then the modeled HMM can be represented as $(\pi, \mathcal{A}, \mathcal{B})$, which can be determined by empirical data [30].

Since we aim to find the most probable sequence of instant mean network traffic given an observation sequence, $\mathcal{X}$ is determined recursively by using the Viterbi algorithm given $\mathcal{Y}^{(q)}$ and HMM with appropriate parameter matrices $(\pi, \mathcal{A}, \mathcal{B})$ [31]. In particular, each state at time t has a partial probability and a partial best path, which indicate the probability and best path of reaching a particular intermediate state in the trellis, respectively. The overall best path is obtained by choosing the state with the maximum partial probability and choosing its partial best path. We first calculate the partial probability of being in the state given $t = 0$ and the observable state q_0, i.e.,

$$\delta_0(n) = \pi(n) b_{nq_0} \qquad (12.14)$$

Then, for δ_t at time $t \geq 1$, the probability of the partial best path to a state n with the observable state q_t is calculated as

$$\delta_t(n) = \max_m (\delta_{t-1}(m) a_{nm} b_{nqt}) \qquad (12.15)$$

We also need to know in which state the system must have been at time $t - 1$ if it is to arrive optimally at state n at time t. This recording is done by holding for each state a back pointer, which points to the predecessor that optimally provokes the current state. The back pointer ϕ_t is formulated as

$$\phi_t(n) = \mathrm{argmax}_m (\delta_{t-1}(m) a_{nm}) \qquad (12.16)$$

Thus we can determine which state at system completion ($t = T$) is the most probable as $x_t = \mathrm{argmax}(\delta_T(x))$. For $t = \{T - 1, T - 2, \ldots, 0\}$, let $x_t = \phi_{t+1}(x_{t+1})$ be backtracking through the trellis following the most probable route. On completion, the sequence $x_0, x_1, \ldots, x_T$ will hold the most probable sequence of hidden states, i.e., the sequence of mean network traffic.

12.3.4 Self-organized VSN forming protocol

We consider the formation of VSNs by following two principles: (i) BSs join a VSN if it makes the average traffic of the VSN approach the mean network traffic. After convergence, all the VSNs' average traffic will be balanced toward the mean network traffic; (ii) BSs join the VSNs so that the average traffic of VSNs lies in different traffic categories, e.g., high, middle, and low. However, in the distributed manner, the challenge is that BSs cannot simply decide to group together without some measurements, e.g., the similarity measures. Thus we proposed a BSVF algorithm based on firefly algorithm (FA). The FA is a heuristic optimization algorithm, inspired by the bioluminescence behavior of fireflies in nature [32], in which the less bright firefly will move to the brighter one, and the brightness is an inverse ratio of the distance different from traditional FA. In the cellular network, BSs are deployed in a hexagonal cell layout and are at a standstill rather than randomly distributed and movable like fireflies. Also, due to the constraint that a BS can only share information with its neighboring BSs, if the VSN controller tries to communicate with its member BS, the communication distance is counted as the number of hops, h_{ij}, rather than Euclidean distance r in the classic FA. Therefore we consider the brightness as an inverse ratio of the communication distance. In our proposed model, the derived BS realistic traffic level l and the estimated mean network traffic $\bar{l}$ are considered as the similarity measures.

Note that we have defined all BSs from a grand coalition $\mathcal{N}$, and at the initial stage, there are K VSNs, $\{\mathcal{C}_k\}$, where $k = \{1, 2, \ldots, K\}$ and $K \geq 1$. Each VSN consists of $\mathcal{C}_k$ BSs. To facilitate the application of BS-off strategies and simplify the VSN forming process, K should be set as a small number so that $\mathcal{C}_k$ can be large enough to apply BS-off strategies. In practice, K can be determined based on the number of hotzones in the network. If there are total Z hotzones, the initial number of VSNs can be $K = Z + 1$. At the initial stage, the VSN controller, $\mathcal{H}_k$, is selected randomly, and each BS selects the nearest $\mathcal{H}_k$ to join (in this case, header selection is similar to the traditional wireless sensor networks [33]).

DEFINITION 12.3 Fringe BS. BS j in VSN $\mathcal{C}_k$ is a fringe BS, if at least one of its neighboring BS i belongs to a different VSN $\mathcal{C}_{k'}$, $k' \neq k$. BS j is confirmed to be a *fringe BS* if it satisfies the following criterion

$$w_{j,k}(t) + \sum_{i=1}^{nj} w_{i,k}(t) \leq n_j, \forall j \in \mathcal{C}_k \tag{12.17}$$

where n_j is the number of BS j's neighboring BSs. Note that $w_{j,k} \in \{0, 1\}$ is a binary variable, which stands for the association between BS j and VSN $\mathcal{C}_k$. If $w_{j,k} = 1$, it means BS j is a member of VSN $\mathcal{C}_k$. Otherwise, $w_{j,k} = 0$. Since each BS has only one associated VSN at any time t, we have

$$\sum_{k=1}^{K} w_{j,k}(t) = 1, \forall j \in \mathcal{N}. \tag{12.18}$$

During the VSN formation period, only fringe BSs are allowed to join one of the neighboring VSNs. Such limitation can avoid isolated BSs or a BS that is never separated from its associated VSN.

For two principles under consideration, brightness β can be defined respectively as (i) the BS realistic traffic of network, i.e., $\beta = l$ and (ii) the absolute value of difference between BS realistic traffic level and the estimated mean network traffic, i.e., $\beta = |l - \bar{l}|$.

Algorithm 12.1 Pseudo code of BSVF algorithm

```
Input:      N, T, K, C_k, H_k
Output:     C_k, H_k
Begin
1:    while (τ < MaxIteration)
2:        for i ∈ N all BSs in N
3:            if i is a fringe BS then
4:                for j = 1 to n_i (all n_i neighboring BSs of BS i)
5:                    Find maximum attractiveness α_ij
6:                    i join j's associate VSN, C_k^j ← C_k^j ∪ {i}
7:                    attractiveness α_ij changes with γ_ij and ĥ_ij
8:                    update BS j's brightness β_j
9:                end for j
10:           end for i
11:       % revise the location of VSN controller H_k
12:       for k = 1 to K (all K VSNs)
13:           for i = 1 to c_k (all c_k BSs)
14:               find maximum h_ik
15:               move H_k towards BS i
16:           end for i
17:       end for k
18:   end while
19:   evaluate average traffic level of VSNs
End
```

Case I: $\beta = l$. In this case, the objective function for forming VSNs is formulated as

$$\min \mathcal{U}^1_{c_k \leq N} = |\varepsilon[\mathcal{L}(t)] - \bar{\mathcal{L}}_k(t)|, \forall k \in \mathcal{K} \quad (12.19)$$

where $\varepsilon[\mathcal{L}(t)]$ is the average network traffic. Note that it is different from the estimated mean network traffic $\bar{l}$ calculated by each individual BS. $\mathcal{L}_k(t)$ is the average traffic of VSN C_k. However, since $\varepsilon[\mathcal{L}(t)]$ is unknown in distributed network scenario, we reformulate the objective function as

$$\min \mathcal{U}^1_{c_{k1}, c_{k2} \leq N} = |\bar{\mathcal{L}}_{k_1}(t) - \bar{\mathcal{L}}_{k_2}(t)|, \forall k_1, k_2 \in \mathcal{K} \quad (12.20)$$

As the BS attractiveness is proportional to the intensity of brightness β observed by the neighboring BSs, the attractiveness function between BSs i and j can be formulated as

$$\alpha_{ij}^1 = \alpha_0 \exp\left(\frac{\gamma_{ij}}{\hat{h}_{ij}}\right) \quad (12.21)$$

where γ_{ij} is defined as a traffic coefficient with respect to the practical average network traffic $\varepsilon[\mathcal{L}(t)]$ and brightness of BSs i and j. In this case, higher attractiveness is gained when the traffic level of compared BSs has a bigger gap. Thus we define the traffic coefficient as $\gamma = \frac{1}{\sqrt{\varepsilon[\mathcal{L}(t)]}}|\beta_i - \beta_j|$ so as to balance the traffic levels, i.e., the average traffic of formed VSN is approaching $\varepsilon[\mathcal{L}(t)]$. Since the average network traffic is not available in the distributed cellular network scenario, we consider a general average value for $\mathcal{L}$ for any BS, i.e., $\varepsilon[\mathcal{L}(t)] = \frac{1}{2}$ for normalized $\mathcal{L} \in [0, 1]$. If BS j is a member of VSN $\mathcal{C}_k$, then $\hat{h}_{ij}$ represents the number of hops from BS i to BS j's associated VSN controller $\mathcal{H}_k$. Thus we have $\hat{h}_{ij} = h_{jk} + h_{ij} \cdot \alpha_0$ is the attractiveness at $h = 0$, and usually we set $\alpha_0 = 1$ for simplicity. After comparison, BS i joins BS j with higher attractiveness. After that, the brightness of BS j from iteration τ to iteration $\tau + 1$ is updated as

$$\beta_j(\tau + 1) = \beta_j(\tau) + \frac{\beta_i}{c_k + 1} \quad (12.22)$$

When the convergence of VSN formation is completed, the average traffic in each VSN will be balanced.

THEOREM 12.1 *VSNs with balancing average traffic. If the brightness is set to be the BS realistic traffic level, after running the BSVF algorithm, for any VSNs, k1 and k2, we have*

$$\lim_{n_{k_1}, n_{k_2} \to \infty} \Pr(|\bar{\mathcal{L}}_{k_1} - \bar{\mathcal{L}}_{k_2}| > \varepsilon) = 0, \forall k1, k2 \in \mathcal{K} \quad (12.23)$$

Proof: Let BSs realistic traffic levels $l_1, l_2, \ldots, l_{n_k}$ in any VSNs $\mathcal{C}_k, l_1, l_2, \ldots, l_{n_k}$ be a sequence of i.i.d random variables with average value $\bar{\mathcal{L}}_k = \frac{1}{n_k}(l_1 + l_2 + \cdots + l_{n_k})$. Based on Chebyshev's inequality on $\bar{\mathcal{L}}_k$, we have

$$\Pr(|\bar{\mathcal{L}}_k - \mu| \geq \varepsilon) \leq \frac{\sigma^2}{n_k \varepsilon^2} \quad (12.24)$$

where μ is the common mean with $\varepsilon[\bar{\mathcal{L}}_k] = \mu$ and σ^2 is the variance of random variable sequence $l_1, l_2, \ldots, l_{n_k}$. From (12.24), we have

$$\Pr(|\bar{\mathcal{L}}_k - \mu| < \varepsilon) = 1 - \Pr(|\bar{\mathcal{L}}_k - \mu| \geq \varepsilon) \geq 1 - \frac{\sigma^2}{n_k \varepsilon^2} \quad (12.25)$$

As n_k goes to infinity, expression (12.25) approaches 1. Thus we have $\Pr(|\bar{\mathcal{L}}_k - \mu|) \to 0$ as $n_k \to \infty$. For any VSNs $k1$ and $k2$, we can obtain $\bar{\mathcal{L}}_{k1} \to \mu$ as $n_{k1} \to \infty$ and $\bar{\mathcal{L}}_{k2} \to \mu$ as $n_{k2} \to \infty$. Hence, expression (11.23) holds. □

Case II: $\beta = |l - \bar{l}|$. In this case, the VSNs are formed based on different categories of traffic levels. The objective function can be formulated as

$$\max \mathcal{U}^2_{c_k \leq N} = |\mathcal{L}(t) - \bar{\mathcal{L}}_k(t)|, \forall k \in \mathcal{K}. \tag{12.26}$$

The attractiveness function between BSs i and j can be formulated as

$$\alpha^2_{ij} = \alpha_0 \exp(-\gamma_{ij} \hat{h}_{ij}) \tag{12.27}$$

In this case, higher attractiveness is gained when two BSs have similar traffic levels. The brightness update function is formulated in the same way as Equation (12.22). When VSN formation procedure converges, some VSNs will gather the BSs with high traffic levels, while others collect the BSs with low or moderate traffic levels.

THEOREM 12.2 *VSN with heterogeneous average traffic. When the brightness is set to be the absolute value of the difference between BS realistic traffic level and estimated mean network traffic, for any VSNs k_1 and k_2, if $l_i > \bar{l}_i, l_i \in (l_1, l_2, \ldots, l_{n_{k_1}})$ in VSN $\mathcal{C}_{ki}$, and $l_j < \bar{l}_j, l_j \in (l_1, l_2, \ldots, l_{n_{k_2}})$ in VSN $\mathcal{C}_{k2}$, we have*

$$\lim_{nk_1, nk_2 \to \infty} \Pr(\bar{\mathcal{L}}_{k_1} - \bar{\mathcal{L}}_{k_2}) > 0, \forall k1, k2 \in \mathcal{K} \tag{12.28}$$

Proof: It is obvious that the BSs with realistic traffic levels under or over the BSs estimated mean network traffic are separately grouped in terms of the brightness and attractiveness functions in Case II. It is the necessary and sufficient conditions for the assumption that $l_i > \bar{l}_i, l_i \in (l_1, l_2, \ldots, l_{n_{k_1}})$ in VSN $\mathcal{C}_{k1}$, and $l_j < \bar{l}_j, l_j \in (l_1, l_2, \ldots, l_{n_{k_2}})$ in VSN $\mathcal{C}_{k2}$. Since $\varepsilon[\mathcal{L}] \approx \frac{1}{n_{k_1}}(\bar{l}_1 + \bar{l}_2 + \cdots + \bar{l}_{n_{k_1}}) \approx \frac{1}{n_{k_2}}(\bar{l}_1 + \bar{l}_2 + \cdots + \bar{l}_{n_{k_2}})$ as $n_{k_1}, n_{k_2} \to \infty$ we have $\bar{\mathcal{L}}_{k_1} - \frac{1}{n_{k_1}}(l_1 + l_2 + \cdots + l_{n_{k_1}}) > \varepsilon[\mathcal{L}]$ as $n_{k_1} \to \infty$ and $\bar{\mathcal{L}}_{k_2} = \frac{1}{n_{k_2}}(l_1 + l_2 + \cdots + l_{n_{k_2}}) < \varepsilon[\mathcal{L}]$ as $n_{k_1}, n_{k_2} \to \infty$. This completes the proof. $\square$

COROLLARY 12.1 *The formation of virtual small networks (VSNs) leads to better power saving results than the case without VSNs under base station off (BS-off) schemes.*

Corollary 12.1 can be proved by an example. Let a cellular network be partitioned into three VSNs, $\{\mathcal{C}_{k_1}, \mathcal{C}_{k_2}, \mathcal{C}_{k_3}\}$. The average traffic of the VSNs is set to be very different with $\bar{\mathcal{L}}_{k_1} > \bar{\mathcal{L}}_{k_2} > \bar{\mathcal{L}}_{k_3}$. According to the implementation of BS (v, u)-off strategy, we assume the best BS-off schemes for each VSN are (v, u)-off, $(v + 1, u)$-off, and $(v + 2, u)$-off, respectively, and $u > v + 2$. The network power saving ratio can be simply derived as $\frac{1}{3}\left(\frac{v}{u} + \frac{v+1}{u} + \frac{v+2}{u}\right) = \frac{v+1}{u}$. In fact, this scenario is similar to Case II, which groups the BSs into different traffic categories. In Case I, after the convergence, all the VSNs have a similar average traffic level, which is approximately $\bar{\mathcal{L}}_{k_2}$. Thus with the corresponding best BS-off scheme $(v + 1, u)$-off, the network power saving ratio is approximately $\frac{v+1}{u}$. However, in the traditional network without VSNs, the unified BS-off scheme is adopted. In order to satisfy the entire network demand, the best BS-off scheme is obviously (v, u)-off with the network power saving ratio of $\frac{v}{u}$.

At the end of each iteration, the location of VSN controller $\mathcal{H}_k$ should be updated to guarantee that the VSN controller always stays in the geological center of the VSN. $\mathcal{H}_k$ needs to maintain a link table, which keeps a record of hopping topologies for the member BSs. Based on the link table, $\mathcal{H}_k$ will move toward BS i with maximum h_{ik}. The detailed process of BSVF algorithm is presented in Algorithm 12.1.

12.3.5 BS-off matching scheme

After the VSN formation process, each VSN controller selects the best BS-off strategy for its VSN. The selection can be formulated as an optimization problem, which aims to minimize total system power consumption at time slot t, $\mathcal{P}_{nw}^t$. We can reformulate equation (12.4) by focusing on the selection of BS-off scheme.

$$\min \mathcal{P}_{nw}^t = \sum_{k=1}^{K} \sum_{j=1}^{c_k} \mathcal{P}_{j,k}^{(\Omega_k)}(t)$$

$$= \sum_{k=1}^{K} \sum_{j=1}^{c_k} ((1+\rho)\mathcal{P}_{sta}^{(\Omega_k)} + \mathcal{P}_{dyn}^{(\Omega_k)}(t))_{j,k} \quad (12.29)$$

$$= \sum_{k=1}^{K} \sum_{j=1}^{c_k} \left((1+\rho)\mathcal{P}_{sta}^{(\Omega_k)} + \sigma \cdot w_{j,k} \cdot \lambda_{j,k} \cdot f_{j,k}^{(\Omega_k)}(t) \right) \quad (12.30)$$

$$\text{s.t.} \sum_{k=1}^{K} w_{j,k} = 1, \forall j = 1, 2, \ldots, N. \quad (12.31)$$

$$\sum_{j=1}^{u^{(\Omega_k)}} f_{j,k}^{(\Omega_k)}(t) \leq (u-v)^{(\Omega_k)} f_{max}. \quad (12.32)$$

$$u^{(\Omega_k)} \leq \sum_{j=1}^{c_k} \lambda_{j,k} \leq c_k, \forall u = 1, 2, \ldots, n_j^{max} + 1. \quad (12.33)$$

where σ stands for the coefficient of BS dynamic power consumption P_{dyn}. $\lambda_{j,k} \in \{0, 1\}$ indicates BS on/off state, in which $\lambda_{j,k} = 1$ means a BS j in VSN $\mathcal{C}_k$ is active, otherwise $\lambda_{j,k} = 0$. Constraint (12.31) limits that any BS must be associated with only one VSN at any time. Constraint (12.32) assures the remaining active BSs can guarantee the service of sleep BSs under certain BS-off strategy Ω_k, while the aggregated traffic loads of both active BSs and the covered sleep BSs cannot exceed overall predefined maximum user capacity l_{max}, and $l_{j,k}(t)$ indicates the traffic load in BS j of VSN $\mathcal{C}_k$ at time t. Since different BS-off strategy is applied in each VSN, constraint (12.33) represents that the number of active BSs in VSN $\mathcal{C}_k$ has to be larger than u given Ω_k. In practice, since the set of feasible BS-off strategies Ω is finite, exclusive searching can be adopted over all potential candidates to find the best one for implementation.

12.3.6 BS power on/off weight

In the BS-off matching scheme, the number of switched-off BSs can be determined in each VSN. However, for a specific BS, whether it is power on or off is not carefully decided yet. Therefore we need to further determine which BS is better being switched off. Note that too often transitions between the active and sleep modes may lead to more power consumption when considering the power cost in the BS transient period. Also, frequent transitions of BS working modes may affect system stability. Thus we introduce the concept of BS on/off weight, which can be simply integrated to the BSVF algorithm. The BS on/off weight, denoted by $\mathcal{W}_{on}$, is defined as the probability that a BS turns to or stays in active mode. n_{con} is a constant for adaptive control of $\mathcal{W}_{on}$. n_{con} can be either an incremental or decremental value for the BS on/off weight learning procedure. The design principles to prevent a BS in frequent working mode transitions are summarized as: (i) if a BS is in active mode currently, its probability of staying active in the next operation period is increasing; (ii) if a BS is in sleep mode, its probability of switching power on in the next operation period is decreasing. Such principles can make a BS enlarge the probability of keeping its working modes unchanged during each operation period.

At the initial stage, each BS is assigned to a random value of $\mathcal{W}_{on} \in (0,1)$. The value of n_{con} can be set as a very small value. After that, BSs will update the value of $\mathcal{W}_{on}$ as a learning procedure at the end of each operation stage, which is required to obey the following rules.

- If $0 < \mathcal{W}_{on} < 1$, $\mathcal{W}_{on} = \mathcal{W}_{on} + n_{con}$ when a BS is in active mode or $\mathcal{W}_{on} = \mathcal{W}_{on} - n_{con}$ when a BS is in sleep mode.
- If $\mathcal{W}_{on} \leq 0$, $\mathcal{W}_{on} = \max(0, \mathcal{W}_{on} + n_{con})$.
- If $\mathcal{W}_{on} \geq 1$, $\mathcal{W}_{on} = \max(1, \mathcal{W}_{on} - n_{con})$.

Thus BSs with larger value of $\mathcal{W}_{on}$ possess a high probability of being switched on or kept in active mode. On the other hand, a BS is more likely to be powered off or kept in sleep mode with small $\mathcal{W}_{on}$. Note that the BS as a VSN controller should be kept in active mode during an operation period with $\mathcal{W}_{on} = 1$.

12.4 Simulation results

In this section, simulation results are demonstrated to evaluate the performance of the proposed RAC process integrating BSVF algorithm, BS-off matching, and BS power on/off weight schemes in wireless cellular networks. The simulation scenario is based on the evaluation methodology described in [34], while the proposed distributed BS management strategy is compared with the traditional unified BS-off strategy, called unified BS-off [11, 12].

The micro-hexagonal cell layout deployed in an urban environment is considered. Followed by the power consumption characteristic of a typical micro BS, its cell radius is set to $\mathcal{R} = 500$ m, which is corresponding to $\mathcal{P}_{sta} = 237$ watts for a typical micro BS. The power transient cost ratio is set to be $p = 0.05$. The power supply loss is approximately 10% and the efficiency of PA is 20% [35]. The path-loss model is compliant with the micro cell test environment in ITU report with center frequency of 2.655 GHz [34] and receiver sensitivity of MUs is –120 dB. The maximum number of MUs that can be served by a BS is $f_{max} = 25$, which corresponds to assigning 5 MHz bandwidth in an LTE system. For the traffic distribution model, the minimum number of hotspots is set to

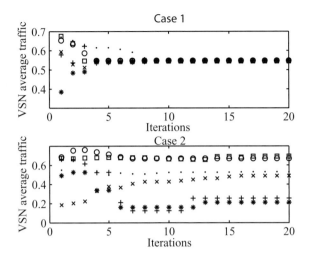

Figure 12.6 Convergence on VSNs average traffic in two cases of proposed BSVF algorithm, with $N = 300$, $K = 6$. Different marks mean the average traffic of VSNs.

$Z = 4$ with abruptness factor $v = 3, 5$ in (12.3), while other regions with $v = 1$ represent the low-traffic zones. The arrival process of MUs follows Possion distribution.

Figure 12.6 shows the convergence on average traffic levels of VSNs when the proposed BSVF algorithm is running under two cases with different brightness setup. We can observe that, in Case I, the average traffic of all VSNs approaches the same approximate point as the iteration proceeds, i.e., all VSNs' average traffic tends to the global average network traffic, which satisfies Theory (12.23). Also, in Case I, the VSNs may choose the same BS-off strategy due to similar average traffic. However, in Case II, the average traffic of VSNs is obviously variable, which indicates that the VSNs form in different traffic categories, i.e., low, moderate, and high traffic levels. In Case II, each VSN may select the most aggressive BS-off strategy to match individual traffic level. Another observation for both cases is the time spent on VSN formation is small because commonly the convergence of each VSN can be completed in under ten iterations.

Figure 12.7 compares power consumption of the proposed BSVF algorithm and traditional unified BS-off strategy. Simulation results are normalized with respect to the network power consumption without any VSN formation and BS-off mechanisms. Compared to the unified BS-off strategy, it is apparent that the total consumed power of the network can be significantly reduced by the proposed protocol, particularly when the network lies within extremely high and low traffic levels, which may relate to office hours and night time, respectively. We also observe that Case II is slightly superior to Case I. This is because Case II can provide a better fit for the traffic fluctuations, especially when the traffic in the whole network is more unevenly distributed. In this scenario, it is easier to form VSNs in different traffic categories than to balance the traffic levels of all VSNs. In short, the proposed protocol can save 60% power consumption compared to that without any VSN formation and BS-off mechanisms, and both proposed cases outperform the traditional unified BS-off strategy with 10% power saving enhancement.

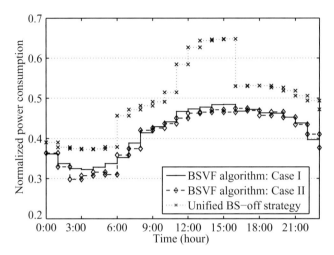

Figure 12.7 Power consumption comparison of Case I and II with the unified BS-off strategy, $N = 200$, $K = 5$.

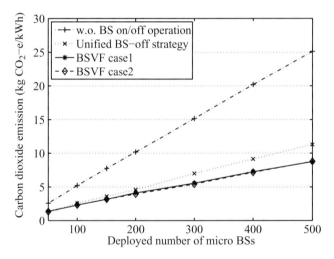

Figure 12.8 $CO_{2\text{-}e}$ varies with increasing number of deployed micro BSs, $K = 5$.

We further show CO_2-e in daily statistics with different numbers of BSs in Figure 12.8. CO_2-e equals 0.457 kg CO_2 per kWh in terms of the report on electricity emission factors [36]. As the number of BSs increases, the emitted CO_2 per kWh of the proposed protocol is obviously less than the unified BS-off strategy and far less than the case without BS on/off operation. When there exist $N = 500$ BSs in the network, the proposed protocol can reduce emissions by 16.5 kg CO_2 daily.

In Figure 12.9, we present the power saving ratio with the choice of initializing different numbers of VSNs, e.g., $N = 200$. Our assumption has proved that if we have a total of Z hotspots in the network, the optimal number of partitioned VSNs is $K = Z + 1$. A smaller number of VSNs is not able to adaptively cover the heterogeneous traffic

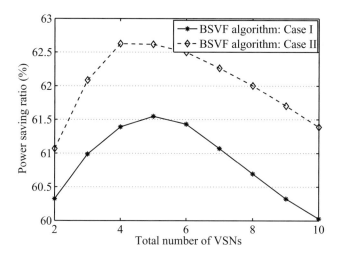

Figure 12.9 The variation tendency of power saving ratio under Cases I and II with forming different number of VSNs, $N = 200$.

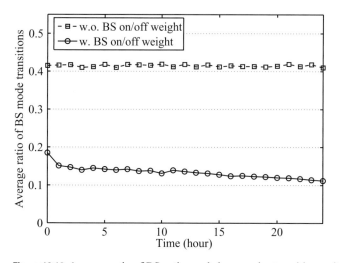

Figure 12.10 Average ratio of BS active and sleep modes transitions with/without BS on/off weight, $N = 300$, $K = 5$.

distribution, while too many VSNs will cause performance degradation due to more aggregated power cost in BS transient period.

Figure 12.10 shows the average ratio of BS mode transitions between the active and sleep modes following the operation time slot with $n_{con} = 0.01$. Note that at each operation period, an average 42% BSs have to change between active and sleep modes without BS on/off weight. However, by taking into consideration the BS on/off weight, the average BS mode transition rate is reduced significantly at 13.5%. Such results validate that the BS can avoid frequently shifting between the active and sleep modes with the BS on/off weight learning procedure.

12.5 Conclusions

In this chapter, we aim to develop a distributed small cell management strategy so as to reduce power consumption in wireless networks. The self-organized VSNs are formed based on the heterogeneous traffic distribution and the underutilized BSs can be switched off during a certain period. In the proposed RAC process, each BS first applies the FIS algorithm to calculate the realistic traffic level based on the collected information, including the number of MUs, the average distance between MUs and BS, and the timeline. Then we formulate HMM for BSs to estimate the mean network traffic distributively at an observation time by applying collected traffic information from the neighboring BSs. After that, the BSVF algorithm is proposed, which targets to adaptively group BSs based on traffic distribution in each operation period. Instead of running a unified BS-off strategy in the entire network, the BS-off matching scheme is applied to make a selection of the best BS-off strategy for each VSN separately. Moreover, the BS on/off weight is introduced as a learning process to revise the probability of BS mode transition, which can avoid a BS making frequent transitions between active and sleep modes. The simulation results prove that our proposed mechanisms can improve power saving in wireless networks significantly.

References

[1] *SMART 2020: Enabling the Low Carbon Economy in the Information Age*, The Climate Group and Global e-Sustainability Initiative (GeSI), 2008, www.smart2020.org.

[2] Z. Hasan, H. Boostanimehr, and V. K. Bhargava, "Green cellular networks: a survey, some research issues and challenges," *IEEE Communications Surveys & Tutorials*, vol. 13, no. 4, pp. 524–540, Nov. 2011.

[3] A. Amanna, *Green Communications*, Wireless@Virginia Tech., Tech. Rep., Feb. 2010.

[4] A. P. Bianzino, C. Chaudet, D. Rossi, and J.-L. Rougier, "A survey of green networking research," *IEEE Communications Surveys & Tutorials*, vol. 14, no. 1, pp. 3–20, Feb. 2012.

[5] C. Han, T. Harrold, S. Armour, *et al.*, "Green radio: radio techniques to enable energy-efficient wireless networks," *IEEE Communications Magazine*, vol. 49, no. 6, pp. 46–54, Jun. 2011.

[6] G. Fettweis and E. Zimmermann, "ICT energy consumption-trends and challenges," *Proc. of the 11th International Symposium on Wireless Personal Multimedia Communications*, vol. 2, no. 4, Sep. 2008, pp. 1–6.

[7] G. Fettweis and E. Zimmermann, *Green Radio: NECs Approach Towards Energy-efficient Radio Access Networks*, NEC Corporation, Tech. Rep., Feb. 2010.

[8] K. J. Christensen, C. Gunaratne, B. Nordman, and A. D. George, "The next frontier for communications networks: power management," *Computer Communications*, vol. 27, no. 18, pp. 1758–1770, Dec. 2004.

[9] M. A. Marsan, L. Chiaraviglio, D. Ciullo, and M. Meo, "Optimal energy savings in cellular access networks," *Proc. of IEEE International Conference on Communications Workshops (ICC'09, GreenCom Wksp.)*, Jun. 2009, pp. 1–5.

[10] E. Oh, B. Krishnamachari, X. Liu, and Z. Niu, "Toward dynamic energy-efficient operation of cellular network infrastructure," *IEEE Communications Magazine*, vol. 49, no. 6, pp. 56–61, Jun. 2011.

[11] Z. Niu, Y. Wu, J. Gong, and Z. Yang, "Cell zooming for cost-efficient green cellular networks," *IEEE Communications Magazine*, vol. 48, no. 11, pp. 74–79, Nov. 2010.

[12] X. Weng, D. Cao, and Z. Niu, "Energy-efficient cellular network planning under insufficient cell zooming," *Proc. of IEEE 73rd Vehicular Technology Conference (VTC-Spring'11 Greenet Wksp.)*, May 2011, pp. 1–5.

[13] S. Bhaumik, G. Narlikar, S. Chattopadhyay, and S. Kanugovi, "Breathe to stay cool: adjusting cell sizes to reduce energy consumption," *Proc. of 1st ACM SIGCOMM Workshops on Green Networking*, Sep. 2010, pp. 41–46.

[14] W. T. Wong, Y. J. Yu, and A. C. Pang, "Decentralized energy-efficient base station operation for green cellular networks," *Proc. of IEEE Global Communications Conference (GLOBECOM'12)*, Dec. 2012, pp. 5194–5200.

[15] Y. Chen, S. Zhang, S. Xu, and G. Y. Li, "Fundamental trade-offs on green wireless networks," *IEEE Communications Magazine*, vol. 49, no. 6, pp. 30–37, June 2011.

[16] J. Hoydis, M. Kobayashi, and M. Debbah, "Green small-cell networks," *IEEE Vehicular Technology Magazine*, vol. 6, no. 1, pp. 37–43, Mar. 2011.

[17] J. Hoydis, M. Kobayashi, and M. Debbah, *LTE Advanced: Heterogeneous Networks*, Qualcomm Inc., Tech. Rep., Jan. 2011, www.qualcomm.com/media/documents/lte-heterogeneous-networks.

[18] C. Prehofer and C. Bettstetter, "Self-organization in communication networks: principles and design paradigms," *IEEE Communications Magazine*, vol. 43, no. 7, pp. 78–85, Jul. 2005.

[19] H. ElSawy, E. Hossain, and D. I. Kim, "HetNets with cognitive small cells: user offloading and distributed channel access techniques," *IEEE Communications Magazine*, vol. 51, no. 6, Jun. 2013.

[20] D. Lopez-Perez, I. Guvenc, G. De LaRoche, M. Kountouris, T. Q. Quek, and J. Zhang, "Enhanced intercell interference coordination challenges in heterogeneous networks," *IEEE Wireless Communications*, vol. 18, no. 3, pp. 22–30, Jun. 2011.

[21] O. Aliu, A. Imran, M. Imran, and B. Evans, "A survey of self organisation in future cellular networks," *IEEE Communications Surveys & Tutorials*, vol. 15, no. 1, pp. 336–361, Feb. 2012.

[22] L. Chiaraviglio, D. Ciullo, M. Meo, M. A. Marsan, and I. Torino, "Energy-aware UMTS access networks," *Proc. of 11th International Symposium on Wireless Personal Multimedia Communications (W-GREEN'08)*, Sep. 2008, pp. 1–8.

[23] 3GPP TS 36.141 V11.2.0, "Technical specification group radio access network, E-UTRA base Station conformance testing, Release 11," Nov. 2012.

[24] M. F. Hossain, K. S. Munasinghe, and A. Jamalipour, "A protocooperation-based sleep-wake architecture for next generation green cellular access networks," *Proc. of 4th International Conference on Signal Processing and Communication Systems (ICSPCS'10)*, Dec. 2010, pp. 1–8.

[25] F. Richter, A. J. Fehske, and G. P. Fettweis, "Energy efficiency aspects of base station deployment strategies for cellular networks," *Proc. of 70th Vehicular Technology Conference Fall (VTC-Fall'09)*, Sep. 2009, pp. 1–5.

[26] H. Shu, Q. Liang, and J. Gao, "Wireless sensor network lifetime analysis using interval type-2 fuzzy logic systems," *IEEE Transactions on Fuzzy Systems*, vol. 16, no. 2, pp. 416–427, Apr. 2008.

[27] J. S. Lee and W. L. Cheng, "Fuzzy-logic-based clustering approach for wireless sensor networks using energy predication," *IEEE Sensors Journal*, vol. 12, no. 9, pp. 2891–2897, Sep. 2012.

[28] L. A. Zadeh, "Fuzzy algorithms," *Information and Control*, vol. 12, no. 2, pp. 94–102, Feb. 1968.

[29] E. Cox, "Fuzzy fundamentals," *IEEE Spectrum*, vol. 29, no. 10, pp. 58–61, Oct. 1992.

[30] D. Burshtein, "Robust parametric modeling of durations in hidden markov models," *IEEE Transactions on Speech and Audio Processing*, vol. 4, no. 3, pp. 240–242, Aug. 1996.

[31] H. L. Lou, "Implementing the Viterbi algorithm," *IEEE Signal Processing Magazine*, vol. 12, no. 5, pp. 42–52, Sep. 1995.

[32] X. S. Yang, *Nature-inspired Metaheuristic Algorithms*. Luniver Press, 2010.

[33] J. Fang and H. Li, "Power constrained distributed estimation with cluster-based sensor collaboration," *IEEE Transactions on Wireless Communications*, vol. 8, no. 7, pp. 3822–3832, Jul. 2009.

[34] ITU-R, *Guidelines for Evaluation of Radio Interface Technologies for IMT-Advanced*, Geneva, Switzerland, Rep. ITU-R M.2135-1, Tech. Rep., Dec. 2009, www.itu.int/pub/R-REP-M.2135/en.

[35] O. Arnold, F. Richter, G. Fettweis, and O. Blume, "Power consumption modeling of different base station types in heterogeneous cellular networks," *Proc. of 19th Future Network and Mobile Summit,* Jun. 2010, pp. 1–8.

[36] *2012 Guidelines to DEFRA/DECC's GHG Conversion Factors for Company Reporting.* Department for Environment, Food and Rural Affairs (DEFRA), UK, Tech. Rep., Aug. 2012, www.defra.gov.uk/publications/2012/05/30/pb13773-2012-ghg-conversion.

13 Mobility management in small cell heterogeneous networks

Peter Legg and Xavier Gelabert

13.1 Introduction

In cellular networks, handover refers to the mechanism by which the set of radio links between an *active mode* mobile device and base station cells is modified. Mobility in the *idle mode* (when the mobile has no data bearers established and is not transmitting or receiving user plane traffic), termed cell selection/reselection, typically ensures that the UE selects the strongest available cell in preparation for an outgoing or incoming call/data session. Handover reliability is a key performance indicator (KPI) since it directly impacts the perceived quality of experience (QoE) of the end user. In contrast, cell reselection is less important since no bearers are established and suboptimal performance is apparent only on call establishment and as a signaling cost to the network operator. For this reason, the remainder of this chapter focuses on handover.

In GSM and LTE the mobile supports only a single radio link such that the handover swaps this link from one cell (the serving cell) to another (the target cell). In WCDMA, however, multiple links (on the same frequency) may be established (known as "soft handover"). Handovers can be classified as:

- intra-RAT, meaning within the same radio access technology (RAT), for example, LTE to LTE
  - intra-frequency (serving and target cells are on the same frequency)
  - inter-frequency (serving and target cells are not on the same frequency)
- inter-RAT
  - between cells of different RATs.

Handover may be triggered for a number of reasons:

- to maintain the connectivity of the mobile and support data transfer (often called a "coverage handover")
- to balance the loading of cells with overlapping coverage or to handover a mobile between overlapping cells to ensure data rates demanded by an ongoing service are met (often called a "vertical handover").

Vertical handovers target stationary mobiles, implying that the radio conditions of links to serving and target cells are relatively stable. More challenging are coverage handovers that result from the motion of the mobile, leaving the coverage of the serving cell and entering that of the target cell. Since indoor users are usually stationary, the

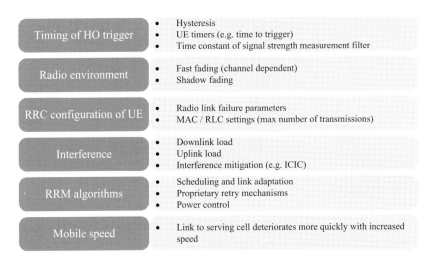

Figure 13.1 Factors impacting coverage handover reliability.

focus of coverage handovers is on outdoor users, on foot or in vehicles. The coverage handover requires a complex signaling exchange between the mobile and the two cells involved, along with signaling over the wired transport or backhaul network. The LTE intra-frequency handover is described in detail below, but there are some factors that in general impact the success of a coverage handover (see Figure 13.1).

In many cases, handover success rates should be maintained at a level of 97 to 99%. As indicated above, the network is unable to control all aspects impacting the handover reliability. For example, shadow fading results from obstacles such as trees and buildings lying between the mobile and the serving or target cells. The operator can influence the timing of the handover through configurable parameters (such as hysteresis) and this is the normal practice in the field. However, many studies have shown that while the configuration of an early handover improves reliability this is at the expense of an increase in the handover count or occurrence frequency [4, 2]. This may lead to small interruptions to the service to the mobile and increased signaling over the air interface and within the RAN. An unnecessary handover or handover sequence is one for which there would be no dropped call or radio-link failure if the handover had not taken place. Ping-pong handovers are a special case where the mobile returns to its original cell shortly after a handover out of it.

While from a capacity point of view the deployment of small cells (Scell), also called pico cells[1], to create a HetNet leads to improved capacity, it poses a large number of new challenges for handover [5]. The topic was also studied by 3GPP for an LTE HetNet [6], identifying the need for enhancements to existing handover mechanisms to give improved reliability and quality (e.g., interruption time and packet loss) for handovers between macro cells (Mcells) and Scells, on the same frequency, for UE speeds above 30 km/h [7]. Moreover, mobility should be supported in diverse coverage overlapping

---

[1] In the remainder of this chapter we will use small cell and pico cell interchangeably.

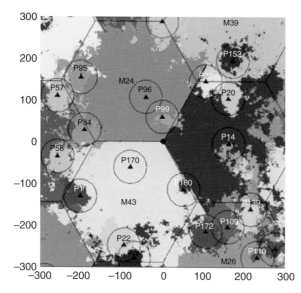

Figure 13.2 Coverage map (gray-scale representing association to a particular cell). Macro base station is represented by • and pico base stations by ▲▴.

conditions, while allowing offloading of traffic from Mcells to Scells. Figure 13.2 depicts the coverage map of a typical use-case of four Scells per Mcell. The issues with HetNet handovers are caused by the different radio propagation and interference scenarios compared to an Mcell-only network, and will be discussed at length later in this chapter.

Besides the LTE HetNet, the co-deployment of WiFi and WCDMA or LTE is popular today. Other cases, such as the mixed RAT usage of small cells and WCDMA HetNet, are less common and will not be addressed further.

Next we address the existing handover mechanism in LTE, namely the backward handover, in a single-carrier scenario. Failure mechanisms are identified and results are presented for different use-cases. We later explain the challenges behind the inter-frequency HO, and subsequently outline some methods to enhance the handover performance in Scells.

13.2 Mobility in LTE small cell HetNets

13.2.1 LTE handover

Handover takes place whenever a mobile user is moving away from its serving (or source) cell (s-cell) and sufficient signal quality can no longer be maintained. A better choice of s-cell must be found, namely the target cell (t-cell). In LTE, the architecture supporting mobility is given in Figure 13.3. Besides the obvious involvement of both the s-cell (provided by the s-eNB) and the t-cell (provided by the t-eNB, connected to s-eNB by the X2 interface), two main network elements are considered: the mobility management

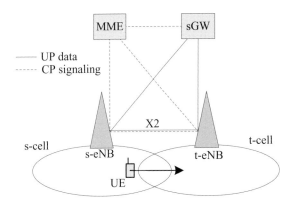

Figure 13.3 Network entities involved in handover procedures.

entity (MME) and the serving gateway (SGW). The MME handles, among other features, the control plane (C-plane) core network functions related to user mobility in LTE. On the other hand, the SGW routes and forwards user plane (U-plane) data packets between the core network and the radio access network. The SGW also acts as a mobility "anchor" for the U-plane during handover. In LTE, a UE-assisted/network-directed handover is defined [8]. This implies that the UE provides measurement(s) to the network (eNB) to assist it to determine when handover should take place (decision lies with the network). Since measurements are sent to the serving cell the handover is "backward," rather than forward when measurements are sent by the UE to the identified target cell.

The UE is configured to perform measurements over particular cell-specific reference symbols (RS) from the s-cell and neighbor cells. In LTE, a UE may measure two parameters: RSRP (reference signal received power) and RSRQ (reference signal received quality). For each RS, the UE measures the RSRP, which includes path loss, antenna gain, log-normal shadowing, and fast fading averaged over all the reference symbols within the measurement bandwidth BW_m. In addition, a time domain averaging (e.g., sliding window) by Layer 1 (physical layer) may provide several averaged measurement samples per measurement period (200 ms) to Layer 3. The filtered measurement of RSRP, is updated at the UE as the output of a first-order infinite impulse response (IIR) filter given by [9]:

$$F_n = \left(1 - 0.5^{\frac{K}{4}}\right) \cdot F_{n-1} + 0.5^{\frac{K}{4}} \cdot M_n, \qquad (13.1)$$

with F_n the updated filter measurement result, F_{n-1} the old filtered measurement result, and M_n is the latest received measurement result from the physical layer. Finally, K is a filter coefficient that is configured by RRC signaling. The formula assumes one new M_n is taken per measurement period. This averaging is performed in Layer 3 (RRC), and is known as L3 filtering.

After updating F_n, if an "entry condition" is fulfilled and maintained for a predefined duration of time (time to trigger, TTT), see [9], a measurement report (measReport) is triggered to the s-cell. Reception of the measurement report may then result in the

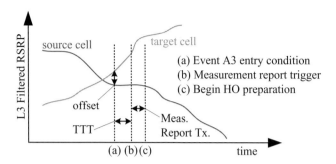

Figure 13.4 Intra-frequency handover trigger mechanism in LTE.

initiation of the handover (the behavior of the eNB is not specified by 3GPP in this respect).

In LTE, there are seven types of measurement report events each with a defined entry condition (see [9]).

- event A1: serving becomes better than threshold
- event A2: serving becomes worse than threshold
- event A3: neighbor becomes offset better than serving
- event A4: neighbor becomes better than threshold
- event A5: serving becomes worse than threshold 1 and neighbor becomes better than threshold 2
- event B1: inter-RAT neighbor becomes better than threshold
- event B2: serving becomes worse than threshold 1 and inter-RAT neighbor becomes better than threshold 2.

LTE intra-frequency handovers are typically triggered upon reception of an event A3 measurement report using RSRP, i.e., that the filtered RSRP of the target cell (t-cell) is better than that of the s-cell by a hysteresis margin (aka offset).

The LTE inter-frequency handover may be used when there is a gap in the coverage of the frequency layer of the source cell (coverage handover, using event A5 or A2 together with A4). Another possibility is when there are benefits in pushing the UE to another frequency layer (vertical handover, use event A4 or A3 with RSRQ triggering [10]).

Figure 13.4 illustrates the intra-frequency handover (HO) procedure pointing out its relevant aspects.

Once the measReport is correctly received at the s-cell, the HO preparation phase starts between the s-eNB and t-eNB, which includes admission control procedures. Upon successful admission, the t-cell prepares for HO and acknowledges the handover request sent by the s-cell. Forwarding of any queued (downlink) data starts between the s-cell and t-cell and a HO command (HOcmd) is sent from the s-cell to the UE. Upon successful reception of the HOcmd the UE accesses the t-cell, by means of a random access channel (RACH) procedure, and delivers a HO confirm (HOconf) message. Figure 13.5 shows the handover procedure.

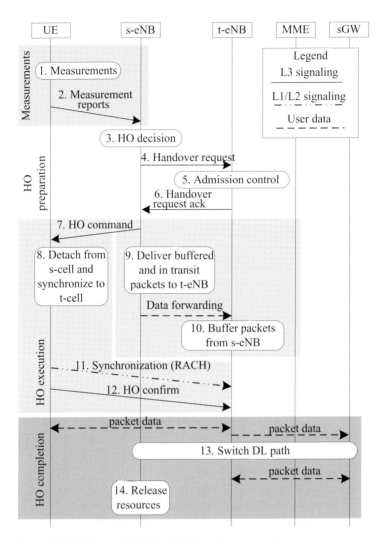

Figure 13.5 LTE intra-MME/intra-SGW handover procedure.

Typically, LTE intra-frequency HO mechanism optimization (see Section 13.3) deals with the adjustment of the offset, the TTT, and the L3 filter coefficient K, in order to achieve a good compromise between HO reliability and frequency [11].

A key aspect of the HO process is to reduce the possibility of radio link failure (RLF), that is, when the UE identifies that the signal quality to the s-cell falls below the minimum required to maintain reliable communication. In LTE, RLF can be detected based on *in-sync* (PDCCH link quality is acceptable) and *out-of-sync* (PDCCH link quality is not acceptable) indications passed from L1 to L3 in conjunction with an L3 timer mechanism (timer T310) [9]. In a typical implementation, RLF is declared if the L3 receives a single out-of-sync indication (N310 = 1) and it does not receive an in-sync indication for the following T310 period (N311 = 1).

In addition to in-sync and out-of-sync indications, an RLF may be declared by the UE upon the indication from the RLC that the maximum number of retransmissions at the link level has been reached.

Once upon detecting an RLF, or upon an HO failure (as described below), a procedure called RRC connection re-establishment is triggered ([9, §5.3.7]). As its name suggests, this procedure aims to quickly re-establish the RRC connection without the UE falling back to the RRC idle state [12].

In LTE Rel-8, T310 is signaled by system information and the same value must be used by all UEs in the cell. In LTE Rel-9, UE-specific signaling of T310 was introduced. This allows the eNB to use different values according to the bearers supported by the UE. T310 is often triggered by the poor SINR conditions during a coverage HO, and needs to be set sufficiently long so that the UE has an adequate time to try and deliver a measurement report to the eNB. Larger T310 values give a useful improvement in HO and it takes longer before recovery (RRC re-establishment) is attempted.

Typically we expect shorter T310 values when GBR bearers are configured – the service interruption is annoying to the end user and it is desirable to initiate the re-establishment procedure quickly, and prevent the UE from remaining in a deeply shadowed/faded cell [13]. This raises the question whether UEs with such GBR bearers should employ different A3 parameters, for example, using a smaller offset to reduce the likelihood of HO failure.

Based on the above, the HO can experience difficulties and fail in several ways. Next we describe the considered failures cases for the intra-frequency HO case.

- F0: RLF (T310 expiry) before A3 measurement report is generated, UE loses coverage in the source cell and does not attempt to send a measurement report.
- F1: RLF (T310 expiry) after A3 measurement report generated but not received, T310 expires before the measurement report is received.
- F2: A3 measurement report RLC transmission failure. Transmission of the measurement report is unsuccessful, including all MAC and RLC retransmissions (subject to maximum attempt limit).
- F3: RLF (T310 expiry) before HO command initiated but after measurement report received. Either during HO preparation or following failed preparation (as illustrated).
- F4: RLF (T310 expiry) before HO command received at UE, after the HO command was initiated.
- F5: RACH failure (called "handover failure" by 3GPP [9]). The HO command is successfully received by the UE and it starts timer T304. However, attempts to complete the random access procedure fail, T304 expires, and failure occurs.
- F6: RLF before HO complete command received, after RACH succeeds. The random access procedure (preamble and the response) succeeds, so T304 is stopped, but the UE falls out-of-sync and eventually RLF is declared.
- F7: HO complete command RLC transmission failure. Figure 13.6 illustrates these eight HO failure cases.

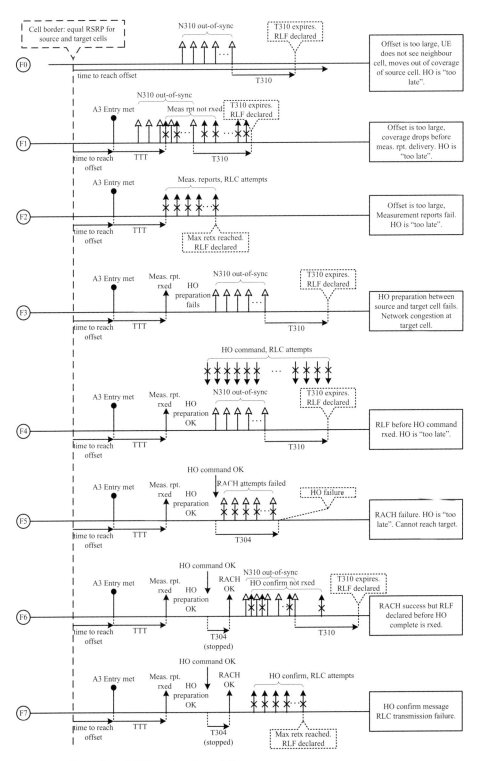

Figure 13.6 Illustration of different handover failure cases.

Table 13.1 System description for performance evaluation.

Feature		Implementation
network layout	macro	3 × 19 cells on hexagonal grid, ISD 350 m
	pico	4 picos per macro cell, randomly placed with min. pico–pico (macro–pico) distance of 40 m (75 m)
antennae	macro	14 dB gain, 3D model with downtilt 15 degrees
	pico	omnidirectional, 5 dB gain
	UE	omnidirectional, 0 dB gain
shadow fading	to macro	$\sigma = 8$ dB
	to pico	$\sigma = 10$ dB
traffic loading	downlink	artificial PRB loading, default is 100%
	uplink	stationary UE with full buffer, default is 1 UE per macro cell
handover model		Includes PDCCH model, all RRC signaling, RLC, MCA, RACH
RLF model		in/out-of-sync thresholds ($Q_{in} = -4.8$ dB/$Q_{out} = -7.2$ dB), N310 = 1, N311 = 1, T310 = 1 s
mobility model		Mobile UEs move in straight lines on heading 045° at constant speed
PDCCH model		Use the latest measured RLF SINR, which is measured every 10 ms without filtering and compared with Q_{out} to model PDCCH decoding error.
receiver model		EESM, 1 × 2 MRC for DL and UL
HO parameters		TTT is 64 ms and $K = 4$ (unless otherwise stated)

13.2.2 Intra-frequency handover: numerical results

In this section, we show some numerical results on the performance of LTE intra-frequency HO scheme in dense small cell HetNets. The main parameters and simulation assumptions are given in Table 13.1.

Figure 13.7 shows the impact of varying the offset on the HO failure rate for different speeds (1, 5, 10, 20, and 50 m/s) and HO types (M2M, M2P, P2M, and P2P[2]). An overall increase of the HO failure rate can be noted for all HO types as the offset is increased. This reflects the fact that delaying the HO by a larger offset causes radio link conditions to degrade in the serving cell and thus a greater potential for RLF to occur. In this case an HO is said to be *too late* (failure cases F0, F1, F2, and F4, see Figure 13.6). Simulations show that most failure events are RLF that occur while the HO command is being transmitted (case F4). This reflects poor downlink SINR conditions in the source cell that result in an early triggering of T310 and difficulty in receiving the HO command. In addition, the HO failure rate degrades with increasing user speed, since radio conditions in the serving cell deteriorate more quickly after triggering of the measurement report at higher speeds. M2M handovers are the most reliable at lower speeds, although less so than if there were no pico cells. Noteworthy, P2M (outbound) HO failure is always more severe than M2P (inbound) HO failure for all speeds. We will address this issue

---

[2] Respectively, macro–macro, macro–pico, pico–macro, and pico–pico.

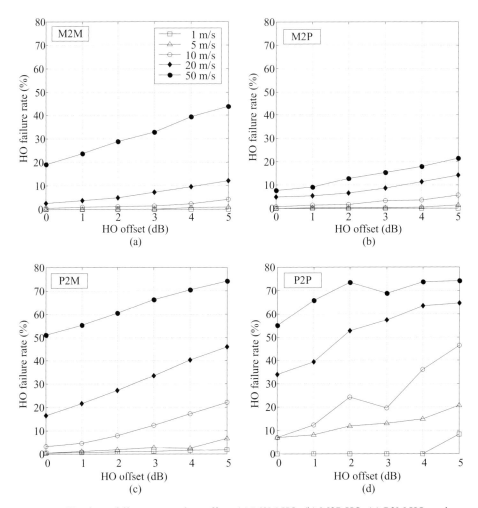

Figure 13.7 Handover failure rate against offset: (a) M2M HO, (b) M2P HO, (c) P2M HO, and (d) P2P HO.

later in this section. Moreover P2P handover is highly prone to failures noting, however, that such HO types are less likely to occur in the described scenario (we consider four Scells per Mcell).

Figure 13.8 shows the mean time between HO (MTBH) against HO offset for different user speeds. The larger the offset the lower the probability of an HO as the UE passes through the cell-edge region where source and target cell strengths fluctuate from shadow fading, hence larger MTBH arise. Thus there is a clear tradeoff between HO reliability (HO failure rate, recall Figure 13.7) and HO frequency (or MTBH). It is difficult to achieve satisfactory KPIs for HO reliability and HO frequency by solely actuating on the HO offset.

A closer look to the comparison between M2P and P2M is presented in Table 13.2. Regardless of the UE speed, and regardless of the pico cell location (at macro cell edge

Table 13.2 Handover reliability performance per type for different speeds and topology considerations. Setup was in this case: HO offset 3 dB, TTT 256 ms, and $K = 4$. Two pico cells were deployed in each macro cell.

Relative performance	10 m/s			20 m/s		
	M2P	P2M	M2M	M2P	P2M	M2M
(HO failure rate, %)						
Pico at macro edge	7.9	21.3	3.5	20.5	52.7	14.5
Pico at macro center	3.9	19.9	3.0	12.7	50.2	12.9

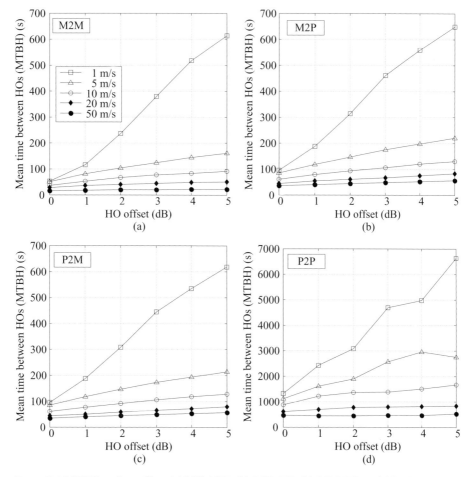

Figure 13.8 MTBH against offset: (a) M2M HO, (b) M2P HO, (c) P2M HO, and (d) P2P.

or center), M2P always has better performance than P2M in terms of HO reliability. The reasons may be found when looking at the downlink SINR at HO command trigger (refer to step 7 in Figure 13.5). Figure 13.9 plots the CDF of the SINR measured at HO command transmission instants. As it can be seen, because M2P HOs always have better

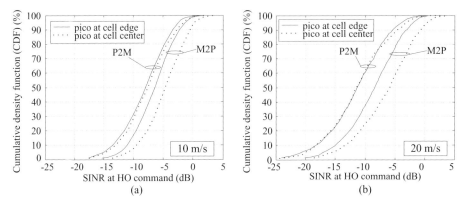

Figure 13.9 Downlink SINR at HO command trigger for (a) 10 m/s UE speed and (b) 20 m/s UE speed. Setup was in this case: HO offset 3 dB, TTT 256 ms, and $K = 4$. Two pico eNBs were deployed in each Mcell.

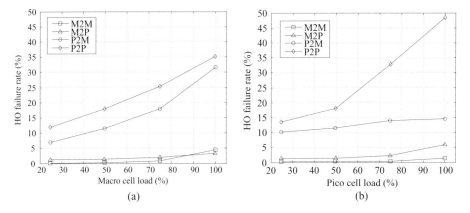

Figure 13.10 Handover failure against (a) macro cell load and (b) pico cell load. UE speed is 20 m/s, offset is 3 dB.

SINR at HOcmd transmission time than P2M, the former experience higher reliability. This has been assessed for different UE speeds and different locations for small cells within a macro cell.

The HO reliability is also sensitive to the load of the system and hence the perceived interference during HO procedures. Figure 13.10a shows the HO reliability against the downlink load level (in %) in the macro cell network (pico load is 100%). Macro load increase causes sharp HO failure rate increase for P2M and P2P, moderate increase for M2M HOs, and slight increase for M2P HOs. M2M and P2M HO degradation is mainly due to increased interference from target cell (plus any third macro cell). On the other hand, P2P HO degradation is due to heavily added background interference from macro cells. HO reliability dependence on increasing the load level in the pico cell network (macro load is 100%) is presented in Figure 13.10c. Pico load increase causes sharp HO failure rate degradation for P2P, moderate degradation to M2P, and slight

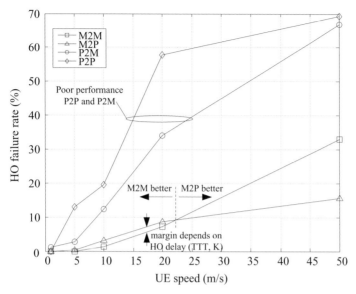

Figure 13.11 Impact of UE speed for offset = 3 dB.

degradation for P2M and M2M HOs. M2M and P2M degradation is mainly because of added background interference, whereas P2P and M2P HO degradation is due to greater interference from target cell.

As we have seen, HO reliability degrades with UE speed. In Figure 13.11 the HO failure rate is plotted against the UE speed for different HO types. The highest impact is for P2P HO and P2M HO. M2P and M2M are less sensitive to UE speed. M2M HO failure rates are a little higher than in a macro-only network, mainly because of DL interference from pico cells. At speeds below 20 m/s, the HO failure rate of M2P is larger than M2M, where such difference depends on the chosen TTT value. Above 20 m/s the M2M HO failure rate is higher than that of M2P. Although P2P exhibits the worst behavior, note that occurrence is rare (2% in total), thus P2M should be, in this case, the major concern.

So far, the Scell deployment has been restricted to a density of four pico cells per Mcell. Noting that further densification is required for satisfying the capacity needs in future systems, Figure 13.12a illustrates the HO frequency against Scell densification (in the horizontal axis) for different user speeds. As expected, both increasing the Scell density and speed has the effect of increasing the number of HOs a user experiences. This is consistent with 3GPP results in [6, 10]. At lower speeds the HO frequency flattens above 10 PBS/Mcell whereas, for higher speeds, a prolonged increasing trend is noted. The HO failure rate, Figure 13.12b, increases when increasing the Scell density (abscissa) for speeds of 10, 20, and 50 m/s. For lower speeds however, HO failure levels off starting from approximately 10 PBS/Mcell. For 10 m/s and above, HO failure rate degrades with increased speed. From Figure 13.12b we also realize that even for a rather pessimistic target of 5% HO failure rate, such a limit is exceeded as we increase speed and Scell density, hence demonstrating the limitations of the LTE handover mechanism.

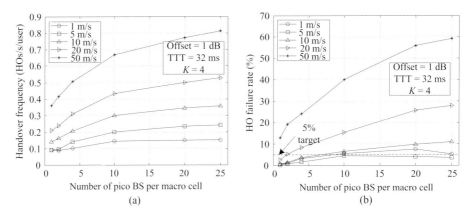

Figure 13.12 User speed and small cell density impact of HO performance. (a) HO frequency against Scell density, (b) HO failure rate against Scell density.

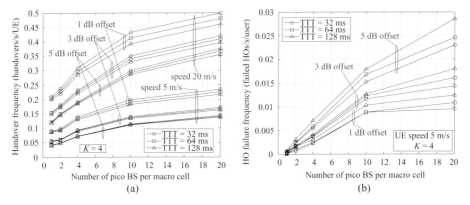

Figure 13.13 Handover parameter and small cell density impact of HO performance (from [1]). (a) Impact of Scell density on HO frequency for different HO parameter setup (offset, TTT) and UE speeds. (b) Impact of Scell density on HO failure frequency for different HO parameter setup (offset and TTT).

Figure 13.13a shows the impact of increasing the Scell density on the HO frequency for different offsets and TTT values. Lowering the offset and TTT also increases the number of HOs since this tuning favors HOs to be triggered at an early stage (recall Figure 13.4). Figure 13.13b depicts the HO failure frequency, noting that failures grow with increased Scell density since interference conditions get more challenging. In this case, low values of offset and TTT translate into lower HO failure since an early HO trigger gives better radio conditions in the source cell.

13.2.3 Inter-frequency handover

LTE inter-frequency coverage HOs can be largely avoided by careful cell planning of each frequency layer, and therefore have received little attention in the literature. Some aspects are addressed in [14]. A significant challenge is the configuration of inter-frequency measurements and measurement reporting at the UE before it leaves the coverage of the serving cell layer. Another risk is that of premature (unnecessary) inter-frequency HOs that occur when keeping the UE on the original layer would have given a successful intra-frequency HO. Since the target cell does not generate interference reliability should be good if the HO is triggered early enough.

Heterogeneous networks consisting of Mcells and Scells deployed on separate carrier frequencies can overcome many of the limitations of the co-channel deployment in terms of interference. This is especially true in the case where no interference coordination mechanisms are implemented (e.g., eICIC as discussed below). There are different views on whether this gives greater capacity, but it is clear that the offloading is dependent on the vertical HO between the layers.

To trigger the inter-frequency HO, the UE needs to identify t-cells operating on different carrier frequencies to the s-cell. In order to do so, inter-frequency measurements are carried out during periodic measurement gaps. LTE defines two measurement gap patterns of length 6 ms, the first one occurring every 40 ms (pattern #0) and the second one occurring every 80 ms (pattern #1) [12]. Full details on inter-frequency measurements in E-UTRAN can be found in [15]. Clearly, shorter measurement gap repetition periods (MGRP) will provide better identification of t-cells in other frequencies but will on the contrary cause more interruption on the s-cell data transmission and reception. Increased UE battery consumption is also a concern.

A number of solutions devoted to optimizing the data offloading potential to Scells subject to these constraints have been studied by 3GPP in [6] and references therein (see §6.4). These range from enabling longer measurement periods, performing measurements without gap assistance, and using UE mobility state estimation (MSE) based measurements among others. The drawn conclusion from the study in [6] is that continuously performing measurements according to existing performance requirements results in very high battery consumption without showing significant impact on Scell offloading potential.

Further work by 3GPP [10] has demonstrated that mobility robustness is better when macro and small cells are placed on different frequency layers, although failure rates are higher than a macro-only deployment. If DRX is used then performance is degraded, and similarly if HO thresholds are chosen to ensure longer stay times on the small cells (for better offloading) then there is a cost in terms of HO failures and ping-pongs.

13.2.4 Mobility enhancers

Current mobility mechanisms in LTE may be enhanced by considering extensions to existing implementations and/or by improving the conditions during the HO process and making it less error prone. Next we discuss some alternatives and their application to the HetNet scenario where improvements are most needed.

13.2.4.1 Forward handover

Forward HO is a technique similar to the cell update procedure used in UMTS to recover from the loss of a radio link or a failed reconfiguration procedure [16]. It can

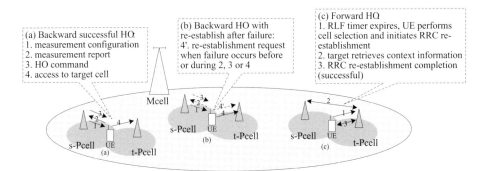

Figure 13.14 Comparison of (a) successful backward HO, (b) backward HO with RRC re-establishment, and (c) forward HO.

be mainly viewed as a recovery mechanism similar to RRC re-establishment in LTE. In a simplified way (refer to §1.2.1 for further details) the successful backward HO can be represented by the four steps given in Figure 13.14a. When the UE is unable to receive an HO command from the source cell and RLF occurs, see Figure 13.14b, it attempts an RRC re-establishment procedure to the target cell. Subsequently, if the target eNB already has the RRC context of the UE and the link from the UE to the target cell satisfactorily supports RRC signaling exchange, the re-establishment will succeed. Forward HO, depicted in Figure 13.14c, operates in a similar way by directly attempting access to the target cell after RLF (RRC re-establishment), but the target cell is able to fetch the UE context information from the source cell (if not currently available). T310 can be set to a small value (e.g., 50 ms) to minimize interruption time.

Forward HO may be successful even if the radio conditions are not good enough for the source cell to be able to decode the measurement report from the UE and prepare the target cell. The success of the HO procedure even with complete failure of signaling with the source eNB makes forward HO robust to rapidly changing signal strength conditions [16]. A drawback of forward HO is that the UE determines which cell to attach to, which excludes smart decisions from the network, for example, exploiting knowledge of cell loading, UE speed, or UE trajectory. The LTE standard does not currently support the context fetch of forward HO.

13.2.4.2 Network controlled handover

In [11], the network controlled handover (NCH) mechanism is presented. This mechanism exploits measurement reports to identify the best target cell and channel quality measurements to realize a handover-specific timing decision. Compared to the backward HO described above, the NCH algorithm only differs in the HO trigger mechanism, whereas the remaining steps in the HO preparation and execution remain unchanged. The method is compliant with LTE Release 8. Recall from Section 13.2.1, the backward HO is triggered as soon as the measurement report is received by the cell serving it. However, the HO success depends on many external factors that are not captured by the measurement report (e.g., cell load, SINR, and UE speed). Network controlled handover

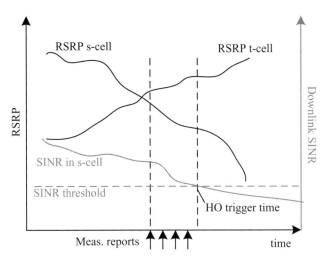

Figure 13.15 Proposed handover trigger mechanism.

exploits the measurement report mechanism to identify candidate target cells, but the timing of the HO is dictated by the network for each individual HO according to other criteria. An implementation of NCH that uses the downlink SINR in the source cell as the criterion is described next and illustrated in Figure 13.15.

1. The eNB configures a "triggered periodic" A3 event with a small offset (say 1 dB) and short TTT (0 ms in practice).
2. Once a report from a UE has been received, the source eNB monitors the downlink SINR of the UE in the source cell.
3. Once the SINR passes below a threshold value the HO preparation is triggered for the strongest target cell in the last measurement report.

Note NCH uses separate signaling mechanisms to identify the target cell and to trigger the HO. The eNB can use downlink measurements taken by the UE to reflect the downlink channel quality. RSRQ is available but only reflects the SINR accurately under full and equal downlink transmission power across the frequency band. One possible approach is to use the filtered wideband channel quality indication (CQI) based on UE CQI reports. These can be configured to be sent periodically on the PUCCH channel. An MRO algorithm based on NCH can optimize HO trigger parameters such as the CQI threshold based on the statistics of the HO performance (see Section 13.3.2).

Since NCH can trigger each HO on demand, on time and only if necessary, this improves greatly the tradeoff relationship of HO reliability and HO frequency. Figure 13.16 shows the performance evaluation of NCH by simulation using four different values of the threshold CQI against the backward HO also for four different settings of the offset and TTT values [11]. For the case of NCH the measurement report configuration used an offset of 0 dB, TTT of 0 ms, and a reporting interval of 240 ms. A HetNet of one pico per macro cell is used. Comparison is shown for the M2M HO

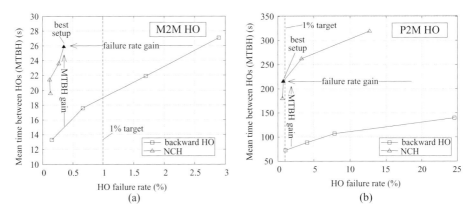

Figure 13.16 Performance gains of NCH with respect to backward HO for (a) M2M HO and (b) P2M HO.

(Figure 13.16a) and for the P2M HO (Figure 13.16b). In both cases, the NCH outperforms the backward HO in terms of lower HO failure rate and higher MTBH (i.e., lower HO frequency).

13.2.4.3 Enhanced ICIC

In order to address the problems caused by the downlink transmit power difference between eNodeBs (eNBs) and pico cell eNBs (PeNBs) in HetNets, cell selection methods allowing UEs to associate with cells that do not provide the strongest downlink RSRP have been proposed. A widely considered approach is cell range expansion (CRE), in which a positive range expansion bias is added to L3 HO measurements at UEs to increase the pico cells' downlink coverage footprint. Figure 13.17 illustrates this concept.

Although CRE is able to mitigate uplink inter-cell interference and provide load balancing in HetNets, it degrades the downlink signal quality of pico cell UEs (PUEs) in the expanded region, since these PUEs are not connected to the cell that provides the strongest RSRP. This is particularly critical for the transmission of the HO command for an outgoing HO (pico to macro), which will suffer severe interference and degrade the overall HO performance. Consequently, in Release 10 of LTE, Enhanced inter-cell interference coordination (eICIC) was introduced to mitigate downlink inter-cell interference for range-expanded PUEs. The scheme uses almost blank subframes (ABS), which are subframes in which no control or data signals but just reference signals are transmitted. Specifically, macro cells schedule ABS, and pico cells schedule range-expanded PUEs in the subframes that overlap with the macro cell ABS, so that their performance can be enhanced.

From the point of view of mobility, our studies revealed the following conclusions:

1. CRE will cause increased HO possibilities, hence increased total HO number.
2. CRE (without eICIC) will degrade the overall HO reliability performance:
    i. CRE will improve M2M HO performance since M2M HOs suffering high interference from pico cells are eliminated

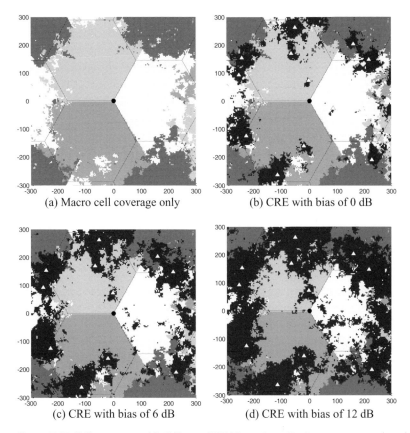

Figure 13.17 Cell coverage with different CRE bias values. Darker coverage regions indicate Pcell coverage.

  ii. CRE will improve M2P HO performance greatly because it is executed earlier (by bias dB)
  iii. CRE will degrade P2M HO performance greatly, because P2M is executed later (by bias dB). Furthermore, there will be more P2M HOs when CRE is used
  iv. CRE will degrade P2P HO performance, because HO happens further away from the source pico antenna, so there is higher background interference; also, P2P HO number will increase greatly when CRE used.
3. CRE with eICIC can improve overall HO reliability compared to the baseline of no CRE:
  i. M2M and M2P HO performance isn't impacted since downlink interference situation does not change, and failures mostly come from downlink problems
  ii. eICIC can help P2M HO performance greatly. While when too large a bias is used (for example, 12 dB), the P2M HO problem still is the bottleneck
  iii. eICIC can also help P2P HO performance because it reduces the background interference greatly when edge pico users are handing over.

13.2.4.4 Dual connectivity

In LTE HetNet, an attractive deployment strategy is to backhaul small cells through a parent macro cell. This opens up the possibility of cooperation between these cells to serve UEs lying within the coverage of both. In 3GPP, different means of cooperation are currently being studied, under the title "dual connectivity" [10]. The motivations for dual connectivity include improving mobility robustness, network capacity, and energy saving. Backhaul is non-ideal (fiber is excluded).

From a mobility perspective, dual connectivity may offer the ability to send the RRC signaling during an HO using either a signaling radio bearer (SRB) established with the macro cell, or an SRB established with the small cell, or indeed both. These constitute different diversity choices and are termed RRC diversity. Furthermore, in dual radio link monitoring (RLM), the UE could declare RLF only if both links are out of sync. Full tx/rx diversity and dual RLM gives large reductions in HO failure rate, particularly with dense small cell deployments [10]. For example, with ten small cells per macro cell, the gain in failure rates is 80%. However, there are many practical difficulties in adding such functionality to LTE.

In a concept known as "shared cell," "soft cell" [17], or "phantom cell" [18], the small cells are not conventional cells because they lack cell-specific channels such as synchronization signals and cell-specific reference symbols, but are part of the macro cell. In [18] dual connectivity furthermore splits the C-plane (SRB, provided by the macro cell) and U-plane (provided by the small "phantom cells"). Within the coverage of a macro cell there can be multiple phantom cells, and the user plane to the UE is switched among these phantom cells as the UE moves. This is not an L3 HO process but is instead an L1/L2 mobility procedure. With dense deployments of phantom cells, the focus shifts to this new mobility procedure, which must rapidly and reliably switch the user plane link of the UE.

13.3 Mobility robustness optimization (MRO)

In the first release of LTE, Release 8, 3GPP introduced support for a self-organized network (SON) by adding self-configuration functionality for the eNB, for example, automatic neighbor relation [8]. Self-optimization, representing the automatic adjustment of the parameters and operation of a live network, was added in Release 9, including the use-case of intra-LTE handover optimization, known as mobility robustness optimization (MRO) [8]. In Release 10, support for inter-RAT MRO was added. Another aspect of MRO, not considered in detail by 3GPP, is the management of idle mode mobility parameters. In particular, it is seen as important to align cell boundaries for idle mode and active mode such that cell reselection (handover) is avoided after an RRC state transition from active to idle (idle to active). Further discussion below is restricted to the active mode case (i.e., handover), which is more important since dropped calls and/or service interruption may occur.

Table 13.3 Handover failure detection mechanisms.

	Failure	Detection mechanism
1	Handover too late	r-eNB sends an RLF indication message to s-eNB. S-cell recognizes that a UE for which it holds the context has suffered RLF and has attempted RRC re-establishment at r-cell.
2	Handover too early (failure during an HO)	In this case detection is trivial since the UE attempts re-establishment in the source cell. No X2 signaling is involved.
3	Handover to wrong cell (failure during an HO)	r-eNB sends an RLF indication message to s-eNB. S-cell recognizes that a UE, which was undertaking an HO to t-cell, has suffered RLF and has attempted RRC re-establishment at r-cell.
4	Handover too early (failure shortly after an HO)	After a successful HO from s-cell to t-cell, RLF occurs and the UE attempts RRC re-establishment in s-cell. s-eNB sends an RLF indication to t-eNB, which identifies the UE recently handed over (time < threshold) from s-cell and sends an HO report to s-eNB. S-cell identifies the failure case.
5	Handover to wrong cell (failure shortly after an HO)	This is similar to case 4. r-eNB sends an RLF indication to t-eNB, which then sends an HO report to s-eNB.

The objectives of MRO are not stipulated by the standard, but one proposal is [2]:

1. to achieve an HO failure rate that is below a target value
2. to minimize the number of HOs while achieving objective 1.

MRO functionality must be placed in both the RAN and the OAM, and the standard allows for different functional splits. In the remainder of this section it is assumed that the RAN identifies adjustments that should be made to the HO timing, for example, by tuning HO parameters such as offset and/or TTT. The OAM system is able to control the scope of the optimization algorithm running in the eNBs, for example, by setting minimum and maximum parameter values [19]. In Release 9, 3GPP introduced two procedures to assist in the identification of HO failure events [20]:

- radio link failure (RLF) indication
- handover report.

These address the general failure cases (see Section 13.2.1 above) in which the cells involved (source cell (s-cell), target cell (t-cell), cell where re-establishment is attempted (r-cell)) belong to different eNBs (s-eNB, t-eNB, r-eNB, respectively). If this is not the case some signaling can be saved but the mechanism is the same.

In all cases the objective is to provide information on the failure to the s-cell such that it can make corrections for subsequent outgoing HOs. Table 13.3 provides a summary of failure types and how they are detected.

Another aid to failure detection is the RLF report [8] that can be generated by a UE and sent to the network. This gives details of the failure event and includes radio

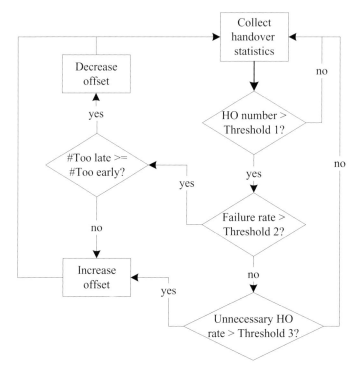

Figure 13.18 Example of MRO flow chart.

measurements prior to the failure. The primary motivation for this report was to pinpoint RLF events occurring in coverage holes lying between two cells, since such failures cannot be corrected by MRO alone.

13.3.1 MRO for macro cell networks

Early work on MRO considered initial LTE deployments consisting of macro cells in dense urban environments within the Socrates research project [21–23]. The algorithms described are driven by KPIs accumulated over some period of time, such as HO failure rate and ping-pong rate. Since LTE allows adjustment of HO hysteresis (aka offset) on a per cell pair basis, KPIs are gathered in a similar fashion. According to the KPIs an adjustment of the hysteresis or the TTT is made, and then further statistics are gathered before the next adjustment is considered. Clear benefits are seen, especially when UE movement is restricted to cross cell boundaries in limited places, as on major roads, because the HO behavior and shadowing environment is then more consistent for a single cell pair. In [24] an MRO algorithm adjusting the HO margin according to the change in user mobility and taking into account the nature of HO failure events is presented. A reduction in both the HO failure and ping-pong rates were noted.

A simple MRO algorithm is drawn in Figure 13.18. After a fixed number of HOs (Threshold 1), the failure rate is compared against the target value (Threshold 2). If

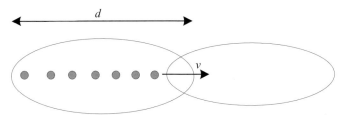

UEs with line density ρ moving at speed v

Figure 13.19 Conceptual UE mobility model.

the rate is too high then the offset is adjusted by 1 dB, upward or downward according to whether the too late or too early failures predominate. Adjustment of the offset used by the UE is possible from the source cell to each target cell individually using the parameter cellIndividualOffset [9]. In contrast, TTT and K are target-cell agnostic making it difficult to use them to tune the timing of one outgoing HO (i.e., to one target cell) alone. Alternatively, if the failure rate is below the target but the unnecessary HO rate is too high then the offset can be increased.

The MRO algorithms described above employ KPIs derived from simple Boolean assessment on each HO event, such as HO failure: true/false; HO ping-pong: true/false. To capture accurate KPIs, the measurement period has to encompass a large number of HOs since the unwanted HO outcomes (failure, ping-pong) are typically rare, representing only a few percent of all HO events. For example, to measure an HO failure rate of 1% over 1,000 HOs should be monitored per cell pair.

Would such an algorithm, driven by hard metrics, be able to respond to environmental changes such as w.r.t. UE speed (vehicle speed increases after clearing of an accident) or network loading (transition to the busy traffic hour)? To achieve 1,000 HOs between two cells in one hour, the HO frequency f is 0.28 handovers/s, or a mean time between handovers (MTBH) of 3.6 s. If we assume that UEs move from one cell to another at speed v, and the line density of users is ρ then (Figure 13.19):

$$f = \rho \cdot v. \tag{13.2}$$

If the cell size is d, then the number of users in each cell, m, is given by

$$m = d \cdot \rho = \frac{d \cdot f}{v}. \tag{13.3}$$

If we substitute f as 0.28 s^{-1}, d as 500 m, v as 1 m/s (3.6 km/h), then we get $m = 140$. So with pedestrian users, we need 140 users in the cell moving to each neighbor cell. With six neighbor cells we require $6 \times 140 = 840$ RRC connected UEs in the cell. Numbers as large as these are possible in a mature heavily loaded deployment but are high for early LTE systems. However, since the focus of MRO is on addressing high-speed users (vehicles) we could expect the HO failure rate driven algorithm (the baseline algorithm described above) to be able to judge the failure rate over an hour. For example, with a vehicle speed of 20 m/s we only require 42 RRC connected users in each cell. Thus reactivity on a one-hour basis is feasible under high-load and high-mobility cases, but load and mobility can fluctuate more rapidly than this.

A more reactive MRO algorithm, driven by so-called "soft" metrics, has been shown to outperform the hard-metric driven conventional algorithm of Figure 13.18 [25]. Soft metrics are measurements taken on each HO that demonstrate some correlation with the HO reliability. Examples of soft metrics are measures of the HO command transmission time or the number of HARQ transmissions to send the HO command to the UE. After only 100 HOs, statistical measures of soft metrics (e.g. the 90-percentile value) stabilize and can be compared to thresholds to determine HO offset adjustment. Thus "soft MRO" is an order of magnitude more reactive than the hard-metric algorithm – the operating point of the HO can be adjusted quickly to the optimum point as the environment changes, saving HO events and failures [25].

13.3.2 MRO for HetNet

The above sections have highlighted the increase in HO frequency and the deterioration in HO performance when we deploy outdoor small cells into a macro network to address high-traffic clusters. Small cells capture fewer users than macro cells, the drop in statistical multiplexing leads to greater variations in their load and the interference generated. Furthermore, small cells are likely to be deployed in increasing numbers over time, adding new neighbor relationships and HO sequences. Energy saving practices may put small cells into "sleep modes" for short time periods (of the order of minutes) giving rapid birth and death of neighbor relations. All these factors make MRO for Het-Net more challenging and more important for HetNet than for macro-only deployments. Literature on MRO for HetNet is scarcer than for homogeneous (macro-only) networks. An example is [26], where the proposed MRO algorithm considers a HetNet deployment and adjusts the cell individual offset (CIO) according to the rate of ping-pong HOs and HO failures.

The network controlled handover (NCH) described above provides UE level HO triggering that auto-corrects for variations in the radio environment and interference levels. However, the best NCH settings are dependent on the UE speed [11]. A conventional MRO algorithm, similar to Figure 13.18, can tune NCH parameters but reactivity is slow, for reasons described above. What gains would be realized by using soft MRO? This question has been evaluated using a system simulator to model a number of small cells deployed to cover a large square in central London, Figure 13.20 [2]. In this case, the placement of small cells is not random but carefully deployed to cover a particular location.

Test UEs travel around the square perimeter at 10 km/h (severe congestion), and at a point in time the congestion is suddenly ended and speeds rise to 20 km/h. Looking at the HO from cell 68 to cell 66, the HO failure rate increases at this point (trees shadow this cell border) but is then brought back to the target value by the MRO algorithm. In Figure 13.21, it has demonstrated that recovery is much faster for soft MRO (case 2) than conventional MRO (case 1): 50 s compared to 500 s.

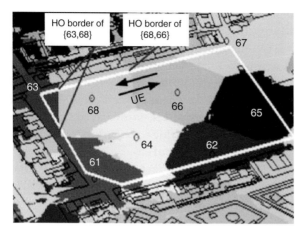

Figure 13.20 Network coverage map for HetNet MRO test [2].

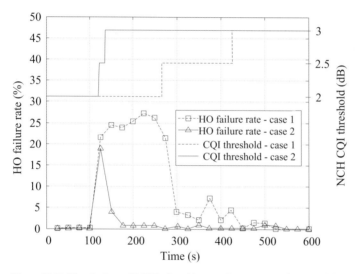

Figure 13.21 Simulation of MRO algorithm performance under speed increase [2].

13.4 Inter-system mobility: LTE to WiFi

Although at first sight WiFi and 3GPP cellular technologies appear as competing solutions, especially in indoor environments, soon it was realized that both technologies could actually coexist resulting in potential benefits to the end-users, operators, service providers, as well as technology providers [27]. A main benefit for integrating both technologies is traffic offloading of cellular to WiFi, by which cellular systems can be relieved from both high load and excessive interference.

The wide market adoption of the IEEE WiFi standard in the early 2000s instigated mobile operators and vendors to undertake studies for the potential integration of WiFi within cellular networks. As a result, the WLAN Inter-working Task Force was created, which led to the definition of various use cases and architectures with different degrees of inter-working [28]. These scenarios led to different architecture implementation alternatives, ranging from "loosely" to "tightly" coupled interworking architectures according to how network entities were reused and interconnected between WLAN and the cellular network. A good overview of such alternatives is provided in [29].

Following the outcome of the WLAN Interworking Task Force, the 3GPP started developing a number of standards to support interworking between WiFi and cellular networks. This first integration was to the UMTS core network, which gave rise to the IWLAN (integrated/interworked WLAN) set of standards [30]. Later, WiFi integration was brought to the evolved packet core (EPC), being referred to as EPC standards [31]. In the remainder of this section we will refer to the latter set, with particular emphasis on the radio access network procedures.

WLAN interworking and integration with LTE has been addressed by 3GPP specifications both in the core network (CN) and in the radio access network (RAN). In particular, a new class of non-3GPP access network is defined, being coined as "trusted," denoting that access points within the trusted network are deployed and managed by the operator, so that a UE can connect to it without requiring any additional security measures. Trusted non-3GPP accesses can interact directly with the EPC. Different network architectures arise supporting forms of interworking between 3GPP and non-3GPP networks (both trusted and non-trusted), which are documented in [31]. In addition, in order to handle the increased HO complexity in these scenarios, a new functionality has been added to the EPC, namely the access network discovery and selection function (ANDSF) [31]. The role of the ANDSF is to exchange discovery information and policies with the UE according to some operator requirements. In particular, it provides the inter-system mobility policy, access network discovery information (such as the SSID in the case of WLAN) and also the inter-system routing policy. Information obtained through the ANDSF is stored in the UE.

RAN-level enhancements for WLAN/3GPP interworking are currently being investigated in Release-12 [3]. These enhancements intend to mitigate a number of well-known problems. First, the fact that operator-deployed WLAN networks are often underutilized. Second, the user experience is suboptimal when the UE connects to an overloaded WLAN. And, third, the UE battery drains caused by excessive and unnecessary WLAN scanning. Among the required solutions are those of providing improved bidirectional load balancing between WLAN and 3GPP radio access in order to increase system capacity, and also to improve the utilization of WLAN when available and not congested. Scenarios considered so far focus on WLAN nodes deployed and controlled by operators, thus falling into the trusted non-3GPP network case.

We now present a possible solution, as suggested in [3], for traffic steering in RRC connected mode between an eNB and a WLAN AP. In this solution, traffic steering is controlled by the network, and user preference for either cellular or WLAN access always takes precedence over rules by the ANDSF (if present). The traffic-steering procedure comprises the steps indicated in Figure 13.22.

In the measurement control, the eNB configures the UE measurement procedures including the identity of the target WLAN to be measured. Measurement events to trigger

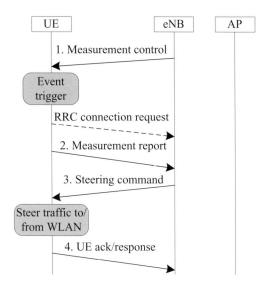

Figure 13.22 Traffic steering for UEs in RRC connected state (from [3]).

reporting can be based, for example, on whether WLAN becomes better (worse) than some threshold value. The identity of the WLAN to be measured can be obtained via the eNB conveying the basic service set identifier (BSSID), the service set identifier (SSID), or others (see [3]). The UE is triggered to send a measurement report by the rules set by the measurement control. Measurements to report may include the received channel power indicator (RCPI), the received signal-to-noise indicator (RSNI), and others. Once the measurement report has been received the eNB sends the steering command message to the UE to perform traffic steering based on the reported measurements.

13.5 Summary

In this chapter we have covered the essential procedures supporting active mode user mobility in small cell HetNets. Using the LTE HO mechanism as a baseline, we have outlined the main principles behind its operation, starting from the measurement gathering and reporting system, and followed by the signaling exchange between the UE and both the source and target cells. The main causes for HO failure have been outlined, and numerical results have shown the dependency of such failures with respect to user speed, small cell density, and other HO design parameters. The introduction of small cells not only brings about more HOs but also reduces the reliability of each one. Several ways to enhance existing HO schemes were outlined with emphasis on small cell HetNets, including forward HO, network-controlled HO, and the use of eICIC, among others. The use of self-optimizing mechanisms for mobility, or MRO, has proven to be an efficient way to achieve low HO failure rate in addition to minimizing the number of HOs. Using additional context information such as, for example, user speed and/or position could be used to help in optimizing mobility procedures. However, such information must be reliably available in real time. Finally, the importance of enabling inter-system mobility with widespread technologies such as ubiquitous WiFi was also addressed.

References

[1] X. Gelabert, G. Zhou, and P. Legg, "Mobility performance and suitability of macro cell power-off in LTE dense small cell HetNets," *2013 IEEE 18th International Workshop on Computer Aided Modeling Analysis and Design of Communication Links and Networks (CAMAD) (CAMAD 2013)*, Berlin, Germany, Sep. 2013.

[2] G. Zhou and P. Legg, "An evolved mobility robustness optimization algorithm for LTE heterogeneous networks," *Vehicular Technology Conference (VTC Spring), 2013 IEEE 77th*, 2013, pp. 1–5.

[3] 3GPP TR 37.834 V1.0.0, "Study on WLAN/3GPP radio interworking (Release 12)," *Tech. Rep.*, Aug. 2013.

[4] P. Legg, G. Hui, and J. Johansson, "A simulation study of LTE intra-frequency handover performance," *Vehicular Technology Conference Fall (VTC 2010-Fall), 2010 IEEE 72nd*, 2010, pp. 1–5.

[5] D. Lopez-Perez, I. Guvenc, and X. Chu, "Mobility management challenges in 3GPP heterogeneous networks," *IEEE Communications Magazine*, vol. 50, no. 12, pp. 70–78, 2012.

[6] 3GPP TR 36.839 V11.1.0, "Mobility enhancements in heterogeneous networks (Release 11)," *Tech. Rep.*, Dec. 2012.

[7] 3GPP TR 36.932 V12.0.0 (2012–12), "Scenarios and requirements for small cell enhancements for E-UTRA and E-UTRAN (Release 12)," *Tech. Rep.*, Dec. 2012.

[8] 3GPP TS 36.300 V8.12.0, "E-UTRA and E-UTRAN; overall description; stage 2 (Release 8)," *Tech. Rep.*, Apr. 2010.

[9] 3GPP TS 36.331 V8.20.0, "E-UTRA radio resource control (RRC); protocol specification (Release 8)," *Tech. Rep.*, Jul. 2013.

[10] 3GPP TR 36.842 V1.0.0, "Study on small cell enhancements for E-UTRA and E-UTRAN: higher layer aspects," *Tech. Rep.*, Nov. 2013.

[11] G. Zhou, P. Legg, and G. Hui, "A network controlled handover mechanism and its optimization in LTE heterogeneous networks," *2013 IEEE Wireless Communications and Networking Conference (WCNC)*, 2013, pp. 1915–1919.

[12] M. Baker, S. Sesia, and I. Toufik, *LTE: The UMTS Long Term Evolution: From Theory to Practice; [Including Release 10 for LTE-Advanced]*. Wiley, 2011.

[13] R2-094304 RAN2 #67 Qualcomm, "RLF handling for real time services," *Tech. Rep.*, Aug. 2009.

[14] J. Kurjenniemi and T. Henttonen, "Effect of measurement bandwidth to the accuracy of inter-frequency RSRP measurements in LTE," *Personal, Indoor and Mobile Radio Communications, 2008. (PIMRC 2008). IEEE 19th International Symposium*, 2008, pp. 1–5.

[15] 3GPP TS 36.133 V9.15.0, "Requirements for support of radio resource management (Release 9)," *Tech. Rep.*, Mar. 2013.

[16] Qualcomm, "LTE mobility enhancements," *White Paper*, Feb. 2010.

[17] S. Parkvall *et al.*, "Heterogeneous network deployments in LTE: the soft-cell approach," *Ericsson Review*, Feb. 2011.

[18] H. Ishii, Y. Kishiyama, and H. Takahashi, "A novel architecture for LTE-B: C-plane/U-plane split and phantom cell concept," *Globecom Workshops (GC Wkshps), 2012 IEEE*, 2012, pp. 624–630.

[19] 3GPP TS 32.522 V10.6.0, "Self-organizing networks (SON) policy network resource model (NRM) integration reference point (IRP); information service (IS) (Release 10)," *Tech. Rep.*, Sep. 2013.

[20] 3GPP TS 36.423 V9.6.0, "X2 application protocol (X2AP) (Release 9)," *Tech. Rep.*, Apr. 2011.

[21] T. Jansen *et al.*, "Handover parameter optimization in LTE self-organizing networks," *Vehicular Technology Conference Fall (VTC 2010-Fall), 2010 IEEE 72nd*, Sep. 2010.

[22] T. Jansen *et al.*, "Weighted performance based handover parameter optimization in LTE," *Vehicular Technology Conference (VTC Spring), 2011 IEEE 73rd*, May 2011.

[23] S. Hämäläinen, H. Sanneck, and C. Sartori, *LTE Self-Organising Networks (SON): Network Management Automation for Operational Efficiency*. Wiley, 2011.

[24] K. Kitagawa, T. Komine, T. Yamamoto, and S. Konishi, "A handover optimization algorithm with mobility robustness for LTE systems," *Personal Indoor and Mobile Radio Communications (PIMRC), 2011 IEEE 22nd International Symposium on*, Sept 2011, pp. 1647–1651.

[25] G. Hui and P. Legg, "Soft metric assisted mobility robustness optimization in LTE networks," *Wireless Communication Systems (ISWCS), 2012 International Symposium on*, 2012, pp. 1–5.

[26] Y. Watanabe, H. Sugahara, Y. Matsunaga, and K. Hamabe, "Inter-eNB coordination-free algorithm for mobility robustness optimization in LTE Het-Net," *Vehicular Technology Conference (VTC Spring), 2013 IEEE 77th*, June 2013.

[27] Small Cell Forum, "Integrated femto–WiFi networks," *Tech. Rep.*, Feb. 2012.

[28] 3GPP TR 22.934 V1.2.0, "Feasibility study on 3GPP system to WLAN inter-working," *Tech. Rep.*, May 2002.

[29] A. K. Salkintzis, C. Fors, and R. Pazhyannur, "WLAN-GPRS integration for next-generation mobile data networks," *Wireless Communications, IEEE*, vol. 9, no. 5, pp. 112–124, 2002.

[30] 3GPP TS 23.234 V11.0.0, "3GPP system to wireless local area network (WLAN) interworking; system description (Release 11)," *Tech. Rep.*, Sep. 2012.

[31] 3GPP TS 23.402 V12.2.0, "Architecture enhancements for non-3GPP accesses (Release 12)," *Tech. Rep.*, Sep. 2013.

14 The art of deploying small cells: field trial experiments, system design, performance prediction, and deployment feasibility

Doru Calin, Aliye Özge Kaya, Amine Abouliatim, Gonçalo Ferrada, and Ionel Petrut

We disseminate a set of small cells' field trial experiments conducted at 2.6 GHz and focused on coverage/capacity within multi-floor office buildings. LTE pico cells deployed indoors as well as LTE small cells deployed outdoors are considered. The latter rely on small emission power levels coupled with intelligent ways of generating transmission beams with various directivity levels by means of adaptive antenna arrays. Furthermore, we introduce an analytical three-dimensional (3D) performance prediction framework, which we calibrate and validate against field measurements. The framework provides detailed performance levels at any point of interest within a building; it allows us to determine the minimum number of small cells required to deliver desirable coverage and capacity levels, their most desirable location subject to deployment constraints, transmission power levels, antenna characteristics (beam shapes), and antenna orientation (azimuth, tilt) to serve a targeted geographical area. In addition, we disseminate specialized solutions for LTE small cells' deployment within hotspot traffic venues, such as stadiums, through design and deployment feasibility analysis.

14.1 Introduction

Small cells are low-cost, low-power base stations designed to improve coverage and capacity of wireless networks. By deploying small cells on top and in complement to the traditional macro cellular networks, operators are in a much better position to provide the end users with a more uniform and improved quality of experience (QoE). Small cells' deployment is subject to service delivery requirements, as well as to the actual constraints specific to the targeted areas. For a good uniformity of service, in populated areas where the presence of buildings is the main reason for significant radio signal attenuation, small cells may need to be closely spaced, e.g., within a couple of hundred meters from each other. Naturally, the performance of small cells is highly dependent on environment-specific characteristics, such as the materials used for building construction, their specific propagation properties, and the surroundings. It is particularly important to have a proper characterization of an environment where small cells are deployed.

This chapter focuses on in-building performance and feasibility of LTE small cells through measurements, taking as reference both outdoor small cell and indoor pico cell deployments. We created scenarios where wireless connectivity within a target building is offered either by small cells located on the exterior of other buildings (small cells with outdoor characteristics) or simply by small cells located within the target building (pico cells with indoor characteristics). We conducted several LTE small cells' trials at 2.6 GHz in various European cities for in-building coverage. We disseminate various measurements results obtained through these LTE trials.

The chapter consists of three parts. In the first part we analyze the measurement results obtained through LTE small cell measurements campaigns at 2.6 GHz. In the second part we introduce an analytical 3D coverage and capacity prediction framework, which we calibrate and validate against field measurements. This framework could be used to assess the performance and feasibility of small cell deployments for various what-if scenarios. Our framework utilizes the ray-tracing tool WiSE [1] for 3D modeling of environments where measurements have been taken. Further, we establish confidence that a well-designed environment simulator can be used reliably in system performance studies. In particular, this framework can determine detailed performance levels or channel characteristics at any point of interest within a building and allows for fast and easy what-if scenarios testing. It can be used to create rules of thumb for deploying small cells, and can be applied to large-scale small cell deployments. In the last part, we show specialized solutions for LTE small cells' deployment within hotspot traffic venues, such as stadiums, through deployment feasibility analysis using this framework.

14.2   LTE small cell field trials

The goal of the field trials was to investigate the indoor performance and feasibility with LTE small cells placed outdoors and indoors. The measurements are performed at 2.6 GHz over 10 MHz bandwidth, in dedicated small cell carrier (no presence of macro cell signal at 2.6 GHz). Measurements are done for a single small cell and a single test mobile or user equipment (UE) to understand the fundamental aspects of outdoor–indoor and indoor–indoor propagation. We varied the location of the UE throughout the buildings and measured several key performance indicators (KPIs), as reported by test mobiles. For the analysis in this chapter we are addressing the reference signal received power (RSRP), signal-to-noise ratio (SNR), and physical downlink shared channel (PDSCH) data rate at the selected UE locations. In LTE, RSRP is the average of the power of all resource elements that carry cell-specific reference signals over the entire bandwidth [2].

In the following, we show results from measurement trials in two different buildings of different sizes, labeled as Building A and Building B. We have conducted trials in many other buildings and found similar results.

Building A: a small office building with large rooms. This building is of size 66 m × 24 m and has three floors as depicted in Figure 14.1. It has an open floor plan with lots

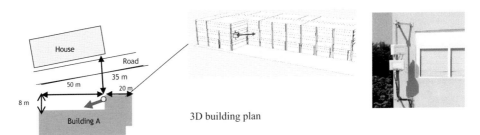

Figure 14.1 Layout and 3D building plan of Building A and its surroundings. The small cell location and antenna orientation is marked with an arrow toward the middle of the long edge of the building. Outdoor small cells are attached to the exterior of the building at the height of the floor where measurements have been taken (right).

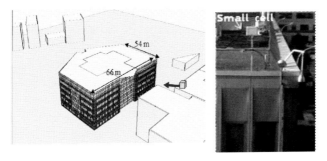

Figure 14.2 3D building plan of Building B and its surroundings (left). The small cell location is marked with a cube and its direction with an arrow toward the middle entrance of the building. The small cell is placed on top of a 1.5 m mast (right).

of furniture and equipment in the hallways. We report the measurement campaign taken on the middle floor of this building.

Services from outdoor small cell: the outdoor small cell was attached on the exterior of the building on the same level as the middle floor. The arrow in Figure 14.1 shows the main direction of the beam formed by the small cell. We refer to this beam orientation as the 0° beam. During measurements, this beam has been also electronically steered by +/−30° in the azimuth plane. For each instance of the beam we have taken independent measurements within the building (only one beam instance was active during measurements).

Services from indoor small cells: we show results also for the cases where multiple indoor small cells are used for coverage.

Building B: a mid-size office building with small offices. The building has five floors and is of size 66 m × 54 m. Figure 14.2 shows the 3D building plan and the location of the small cell.

Services from outdoor small cell: this is the only type of deployment considered for this office building. The small cell was placed on the rooftop of the building in front of the measurement building. The rooftop of the building hosting the small cell is on the same level as the windows on the fourth floor of the measurement building. It is

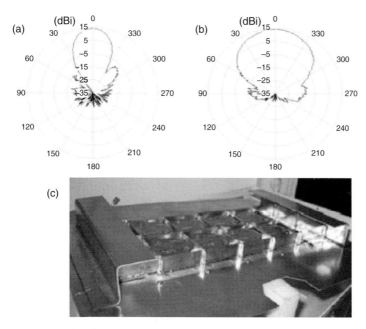

Figure 14.3 Directive antenna pattern for the outdoor small cell: (a) horizontal plane (14.2 dBi gain, 25° horizontal half power beamwidth); (b) vertical plane (14.2 dBi gain, 49.6° vertical half power bandwidth); (c) eight-element smart prototype antenna panel for outdoor deployment to improve the indoor antenna coverage.

placed 13 m above the ground. The antenna was mechanically steered by 30° in the azimuth plane toward the middle entrance of the measurement building, as indicated by the arrow. The street width between the building where the small cell is located and the building to be covered is 22 m. The UE equipment is an LG G7 USB dongle attached to a laptop placed on a measurement table on a trolley at 80 cm height above the floor. We performed measurements on every floor.

For outdoor to indoor coverage in both buildings we used the same directive beam with 14.2 dBi gain. The beam is formed with an eight-element smart antenna panel (Figure 14.3). Its horizontal and vertical patterns corresponding to the 0° beam direction are illustrated in Figure 14.3a and Figure 14.3b, respectively. The antenna pattern has 25° horizontal half power beamwidth and 49.6° vertical half power beamwidth. The narrow horizontal beamwidth is designed with the purpose to reduce the interference with any potential adjacent transmitters, while larger vertical beamwidth enables coverage across several floors of a building. All measurements using the outdoor small cell were made with the LTE transmit diversity mode or transmission mode 2 (TM2) [3–5], which can yield a maximum measurable throughput over-the-air around 30 Mbps (single data stream) over a 10 MHz channel.

For indoor-to-indoor measurements we used a less directive pico cell pattern [6]. The antenna pattern has a directivity of 7 dBi and has a 60° horizontal and 62° vertical half power beamwidth. All measurements with the pico cells were made using the LTE closed

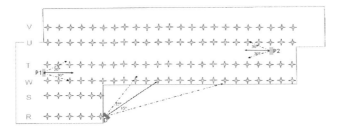

Figure 14.4 Grid of measurement points and small cell locations. The outdoor small cell is located at D. The pico cells are located indoors at *P1* and *P2*.

loop spatial multiplexing mode or Transmission mode 4 (TM4) [3–5], which can yield a maximum measurable throughput over-the-air around 60 Mbps (two data streams) over a 10 MHz channel.

14.2.1 Outdoor–indoor and indoor–indoor LTE small cell trials in Building A

Figure 14.4 shows the grid of measurement points in Building A, where KPIs are collected along with the small cell locations on the middle floor. We grouped the measurement points as rows labeled with the letters *R, S, W, T, U,* and *V*. The rows *R* and *S* are close to the outdoor small cell located at the point labeled with *D*, though they are not in the main beam direction. Most of the measurement points on row *W* are near the windows. The rows *V* and *U* are the furthest away in the interior of the building. There are up to 23 measurement points on each row. The straight line at point *D* shows the 0° beam direction of the small cell. The directions of the beams resulting from electronic beam-steering by +/–30° of the 0° beam in the azimuth plane are also represented by the dashed lines coming out of the location *D*.

The pico cells are located at the points *P1* and *P2*, attached to the ceiling, with antenna orientation pointing at each other. The corresponding arrows indicate the main directions of the pico cells' antennas.

Two-dimensional (2D) maps visualize the variation of KPIs of interest over an entire area and help identifying potential zones of a building that need improvement in terms of coverage and data rates. The 2D maps for Building A are obtained by interpolating the KPIs collected at the measurement points shown in Figure 14.4.

14.2.1.1 Coverage and data rates using the outdoor small cell

Figure 14.5 shows the 2D maps for RSRP, SNR, and PDSCH data rate maps for the 0° beam of the outdoor small cell. The total transmit power was set to 125 mW. Although the transmit power is quite small, the significant small cell antenna gain enables penetration of the signals into the building and yields good coverage over a large area of the building. We measured RSRP values up to –68 dBm near windows. The maximum possible downlink (DL) data rate of 30 Mbps was achieved at measurement points falling within the span of the 0° beam. Further, we were able to ensure data rates higher than 15 Mbps for 85% of receiver locations. The lowest measured data rate was

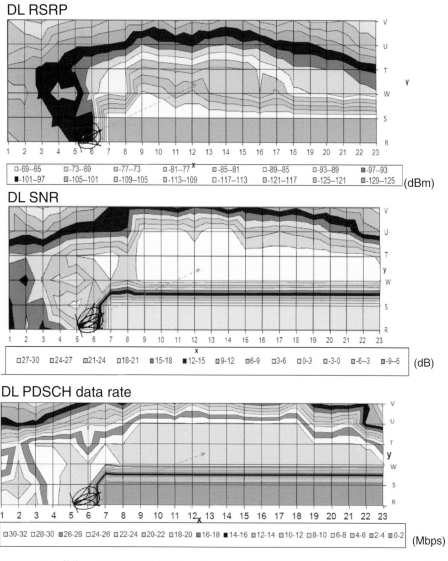

Figure 14.5 Building A: 2D RSRP, SNR, and measured PDSCH data rate maps using the outdoor small cell (0° beam).

7.4 Mbps, corresponding to the right uppermost corner of the building. Higher data rates would be possible through spatial multiple input multiple output (MIMO), which allows for multiple data streams to be delivered simultaneously.

The SNR shown in the 2D map is capped at 29 dB. The SNR was above 20 dB at 52% of the locations. Only 5% of measured locations have a SNR below 7 dB. Outdoor–indoor penetration depends highly on the type of the windows and exterior wall. Building A has regular windows without any metallic shielding, which was favorable for outdoor–indoor propagation. On the other hand, its exterior wall is made of a thick concrete layer and the floors were filled with lossy office furniture and metallic equipment on the shelves. To quantify the effective losses through the exterior we have placed the UEs outdoors, 30 cm away from the window at 140 cm height (this is the level that corresponds to the base of the window on this floor). In addition, we have also placed UEs indoors, 20 cm away from the exterior wall, on a table at 80 cm height (40 cm below the base of the window). The outdoor locations were in line of sight of the outdoor antenna. We repeated this experiment at measurement points parallel to row W in Figure 14.4, which are associated with six adjacent windows within the span of the 0° beam. The windows are around 6 m apart from each other. The difference between measured outdoor and indoor RSRP values in these experiments varied between 10 dB and 21 dB. This corresponds to the penetration loss caused by the front wall of this building.

14.2.1.2 Small cell footprint within the building

Figure 14.6 displays the measured received signal strength indicator (RSSI) values for the beams directed at three different azimuths [30°, 0°, and −30°], each active one at a time. For each azimuth value we see that the coverage footprint within the building is consistent with the orientation and directivity characteristics of the antenna, yielding the low angular spread at the transmitter side. This is an important result, which can be further exploited by enabling multiple simultaneous active beams with narrow horizontal beamwidths, without causing severe interference to each other.

14.2.1.3 Indoor small cell measurements

Pico cell transmit power was set to 250 mW for these experiments with the indoor small cells. Figure 14.7 shows the RSRP, SNR, and DL PDSCH data rate for the scenario where only the pico cell at *P1* marked in Figure 14.4 is active. With a single pico cell the transmitted signals penetrate well up to the measurement column 16. The cell edge forms on the opposite side of the building. In Figure 14.8, both pico cells located at *P1* and *P2* are active. This improves the RSRP on the side of the building close to *P2*, but reduces the SNR and PDSCH data rates in the middle of the building.

Figure 14.9 shows the comparison of the cumulative distribution function (CDF) of PDSCH data rates between one pico cell (only *P1* active) and two pico cell deployments (both *P1* and *P2* are active) in Building A. The CDFs we show in this chapter are the empirical CDFs of collected KPIs. Let **x** be the vector including the N collected KPI values with the ith element denoted as x_i. The empirical CDF $F(x)$ is defined as the proportion of x_i values less than or equal to **x**.

Although RSRP distribution is improved by adding the second small cell at *P2*, as shown in Figure 14.4, there are no data rate gains in comparison to the single pico cell scenario due to the reduced SNR in the middle of the building. On the other hand, each small cell is an access point, serving users in its proximity. Through these experiments, one can interpret that the small cells' densification is possible.

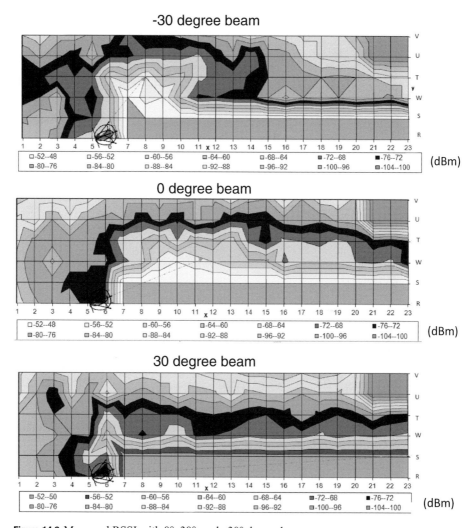

Figure 14.6 Measured RSSI with 0°, 30°, and –30° degree beams.

14.2.2 Outdoor–indoor LTE small cell trials in Building B

We performed measurements on every floor of Building B while walking along the corridors parallel to the building edges. The outdoor small cell transmit power was set to 250 mW. Figure 14.10a shows the 2D RSRP maps on each floor along the measurement trajectory. Similarly, Figure 14.10b displays the 2D PDSCH data rate maps for the same set of measurement locations. For some floors we were able to maintain wireless connectivity while penetrating deep inside the building. For some other floors, measurements were not possible when the received signal was too much attenuated through propagation losses. The highest RSRP and data rates were measured along the orientation of the beam. We also noticed high RSRP and data rates values

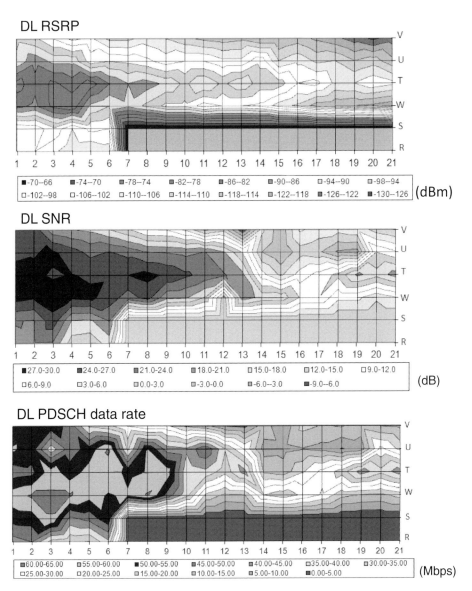

Figure 14.7 2D RSRP, SNR, and measured PDSCH data rate maps with only the pico cell at P1 active.

on the side of the measurement building that falls outside the beam footprint. This phenomenon is primarily attributed to rays' reflections from the building in front of the measurement building (canyon effect), which creates indirect radio coverage. This building acts as a reflector, causing canyon effects; it induces indirect radio coverage, improving significantly radio conditions in the proximity of the window area of the target measurement building in particular.

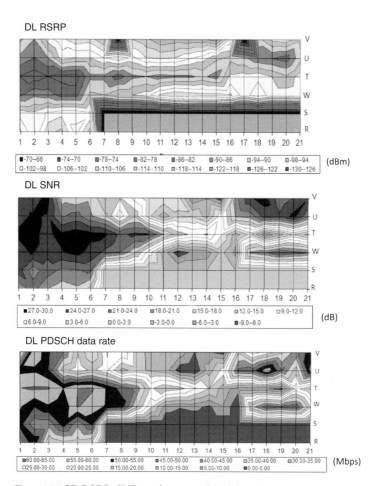

Figure 14.8 2D RSRP, SNR, and measured PDSCH data rate maps with both pico cells at P1 and P2 active.

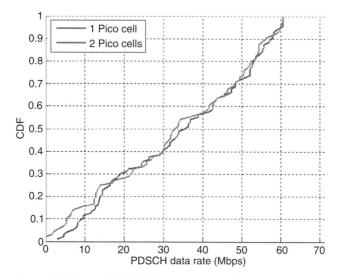

Figure 14.9 Measured PDSCH data rate comparison between single pico cell and two pico cell deployments.

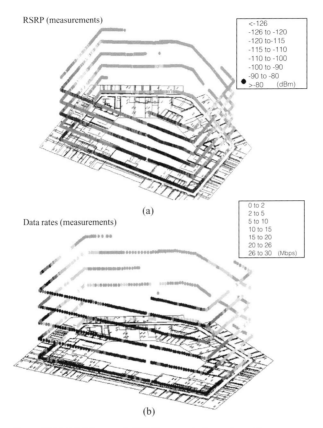

Figure 14.10 (a) Measured RSRP on each floor along the measurement path; (b) measured PDSCH data rate on each floor along the measurement path.

This observation is well supported through experiments consisting of comparing the coverage with and without the reflector building, and in varying the reflection properties of the reflector building.

Figure 14.11 shows the average RSRP per 0.05 dB log-distance bin on each floor.

Here we only included the points on the corridors close to windows. d_0 is the reference distance set to 1 m, while d is the distance between the transmitter and the receiver. The highest RSRP values are measured on the first floor, although the outdoor small cell is placed at the height of the fourth floor. The first two floors' windows are known to be much less lossy, as also noted by previous radio frequency (RF) characterization campaigns in the same building.

The different size and material used for windows are the cause for this variation among measurements per floor. Especially with the architectural choices and various materials used in the construction of modern buildings, such variations are more likely to occur. In measurement campaigns for older, more uniformly structured buildings, we did not see such counter-intuitive performance variations. The performance variation observed in

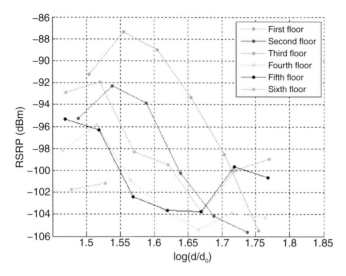

Figure 14.11 Average RSRP versus log distance per floor.

this building is in contrast with commonly assumed stochastic models, which considers uniform penetration loss per floor.

14.3 LTE performance prediction framework validation with measurements

3D ray-tracing based techniques simulate the physical environment and use wave propagation physics to predict the radio signal produced at any receive location from any transmit location. They also account for transmission through walls and diffraction around walls. When the physical environment is well specified, for example the layouts and materials of walls, floors, and ceilings are known, 3D environment simulations can be deployed on a very large scale and with limited effort.

Our experience has indicated that after an initial training and calibration phase with measurements, 3D environment simulations can be used with a high degree of confidence in complement of measurements, which are known to be labor-intensive and costly [7]. When measurements are not available for whatever reason, 3D environment simulations can be used instead, as an efficient tool to understand performance of wireless technologies, and in particular to get a good sense of how some specific product features, such as power levels and/or antenna characteristics, would impact product performance [8, 9]. They can also be used to assist RF network design, and to derive rules of thumb for wireless products deployment in real environments. Their prediction accuracy can be only as good as the environment modeling assumptions are.

Ray tracing is already established as a promising environment emulator for studying channel characteristics in urban environments. There are several studies confirming ray-tracing predictions with experimental results. Ray-tracing predictions in urban environments have been favorably compared to field measurements for received power, RMS delay spread, cross-pol behavior, angle of arrival, and link performance in many studies conducted by independent research groups e.g., [10–14].

The ray-tracing tool we use is WiSE (for wireless system engineering), an environment simulator developed by Bell Labs [1]. WiSE takes into account the geometry of buildings, type of materials, angle of incidence, dielectric coefficients, wall thickness, frequency, and 3D antenna patterns. It has been validated for both indoor and outdoor environments in previous studies [14–17]. The distributions of arrival times and angular spreads generated with WiSE agree quite well with those of an empirical model based on measurements in [17].

Given a building plan and transmitter and receiver locations, WiSE measures radio-signal performance at any point in the building. WiSE simulates the ray traces at the receiver both in elevation and azimuth plane. For each received ray, WiSE computes the angle of arrival (AOA), the delay, and the power. The power of a particular ray is given by:

$$P_R = G_t G_R \left(\frac{\lambda}{2\pi d}\right)^2 \prod_{m=1}^{M} t_m \prod_{k=1}^{M} r_k P_T \qquad (14.1)$$

where d is the unwrapped distance traveled by the ray, G_t is the transmit antenna gain, G_r is the received antenna gain, P_t is the transmit power, and λ is the wavelength. We assume the ray undergoes M transmissions and K reflections before reaching the receiver; t_m denotes the transmission coefficient for the mth transmission; and r_k denotes the reflection coefficient for the k_{th} reflection. The reflection and transmission coefficients are computed by each interaction with a wall taking into account the frequency and angle of incidence. We forgo in the scope of the chapter explaining further details of the WiSE ray-tracing tool. For more details on WiSE ray-tracing algorithm we refer the readers to the reference [1]. We are aware of other ray-tracing emulators for indoor–indoor or outdoor–indoor simulations (e.g., [18]) available commercially but we did not compare their performance with WiSE, as our main goal did not consist in a cross-ray tracing tools performance comparison.

We modeled the buildings where the field trials have been held. We had the exact detailed information on the geometry of the buildings, and some prior information on the electromagnetic properties of the materials based on propagation loss studies conducted before in the same buildings. We set these parameters after some trial-error-steps, where the simulations are compared against the field measurements and adequate tuning of in-building propagation losses was made to match experimental results.

For prediction we used the same 3D antenna patterns and power levels as in the field trial experiments. During measurements we recorded the measured metrics at specific locations. We exported these locations to WiSE and predicted the corresponding RSRP and data rates. The validation consists in comparing the measured RSRP and

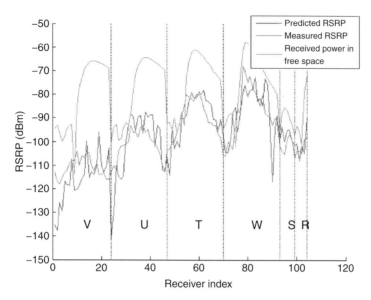

Figure 4.12 Building A: comparison between measured/predicted RSRP and the received power in free space.

measured data rate against corresponding predicted values, through a point-by-point comparison.

Figure 14.12 shows the predicted versus measured RSRP on the rows R, S, W, T, and U in Building A. The dark line corresponds to the predicted RSRP without any losses through the building, which corresponds to the free space propagation. We determined it by removing the building from the simulation set-up, but keeping all other parameters unchanged. Comparing the free space with the measured RSRP we observe that the measurement pattern follows the antenna directivity characteristics. This is an indicator of low angular spread, which encourages the use of directive antennas for outdoor small cell deployments.

Figure 14.13 shows a direct comparison of measured and predicted values for RSRP and PDSCH data rate along the measurement path on the fourth floor of Building B. We have observed similar matches between predicted and measured values on the other floors of the building.

Figure 14.14a illustrates the comparison of the measured and predicted RSRP CDFs for all of the measurement locations in the whole building. Noticeably, the measured RSRP values are higher than the predicted ones when the RSRP signal is very weak. This is due to the measurement noise, which is not captured via predictions. A correction factor may be introduced in the predictions to account for the noise effect at low RSRP values. In these statistics we have included the locations with lower RSRP values, which would normally be served by other small cells through a complete wireless network deployment.

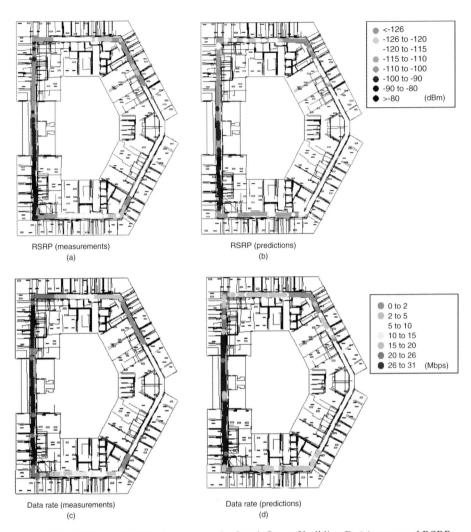

Figure 14.13 RSRP and PDSCH data rate on the fourth floor of building B: (a) measured RSRP; (b) predicted RSRP; (c) measured PDSCH data rate; and (d) predicted PDSCH data rate.

Figure 14.14b indicates that the predicted PDSCH data rate is very close to the measured one. We measured indoor data rates up to the maximum achievable of 30 Mbps (over a 1 MHz LTE channel), through the LTE transmission mode TM2. We were able to ensure that 60% of the receiver locations within the building can be served at more than 10 Mbps with only one small cell. The predicted data rate is estimated by mapping the predicted SNR values to data rates.

Overall, the power predictions agree well with measurements. The agreement is encouraging: 3D prediction analysis can be used as an alternative to or simply in supplement of labor-intensive measurements. As an example, Figure 14.15 shows the predicted RSRP on the entire fourth floor of the measurement building, after calibrating the tool with the RSRP values on the measurement path.

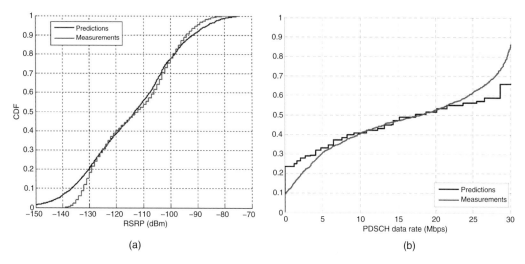

Figure 14.14 Building B: (a) CDF comparison of measured and predicted RSRP values; (b) CDF comparison of measured and predicted PDSCH data rates.

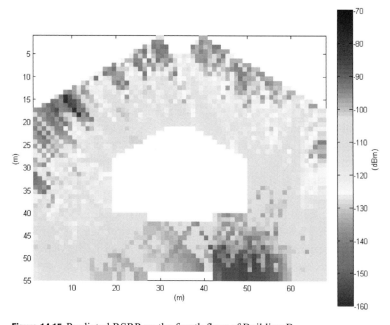

Figure 14.15 Predicted RSRP on the fourth floor of Building B.

In the course of our investigations, we realized that the main reason that compromises prediction accuracy is the environment model, as it is subject to material properties of walls, floors, and ceilings not well specified. One can use a set of preliminary measurements to adjust the material propagation properties and fine tune the prediction model. This technique has the advantage of minimizing the prediction errors.

14.4 High density small cells' design for stadiums using the LTE performance prediction framework

Offering good QoE in stadiums poses unprecedented challenges to wireless operators, due to extreme traffic conditions. During popular sporting events, there could be tens of thousands of active users packed into a relatively small area, sharing pictures and video clips through smart phone applications. This creates high traffic densities and drives requirements for high capacity, and yet economically feasible solutions for stadiums. The main issue for an ultra-hotspot traffic stadium is providing sufficient capacity for peak traffic conditions. One must be able to deploy many sites within a short distance of each other and yet to robustly mitigate the resulting interference. Therefore determining the number of required small cells, their desirable location, desirable antenna patterns and their orientation, and desirable power levels – all subject to a given traffic volume and to a target data rate per subscriber – are key design criteria.

Currently, wireless operators are using a combination of several solutions to tackle this problem: distributed antenna systems (DAS), traditional macrocellular networks, WiFi, and LTE small cells. These technologies are often used in complement of each other to increase the total available capacity in hotspot traffic scenarios. DAS is a network of spatially separated antenna nodes connected to a common source via a transport medium that provides wireless service within a geographic area or structure. The DAS system is still constrained to the capacity offered by a macro base station sector, which limits its potential to scale up with heavy traffic demands. To increase the capacity of DAS systems, new macro sectors are needed to be deployed. An analysis on comparison of DAS and LTE small cell deployments in a stadium could be found in [9].

In this chapter, we investigate the ability of low-power small cells to respond to heavy traffic demands in stadiums using the LTE performance prediction framework. The analysis solely addresses the seating area of the stadium. We show that the outdoor seating area could be served with small cells radiating power levels as low as 10 mW, which makes the interference from the outdoor small cells (deployed in the outdoor seating area) to the small cells deployed in indoor concession area negligible. Therefore small cell deployments for interior concession area and outdoor seating area of the stadium can be planned independently, as long as appropriate power levels are selected. For the interested reader, we refer to the reference [9] for a detailed coverage and capacity evaluation over the outdoor seating area as well as over the indoor concession area of the stadium.

14.4.1 3D stadium model

It is particularly important to have a proper characterization for a stadium environment, including walls, construction materials, and corresponding propagation characteristics, which allow to account for modeling of realistic propagation conditions. We have taken as a reference a large football stadium and have constructed a realistic 3D stadium model in our ray-tracing framework, as represented in Figure 14.16. The stadium has

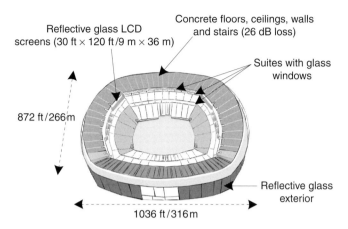

Figure 14.16 Simulated 3D stadium model.

the following dimensions: 316 m length, 266 m width, and 52 m height. The exterior of the stadium is covered with reflective glass, which has a 5 dB propagation loss at 2.1 GHz. Floors, ceilings, walls, and stairs are built of concrete, which introduces a 26 dB propagation loss at 2.1 GHz. There are three concourses, with corresponding outdoor seating areas. There are suites with glass windows, which are also modeled, as well as a big reflective glass screen (approximate dimensions 10 m × 40 m), which impacts the propagation conditions.

14.4.2 3D antenna patterns for stadiums

Two types of antenna patterns have been considered throughout the analysis. Their characteristics for the 2.1 GHz carrier frequency are described below.

- **A broad antenna pattern:** the horizontal and vertical patterns are illustrated in Figures 14.17a and 14.17b, respectively. It has a directivity of 7.7 dBi and around 80° horizontal and vertical half power beamwidth.
- **A narrower and more directive antenna pattern:** the horizontal and vertical patterns are illustrated in Figures 14.17c and 14.17d, respectively. It has a directivity of 11 dBi and 24.6° horizontal and 95° vertical half power beamwidth.

14.4.3 Small cell design for stadiums

Several key small cells' deployment criteria have been used together to respond to the high capacity demand in the stadium.

- Several tiers (rings parallel to the circumference of the stadium) of small cells are used for a typical multi-concourse stadium. In this particular case, there are three tiers, one per concourse. The main idea is that each concourse has its own dedicated tier of small cells.

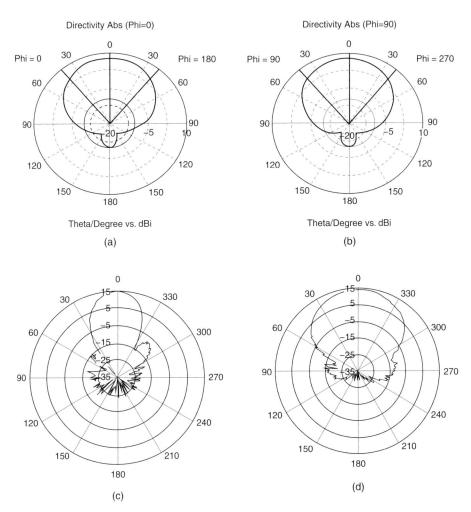

Figure 14.17 Small cells' antenna patterns. (a) Broad antenna – horizontal plane (7.7 dBi gain, 79.3° horizontal half power beamwidth); (b) broad antenna – vertical plane (7.7 dBi gain, 79.2° vertical half power beamwidth); (c) directive antenna – horizontal plane (14.2 dBi gain, 25° horizontal half power beamwidth); (d) directive antenna – vertical plane (14.2 dBi gain, 50° vertical half power beamwidth).

- Small cells' antennas are deployed facing straight down (90° downtilt) along each tier to minimize interference. This technique is used to limit the cross-site interference along the same tier and between adjacent tiers of small cells (referred to in traditional stadium DAS deployments as cross-tier interference leaking).
- The small cells are packed into each tier of the stadium like spokes in a wheel. This is achieved by pointing the main beam of the antenna along the radius (the line between the location of the small cell and the center of the stadium). For simplicity, we assumed a uniform traffic distribution across the stadium. Such a design creates a coordinated antenna orientation across sites and minimizes cross-site interference. Non-uniform

traffic distribution can be accommodated, and the design can be tailored to adapt to such conditions.

14.4.4 Simulation scenarios

Small cells in these scenarios operate in LTE frequency division duplexing (FDD) mode. This assumes paired spectrum for uplink and downlink directions. The analysis focuses in downlink and assumes 10 MHz of bandwidth and a deployment at 2.1 GHz.

Deployment scenarios considered in the analysis are for various RF-carrier frequency reuse, namely frequency reuse 1 and frequency reuse 2. Further frequency reuse factors are possible in LTE, but are not considered in this analysis. In the case of a frequency reuse 1 deployment, each small cell operates over the entire bandwidth of 10 MHz available for downlink transmissions (single frequency, f1, deployed on each site). In the case of a frequency reuse 2 deployment, there are two RF carriers of 5 MHz bandwidth each, deployed alternately across adjacent sites on the same tier (two frequencies, $f1$ and $f2$, deployed alternately following the pattern $f1, f2, f1, f2$, across sites on the same tier). Another simple and efficient way to limit the cross-site interference is by employing a time-division access to the entire available frequency spectrum (10 MHz) across sites. This allows each site to get full access to the entire spectrum for a fraction of time, with a suitable duty cycle. For instance, half time access can be implemented across adjacent sites by reinforcing a pattern $t1, t2, t1, t2$, across sites on the same tier, with $t1 = t2 = T/2$, where T is the desirable duty cycle. This option provides a high degree of design flexibility to control interference and to adapt to various traffic conditions.

Several power levels have been simulated assuming no presence of macro cells' signals. Note that lessons learnt from DAS deployment indicate that it is desirable to direct away the polluting macro cells (interfering with the DAS infrastructure) as much as possible from the stadium. We would advise to apply this learning from the field to stadium small cells' deployments as well. The immediate remark is that such an open environment does not need a lot of power to ensure radio coverage. Certainly, the need for high capacity drives the requirement for many small cells in the stadium. With many small cells, we have noticed high levels of interference despite usage of very small power transmitters. We found out that a power level as low as 10 mW per transmitter is sufficient for this environment. However, one should note that, depending on the potential levels of interference from the macro cells nearby, the desirable small cell power levels may be adjusted accordingly to compensate the pollution levels.

14.4.5 Stadium coverage

This section describes a set of results, in the form of coverage maps. Figures 14.18a and 14.18b illustrate the coverage maps for a deployment of 62 LTE small cells with frequency reuse 1 and frequency reuse 2, respectively. Small cells used in these scenarios are equipped with broad antennas (antenna diagrams in Figures 14.17a and 14.17b). Such broader antenna patterns are found appropriate at lower site density.

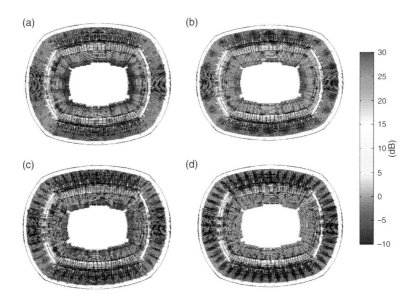

Figure 14.18 SINR with LTE small cells' stadium deployment. (a) 62 small cells frequency reuse 1; (b) 62 small cells frequency reuse 2; (c) 124 small cells frequency reuse 1; (d) 124 small cells frequency reuse 2.

It is obvious in both cases that the area in the proximity of each small cell enjoys a much better quality than the rest of the area. Indeed, such a small cells' deployment creates zones of high "local" capacity. Further improvement in radio quality conditions is also well visible through a frequency reuse 2 deployment.

Figures 14.18c and 14.18d illustrate the coverage maps for a deployment of 124 LTE small cells with frequency reuse 1 and frequency reuse 2, respectively. Small cells used in these scenarios are equipped with directive antennas (antenna diagrams in Figures 14.17c and 14.17d). Such narrower antenna patterns are appropriate at higher site density. Indeed, we have noticed that following design criteria introduced in the previous section, a directive antenna provides higher protection to cross-site interference compared to a less directive antenna.

Similar to the 62 small cells' deployment case, it is also visible in both cases that the area in the proximity of each small cell enjoys much better quality than the rest of the area. The further improvement in radio quality conditions is visible for a frequency reuse 2 deployment in this case as well. Noticeable, by doubling the number of small cells, the zones of high "local" capacity in the proximity of small cell locations are well preserved, thanks to the narrower antenna pattern, which creates a visible narrower coverage footprint.

14.4.6 Small cells' stadium capacity and user application layer throughput

We introduce a framework to predict the QoE per user per application over the stadium; the distribution of user application layer throughput is compared for different deployment

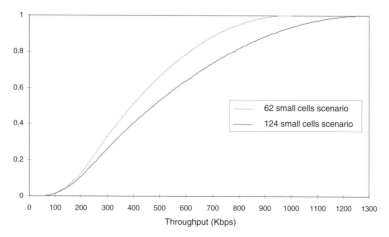

Figure 14.19 CDF comparison of the FTP download application layer throughput per user.

scenarios. The results were obtained using an end-to-end LTE application performance simulation, since the ultimate goal for a quality system is to ensure that the end users experience a good quality of service. An eNodeB scheduler was embedded into an OPNET [19] simulation platform. LTE transmission mode 2 (TM2) is selected at lower SINR while LTE transmission mode 3 (TM3) is selected at higher SINR [3]. A mix of applications was modeled on top, allowing each user to generate several simultaneous applications.

The geometry of users obtained from the 3D propagation models for the 62 and 124 small cells' deployment scenarios are used in the end-to-end LTE application performance simulator, which includes TCP/IP and various application protocols, such as file transfer protocol (FTP).

The following configurations are used in this analysis:

- 62 LTE small cells, each employing broad antenna
- 124 LTE small cells, each employing directive antenna
- Frequency reuse 2, since it shows an overall improvement in radio quality conditions over the frequency reuse 1 deployment
- 2 MB file download (FTP)
- 20 simultaneous active users connected to each small cell

Figure 14.19 displays the FTP download throughput CDF for both 62 and 124 small cells' deployment scenarios. It shows that doubling the number of small cells, while narrowing the antenna patterns (to increase the number of supporting users in the stadium), does not degrade the user level application performance, although the distance between adjacent small cells is halved by migrating from 62 to 124 small cells. In fact, the average throughput per user is even increased by 15.7% (486 kbps, up from 420 kbps), and the entire range of the user throughput experience (including cell edge) is increased as follows.

- 5% CDF of user throughput: 141 kbps with 62 LTE small cells versus 149 kbps with 124 LTE small cells
- 50% CDF of user throughput: 395 kbps with 62 LTE small cells versus 477 kbps with 124 LTE small cells
- 95% CDF of user throughput: 802 kbps with 62 LTE small cells versus 1040 kbps with 124 LTE small cells
- Average user throughput: 420 kbps with 62 LTE small cells versus 486 kbps with 124 LTE small cells

This indicates good capacity scalability of the small cells, with the increased number of users. Saturation in user throughput is expected at some point with further small cells' densification.

14.5 Summary

We disseminated a set of small cells' field trial experiments conducted at 2.6 GHz and focused on coverage/capacity within multi-floor office buildings. LTE pico cells deployed indoors as well as LTE small cells deployed outdoors are considered. The latter rely on small emission power levels coupled with intelligent ways of generating transmission beams with various directivity levels by means of adaptive antenna arrays. With a single small cell, we were able to establish good indoor coverage and to deliver high data rates for large parts of a mid-size six-floor office building. Using a 250 mW total transmit power for a small cell equipped with a directive antenna, and while operating over a 10 MHz LTE channel at 2.6 GHz in dedicated carrier and enabling the LTE transmit diversity mode (TM2), we were able to ensure throughputs higher than 10 Mbps for 60% of receiver locations within the building. The feasibility of such small cells' deployments is in line with other LTE measurement campaigns associated with other buildings.

Furthermore, we introduced an analytical 3D performance prediction framework, which we calibrated and validated against field measurements. The framework provides detailed performance levels at any point of interest within a building; it allows us to determine the minimum number of small cells required to deliver desirable coverage and capacity levels, their most desirable location subject to deployment constraints, transmission power levels, antenna characteristics (beam shapes), and antenna orientation (azimuth, tilt) to serve a targeted geographical area. Our experience has indicated that after an initial training and calibration phase with measurements, environment simulations can be used with a high degree of confidence in complement of measurements, which are known to be labor-intensive and costly. The good match between measurements and predictions encourages the use of the 3D performance prediction framework, in complement to field measurements, to support small cell deployments.

In addition, we showed specialized solutions for LTE small cells' deployment within hotspot traffic venues such as stadiums, through a deployment feasibility analysis conducted at 2.1 GHz. Via end-to-end LTE application performance simulations, we also demonstrated scalability of capacity based on the end user perceived QoE. A system configured with 62 LTE small cells, frequency reuse 2, and broad antenna can deliver an average 420 kbps per user for 20 simultaneous active users per small cell (1,240 simultaneously active users over the entire outdoor seating area). On the other hand, a system configured with 124 LTE small cells, frequency reuse 2, and directive antenna can deliver an average 486 kbps per user for 20 simultaneous active users per small cell (2,480 simultaneously active users over the entire outdoor seating area).

Through analysis and design choices presented in this chapter we demonstrated that the small cells could offer scalable solution in response to the high capacity requirements for stadiums and sports venues in general.

Acknowledgment

We would like to thank our colleagues Florian Pivit and Denis Rouffet for their contributions to the early definition of the small cell products. We would like to thank also our colleagues Titos Kokkinos and Jean-Pierre Harel for their support with antenna patterns.

References

[1] S. Fortune, D. Gay, B. Kernighan, O. Landronand, R. Valenzuela, and M. Wright, "WISE design of indoor wireless systems: practical computation and optimization," *IEEE Computational Science and Engineering*, vol. 2, no. 1, pp. 58–68, 1995.

[2] 3GPP, "Evolved universal terrestrial radio access (E-UTRA); physical layer: measurements," 3rd Generation Partnership Project (3GPP), TS 36.214, Dec. 2009. www.3gpp.org/ftp/Specs/html-info/36214.htm.

[3] S. Sessia, S. Toufik, and M. Baker, *LTE: The UMTS Long Term Evolution: FromTheory to Practice: 1st Edition*. John Wiley & Sons, Incorporated, 1987.

[4] 3GPP, "Evolved universal terrestrial radio access (E-UTRA); physical layer: procedures," 3rd Generation Partnership Project (3GPP), TS 36.213, Dec. 2009. www.3gpp.org/ftp/Specs/html-info/36213.htm.

[5] 3GPP, "Evolved universal terrestrial radio access (E-UTRA); long term evolution (LTE) physical layer; general description," 3rd Generation Partnership Project (3GPP), TS 36.201, Dec. 2009. www.3gpp.org/ftp/Specs/html-info/36201.htm.

[6] "Kathrein antenna type: 80010677," www.kathrein.de.

[7] D. Calin, A. O. Kaya, A. Abouliatim, G. Ferrada, P. Richard, and A. Segura, "On the feasibility of outdoor-to-indoor LTE small cell deployments: field trial experiments and performance prediction," *IEEE GLOBECOM*, 2013.

[8] A. O. Kaya and D. Calin, "Modeling three dimensional 3D channel characteristics in outdoor-to-indoor LTE small cell environments," *IEEE MILCOM*, 2013.

[9] D. Calin, A. O. Kaya, B. Kim, K. Yang, and S. Yiu, "On the high capacity light-radio metro cell design for stadiums," *Bell Labs Technical Journal*, vol. 18, no. 2, pp. 77–97, 2013.

[10] S.-C. Kim, J. Guarino, B. J. Willis, *et al.* "Radio propagation measurements and prediction using three-dimensional ray tracing in urban environments at 908 MHz and 1.9 GHz," *Vehicular Technology, IEEE Transactions*, vol. 48, no. 3, pp. 931–946, May 1999.

[11] V. Erceg, S. Fortune, J. Ling, J. Rustako, and R. Valenzuela, "Comparisons of a computer-based propagation prediction tool with experimental data collected in urban microcellular environments," *Selected Areas in Communications, IEEE Journal*, vol. 15, no. 4, pp. 677–684, May 1997.

[12] G. Athanasiadou, A. Nix, and J. McGeehan, "A microcellular ray-tracing propagation model and evaluation of its narrow-band and wide-band predictions," *Selected Areas in Communications, IEEE Journal*, vol. 18, no. 3, pp. 322–335, March 2000.

[13] C. Oestges, B. Clerckx, L. Raynaud, and D. Vanhoenacker-Janvier, "Deterministic channel modeling and performance simulation of microcellular wide-band communication systems," *Vehicular Technology, IEEE Transactions*, vol. 51, no. 6, pp. 1422–1430, Nov. 2002.

[14] A. O. Kaya, L. J. Greenstein, and W. Trappe, "Characterizing indoor wireless channels via ray tracing combined with stochastic modeling," *Wireless Communications, IEEE Transactions*, vol. 8, no. 8, pp. 4165–4175, August 2009.

[15] R. Valenzuela, D. Chizhik, and J. Ling, "Measured and predicted correlation between local average power and small scale fading in indoor wireless communication channels," *Vehicular Technology Conference, 1998. VTC 98. 48th IEEE*, vol. 3, pp. 2104–2108, May 1998.

[16] V. Erceg, S. Fortune, J. Ling, A. J. Rustako, and R. Valenzuela, "Comparisons of a computer-based propagation prediction tool with experimental data collected in urban microcellular environments," *Selected Areas in Communications, IEEE Journal*, vol. 15, no. 4, pp. 677–684, May 1997.

[17] G. German, Q. Spencer, L. Swindlehurst, and R. Valenzuela, "Wireless indoor channel modeling: statistical agreement of ray tracing simulations and channel sounding measurements," *Proc. IEEE International Conference on Acoustics, Speech, and Signal Processing*, 2001, pp. 2501–2504.

[18] "Ibwave design," www.ibwave.com.

[19] "OPNET technologies," www.opnet.com.

15 Centralized self-optimization of interference management in LTE-A HetNets

Yasir Khan, Berna Sayrac, and Eric Moulines

In this chapter we address interference mitigation in LTE-A co-channel Het-Net deployments, a major issue for reaching substantial capacity enhancements. We provide an extensive literature survey of the existing co-channel interference mitigation methods involving optimization of the related network parameters, resulting in corresponding improvements in quality of service (QoS). Since the number of base stations (macro, micro, pico, and femto) increases considerably in a HetNet deployment, optimization of network parameters with such a high number of nodes becomes complex and costly, calling for the inevitable need for self-optimization. We propose a self-optimization framework based on efficient statistical modeling combined with robust sequential optimization, which is very suitable for implementation in a centralized manner at the operator's management plane. In particular, the proposed methodology is based on the following two concepts, which will be described in detail.

1. Modeling of the functional relationships between network parameters and relevant key performance indicators (KPIs) by using a statistical modeling technique called *Kriging*.
2. Optimization of the KPIs through these statistical relationships by using a pattern search algorithm.

The proposed methodology is applied to two interference mitigation techniques, namely enhanced inter-cell interference coordination and active antenna systems-based interference mitigation, and their comparative performances are analyzed.

15.1 Introduction

Mobile usage and data traffic is increasing globally at a remarkably fast rate. Ericsson reported data traffic to almost have doubled between 2012 and 2013 and the industry is preparing for a 1,000 times increase by 2020 [1]. Given the rise in traffic demand, the industry is looking for solutions to challenges both in technological as well as in managerial aspects of a network.

Until recently, a part of this challenge was addressed by increasing the number of sites that could be deployed in a given area and on a given frequency. However, the rise in interference has prevented further densification in a given radio access network (RAN). Furthermore, installations of new base stations transmitting on high power has become a societal issue. Of course, the bandwidth can be expanded by adding additional spectrum, but this solution again involves economical factors apart from the issue of limitations in the amount of bandwidth available for bracing a challenge of 1,000 times traffic rise.

One of the solutions increasingly adopted by operators is the co-channel deployment of macro and low-power nodes (typically pico cells). However, in such a heterogeneous (HetNet) deployment wherein all the layers (i.e., macro and pico layers) are operating on the same frequency, inter-layer interference limits drastically the capacity of the network. Frequency reuse scheme of interference management becomes less relevant due to the significant interference between the high-power macro transmissions and the low-power pico cells.

HetNets involve several co-existing cellular operations. With the increased number of nodes, the number of parameters needed to be optimized by the operators at the operations and maintenance center (OMC) increases drastically. For a long time, the problem of efficient optimization has been the primary concern for telecom operators. Increase in competition accompanied by user demands for lower subscription prices have resulted in operators attempting measures to cut down operational expenditure (OPEX). Savings in OPEX over longer time periods can greatly impact capital expenditures (CAPEX) as the infrastructure investments are delayed.

Historically operators continuously monitored and manually adjusted network parameters (NPs) at the OMC so as to bring about a desired quality of service (QoS) improvements in a specific part of their network. With the evolution in the complexity of the networks and automation solutions, concepts such as self-organizing networks (SON) have recently emerged in the literature, which promises automation of the optimization tasks thereby reducing OPEX and delaying CAPEX e.g., [2, 3].

This chapter presents some of the literature in this framework. More specifically, the chapter is a guide to engineers and researchers on how to build a framework for optimizing network based on QoS specific objectives. Among pioneering frameworks on centralized self-optimizing algorithms there are the works of [4] wherein the authors present a centralized load balancing algorithm based on changes to mobility parameters. In [5], the authors propose a centralized capacity and coverage optimization approach through adaptation of antenna system parameters using *case-based reasoning* (CBR) algorithm. In [6, 7] the authors define an iterative self-optimization methodology, which involves relating the network key performance indicators (KPIs) and the network parameters (NPs) using linear and logistic models followed by numeric optimization on the network function.

Our methodology and problem analysis is important from both practical and theoretical perspectives. From an engineering point of view, we propose a practical centralized self-optimization methodology for a given QoS objective. From a mathematical point of view, we model the network QoS objectives by Gaussian processes which have properties suitable for global active optimizations.

In this chapter we provide a detailed survey of: (1) interference management in HetNets, as well as (2) SBO techniques as a practical and effective means to put in practice centralized self-optimization of interference management in HetNets. As an example,

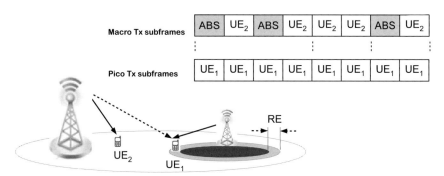

Figure 15.1 eICIC concept indicating the subframe synchronization between the macro and pico layers.

we propose an SBO-based centralized self-optimization solution for two interference mitigation techniques in a HetNet hotspot scenario. The surrogate based optimization (SBO) framework proposed in this chapter for a centralized self-optimization implementation is more precise and efficient than manual process and thus also guarantees improved QoS. With the evolution of mobile networks from a simple single radio access technology (RAT) to multi-RAT, multi-vendor, multi-technology, and highly complex entities, a simple yet efficient optimization framework is sought for implementation by the operators at the OMC level in a centralized manner.

This chapter is structured into the following parts. The first part involves a survey of various interference problems existing in a typical HetNet and also the various interference mitigation techniques. In the second part we provide an extensive literature survey on the Gaussian process, surrogate-based optimization (SBO) techniques and their applications in several fields of engineering. In the third part, we develop a generic SBO framework for application to centralized network optimization. The methodology involves an iterative SBO technique, which models functional KPI-NP relationships and further uses pattern search algorithms to optimize the modeled surrogate. Within the self-optimizing context for co-channel LTE-A HetNets, we apply our methodology to two scenarios. Our results indicate the effectiveness of the SBO algorithm and also demonstrate the superiority of enhanced inter-cell interference coordination (eICIC) for interference management (Figure 15.1).

15.2   Interference management in HetNets

With the rapid explosion in mobile data, operators have to find solutions to increase the network capacity. One of the straightforward solutions is to add more macros in the network. However, this process is expensive and complicated in a dense urban area due to the scarcity and costs of new site acquisitions. To increase capacity in regions with high mobile utilizations, a more cost effective and efficient model for the operators is to deploy low-power pico cells. Such concentrated deployments bring networks closer to the users thereby increasing the user QoS and also improving the overall network coverage and/or capacity. The macro base stations are typically transmitting at a higher power (5 W to 40 W) while the picos usually at a much lower level (100 mW to 2 W). If the macro and pico layers are both operating on the same carrier frequencies (co-channel deployment), the users in the pico are interfered by the high-power macro transmissions. Such a deployment increases the overall capacity but at a cost of degraded channel conditions for pico cell-edge users to a larger extent and the macro cell-edge users to a smaller extent due to lower transmission powers of the pico cells. In order to improve the pico cell-edge users' QoS while at the same time not compromising on the overall capacity improvement, several techniques for managing interference have been proposed. These techniques can be classified broadly along four categories given below.

1. Time domain techniques
2. Frequency domain techniques
3. Space domain techniques
4. Other techniques.

15.2.1 Time domain techniques

Such techniques are those wherein the pico cell-edge users are protected by scheduling the macro and pico users in time domain. This method known as time domain multiplexing eICIC typically involves complete/partial muting of the interfering layer subframes while scheduling/prioritizing the cell-edge users of the interfered layer for reducing the overall network interference [2, 8, 9]. Two of these techniques are discussed in sections below.

15.2.1.1 Pico cell range extension

To optimize system capacity, it is primordial to push more users of the macro to the more spectrally efficient pico cells. This is typically carried out by range extension (RE) for the pico eNBs. During the cell selection process, a user equipment (UE) gets connected to that cell which demonstrates a higher reference signal received power (RSRP) measurement. To increase the pico cell coverage area, a cell-specific positive bias (offset) is applied to UE pico measurements thereby *biasing* the UE attachments to the pico. The optimum bias for each pico cell is typically unknown and can be optimized depending on the traffic characteristics. The macro–pico cell selection dilemma can be given by:

$$Cell\ ID_{serving} = \arg\max_{i}[RSRP_i \pm RE_i]$$

where $RE_i = 0$ dBm for macro and $RE_i \geq 0$ dBm for pico. Although favoring pico selection increases the HetNet system capacity in most cases, too much bias results in pico cell-edge UEs experiencing higher interference from the macro layer and thus a reduced pico SINR. Thus a need arises to find efficient interference management techniques to improve the capacity of the cell-edge users.

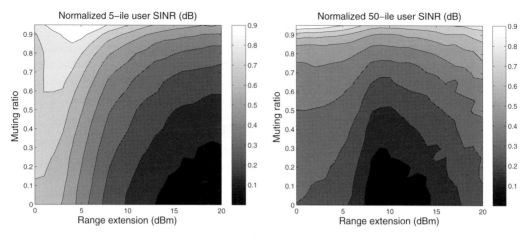

Figure 15.2 Surface plot of cell-edge and cell-center UE SINR.

15.2.1.2 Time domain (TDM) eICIC

During a co-channel operation, both the macro and pico transmit data at the same time. Higher transmission powers of the macro cells result in higher pico cell-edge UE interference, which is even more exaggerated by pico range extension. To protect the pico cell-edge UEs by providing interference mitigation, time domain (TDM) eICIC has been proposed [2, 8].

TDM eICIC involves restricting the macro layer subframe transmission to a fraction of the total subframes. During the fraction of subframes when the macro is not transmitting (muted), preference is given to the pico cell-edge UEs to transmit. The restricted macro subframes are called "almost blank subframes" (ABS) as the control signals necessary for maintaining UE connections are still transmitted during the *muting*. Strict time and phase synchronization and ABS pattern information exchange is ensured being made possible over the X2 interface between the pico and the macro cells employing TDM eICIC.

The influence of range extension and muting ratio on cell-edge SINR and mean SINR in a typical HetNet is as shown in Figure 15.2.

15.2.2 Frequency domain techniques

These techniques involve users' scheduling in the frequency domain i.e., the bandwidth is split into several disjoint sets, which are then assigned to individual layers such as the different types of inter-cell interference coordination (ICIC) schemes [10], dynamic frequency partitioning [11], fractional frequency reuse [12], and soft frequency reuse [13].

15.2.3 Space domain techniques

These techniques use inter-cell coordination for spatially multiplexing the users as in coordinated multi-point (CoMP) technique [14, 15], which involves interference

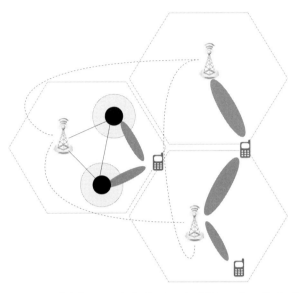

Figure 15.3 CoMP scenario indicating the X2 interface (dashed line) and the fiber backhaul (continuous line) for coordination.

cancellation through coordinated/joint-coordinated beamforming or joint processing/ joint transmission [16, 17].

15.2.3.1 Coordinated multi-point (CoMP)

CoMP, which is also called multi-cell multiple input multiple output (MIMO), is an inter-cell interference coordination technique wherein multiple base stations cooperate in a way to mitigate the inter-cell interference to serve multiple users simultaneously [18]. Intra-site CoMP algorithms (for example those operating between different sectors of the same base station) are faster as the information exchange is within a site. In contrast, inter-site CoMP involves coordination of multiple sites by exchange of information via the backhaul. Depending upon how the cells coordinate, CoMP can be classified into coordinated beamforming (CB)/coordinated scheduling (CS) and coordinated joint transmission (CJT)/coordinated joint processing (CJP). In CB/CS the base stations are equipped with multiple antennas and remote receivers and communications to multiple users of each cell are carried out simultaneously via a mutual dynamic coordination of beamforming/scheduling functionality between the cells. In CJP/CJT, transmission to a single user is carried out from multiple points. The user receives transmission data from geographically separated antennas (multi-points) coordinated as a single transmitter. However inter-site coordinated JP/JT involves a stringent requirement on backhaul capacity. Figure 15.3 shows a typical CoMP scenario indicating coordination over the X2 interface between the macro cells and the macro–pico fiber backhaul. In the literature, CoMP studies have been carried out for distributed antenna systems (DAS) scenario where cooperative transmission is used between distributed transmission points and relative advantages/disadvantages have been studied [19]. Studies have been carried out

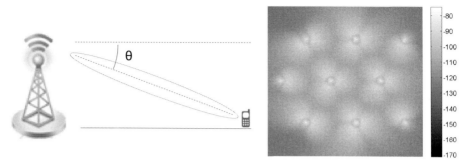

Figure 15.4 (a) AAS tilt angle in degrees; (b) spatial path-loss map in dB.

on CoMP deployment scenarios, practical implementation, and operational challenges. Investigations to study the performance of CoMP in terms of QoS with unreliable backhaul have been carried out and it was concluded that using a fully distributed architecture, the coherent joint transmission scheme was the most robust [20]. A performance comparison between CoMP and eICIC for a HetNet scenario has been investigated to provide the relative advantages/drawbacks of each [18, 21].

15.2.4 Other techniques

These involve optimization of transmission characteristics such as power control methods wherein the total transmit power of the cells is optimized to minimize interference [22–24] or antenna-tilt based methods wherein the characteristics of antenna beams are optimized such as those in active antenna systems (AAS) [25].

15.2.4.1 AAS tilt-based interference management

An AAS involves arrays of antenna elements each of which can be electronically phased to obtain a unique radiation pattern for the array (Figure 15.4). The advantage of the AAS lies in having the flexibility to electronically form the antenna beams by changing antenna tilt, azimuth, or beam shape, thus providing flexibility and control over the interference and the performance of the network in terms of coverage, capacity, and quality of service (QoS) [3, 25] for varying traffic characteristics. The 3rd Generation Partnership Project (3GPP) introduces capacity and coverage optimization (CCO) as an important use case for LTE and LTE-A networks [3] and antenna tilt is an important parameter that can be used to carry out CCO by reducing inter-cell interference [26, 27]. Traditionally, once deployed, optimization engineers used to recommend changes to the antenna tilts, which were changed by technicians mechanically, which thus required a site visit and utilization of costly human resource. Later, remote electrical tilts (RETs) were introduced, which involved electrically changing the antenna tilts via an actuator and thus moved the on-site human intervention out of the optimization loop [26–28]. Both the mechanical or electrical tilt involved beams that were always fixed to the antenna plane, requiring tilting of the entire antenna transmission assembly to achieve a given tilt objective and thus were not flexible in themselves. With the advent of AAS, operators now have a better control and a possibility of custom tilt, azimuth, in the number of beams that can be defined leading to sectorization. One important contrast from AAS is that in RET, the entire radiation plane including the side lobe moves with each tilt of the main beam, while in AAS the side lobe can be controlled dynamically. AAS involve built in redundancy and improved thermal performance and AAS based antenna beams can be adapted electronically if one or more of its antenna elements fails, thus providing an improved operational efficiency for the operator. Vertical sectorization, carrier-based tilt, separate tilts for transmissions and receptions, and tilt by standards (LTE,LTE-A,HSDPA) are other benefits.

15.3 Surrogate-based optimization (SBO)

Engineering problems involve real operations or complex numerical simulations to replicate real-world behavior. Many problems are highly complex and theoretical models to replicate and subsequently optimize their performance are not available. Offline design optimization problems may involve time-consuming simulations (typically numerical function evaluations on a computer) while online optimization problems are costly due to the quantitativeness of the evaluations themselves.

In both problems a direct attempt to solve the high-fidelity (fine) models by incorporating the offline simulator or the online scenario directly in a close-loop optimization engine may be impractical. Conventional optimization algorithms (e.g., gradient-based schemes with numerical derivatives) in such an optimization engine require a very large number of objective functional evaluations for a single run, thereby rendering such an optimization approach costly. Moreover offline real-world problems involve inherently noisy, discontinuous, and complex responses, which are difficult to handle by conventional optimizers.

A solution to the above problem, and one that has been widely studied in literature, is surrogate-based optimization in which the computation/evaluation-expensive, analytically intractable high-fidelity black-box model is replaced by an iterative process, which involves creation, optimization, and update of a tractable, low-fidelity (coarse) surrogate model [29]. This iterative process involves low-fidelity model optimization by evaluation of high-fidelity data and its subsequent verification on the low-fidelity model. Once verified, the high-fidelity data is used to update the surrogate thereby allowing a gradual refinement of the low-fidelity model at each iteration [29].

Surrogate-based optimization has been widely studied in the literature for design optimization problems such as to design supersonic turbine blades [30], aerofoil shapes [31], helicopter rotor blade [32, 33], and antenna design problems [34]. Substantial energy efficiency improvements of buildings has been successfully demonstrated by using SBO to solve design of low-energy buildings [35, 36].

Telecommunication networks are a complex set of systems, which, when viewed from the system-level perspective, involve noisy network data and complex interactions that are difficult to model thereby making SBO a suitable candidate for optimization. The KPIs at the OMC plane are very noisy. A lot of this noise can be overcome by filtering over a certain period of time. Surrogate-based approximation for KPIs, which are filtered over acceptable periods of time, can be a powerful tool to reveal overall trends in data, and subsequent optimization [37].

The network optimization problem can be formulated as a generic minimization problem of the following form

$$\mathbf{x}^* = \arg\min_{\mathbf{x}} Y(\mathbf{x}) \tag{15.1}$$

where $Y(\mathbf{x})$ is the network objective function (typically a KPI or a combination of KPIs) to be minimized at the point $\mathbf{x}$, which is the network variable vector, $\mathbf{x}^*$ denotes the optimal network variable vector.

There are several reasons why an optimization involving the high-fidelity network model directly in the form of a closed loop using traditional optimization algorithms will be impractical. First, the network evaluations are costly from a user QoS perspective. An operator may not wish to perform extensive evaluations at those network variable settings where the user QoS are detrimental. Secondly, $Y(\mathbf{x})$ is indeterministically noisy due to several factors such as randomness in user arrivals and departures, mobility, network usage, etc. The objective function can even be discontinuous in certain regions.

The problem of network management is critical when more than one RAT is deployed in a given area as in the case of a HetNet, wherein interference mitigation is a crucial management issue. With the evolving traffic and changes in user distribution, the network needs to adapt itself to the changes and enable operations with the least interference and the maximum capacity for its QoS objectives. As discussed in all the HetNet interference management techniques described above, these techniques involve changes to network parameters, i.e., to those of the macro and pico cells, so as to ensure optimum cell-edge and cell-user performance for both macro and pico users. The objective poses a challenge in that the optimum parametric setting needs to be found using the least network evaluations as possible.

15.3.1 Optimization methodology

One approach to solve this problem is to optimize a low-fidelity model (surrogate), which is comparatively faster and an approximative of the high-fidelity model. This surrogate can be a good representative of $Y(\mathbf{x})$ while regressing through noise and also providing continuity in regions of discontinuity. The representativeness by the surrogate can be verified through each step of the optimization by high-fidelity model (network) evaluations at the predicted model minimizer of the surrogate. For the purpose of clarity, in the following subsections we will call the underlying high-fidelity model ($Y(\mathbf{x})$) by *network evaluations* and the representative low-fidelity model by *surrogate*. Assuming that the NP–KPI relationship between the cell-edge/center-user quality (KPIs) and the macro/pico parameters (NPs) of a HetNet can be represented by a surrogate, we propose an SBO framework for HetNet optimization. In the general scope of SBO, a surrogate or a meta-model can be constructed, which is an approximation of an underlying NP–KPI response. The response is typically associated with a domain within which the meta-model is valid. A meta-model can be constructed as an approximation of a few network

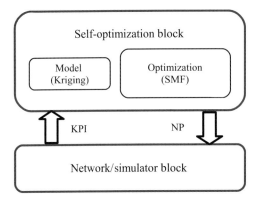

Figure 15.5 Functional block diagram of the proposed automated optimization.

simulations/evaluation, each of which provides a data point for building the NP–KPI surrogate.

The SBO process for a typical network optimization can be summarized as follows:

1. Generate the initial surrogate model using an initial design of n network observations.
2. Optimize the surrogate to obtain an approximate solution of equation (15.1).
3. Observe network at the approximate solution computed in Step 2.
4. Update the surrogate model using the network observation from Step 3.
5. Go to Step 2 if the termination condition is not satisfied.

A stepwise flow chart for SBO is given in Figure 15.6. The block diagram of such an iterative procedure is depicted in Figure 15.5.

Starting with an initial design of n network observations, the optimization problem at any iteration k during the iterative SBO can be given by changing equation (15.1) as:

$$\mathbf{x}^{k+1} = \arg\min_{\mathbf{x}} \hat{Y}_n^k(\mathbf{x}) \tag{15.2}$$

The solution to the above equation gives a network parameter setting $\mathbf{x}^{k+1}$, which is optimal for the surrogate model $\hat{Y}_n(\mathbf{x})$ constructed with an initial n network observations and at k iterations of the optimization process. Given that a right choice of initial design, a good representative surrogate modeling technique, and an efficient optimization algorithm is used, successive solutions of equation (15.2) will hopefully converge towards the optimum NP setting of the real network represented by $\mathbf{x}^*$. Since optimum search is carried out over the surrogate model $\hat{Y}_n(\mathbf{x})$ rather than over the network evaluations $Y(\mathbf{x})$, with the network evaluations carried out only at $\mathbf{x}^{k+1}$, the optimization process is less costly in terms of the number of iterations needed to reach the global optimum.

As discussed above, the convergence guarantee for the solution to equation (15.1) using SBO depends on choosing the right initial design, correct surrogate modeling technique, and an efficient optimization algorithm. For a complex NP–KPI space with several local minima, a well filling initial sampling design is needed before the SBO process can be initiated. A representative surrogate model must be just flexible enough to capture the KPI variations, while avoiding overfitting and interpolating noise. Choosing

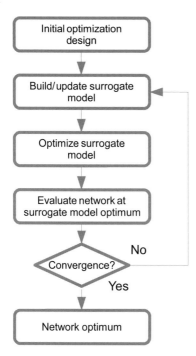

Figure 15.6 Flow chart depicting iterative process flow in SBO.

the right initial design and discussions on different surrogate modeling techniques with a detailed explanation for *Kriging* will be given in Section 15.3.2. Optimization algorithms play a crucial role in determining convergence in SBO techniques. Convergence for some algorithms, such as pattern search in surrogate management framework, is based on the guarantee that a space-filling sampling plan (such as a Latin hypercube sampling) has been chosen, which represents sufficiently the entire search space. Others, such as those based on expected improvement criteria, use the exploration–exploitation dilemma to arrive at the global optimum. These algorithms will be discussed in detail in Section 15.3.3. However, the optimization performance doesn't depend on the initial starting point but rather on the initial surrogate built using the space-filling sampling plan. In all cases, the objective of the optimization is to move to an NP setting, which even if it represents a local optimum, provides an improvement in terms of the current state of network QoS (exploitation step) or in terms of knowledge base of the optimization framework (exploration step).

15.3.2 Surrogate models

The first step in the SBO process is the construction of an initial surrogate model. This can be done using network observations through sampling in the design space. Sampling is typically carried out by specific strategies that are used to allocate samples efficiently through the design space so that the maximum space information can be retrieved using the least number of observations. Based on this design of experiment, observations are carried out in the network and an initial low-fidelity model is constructed using an appropriate surrogate modeling technique. This needs to be followed by model validation, which typically involves evaluating the representativeness of the network evaluations (observations) on surrogate predictions. This is typically carried out by dividing the observations into a training set (for surrogate model construction) and a testing set (for surrogate model validation) and carrying out cross-validation tests. Model validation will give us confidence in the use of a specific sampling strategy, surrogate modeling technique, or the size of the initial sampling design.

In this section, we first describe the basic steps needed for generating initial surrogates. Sampling strategies and the importance of a right strategy are discussed. The subsequent sections discuss the intricacies of network observation collection and the various modeling techniques.

15.3.2.1 Design of experiments

Design of experiments (DoE) involves allocating samples in the design space so as to capture maximum amount of information using a fixed number of sampling points. Based on this design, network evaluations are carried out at these samples to construct the initial observation set to be used to construct the surrogate model. Since the representation by the surrogate modeling technique and the optimization algorithm performance depend on the initial design, the samples must be spread and be sufficiently *space filling* to represent the global trend of the search space. A *space-filling* optimum sampling plan for a certain domain of evaluation is one that contains as few data points as to sufficiently construct a reliable surrogate function thereby allowing the best representation of the functional response. Sampling plans should not be so large that they are overrepresented in the surrogate, neither too small that they construct an unreliable surrogate.

Factorial designs are classical DoE techniques that when applied to discrete design variables, explore a large region of the search space. Continuous variables, when discretized, can be easily analyzed through factorial designs. A straightforward way of sampling a design space uniformly is by means of a rectangular grid of points called a full factorial design. Fractional factorial designs are used when model evaluation is expensive and the number of design variables is large. Central composite design, star design and Box–Behnken design are some commonly used factorial designs [38].

In situations when no prior knowledge of the objective function is available while constructing the initial surrogate, a variety of *space-filling* sampling designs have been recommended. Design of experiment methods such as orthogonal array sampling [39], quasi-Monte Carlo sampling [40], and Hammersley sampling [40] have been discussed extensively in the literature. The most popular and commonly used DOE for uniform sampling distributions is the Latin hypercube sampling (LHS) first introduced by McKay *et al.* in 1979 [41]. This family of sampling designs involves splitting the range of each of the d multi-variable dimensions into a large number of equally sized b bins, thereby yielding a total number of b^d bins in the design space [37]. The samples are then randomly selected in the design space such that (i) each sample is randomly placed inside a bin, and (ii) for all one-dimensional projections of the b samples and bins, there is exactly one sample in each bin. These two conditions yield a randomized sampling plan, whose dimensional projections are uniformly spread. However, this does not guarantee that the plan will be *space filling* as placing all the samples on the main diagonal of the design space will still fulfill (i) and (ii) but will create a non-uniform sampling distribution. Therefore there is a need to find the best *space-filling* LHS design based on a quantitative measure of uniformity. Such a design can be found by incorporating optimization techniques that minimize a specific non-uniformity measure called the Morris and Mitchell criteria given by [42]

$$\Phi_q(X) = \left(\sum_{j=1}^{m} J_j l_j^{-q} \right)^{1/q}$$

where $l_1, l_2, \ldots, l_j, \ldots, l_m$ is the list of the unique values of distances between all possible pairs of points in a sampling plan X, sorted in ascending order, $J_1, J_2, \ldots, J_j, \ldots, J_m$, are defined such that J_j is the number of pairs of points in **X** separated by the distance l_j, q is an exponent whose value is typically equal to 2.

The distance l between any two points $(\mathbf{x}, \mathbf{x}')$ is defined by *p-norm* of the space given by

$$l_p(\mathbf{x}, \mathbf{x}') = \left(\sum_{j=1}^{d} |x_j - \mathbf{x}'_j|^p \right)^{1/p}$$

where $p = 1$ represents the rectangular distance and $p = 2$ yields the Euclidean norm.

Other measures to evaluate the uniformity of a plan include the maximin metric introduced by [43].

15.3.2.2 Collection of observations

Once the sampling design has been defined, the next step is to obtain observations for the samples. Observations in a distributed SON architecture are fairly noisy on account of faster KPI reporting and updates when compared to the centralized architectures. Filtering involves averaging KPIs over specific intervals of times before they can be used by the SBO process. In all cases, filtering is essential to remove the noise to a larger extent and make the inherent functional patterns more explicit.

Centralized SON algorithms have slower updates (in the order of hours, days, or weeks) compared to distributed SON algorithms, which update at a faster rate (typically seconds to minutes). Thus centralized SON can obtain and filter KPIs over longer periods of time. Filtering KPIs typically involves averaging over a fixed period and can be used as a good pre-conditioning tool to obtain a better surrogate quality and faster optimization.

15.3.2.3 Modeling techniques

Once the DoE has been carried out and the observations collected, the next step is to build a surrogate model. The optimization will not only be started on this model but at each iteration of the SBO, the surrogate will be validated with network observations and the surrogate model updated with the new observations.

Given n noisy observations of the KPI $Y(\mathbf{x})$, $\mathbf{Y}^n = [y_1 y_2, \ldots, y_n]^T$ at the $\mathbf{x}$ design points $B^n = [\mathbf{x}_1 \mathbf{x}_2, \ldots, \mathbf{x}_n]^T$, $\mathbf{x}_i = \{x_{iq}\}_{q=1}^d$ for all $i = 1, \ldots, n$, where $y_i = Y(\mathbf{x}_i)$, d is the number of network parameters, $\mathbf{x}_i$ being a combination set of network parameters for all $i = 1, \ldots, n$, a surrogate modeling technique can be used to give a prediction $\hat{Y}_n(\mathbf{x})$ at any unobserved design point $\mathbf{x}$. This section describes surrogate modeling techniques that have been used within the context of SON.

Multiple linear regression

In linear regression, we try to explain any observation y_i, through a design point $\mathbf{x}_i$ using a dependency between y_i and each of the d explanatory variables in $\mathbf{x}_i$. We can write the linear expression for y_i and $\mathbf{x}_i$ inter-dependence as follows:

$$y_i = [1 \mathbf{x}_i]\beta + \epsilon_i \tag{15.3}$$

where $\beta = [\beta_0, \beta_1, \ldots, \beta_d]^T$ denotes the regression coefficients and e represents the regression error.

The best-fit regression coefficients β, can be found using the least-squares criterion where the sum of the squares of the Euclidean distance between each sample y_i and its estimate $\hat{y}_i$ is minimized:

$$\beta^* = \arg\min_{\beta} \sum_{i=1}^{n} (y_i - \hat{y}_i)^2$$

Considering all the n training samples, equation (15.3) can be written in a matrix form as

$$\begin{aligned}\mathbf{Y}^n &= [1^n \mathbf{x}^n]\beta + \epsilon \\ &= \hat{\mathbf{Y}}^n + \epsilon\end{aligned}$$

Linear regression is crucial in developing robust and efficient optimization algorithms for scenarios in which the operator has sufficient confidence on the linearity of the NP–KPI relationship [6].

Regression Kriging

Kriging was coined by Matheron in 1963 [44], in honor of Danie G. Krige who first developed the method and tested it on mining data [45]. The method was formalized within the context of computer experiments by Sacks *et al.* [46] and of optimization by Jones *et al.* [47]. *Kriging* assumes that $Y(\mathbf{x})$ may be expressed as follows:

$$Y(\mathbf{x}) = g(\mathbf{x})^T \beta + Z(\mathbf{x}) + \epsilon(\mathbf{x})$$

where $g(\mathbf{x}) = [g_1(\mathbf{x}) g_2(\mathbf{x}) \ldots g_K(\mathbf{x})]^T$ are known functions, $\beta = [\beta_1 \beta_2 \ldots \beta_K]^T$ are the unknown model parameters, $Z(\mathbf{x})$, $\mathbf{x} \in \mathbb{R}^d$ is the underlying stochastic model, which is a Gaussian process with zero mean and transition-invariant co-variance kernel $\phi(\mathbf{x}, \mathbf{x}')$ for all $\mathbf{x}, \mathbf{x}' \in \mathbb{R}^d$. $\epsilon(\mathbf{x})$ is the additive noise term with zero mean and covariance kernel $\tau^2 \delta(\mathbf{x}, \mathbf{x}')$ for all $\mathbf{x}, \mathbf{x}' \in \mathbb{R}^d$, $\delta(\mathbf{x}, \mathbf{x}')$ is the Kronecker delta function and τ^2 is the noise variance, and $\epsilon(\mathbf{x})$ is assumed to be independent from $Z(\mathbf{x})$. The regression term $g(\mathbf{x})^T \beta$

takes into account the global trend of the function Y, and $Z(\mathbf{x})$ takes into account localized variations. The Kriging predictor is defined by its estimated mean $\hat{m}Y_n(\mathbf{x})$ and variance $\hat{s}^2_{Y_n}(\mathbf{x})$ as [48]:

$$\hat{m}_{Y_n}(\mathbf{x}) = g(\mathbf{x})^T \hat{\beta}_n + \phi_n(\mathbf{x})^T \Phi_n^{-1}(\mathbf{Y}^n - \mathbf{G}_n \hat{\beta}_n) \tag{15.4}$$

$$\hat{s}^2_{Y_n}(\mathbf{x}) = \hat{\sigma}^2_n - \phi_n(\mathbf{x})^T \Phi_n^{-1} \phi_n(\mathbf{x}) + \frac{\left(1 - 1_n^T \Phi_n^{-1} \phi_n(\mathbf{x})\right)^2}{1_n^T \Phi_n^{-1} 1_n} \tag{15.5}$$

where

- the vector of model parameters $\hat{\beta}_n$ is computed as

$$\hat{\beta}_n = \left(\mathbf{G}_n^T \Phi_n^{-1} \mathbf{G}_n\right)^{-1} \mathbf{G}_n^T \Phi_n^{-1} \mathbf{Y}^n$$

- $\mathbf{G}_n$ is a $n \times K$ matrix with $\mathbf{G}_n = g_j(\mathbf{x}_i)$, for all $i = 1, \ldots, n, j = 1, \ldots, K$
- Φ_n is the $n \times n$ correlation matrix whose $(i, j)^{th}$ entry is equal to:

$$\Phi_n(i,j) = \begin{cases} \phi(\mathbf{x}_i, \mathbf{x}_j) + \hat{\tau}^2_n & i = j \\ \phi(\mathbf{x}_i, \mathbf{x}_j) & i \neq j \end{cases} \quad \text{for } i, j = 1, \ldots, n$$

- $\hat{\tau}^2_n$ is the estimated noise variance

Also $g(\mathbf{x})^T \hat{\beta}_n$ is the best unbiased global estimate of $g(\mathbf{x})^T \beta$ using n observations and $\phi_n(\mathbf{x}) = [\phi(\mathbf{x}, \mathbf{x}_1) \phi(\mathbf{x}, \mathbf{x}_2) \ldots \phi(\mathbf{x}, \mathbf{x}_n)]^T$ is the correlation of the unobserved point with the existing (observed) points. Isotropic covariance kernel given by $\phi(\mathbf{x}, \mathbf{x}) = \sigma^2 \exp(-\sum_{j=1}^{d} \theta_j |\mathbf{x}_j - \mathbf{x}'_j|^{p_j})$ for all $x_j, x'_j \in \mathbb{R}$ is assumed in Kriging. While Gaussian kernel has $1/\sigma^2$, the Kriging kernel has a vector $\theta_j = \{\theta_1, \theta_2, \ldots, \theta_d\}^T$, allowing the width of the kernel to vary from variable to variable. Also while in Gaussian kernel, the exponent is fixed at 2, giving a smooth function through the point $\mathbf{x}$, Kriging allows this exponent $P_j = \{p_1, p_2, \ldots, p_d\}^T$ to vary (typically $P_j \in [1, 2]$) for each dimension in $\mathbf{x}$. If we fix pat $p_{(1,2,\ldots,d)} = 2$ and a constant θ_j for all dimensions, the Kriging kernel is the same as the Gaussian.

The estimations of the unknown covariance kernel parameters $[\hat{\sigma}^2, \hat{\beta}, \hat{\tau}, \hat{p}_1, \ldots, \hat{p}_d, \hat{\theta}_1, \ldots, \hat{\theta}_d]$ are based on the maximum likelihood estimation (MLE), which maximizes the *concentrated In-likelihood function L*. Optimization of L is performed with respect to the whole vector of parameters [37]:

$$[\hat{\sigma}^2, \hat{\beta}, \hat{\tau}, \hat{p}_1, \ldots, \hat{p}_d, \hat{\theta}_1, \ldots \hat{\theta}_d] = \arg \min L(\sigma^2, \beta, \tau, p_1, \ldots, p_d, \theta_1, \ldots, \theta_d)$$

Other techniques

Other techniques such as polynomial regression, in which several polynomial basis functions are used to build the surrogates, or radial basis functions, which use several radially symmetric functions and support vector regression (SVR) have also been discussed in the literature [37]. SVR comes from the theory of support vector machines (SVM), developed at AT&T Bell Laboratories in the 1990s [49]. In fact SVRs are considered as an extension to radial-basis function methods and allow us to calculate a margin within which we are willing to accept errors in the measurements [37]. Others include neural networks, which are systems of interconnected "neurons." Using an appropriate neural architecture, the network can be trained sufficiently to approximate a smooth regressor. More details on this can be found in [50].

15.3.3 SBO techniques

In this section, we will introduce and discuss in detail two of the most relevant optimization strategies namely, surrogate management framework and expected improvement based technique, which can be used in the SBO framework to exploit the surrogate model.

15.3.3.1 Expectation of improvement (EI) based infill criterion

A very useful criterion for optimization is the 'expected improvement' criterion, which balances the local exploitation with global exploration. The concept of using this criterion for black-box optimization was first proposed in [47] and has been subsequently studied widely in the literature [37]. The concept models the uncertainty at any given unobserved point x by treating it as the realization of a normally distributed random variable with a certain mean and standard deviation. Incorporating Kriging modeling with this optimization technique now makes perfect sense, as the mean and standard deviation information at any given point can be provided by the Kriging technique. When combined within the Kriging framework, expected improvement is a powerful optimization tool as:

1. The exploration–exploitation dilemma prevents the optimization algorithm from getting stuck in a local minimum.
2. The technique provides a quantification of the improvement expected at any given point as the optimization process is progressing.
3. By considering a multi-dimensional Gaussian problem, the criterion can be used in a multi-objective optimization context or by multiplying with probability of constraint satisfaction for a constrained optimization scenario.

Within the context of optimization based on improvement, the probability that one can achieve an 'improvement' at any point on the surrogate can be mathematically computed. We first define the 'improvement' achieved by the Kriging prediction $\hat{F}_n(\mathbf{x})$ over the n observed values:

$$I_n(\mathbf{x}) \triangleq f_n^{\min} - \hat{F}_n(\mathbf{x}) \tag{15.6}$$

Typically for a Kriging model interpolating over noise-free observations, the choice of $f_n^{\min}$ is such that $f_n^{\min} = \min_{\mathbf{x}} F(\mathbf{x})$, which denotes the currently known minimum at the nth iteration. However, in the framework of noisy observations, this plug-in lacks robustness since a single noisy observation with coincidentally a very low value can severely underestimate $f_n^{\min}$ over the rest of the optimization process. One possibility to deal with this issue is to replace it by some arbitrary target T. This is called the *improvement with "plug-in"* [48] and can be given by modifying equation (15.6) as:

$$I_{T,n}(\mathbf{x}) \triangleq T - \hat{F}_n(\mathbf{x}) \tag{15.7}$$

The choice of T is important as too high or too low values can have significant influence on the shape of $I_{T,n}$ and thus can influence the optimization behavior. A choice that has been proposed in [51] can be $T = \min_{\mathbf{x}} \hat{F}_n(\mathbf{x})$ i.e., we would consider the improvement above the maximum of the predictor instead of the maximum of the noisy observations. In other words, we surrogate the unknown $f_n^{\min}$ by the minimum value of the Kriging mean evaluated over all the observed points $\mathbf{X}^n$.

Probability and expectation of improvement with "plug-in"
It is clear from equation (15.7) that $I_{T,n}(\mathbf{x})$ is also Gaussian with mean $T - \hat{m}_{F_n}(\mathbf{x})$ and variance $\hat{s}_{F_n}^2(\mathbf{x})$. Being a Gaussian process, the probability of an improvement upon T is given by

$$\Pr[I_{T,n}(\mathbf{x})] = \frac{1}{\hat{s}_{F_n}(\mathbf{x})\sqrt{2\pi}} \int_{-\infty}^{0} e^{-\frac{u^2}{2}} dI \tag{15.8}$$

where

$$u = \frac{T - \hat{m}_{F_n}(\mathbf{x})}{\hat{s}_{F_n}(\mathbf{x})}.$$

Instead of defining a probability of improvement, the amount of improvement at a given point can be used as a more precise measure for deciding the next step in the optimization process and is analytically tractable. This measure is the *expectation of improvement with "plug-in"* at a given point x and can be computed by

$$\mathrm{E}[I_{T,n}(\mathbf{x})] = (T - \hat{m}_{F_n}(\mathbf{x}))\Phi(u) + \hat{s}_{F_n}(\mathbf{x})\phi(u) \tag{15.9}$$

where $\Phi(.)$ and $\phi(.)$ are the cumulative distribution function and probability density function respectively. The optimum x value, $\mathbf{x}^*$, which minimizes $\hat{F}_n(\mathbf{x})$ is that value which maximizes our expectation of having an improvement over all the n observed values. The optimization solution can thus be given by:

$$\mathbf{x}^* = \arg\max_{\mathbf{x}} \mathrm{E}[I_{T,n}(\mathbf{x})] \tag{15.10}$$

Augmented expectation of improvement
Augmented expectation (AE) of improvement criterion was first proposed for noisy framework. Instead of considering $T = \min_{\mathbf{x}} \hat{F}_n(\mathbf{x})$, T is taken as $\hat{F}_n(\mathbf{x}')$ where $\mathbf{x}'$ is called the *effective best solution* and is obtained by minimizing $\hat{F}_n(\mathbf{x}) + \alpha \hat{s}_{F_n}(\mathbf{x})$, over all the observed points in order to have a more robust estimate of the plug-in. $\alpha = 1$ has been proposed by [48]. Thus T is the Kriging mean value at the design point with lower

β-quantile, where $\Phi(\beta) = \alpha$. Replacing equation (15.7) with the new value of T and rewriting equation 15.9, we get

$$\text{AE}[I_{T,n}(\mathbf{x})] = (\hat{F}_n(\mathbf{x}') - \hat{m}_{F_n}(\mathbf{x}))\Phi(u') + \hat{s}_{F_n}(\mathbf{x})\phi(u')$$

where $u' = \frac{\hat{F}_n(\mathbf{x}') - \hat{m}_{F_n}(\mathbf{x})}{\hat{s}_{F_n}(\mathbf{x})}$. Note that in the above equation, when $\mathbf{x}' = \mathbf{x}$, $\text{AE}[I_{T,n}(\mathbf{x})]$ does not reduce to zero as would have been the case for $\text{E}[I_{T,n}(\mathbf{x})]$ with $T = \min_\mathbf{x} \hat{F}_n(\mathbf{x})$ in equation (15.9).

One drawback of expectation of improvement-based optimization discussed earlier is that it does not take into account the noise variance of the future observation: everything is calculated as if the next evaluation would be deterministic. Augmented expectation of improvement addresses this issue by adding a multiplicative penalty to $\text{E}[I_{T,n}(\mathbf{x})]$, to obtain a penalized AE of improvement, which penalizes the points whose Kriging variance is small compared to the noise level [48] and is given by:

$$\text{AE}_p[I_{T,n}(\mathbf{x})] = \text{AE}[I_{T,n}(\mathbf{x})] \times \left(1 - \frac{\hat{\tau}_n}{\sqrt{\hat{s}_{F_n}^2(\mathbf{x}) + \hat{\tau}_n^2}}\right)$$

The penalty term is one if $\hat{\tau}_n = 0$, thus reducing to the original $\text{E}[I_{T,n}(\mathbf{x})]$ function and decreases towards zero when $\hat{\tau}_n$ increases. It can be seen that for designs with smaller prediction variance $\hat{s}_{F_n}^2(\mathbf{x})$, the penalty penalizes designs thereby enhancing exploration.

Constrained expectation of improvement
For a constrained optimization scenario, apart from computing the expectation of improvement, the probability of constraint satisfaction must also be incorporated in the optimization problem. It must be noted that $\mathbf{x}^*$ must also satisfy the constraints $c(\mathbf{x})$. In other words, those values of x that do not satisfy these constraints must be discarded. Given a constraint threshold Th_c, the probability that a constraint is satisfied (feasibility of a solution) can be computed in a similar manner as the expectation of improvement. Rather than computing the probability that an improvement is achieved over T, we define the probability that the prediction is greater than a given (user-defined) constraint limit Th_c, i.e., the probability that a constraint is met. Thus we restrict the solution space to those **x** values that satisfy the constraint by defining a measure of "feasibility," given by:

$$H_n(\mathbf{x}) = \Pr[\hat{C}_n(\mathbf{x}) \leq Th_c]$$

Using the same logic as that of the probability of improvement, equation (15.8) can be adapted for the Kriging model of the constraint function $\hat{C}_n(\mathbf{x})$ and written in the form of the error function as [37]:

$$H_n(\mathbf{x}) = 1 - \frac{1}{2}\text{erfc}\left[\frac{Th_c - m_{c_n}(\mathbf{x})}{\sqrt{2}s_{c_n}(\mathbf{x})}\right]$$

The measure of the improvement that the new infill point offers and with the probability that it is also constraint feasible can thus be given by incorporating the probability of constraint feasibility as a multiplicative term, which renders the objective function zero for those x values that do not satisfy the constraints. Thus the optimum x value can be obtained by:

$$\mathbf{x}^* = \arg\max_{\mathbf{x}}\{E[I_{T,n}(\mathbf{x})]H_n(\mathbf{x})\} \quad (15.11)$$

or by using the augmented expectation of improvement criterion:

$$\mathbf{x}^* = \arg\max_{\mathbf{x}}\{AE[I_{T,n}(\mathbf{x})]H_n(\mathbf{x})\} \quad (15.12)$$

Other methodologies to incorporate the constraint within the optimization process have been proposed, such as those by incorporation of the penalty functions. The approach involves a penalization of the objective function whenever a design is considered that violates (or perhaps nearly violates) one or more constraints. Others include the multi-objective Pareto front optimization where in one of the objective functions to be minimized is the constraint function. For more information on these techniques, please refer to [37].

Minimum quantile criterion
A simple criterion that can provide a tradeoff between exploration (high Kriging variance) and exploitation (low Kriging variance) consists in a Kriging percentile, i.e., a weighted sum of $\hat{m}_{F_n}(\mathbf{x})$ and $\hat{s}^2_{F_n}(\mathbf{x})$ given by [48]

$$\mathbf{x}^* = \arg\min_{\mathbf{x}}\{\hat{m}_{F_n}(\mathbf{x}) + \alpha\hat{s}_{F_n}(\mathbf{x})\}$$

Although this method is considered less efficient than the expectation of improvement-based criteria, it has shown successful applications in Kriging-based multi-objective optimizations.

15.3.3.2 Surrogate management framework

The surrogate management framework (SMF) incorporates surrogates within the pattern search algorithmic framework [32]. Pattern search algorithms are a class of direct search methods that are capable of solving global optimization problems of irregular, multimodal objective functions and without the need for calculating any gradient information. A formal definition of pattern search was proposed in [52]. Pattern search algorithms compute a sequence of points that approaches eventually to the globally optimum point.

Pattern search algorithms are characterized by a sequence of *meshes* and a list of *polling conditions*. A mesh is a lattice to which an iterate is restricted during the search. A set of *polling conditions* govern when the current mesh must be refined, ensuring that the algorithm satisfies the demands of the convergence theory for pattern search methods.

Convergence in these methods is typically ensured through a polling condition, which can be the set of vectors formed by taking the differences between the set of trial points at which the objective function is to be evaluated and the current iterate $\mathbf{x}_k^*$ must contain a positive basis for $\Re^d$. At least $d+1$ and at most $2d$ positive basis are required for convergence. A positive basis can be constructed as follows. We let V be the matrix whose columns are the basis elements. Then construct $D = [V, -V \cdot e]$, where e is the vector of ones and $-V \cdot e$ is the negative sum of the columns of V. The columns of D form a $d + 1$ positive basis for $\Re^d$. For example, in two dimensions such a basis could be given by $(1, 0), (0, 1), (-1, -1)$ is a set of vectors.

An advantage of these methods is that constraint feasibility conditions can be incorporated easily in the pattern-search framework during the searches. For unconstrained problems a minimal positive basis $(d + 1)$ is sufficient to guarantee convergence; however, for constrained problems a maximal positive basis $(2d)$ is recommended [32]. A special case of the pattern-search methods is the generalized pattern search (GPS) algorithm with the maximal positive basis set $2d$ vectors. The algorithm uses fixed-direction vectors to compute the set of points forming the mesh. Each run of a GPS algorithm is characterized by a SEARCH and a POLL step.

The surrogate management framework (SMF) is a set of strategies for using approximations in the SEARCH and POLL steps of the GPS algorithm and incorporates an additional EVALUATE/CALIBRATE step. SMF can be used to guarantee local optimum given that the surrogate to start with is a good representative of the underlying model. The convergence proof of the SMF algorithm for noiseless data is given in [32].

Let M_0 denote a mesh on $\mathfrak{B} \equiv \{\mathbf{x} | a \le \mathbf{x} \le b\}$ where $a, b \in \Re^d$ and $a \le b$ means that each coordinate satisfies $a_i \le b_i$. Suppose we have built $\hat{F}_0(\mathbf{x})$, an initial approximation of objective function $f(\mathbf{x})$ on $\mathfrak{B}$, and $\mathbf{x}_0 \in M_0$, let $P_0 \subset M_0$ contain $\mathbf{x}_0$ and any $2d$ points adjacent to $\mathbf{x}_0$ for which the differences between those points and $\mathbf{x}_0$ form a maximal positive basis for $\Re^d$. As the algorithm generates $\mathbf{x}_k \in M_k$, let $P_k \subset M_k$ be defined in the same way. For a constrained optimization, let $\hat{C}_0(\mathbf{x})$ be an initial approximation of constraint function $c(\mathbf{x})$ on $\mathfrak{B}$. Let statements within () define the additional steps to be taken for constrained optimization. Then for $k = 0, 1, \ldots$, do:

1. SEARCH: Use any method to choose a trial set $T_k \subset M_k$. If $T_k \ne \emptyset$ is chosen, then it is required to contain at least one point at which $f(\mathbf{x})$ is not known. (For constrained optimization use a method to choose T_k, such that $\hat{C}_k(\mathbf{x}) \le Th_c$ for all points in T_k.) If $T_k \ne \emptyset$, then go to POLL.
2. EVALUATE/CALIBRATE: Evaluate $f(\mathbf{x})$ on elements in T_k until either it is found that $\mathbf{x}_k$ minimizes $f(\mathbf{x})$ on T_k or until $\mathbf{x}_{k+1} \in T_k$ is identified for which $f(\mathbf{x}_{k+1}) < f(\mathbf{x}_k)$ (and $c(\mathbf{x}_{k+1}) \le Th_c$). If such an $\mathbf{x}_{k+1}$ is found then declare the SEARCH successful. Recalibrate $\hat{F}_k(\mathbf{x})$ (and $\hat{C}_k(\mathbf{x})$ for constrained optimization) with the new values of $f(\mathbf{x})$ (and $c(\mathbf{x}_k)$) computed at points in T_k.
3. If SEARCH was successful, then set $\hat{F}_{k+1}(\mathbf{x}) = \hat{F}_k(\mathbf{x})$ (and $\hat{C}_{k+1}(\mathbf{x}) = \hat{C}_k(\mathbf{x})$), $M_{k+1} = M_k$, and increment k.
    else return to SEARCH with the recalibrated $\hat{F}_k(\mathbf{x})$ (and $\hat{C}_k(\mathbf{x})$), but without incrementing k.
4. POLL: If $\mathbf{x}_k$ minimizes $f(\mathbf{x})$ for $\mathbf{x} \subset P_k$, then declare the POLL unsuccessful, set $\mathbf{x}_{k+1} = \mathbf{x}_k$, and set $M_{k+1} = M_k/2$; else declare the POLL successful, set $\mathbf{x}_{k+1}$ to a point in P_k at which $f(\mathbf{x}_{k+1}) < f(\mathbf{x}_k)$ (and $c(\mathbf{x}_{k+1}) \le Th_c$), and set $M_{k+1} = M_k$.

Recalibrate $\hat{F}_k(\mathbf{x})$ (and $\hat{C}_k(\mathbf{x})$) with the new values of $f(\mathbf{x})$ (and $c(\mathbf{x}_k)$) computed at points in P_k. Set $\hat{F}_{k+1}(\mathbf{x}) = \hat{F}_k(\mathbf{x})$ (and $\hat{C}_{k+1}(\mathbf{x}) = \hat{C}_k(\mathbf{x})$). Increment k.

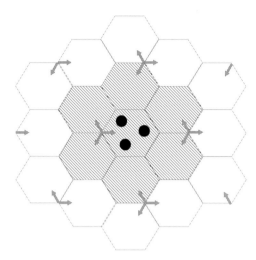

Figure 15.7 HetNet layout indicating directional macro cells (striped) and omni directional pico cells (solid).

15.4 Centralized self-optimization for interference mitigation

The following subsections use the SBO framework developed above to solve and compare the performances of eICIC and AAS-based self-optimization.

15.4.1 eICIC and AAS-based self-optimization

Let us assume a central cluster of *macro* cells represented by the set M. This cluster comprises a central cell surrounded by its first-tier neighbors as shown in Figure 15.7.

A collection of *pico* cells represented by the set P, are located in the central cell of the *macro* cells cluster. A typical self-optimization scenario consists of optimizing a predetermined *KPI* for the network by adjusting either the AAS or eICIC parameters.

15.4.1.1 Objective function definition

1. A typical AAS-based optimization scenario includes adjustments to each of the antenna tilts in M. Here the self-optimization algorithm performs optimization by adjusting each of the antenna tilts and proposes an optimum vector of antenna tilt combination. Let us denote the antenna tilt of cell s as $\theta_s(°)$ and the vector of antenna tilts by $\mathbf{x} = [\theta_1 \theta_2 \ldots \theta_s \ldots \theta_{|M|}]$, $\forall s \in M$. The AAS-based self-optimization objective function can be given by:

$$\mathbf{x}^* = \arg\min_{\mathbf{x}} f(\mathbf{x})$$
$$\text{s.t.} c(\mathbf{x}) \leq Th_c, \theta_{min} \leq \theta_s \leq \theta_{max}, \forall s \in M \quad (15.13)$$

2. A typical eICIC-based self-optimization scenario includes adjustments to each of the *muting ratio* in M and to each of the *range extension* in P. The self-optimization algorithm performs optimization by proposing an optimum joint vector of the *muting*

ratio and *range extension* combination. Let us denote the *muting ratio* of cell s as mR_s, the *range extension* of cell r as RE_r(dBm) and the joint vector of *muting ratio, range extension* combination by $\mathbf{x} = [mR_1 \ldots mR_s \ldots mR_{|M|}, RE_1 \ldots RE_r \ldots mR_{|P|}]$, $\forall s \in M, \forall r \in P$. The eICIC-based self-optimization objective function can be given by:

$$\mathbf{x}^* = \arg\min_{\mathbf{x}} f(\mathbf{x})$$

$$\text{s.t.} c(\mathbf{x}) \leq Th_c, RE_{min} \leq RE_r \leq RE_{max}, \forall r \in P$$
$$\leq mR_s \leq 0, \forall s \in M \quad (15.14)$$

where $f(\mathbf{x})$ and $c(\mathbf{x})$ are the objective function to be optimized and the constraint function respectively. $c(\mathbf{x}) = \max[BCR_1(\mathbf{x}) \ldots BCR_2(\mathbf{x}) \ldots BCR_s(\mathbf{x})]$ where BCR_s is the BCR of cell s, $\forall s \in M$, $|M|$ being the cardinality of M. Th_c is the constraint threshold on BCR, $\theta_{min}, \theta_{max}$ are the minimum and maximum allowable antenna tilt values to prevent coverage issues such as coverage holes, pilot pollution, etc. RE_{min}, RE_{max} are the minimum and maximum allowable range extensions on the pico cells.

Aggregation of KPIs can be used to formulate coverage and capacity objectives $f(\mathbf{x})$. Such an aggregated optimization allows to take into account the effect of all KPIs considered and are thus impacted equally or proportionally by formulating a weighted KPI objective. The optimization in equations (15.13) and (15.14) can be formulated as a minimization problem or as a maximization problem as given below:

1. $f(\mathbf{x}) = f_{\rho5\%}(\mathbf{x}) = R_{max5\%} - \sum_{\forall s \in T \cup P} \rho_s^{5\%}(\mathbf{x})$, for the maximization of cell-edge throughput, where $\rho_s^{5\%}$ is 5%-ile throughput of cell s, and $R_{max5\%}$ is a predetermined value, which is greater than the maximum expected sum of the 5%-ile throughput over T, i.e. $R_{max5\%} > E[\max[\sum_{\forall s \in T \cup P} \rho_s^{5\%}(\mathbf{x}),]]$ where $E[.]$ denotes expectation.
2. $f(\mathbf{x}) = f_{\rho5\%}(\mathbf{x}) = R_{max5\%} - \sum_{\forall s \in T \cup P} \rho_s^{50\%}(\mathbf{x})$, for the maximization of cell center (or median) throughput, where $\rho_s^{50\%}$ is 50%-ile throughput of cell s, and $R_{max50\%}$ is again a predetermined value, which is greater than the maximum expected sum of the 50%-ile throughput over T, i.e. $R_{max50\%} > E[\max[\sum_{\forall s \in T \cup P} \rho_s^{50\%}(\mathbf{x})]]$.

15.4.1.2 Antenna tilt model

The antenna tilt θ for each site determines the boundaries of the cells. Tilts of a group of neighboring cells can be mutually adjusted to reduce interference thereby increasing capacity or to provide coverage to a certain area. For a trisectorial site, at a given location (ψ, φ), where ψ is the elevation angle computed using the antenna height and the ground distance from the antenna to that location; ψ is the azimuth angle computed using the antenna and the location coordinates, 3GPP defines azimuth, elevation, and the total radiation patterns given respectively by,

$$A_H(\varphi) = -\min\left[12\left(\frac{\varphi}{\varphi 3dB}\right)^2, A_m\right]$$

$$Av(\psi) = -\min\left[12\left(\frac{\psi - \theta}{\psi 3dB}\right)^2, SLA_v\right]$$

$$A(\varphi, \psi) = -\min\{-[A_H(\varphi) + Av(\psi)], A_m\}$$

Table 15.1 System-level simulation parameters.

Parameters	Settings
Antenna tilt range	2 to 12 in steps of 2
Antenna height (h)	32 m
Bandwidth	10 MHz
Network layout	500 m macro layer inter-site distance
Macro cell layout	27 macro cells
Pico cell layout	3 pico-eNBs in central macro cell
Path loss macro to UE	128.1 + 37.6 log10(R[km])
Path loss pico to UE	140.7 + 36.7 log10(R[km])
File size	16 M bits
Traffic model	FTP
eNB packet scheduling	Round robin
RE range	0 to 20 dB in steps of 4 dB
Muting ratio range	0 to 1 in steps of 0.2
Shadowing standard deviation	6 dBm
Transmit power	46 dBm (macro), 30 dBm (pico)
Antenna gain	macro: 14 dBi; pico: 10 dBi
PRBs per eNB	50

where A_m and SLA_V are the backward attenuation factors taken as 25 dB and 20 dB respectively, ψ_{3dB} and φ_{3dB} are the half power beam widths of the azimuth and elevation beam patterns, which are taken as 10 dB and 20 dB respectively. The total radiation pattern $A(\varphi, \psi)$ is used to compute the path loss at any given location.

15.4.2 Simulator description

The simulation scenario consists of an LTE-A HetNet consisting of 27 *macro* cells, where $M = \{eNB_1, eNB_2, eNB_3, eNB_{11}, eNB_{17}, eNB_{18}, eNB_{24}\}$, and 3 *pico* cells, where $P = \{P_1, P_2, P_3\}$ are considered. Traffic arrival follows a Poissonian distribution and arrives with a certain intensity in all of the 27 macro and 3 pico cells; however, KPIs are collected from M and P only. Simulations are carried out for the downlink only. A semi-dynamic LTE-A system simulator, which performs correlated Monte Carlo snapshots with a time step of t sec, is used ($t = 1$ sec). Each *eNB* is associated with a fixed number of resource blocks depending on its bandwidth. As the UEs arrive for call admission, the associated resource blocks and coverage are checked for availability. If available, a fixed number of resources are allocated to each UE, and UEs that cannot be served due to non-availability of resource blocks or coverage are either blocked or dropped. During communication UEs download a file of fixed size, those that finish communication leave the system, while those that continue their communication are moved according to a predetermined mobility and mobile-base station attachments are re-calculated. Each simulation lasts for 3,000 s at the end of which statistics of performance metrics (KPIs) are collected. Table 15.1 lists the parameters used in the system simulation.

Table 15.2 *Kriging* prediction quality.

$f(x)$	elCIC			AAS		
	R^2	NMSE	CV	R^2	NMSE	CV
$f_{\rho 5\%}$	93.41	3.9e-4	1.98	88.99	2.7e-4	1.67
$f_{\rho 50\%}$	92	2e-4	1.44	94.92	4e-4	2

15.4.3 Results

15.4.3.1 Observational KPIs

Once simulations are carried out, the results are stored and processed. The following KPIs are observed for evaluating the performance of the centralized self-optimization algorithm:

1. $SINR_{5\%}$ and $SINR_{50\%}$, for the 5%-ile and 50%-ile $SINR$ of the entire network
2. $Th_{P5\%}$ and $Th_{P50\%}$, for the 5%-ile and 50%-ile throughput of the pico cells in P
3. $Th_{M5\%}$ and $Th_{M50\%}$ for the 5%-ile and 50%-ile throughput of the macro cells in M
4. BCR_M and BCR_P, for BCRs in M and P respectively.

15.4.3.2 Prediction quality

Classical prediction quality metrics, such as the coefficient of determination $R^2 = (1 - \frac{SS_{res}}{SS_{tot}}) \times 100$, the normalized mean-squared error $NMSE = \frac{1}{n}\sum_{i=1}^{n}\frac{(y_i - m_{y_i})^2}{\bar{y}\bar{m}_y}$ and coefficient of variation $CV = \frac{\sqrt{\sum_{i=1}^{n}\frac{(y_i - m_{y_i})^2}{n}}}{\bar{y}} \times 100$ have been used as performance indicators, where $SS_{res} = \sum_{i=1}^{n}(y_i - m_{y_i})^2$ is the residual sum of squares, $SS_{tot} = \sum_{i=1}^{n}(y_i - \bar{y})^2$ the total sum of squares, $\bar{y} = \frac{1}{n}\sum_{i=1}^{n}y_i$ is the mean of the observations and $\bar{m}_y = \frac{1}{n}\sum_{i=1}^{n}m_{y_i}$ is the mean of the predictions. R^2 is an indicator of how well the model fits the data ($R^2 = 100\%$ indicating a perfect fit), CV indicates the dispersion of noise around the model and $NMSE$ is an estimator of the overall deviations between predicted and measured values. A total of 400 points (AAS tilt value combinations, elCIC muting ratio, and bias combinations) were set aside for model building and prediction, out of which 205 points were used for building the model and the remaining (195) points were used for evaluating the quality of prediction. The results for prediction quality tests are provided in Table 15.2.

In terms of prediction, both the objective functions tend to show a higher quality with a greater level of prediction confidence. To start the optimization procedure, the default network parameter settings are chosen as $mR = 0$ and $\theta = 6°$ for all the concerned macro eNBs in M, and $RE = 0dB$ for all the pico eNBs in P. Results are obtained for $Th_c = 5\%$.

15.4.3.3 Optimization performance

Optimization for $f_{\rho 5\%}(\mathbf{x})$ results in gains as shown in Figure 15.8. The figure indicates gains for elCIC and AAS optimizations with respect to the default network parameters.

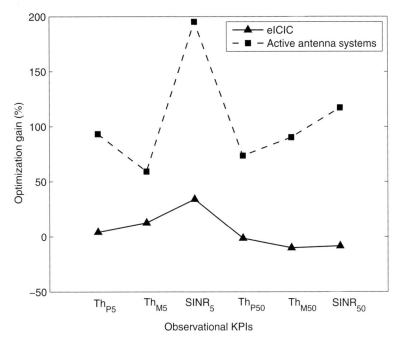

Figure 15.8 Cell-edge optimization performance.

If all the KPIs described in Section 15.4.3 are taken into consideration, it is quite clear that eICIC performs much better than AAS optimizations. Improvements/degradations in other KPIs were achieved by a factor of

- 93% and 4% improvements in eICIC and AAS respectively for $Th_{P5\%}$
- 74% improvement and 1% degradation in eICIC and AAS respectively for $Th_{P50\%}$
- 59% and 13% improvements in eICIC and AAS respectively for $Th_{M5\%}$
- 90% improvement and 10% degradation in eICIC and AAS respectively for $Th_{P50\%}$
- 195% and 35% improvements in eICIC and AAS respectively for $SINR_{5\%}$
- 116% improvement and 8% degradation in eICIC and AAS respectively for $SINR_{50\%}$

Optimization for $f_{\rho 50\%}$ (**x**) results in gains as shown in Figure 15.9. Optimization with eICIC yields positive gains for all cell-center observational KPIs with respect to the default network parameter settings. Performance is degraded for cell-edge UEs using eICIC as is evident from $Th_{P5\%}$, $SINR_5\%$. In contrast AAS performs better for all observational KPIs, though it performs comparatively inferiorly to eICIC for all cell-center KPIs. There were no macro call blocks for either eICIC or AAS optimizations. Improvements/degradations in observational KPIs are summarized below as

- 93% degradation and 8% improvements in eICIC and AAS respectively for $Th_{P5\%}$
- 10% and 3% improvements in eICIC and AAS respectively for $Th_{P50\%}$
- 55% and 10% improvements in eICIC and AAS respectively for $Th_{M5\%}$
- 100% and 33% improvements in eICIC and AAS respectively for $Th_{M50\%}$

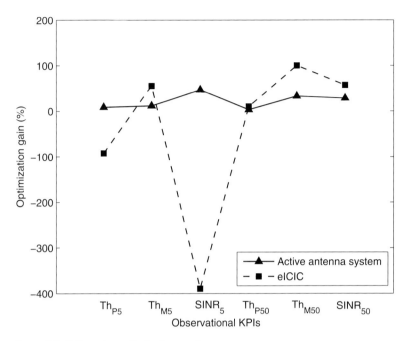

Figure 15.9 Cell-center optimization performance.

- 390% degradations and 46% improvements in eICIC and AAS respectively for $SINR_{5\%}$
- 55% and 28% improvements in eICIC and AAS respectively for $SINR_{50\%}$

15.5 Concluding remarks and open issues

In this chapter, we provided a brief overview of the need for capacity in rapidly evolving mobile networks and the need for interference mitigation techniques in HetNets to meet the traffic demands. We provided an extensive literature survey on the existing interference mitigation solutions. We developed a centralized iterative self-optimization framework based on the concept of surrogate-based optimization. We provided two case scenarios for interference mitigation in HetNets and developed an optimization model for both within a SON framework. The two scenarios were tested on a system-level simulator and the optimized solutions compared. Based on the results obtained from the simulation studies the following conclusions can be drawn:

1. A functional relationship does exist between the NPs and the KPIs, which can be easily modeled by the *Kriging* methodology.
2. Antenna tilt and eICIC have a significant impact on cell-edge and cell-center user quality.
3. For cell-edge optimization performance, the eICIC performs better than the AAS-based optimization for all observational KPIs.

4. For cell-center optimization performance, eICIC once again is superior to the AAS-based optimization; however, only for cell-center observational KPIs. In optimizing cell-center KPIs, it deteriorates drastically the cell-edge KPIs. To optimize cell-center quality while also maintaining a positive mediocre improvement in cell-edge KPIs, AAS-based optimization is preferred over eICIC optimization.

Although Kriging performs well for mapping the KPI–NP relationships, it will be worth comparing the regression performances of Kriging with linear and logistic regression techniques. Given the close similarity between Kriging and SVR, a performance comparison between these two could provide insights into the best regression technique suitable for network optimization. Also investigating the performance of expectation of improvement-based algorithms, which utilize uncertainty information from the Kriging predictions, could be another direction of research. Investigations for the use of constraints within a multi-objective optimization framework, which can provide the operator with a range of competing QoS metrics, are also envisaged as important future work.

References

[1] "Ericsson mobility report" (2013). *Technical report*, Ericsson.
[2] Hmlinen, S., Sanneck, H., and Sartori, C. (2012). *LTE Self-Organising Networks (SON): Network Management Automation for Operational Efficiency*, 1st edn, Wiley Publishing.
[3] Ramiro, J. and Hamied, K. (2011). *Self-Organizing Networks (SON): Self-Planning, Self-Optimization and Self-Healing for GSM, UMTS and LTE*, Wiley.
[4] Suga, J., Kojima, Y. and Okuda, M. (2011). "Centralized mobility load balancing scheme in LTE systems," *ISWCS, IEEE*, pp. 306–310.
[5] Yilmaz, O. N., Hamalainen, J. and Hamalainen, S. (2013). "Optimization of adaptive antenna system parameters in self-organizing lte networks," *Wireless Networks*, 19(6), 1251–1267.
[6] Tiwana, M., Sayrac, B., and Altman, Z. (2009). "Statistical learning for automated RRM: application to EUTRAN mobility," *Communications, 2009. ICC '09. IEEE International Conference*, pp. 1–5.
[7] Tiwana, M. I., Sayrac, B., and Altman, Z. (2010). "Statistical learning in automated troubleshooting: application to LTE interference mitigation," *Vehicular Technology, IEEE Transactions* 59(7), 3651–3656.
[8] Pedersen, K. I., Wang, Y., Soret, B., and Frederiksen, F. (2012). "EICIC functionality and performance for LTE HetNet co-channel deployments," *VTC Fall, IEEE*, pp. 1–5.
[9] Khan, Y., Sayrac, B., and Moulines, E. (2013c). "Surrogate based centralized SON: application to interference mitigation in LTE-A HetNets," *Proc. of IEEE VTC*, Spring, Dresden.
[10] Pauli, V., Naranjo, J. D., and Seidel, E. (2010). "Heterogeneous LTE networks and inter-cell interference coordination," *Technical report*, Nomor Research GmbH, Munich.
[11] Rahman, M., Yanikomeroglu, H., and Wong, W. (2009). "Interference avoidance with dynamic inter-cell coordination for downlink LTE system," *IEEE WCNC, WCNC'09*, IEEE Press, Piscataway, NJ, USA, pp. 1238–1243.

[12] Saquib, N., Hossain, E., and Kim, D. I. (2013). "Fractional frequency reuse for interference management in LTE-Advanced HetNets," *Wireless Communications, IEEE* 20(2), 113–122.

[13] Kosta, C., Imran, A., Quddus, A., and Tafazolli, R. (2011). "Flexible soft frequency reuse schemes for heterogeneous networks (macrocell and femtocell)," *Vehicular Technology Conference (VTC Spring), 2011 IEEE 73rd*, pp. 1–5.

[14] Xia, P., Liu, C.-H., and Andrews, J. G. (2013). "Downlink coordinated multi-point with overhead modeling in heterogeneous cellular networks," *IEEE Transactions on Wireless Communications* 12(8), 4025–4037.

[15] Simonsson, A. and Andersson, T. (2012). "LTE uplink comp trial in a hetnet deployment," *VTC Fall, IEEE*, pp. 1–5.

[16] Dahrouj, H. and Yu, W. (2010). "Coordinated beamforming for the multicell multi-antenna wireless system," *IEEE Transactions on Wireless Communications* 9(5), 1748–1759.

[17] An, H., Mohaisen, M., Han, D., and Chang, K. (2010). "Coordinated transmit and receive processing with adaptive multi-stream selection," CoRRabs/1005.5054.

[18] Geirhofer, S. and Gaal, P. (2012). "Coordinated multi point transmission in 3gpp lte heterogeneous networks," *Globecom Workshops, 2012 IEEE*, pp. 608–612.

[19] Clerckx, B., Kim, Y., Lee, H., Cho, J., and Lee, J. (2011). "Coordinated multi-point transmission in heterogeneous networks: A distributed antenna system approach," *Circuits and Systems (MWSCAS), 2011 IEEE 54th International Midwest Symposium*, pp. 1–4.

[20] Mayer, Z., Li, J., Papadogiannis, A., and Svensson, T. (2013). "On the impact of backhaul channel reliability on cooperative wireless networks," *Communications (ICC), 2013 IEEE International Conference*, pp. 5284–5289.

[21] Barbieri, A., Gaal, P., Geirhofer, S., Ji, T., Malladi, D., Wei, Y., and Xue, F. (2012). "Coordinated downlink multi-point communications in heterogeneous cellular networks," *Information Theory and Applications Workshop (ITA)*, 2012, pp. 7–16.

[22] Yeh, S.-P., Talwar, S., Himayat, N., and Johnsson, K. (2010). "Power control based interference mitigation in multi-tier networks," *GLOBECOM Workshops, IEEE*.

[23] Khan, Y., Sayrac, B., and Moulines, E. (2013b). "Centralized self-optimization of pilot powers for load balancing in LTE," *Personal Indoor and Mobile Radio Communications (PIMRC), 2013 IEEE 24th International Symposium*, pp. 3039–3043.

[24] Lu, Z., Sun, Y., Wen, X., Su, T., and Ling, D. (2012). "An energy-efficient power control algorithm in femtocell networks," *Computer Science Education (ICCSE), 2012 7th International Conference*, pp. 395–400.

[25] Khan, Y., Sayrac, B., and Moulines, E. (2013a). "Centralized self-optimization in lte-a using active antenna systems," *IFIP Wireless Days Conference*.

[26] Yilmaz, O., Hamalainen, S., and Hamalainen, J. (2009). "Comparison of remote electrical and mechanical antenna downtilt performance for 3GPP LTE," *Vehicular Technology Conference Fall (VTC 2009-Fall), 2009 IEEE 70th*, pp. 1–5.

[27] Yilmaz, O., Hamalainen, J., and Hamalainen, S. (2010). "Self-optimization of remote electrical tilt," *Personal Indoor and Mobile Radio Communications (PIMRC), 2010 IEEE 21st International Symposium*, pp. 1128–1132.

[28] Athley, F. and Johansson, M. (2010). "Impact of electrical and mechanical antenna tilt on LTE downlink system performance," *Vehicular Technology Conference (VTC 2010-Spring), 2010 IEEE 71st*, pp. 1–5.

[29] Barthelemy, J.-F. and Haftka, R. (1993). "Approximation concepts for optimum structural design: a review," *Structural Optimization* 5(3), 129–144.

[30] Papila, N., Shyy, W., Griffin, L., and Dorney, D. J. (2002). "Shape optimization of supersonic turbines using global approximation methods," *Journal of Propulsion and Power*.
[31] Rai, M. M. and Madavan, N. K. (2000). "Aerodynamic design using neural networks," *AIAA Journal* 38, 173–182.
[32] Booker, A. J., Dennis Jr., J. E., Frank, P. D., Serafini, D. B., Torczon, V., and Trosset, M. W. (1998). *A Rigorous Framework for Optimization of Expensive Functions by Surrogates*.
[33] Glaz, B., Friedmann, P. P., and Liu, L. (2008). "Surrogate based optimization of helicopter rotor blades for vibration reduction in forward flight," *Structural and Multi-disciplinary Optimization* 35(4), 341–363.
[34] Koziel, S. and Ogurtsov, S. (2012). "Selecting model fidelity for antenna design using surrogate-based optimization," *Antennas and Propagation Society International Symposium (APSURSI), 2012 IEEE*, pp. 1–2.
[35] Wright, J. A., Loosemore, H. A., and Farmani, R. (2002). "Optimization of building thermal design and control by multi-criterion genetic algorithm," *Energy and Buildings* 34(9), 959–972.
[36] Coley, D. A. and Schukat, S. (2002). "Low-energy design: combining computer-based optimisation and human judgement," *Building and Environment* 37(12), 1241–1247.
[37] Alexander, I. J., Forrester, A. S., and Keane, A. J. (2008). *Engineering Design via Surrogate Modelling: A Practical Guide*, John Wiley & Sons Ltd.
[38] Koziel, S. and Leifsson, L. (2013). *Surrogate-Based Modeling and Optimization: Applications in Engineering*, SpringerLink: Bücher, Springer.
[39] Queipo, N., Haftka, R., Shyy, W., Goel, T., Vaidyanathan, R., and Kevintucker, P. (2005). "Surrogate-based analysis and optimization," *Progress in Aerospace Sciences* 41(1), 1–28.
[40] Giunta, A. A., Wojtkiewicz, S. F., and Eldred, M. S. (2003). Overview of modern design of experiments methods for computational simulations.
[41] McKay, M. D., Beckman, R. J., and Conover, W. J. (2000). "A comparison of three methods for selecting values of input variables in the analysis of output from a computer code," *Technometrics* 42(1), 55–61.
[42] Morris, M. D. and Mitchell, T. J. (1995). *Journal of Statistical Planning and Inference* 43(3), 381–402.
[43] Johnson, M. E., Moore, L. M., and Ylvisaker, D. (1990). "Minimax and maximin distance designs," *Journal of Statistical Planning and Inference* 26(2), 131–148.
[44] Matheron, G. (1963). "Principles of geostatistics," *Economic Geology* 58(8), 1246–1266.
[45] Krige, D. G. (1953). "A statistical approach to some basic mine valuation problems on the Witwatersrand," *OR* 4(1).
[46] Sacks, J., Welch, W. J., Mitchell, T. J., and Wynn, H. P. (1989). "Design and analysis of computer experiments," *Statistical Science* 4(4), 409–423.
[47] Jones, D., Schonlau, M., and Welch, W. (1998). "Efficient global optimization of expensive black-box functions," *Journal of Global Optimization* 13(4), 455–492.
[48] Picheny, V., Wagner, T., and Ginsbourger, D. (2013). "A benchmark of Kriging-based infill criteria for noisy optimization," *Structural and Multidisciplinary Optimization*.
[49] Vapnik, V. N. (1995). *The Nature of Statistical Learning Theory*, Springer-Verlag New York, Inc., New York, NY, USA.
[50] Haykin, S. (1998). *Neural Networks: A Comprehensive Foundation*, 2nd edn, Prentice Hall PTR, Upper Saddle River, NJ, USA.

[51] Vazquez, E., Villemonteix, J., Sidorkiewicz, M., and Walter, E. (2008). "Global optimization based on noisy evaluations: an empirical study of two statistical approaches," *Journal of Physics: Conference Series* 135.

[52] Torczon, V. (1997). "On the convergence of pattern search algorithms," *SIAM Journal on Optimization* 7(1), 1–25.

16 Self-organized ICIC for SCN

Lorenza Giupponi, Ali Imran, and Ana Maria Galindo

In recent years the use of data services in mobile networks has notably increased, which requires a higher quality of services and data throughput capacity from operators. These requirements become much more demanding in indoor environments, where, due to the wall-penetration losses, communications suffer a higher detriment. As a solution, short-range base stations (BSs), known as femto cells [1], are proposed. Femto cells are installed by the end consumer and communicate with the macro cell system through the internet by means of a digital subscriber line (DSL), fiber, or cable connection. Due to this deployment model, the number and location of femto cells are unknown for the operators and therefore there is no possibility for centralized network planning. In the case of co-channel operation, which is the more rewarding option for operators in terms of spectral efficiency, an aggregated interference problem may arise due to multiple, simultaneous, and uncoordinated femto cell transmissions. On the other hand, the macro cell network can also cause significant interference to the femto cell system due to a lack of control of position of the femto nodes and their users. In this chapter we focus then on the challenging problem in the area of small cell networks, the inter-cell interference coordination among different layers of the network.

We propose two different self-organizing solutions, based on two smart techniques that operate on the most appropriate configuration parameters of the network for each situation. In particular, we address femto–macro and macro–femto problems. On the one hand, for the femto–macro case, we propose in Section 16.1 a machine learning (ML) approach to optimize the transmission power levels by modeling the multiple femto BSs as a multi-agent system [2], where each femto cell is able to learn transmission power policy in such a way that the interference it is generating, added to the whole femto cell network interference, does not jeopardize the macro cell system performance. To do this, we propose a reinforcement learning (RL) category of solutions [3], known as time-difference learning. On the other hand, for the macro–femto problem, we propose in Section 16.2 a solution based on interference minimization through self-organization of antenna parameters. In homogeneous macro cell networks, BS antenna parameters are configured in the planning and deployment phase and left unchanged for a long time. This approach works well for homogeneous networks, as the topology of the system remains unchanged over a long time. However, this is not the case in heterogeneous networks where different layers of the networks are deployed and configured in an impromptu manner and based on the timely conditions of the environment. Consequently, the density, locations, and activity levels of the small cells may change over space and time. Hence there is need to revisit the macro cell planning from this perspective and to apply a much more appropriate self-organized approach, which is the one proposed in this chapter.

16.1 Femto–macro interference control: a time-difference learning approach

Self-organization techniques based on ML approaches are introduced in this section to perform the radio resource management (RRM) procedures for the coexistence of macro and femto networks. In particular, we propose to map the femto cells onto a multi-agent system [2], where each femto BS is an intelligent and autonomous agent that learns [3] by directly interacting with the environment and by properly utilizing the past experience. Multi-agent systems are characterized by the following: (i) the intelligent decisions are made by multiple and uncoordinated nodes; (ii) the nodes partially observe the overall scenario; and (iii) their input to the intelligent decisions process is different from node to node since they come from spatially distributed sources of information. The reason for proposing a multi-agent system is found in the impossibility for the femto network to be managed by means of a centralized node, due to the number of femto cells and the lack of information to the network operator regarding their location.

The environment in which the multi-agent system is operating is dynamic due to the characteristics of the mobile wireless scenario, e.g., existence of lognormal shadowing, fading, mobility of user, etc., and to the cross-dependencies of actions made by the multiple agents. We model the natural evolution of the environment through states and the multi-agent system, through a stochastic game. A stochastic game is the extension of a Markov decision process (MDP), which is the natural model of a single-agent scenario, to multiple agents.

Real scenarios formed by simultaneously performing multiple agents commonly present highly dynamic and unstable behaviors. This occurs because the policy learnt by an agent at a given moment may not be valid anymore when the environment switches to a new state due to potential actions performed in parallel by other agents in the system. This characteristic does not allow us to define a probabilistic state transition model, therefore we propose to solve the stochastic game through time-difference RL algorithms. The RL paradigm is based on learning from interactions with the environment through actions, where knowledge is built based on the observed consequences when a given action is executed. In this context, we focus on the paradigm of independent learning where each femto cell (agent) learns independently a power-allocation strategy for interference avoidance. The interactive learning dynamics will be evaluated in terms of system performances and speed of convergence.

We solve the interference management problem from femto to macro systems through Q-learning (QL), which is a typical form of time difference (TD) RL. The TD algorithms are based on quantifying, by means of the Q-function, the quality of an action in a certain state. The knowledge of the agent is then represented by the so called Q-values, which are usually stored in a Q-table. Consequently, states characterizing the environmental situation and the available actions have to be represented by discrete values, and therefore the use of thresholds is mandatory. This entails an important intervention of the learning-system designer selecting the mentioned thresholds for the state representation, and setting the amount and the values of the available actions. When designing the learning system, the selection of the amount of states representing the environment and the actions available in each state, plays an important role in the agent behavior. The size of those sets directly affects the system adaptability and therefore its performance. Besides, it is directly related with the feasibility in the knowledge representation, i.e., when the number of state-action pairs is large or the input variables are continuous, the memory requirement to store the Q-table may become impracticable, as well as the required learning time.

A solution to the previously presented weak points is to use a form of continuous state and action representation without the requirement of near-infinite Q-tables. This would allow us to build a system capable of working independently from the scenario and designer criterion, which would be in line with the self-organized requirements of future networks. To this end, we propose to improve the QL algorithm by the introduction of fuzzy inference systems (FISs), in order to represent state and action spaces continuously. This approach is called fuzzy Q-learning (FQL). Fuzzy logic, introduced by Lotfi Zadeh [4], is a way to map a fuzzy input space to a crisp output space by means of membership functions. The combination of FIS and RL was introduced by Berenji in [5] and then extended by Glorennec and Jouffe in [6] and [7]. They also presented an interesting case study, where they applied FQL for navigation system in autonomous robots [8]. Additional to the benefits already mentioned, FQL offers other interesting advantages such as: (1) a more compact and effective expertness representation mechanism, and (2) the possibility of speeding up the learning process by incorporating offline expert knowledge in the inference rules.

This section is then structured as follows. The system model is presented in Section 16.1.1. In Section 16.1.2, we introduce the multi-agent systems. In Section 16.1.3, we briefly present the QL and FQL algorithms. In Section 16.1.4, we present the simulation scenario, the reference algorithms, and describe relevant simulation results.

16.1.1   System model

We consider a heterogeneous wireless network composed of a set of $\mathcal{M}$ macro cells that coexist with $\mathcal{F}$ femto cells. The $M = |\mathcal{M}|$ macro cells form a regular hexagonal network layout with inter-site distance D, and provide coverage over the entire network, comprising both indoor and outdoor users. The $F = |\mathcal{F}|$ femto cells are placed indoors within the macro-cellular coverage area following the 3rd Generation Partnership Project (3GPP) dual-stripe deployment model. Both macro cells and femto cells operate in the same frequency band, which allows us to increase the spectral efficiency per area through spatial frequency reuse.

An orthogonal frequency division multiple access (OFDMA) downlink is considered, where the system bandwidth B is divided into R resource blocks (RBs), with $B = R \cdot B_{RB}$. RB represents one basic time-frequency unit that occupies the bandwidth B_{RB} over time

T. Associated with each macro cell and femto cell are U^M macro and U^F femto users, respectively. The multi-user resource assignment that distributes the R RB among the U^M macro and U^F femto users, is carried out by a proportional fair scheduler.

We denote by $\mathbf{p}_t^n = (p_{1,t}^n, \ldots, p_{R,t}^n)$ the transmission power vector of BS n at time t, with $p_{r,t}^n$ denoting the downlink transmission power of RB r. The maximum transmission powers for femto cells and macro cells are $P_{\max}^F$ and $P_{\max}^M$, with $P_{\max}^F \ll P_{\max}^M$, such that $\sum_{r=0}^{R} p_{r,t}^m \leq P_{\max}^M, m \in \mathcal{M}$ and $\sum_{r=0}^{R} p_{r,t}^f \leq P_{\max}^F, f \in \mathcal{F}$.

We analyze the system performance in terms of signal to interference-plus-noise ratio (SINR) and achieved data rate given in bit/s. Assuming perfect synchronization in time and frequency, the SINR of macro user u who is allocated in RB r of macro cell $m \in \mathcal{M}$ amounts to:

$$\gamma_{r,t}^m = \frac{p_{r,t}^m h_{r,t}^{mu}}{\sum_{m \in \mathcal{M}, n \neq m} p_{r,t}^n h_{r,t}^{nu} + \sum_{f \in \mathcal{F}} p_{r,t}^f h_{r,t}^{fu} + \sigma^2}$$

where $h_{r,t}^{mu}$ accounts for the link gain between the transmitting macro cell m and its macro user u; while $h_{r,t}^{nu}$ and $h_{r,t}^{fu}$ represent the link gain of the interference that BS n and f imposes on macro user v, respectively. Finally, σ^2 denotes the thermal noise power.

Likewise, the SINR of femto user v who is allocated in RB r by femto cell $f \in \mathcal{F}$ is in the form:

$$\gamma_{r,t}^f = \frac{p_{r,t}^f h_{r,t}^{fv}}{\sum_{m \in \mathcal{M}} p_{r,t}^m h_{r,t}^{mv} + \sum_{n \in \mathcal{F}, n \neq f} p_{r,t}^n h_{r,t}^{nv} + \sigma^2}$$

where $h_{r,t}^{fv}$ represents the link gain between the transmitting femto cell f and its femto user v, $h_{r,t}^{nv}$ and $h_{r,t}^{mv}$ indicate the link gain between BS n and m and femto user v, respectively.

The data rate that cell n achieves at slot t is the sum over the individual rates of all RBs, which is upper bounded by the Shannon capacity:

$$C_t^n = \sum_{r=0}^{R} C_{r,t}^n = \sum_{r=0}^{R} \frac{B}{R} \log_2 \left(1 + \gamma_{r,t}^n\right)$$

where $C_{r,t}^n$ is the capacity of RB r at slot t, and $n \in \mathcal{M}$ determines the rate of a macro cell and $n \in \mathcal{F}$ that of a femto cell. Then, the total sum-rate of the network amounts to:

$$C_t^{sys} = \sum_{m \in \mathcal{M}} C_t^m + \sum_{f \in \mathcal{F}} C_t^f$$

16.1.2 Introduction to multi-agent systems

In machine learning literature, the ability of learning new behaviors online and automatically adapting to the temporal dynamics of the system is commonly associated with RL [9]. At each learning iteration, the agent perceives the state of the environment and takes an action to transit to a new state. A scalar cost is received, which evaluates the quality of this transition. In a wireless setting, the single-agent approach can be applied in scenarios characterized by only one decision maker, as is the case for the radio network controller (RNC) or the serving gateway in cellular networks. On the other hand, the multi-agent approach is applied in situations where the intelligence has to be distributed across multiple nodes, as is the case of femto cell networks. In this chapter we refer to the multi-agent system. The theoretical framework can be found in stochastic games, which are non-cooperative games defined by the following quintuple $\{\mathcal{F}, \mathcal{S}, \mathcal{A}, \mathcal{P}, \mathcal{C}\}$, where:

- $\mathcal{F} = \{1, 2, \ldots, F\}$ is the set of agents (femto BS).
- $\mathcal{S} = \{s_1, \ldots, s_k\}$ is the set of possible states and k is the agent's number of available states.
- $\mathcal{A} = \{a_1, \ldots, a_l\}$ is the set of actions and l is the agent's number of available actions.
- $\mathcal{C} : \mathcal{S} \times \mathcal{A} \rightarrow R$ is the cost function, which is fed back to each agent after the execution of a certain action.
- $\mathcal{P}$ is a probabilistic transition function, defining the probability of migrating from one state to another, provided the execution of a certain joint action.

For each independent agent, the state transition function probabilistically specifies the next state of the environment as a function of its current state and the joint action. The cost function specifies expected instantaneous cost as a function of current state and action. The model is a *Markov* model if the state transitions are independent of any previous environment state or agent actions. The objective is to find a policy that minimizes the cost of each state x. As a result, the aim is to find an optimal policy for the infinite-horizon discounted model, relying on the result that, in this case, there exists an optimal deterministic stationary policy [9].

We define the optimal value of state x as the expected infinite discounted sum of costs that the agent gains if it starts in state x and then executes the optimal policy. We indicate with π the complete decision policy, so that the optimal value of state x can be written as:

$$V^*(x) = \min_{\pi} E\left(\sum_{t=0}^{\infty} \delta^t c(t)\right) \tag{16.1}$$

where $0 \leq \delta < 1$ is a discount factor and $c(t)$ is the value of function $\mathcal{C}$ in time t. According to the principle of Bellman's optimality [9], this optimal value function is unique and can be defined as the solution to the equation:

$$V^*(x) = \min_{a} \left(\tilde{c} + \delta \sum_{y \in \mathcal{S}} P_{x,y}(a) V^*(y)\right) \tag{16.2}$$

which asserts that the value of state x is the expected cost $\tilde{c} = E\{c\}$, plus the expected discounted value of the next state, y, using the best available action. Given the optimal value function, we can specify the optimal policy as:

$$\pi^*(x) = \arg\min_{a} \left(\tilde{c} + \delta \sum_{y \in \mathcal{S}} P_{x,y}(a) V^*(y)\right) \tag{16.3}$$

Figure 16.1 Q-table for task r of agent f.

In the literature, two ways have been identified to solve this problem. The first one consists of the knowledge of the state transition probability function $\mathcal{P}_{x,y}(a)$. The second one, conversely, does not rely on this previous knowledge and is based on RL. As a result, RL is primarily concerned with how an agent ought to take actions in an environment so as to minimize the notion of long-term cost, that is, so as to obtain the optimal policy, when the state transition probabilities are not known in advance. In multi-agent settings, where each agent has incomplete information about state-transition probabilities and payoff functions of other agents, we can say that it learns independently from the other agents, therefore we approximate the other agents as part of the environment, and we still can apply the Bellman's criterion introduced above. In this case, the convergence to optimality proof does not hold strictly, but this independent learning approach has been shown to correctly converge in multiple applications [10]. As a result, we rely on RL, and in particular on QL, which has the ability of learning optimal decision policies through real-time interactions with the surrounding environment.

16.1.3 Learning methods

In this subsection we present the proposed QL and FQL learning algorithms. For the sake of nomenclature simplicity, in what follows we do not include the time indicator t.

16.1.3.1 Q-learning

In QL, the state of the surrounding environment and available actions to the agents are commonly represented by discrete sets. The QL algorithm is based on quantifying, by means of the Q-function, the quality of an action in a certain state. Therefore to be able to learn from the past, the Q-values have to be stored in a representation mechanism. The look-up table, represented in Figure 16.1, is the most commonly used and the most direct method when memory requirement is not a problem.

The state vector $\vec{x} \in \mathcal{S}$ is composed by values of representative variables capturing the surrounding environment, being $\mathcal{S} = \{s_1, \ldots, s_k\}$ the set of possible states the agent can perceive, for each RB r. The set of actions $\mathcal{A} = \{a_1, \ldots, a_l\}$ represent the decisions that the agent can make based on the state vector $\vec{x}$. Based on $\vec{x}$ and the corresponding

Q-values, which are stored in the Q-table, the most appropriate action $a \in \mathcal{A}$ is selected. After the execution of a in $\vec{x}$, the agent receives an immediate scalar cost c and the corresponding Q-value $Q(\vec{x}, a)$ is updated according to the rule:

$$Q(\vec{x}, a) \leftarrow Q(\vec{x}, a) + \alpha \left[c + \delta \min_a Q(\vec{y}, a') - Q(\vec{x}, a) \right] \quad (16.4)$$

where α is the learning rate, $Q(\vec{y}, a')$ is the next state Q-value, and a' is the next state optimal action.

The state representation is given by the vector state:

$$\vec{x} = \{Pow^f, C_r^m, C_r^f\}$$

where $Pow^f = \sum_{r=0}^{r=R} p_r^f$ denotes the total transmission power of femto cell f over all RBs. The considered cost for RB r and femto cell f is:

$$c = \begin{cases} K & \text{if } Pow^f > P_{\max}^F \text{ or } C_r^m < C_{Th}^M, \\ K \exp^{(-C_r^f C_r^m)} & \text{otherwise} \end{cases} \quad (16.5)$$

where K is a constant initialized to a high value and C_{Th}^M is the minimum capacity per RB that the macro cell has to fulfill. The rationale behind this cost function is that if the macro cell capacity at RB r is below the threshold C_{Th}^M and the femto cell total transmission power is above the maximum allowed $P_{\max}^F$, the agent will receive a high-cost value K. On the other hand, if both constraints are fulfilled the femto cell will focus on maximizing its own capacity and the macro cell capacity at RB r.

16.1.3.2 Fuzzy Q-learning

When the number of state-action pairs is large or the input variables are continuous, the memory requirements may become infeasible. In addition, the selection of discrete sets for state and action definitions may highly affect the system performance. The FQL scheme allows to solve those drawbacks, by combining the advantages of FIS and QL. In particular, the FIS allows us to generalize the state space and to generate continuous actions [11], which is a big advantage in terms of precision, since femto cells will be able to better adjust their actions.

We consider an input state vector $\vec{x}$, represented by L fuzzy linguistic variables. For each RB r we denote $\bar{S} = \{\bar{s}_1, \ldots, \bar{s}_k\}$ the set of fuzzy state vectors of L linguistic variables. For state $\bar{s}_i$, we denote $\mathcal{A} = \{a_1, \ldots, a_l\}$ the set of possible actions. The rule representation for state $\bar{s}_i$ is:

$$\text{If } \vec{x} \text{ is } \bar{s}_i, \text{ Then } a_1 \text{ with } q(\bar{s}_i, a_1)$$
$$\ldots$$
$$\text{or } a_j \text{ with } q(\bar{s}_i, a_j)$$
$$\ldots$$
$$\text{or } a_l \text{ with } q(\bar{s}_i, a_l)$$

where a_j is the j-*th* action candidate, which is possible to choose for state $\bar{s}_i$, and $q(\bar{s}_i, a_j)$ is the fuzzy Q-value for each state-action pair $(\bar{s}_i, a_j)$. The FQL has two outputs as a result of the defuzzification process. One corresponds to the inferred action after defuzzifying the n rules and the other represents the Q-value for the state-action pair $(\vec{x}, \hat{a})$. They are given by:

$$a = \frac{\sum_{i=1}^{n} w_i \times \hat{a}}{\sum_{i=1}^{n} w_i} \quad Q(\vec{x}, a) = \frac{\sum_{i=1}^{n} w_i \times q(\bar{s}_i, \hat{a})}{\sum_{i=1}^{n} w_i}$$

where w_i represents the truth value (i.e., the fuzzy-AND operator) of the rule representation of FQL for $\bar{s}_i$, and $\hat{a}$ is the action selected for state $\bar{s}_i$.

Q-values have to be updated after the action selection process. Since there is a fuzzy Q-value per each state-action pair, in each iteration, n fuzzy Q-values have to be updated based on:

$$q(\bar{s}_i, \hat{a}) = q(\bar{s}_i, \hat{a}) + \alpha \Delta q(\bar{s}_i, \hat{a})$$

where $\Delta q(\bar{s}_i, \hat{a}) = [c + \delta(Q(\vec{y}, a') - Q(\vec{x}, a))] \times \frac{w_i}{\sum_{i=1}^{n} w_i} \cdot c$ represents the cost obtained applying action a in state vector $\vec{x}$ and $Q(\vec{y}, a')$ is the next-state optimal Q-value defined as:

$$Q(\vec{y}, a') = \frac{\sum_{i=1}^{n} w_i \times q(\bar{v}_i, a^*)}{\sum_{i=1}^{n} w_i}$$

and $a^* = \arg\min(q(\bar{v}_i, a^*_j))$ with $j = (1, \ldots, l)$ is the optimal action for the next state $\bar{v}_i$, after the execution of action $\hat{a}$ in the fuzzy state $\bar{s}_i$.

Figure 16.2 shows the FQL structure for the FQL power, macro and femto capacity-based (FQL-PMFB) algorithm as a four-layer FIS. The functionalities of each layer are the following:

- *Layer 1*: This layer has as input three linguistic variables defined by the term sets: $T(Pow^f) = \{\text{Very Low(VL), Low(L), Medium(M), High(H), Very High(VH)}\}$, $T(C_r^m) = \{L, M, H\}$ and $T(C_r^f) = \{L, M, H\}$. Therefore considering the number of fuzzy sets in the three term sets in Layer 1 we have $z = |T(Pow^f)| + |T(C_r^m)| + |T(C_r^f)| = 11$ term nodes. Every node is defined by a membership function with a bell-shape form (Figure 16.3), so that the output $O_{1,h}$ for a generic component x of $\vec{x}$ is given by:

$$O_{1,h} = \exp^{-\frac{(x - e^h)^2}{(\rho^h)^2}} \quad h = 1, \ldots, z$$

where e^h and ρ^h are the mean and the variance of the bell-shape function associated to node h, respectively. The membership function definitions for the proposed FQL are based on expert knowledge. Different mean and variance values, and the amount of fuzzy sets for the four term sets description, have been tested through simulations. Subjectivity in the term set definitions is absorbed by the learning process in Layer 3, due to the inherent adaptive capability of FQL algorithms.
- *Layer 2*: This is the rule nodes layer. It is composed by $n = |T(Pow^f)| \times |T(C_r^m)| \times |T(C_r^f)| = 45$ nodes. Each node gives as output the truth value of the i-th fuzzy rule. Each node in Layer 2 has three input values, one from one linguistic variable of each of the three components of the input state vector, so that the i-th rule node is represented

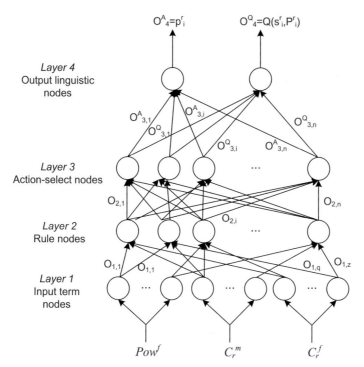

Figure 16.2 Fuzzy Q-learning power, macro and femto capacity-based structure.

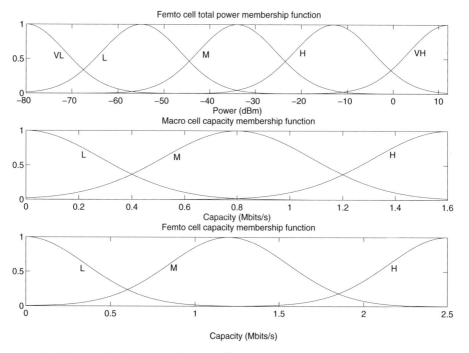

Figure 16.3 Membership functions of the input linguistic variables.

by the fuzzy state vector $\bar{s}_i$. The Layer 2 output $O_{2,i}$ is the product of three membership values corresponding to the inputs. Truth values are represented as:

$$O_{2,i} = \prod_{h=1}^{3} O_{1,h} \quad i = 1, \ldots, n$$

- *Layer 3*: This layer is composed by n nodes each of which is an action-select node based on the ε-greedy policy. Here the set of possible actions for each Layer 3 node are $l = 60$ power levels. Those power levels range from –80 dBm to 10 Bm effective radiated power (ERP). The $q(\bar{s}_i, \hat{a})$ values are initialized based on expert knowledge. The node i generates two normalized outputs, which are computed as:

$$O_{3,i}^A = \frac{O_{2,i} \times \hat{a}}{\sum_{d=1}^{n} O_{2,d}} \quad i = 1, \ldots, n$$

$$O_{3,i}^Q = \frac{O_{2,i} \times q(\bar{s}_i, \hat{a})}{\sum_{d=1}^{n} O_{2,d}} \quad i = 1, \ldots, n$$

- *Layer 4*: This layer has two output nodes, action node O_4^A and Q-value node O_4^A, which represent the defuzzification method. The final outputs are given by:

$$O_4^A = \sum_{i=1}^{n} O_{3,i}^A \quad O_4^Q = \sum_{i=1}^{n} O_{3,i}^Q$$

16.1.4 Evaluation through system-level simulations

In this section we present the reference algorithms we use and some simulation results. The scenario considered to validate the proposed approach is based on 3GPP Technical Specification Group (TSG) Radio Access Network (RAN) WG4 (Working Group 4) simulation assumptions and parameters [12]. Visualized in Figure 16.4, it is deployed in an urban area and operates at 1,850 MHz. We consider $M = 1$ macro cells with radius $D = 500$ m and $BA = 1$ blocks of apartments, based on the dual stripes femto deployment model. We introduce an occupation ratio, which determines whether inside an apartment there is a femto cell or not. Furthermore, each femto cell has a random activity pattern. Each femto cell provides service to its $U^F = 2$ associated femto users, which are randomly located inside the femto cell area. Macro users are also located randomly inside the femto cell block. We consider that macro users are always outdoors and femto users are always indoors. We consider the macro and femto systems to be based on long-term evolution (LTE), therefore the frequency band is divided into RB of width 180 kHz in the frequency domain and 0.5 ms in the time domain. Those RB are composed of $N_{SC} = 12$ subcarriers of width $\Delta f = 15$ kHz and 7 orthogonal frequency division modulation (OFDM) symbols. For simulations, we consider $R = 4$, which corresponds to an LTE implementation with a channel bandwidth of $BW = 0.7$ MHz. The antenna patterns for macro BS, femto BS, and macro/femto users are omnidirectional, with 18 dBi, 0 dBi, and 0 dBi antenna gains, respectively. The shadowing standard deviation is 8 dB and 4 dB, for macro and femto systems, respectively. The macro and femto BS noise figures are 5 dB and 8 dB, respectively. The transmission power of the macro BS

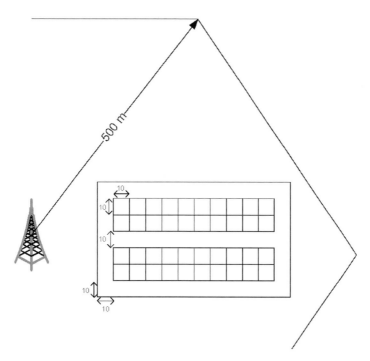

Figure 16.4 System layout.

is 46 dBm, whereas the femto BS adjusts its power through the learning scheme to a value of maximum $P_{max}^F = 10$ dBm.

The considered PL models are based on [12] for the case of urban scenarios and are described with details in [13]. We also consider the 3GPP implementation of frequency-selective fading model specified in [14] (urban macro settings) for macro BS to user propagation, and a spectral block fading model with coherence bandwidth of 750 kHz for indoor propagation.

With respect to the FQL algorithm, the variances ρ^h of the membership functions in Layer 1 are 12, 0.4, and 0.5 for total power, macro cell capacity, and femto cell capacity inputs, respectively. The selected mean values e^h for Pow^f are –80, –55, –34, –13, and 12, for C_r^m are 0, 0.8, and 1.6, and finally for C_r^f are 0, 1.2, and 2.5. Also, we introduce a probability $\varepsilon = 0.07$ of visiting random states in the initial 50% of the FQL iterations and this probability is decreased to $\varepsilon = 0.04$ until the 80% of iterations. The benchmark algorithm is implemented considering $\eta = 0.35$ and $\beta = 0.8$.

16.1.4.1 Reference algorithms

In this subsection we first present the reference algorithms we use to compare the proposed learning methods and then we show some simulation results.

The proposed algorithms are compared to two reference FQL algorithms with lower grade of complexity than the proposed FQL-PMFB and a smart power control (SPC) based on interference measurements proposed in [15].

- **FQL power-based (FQL-PB)**: The algorithm's objective is to maintain only the total femto BS transmission power below a given threshold. The vector state is represented as:

$$\vec{x} = \{Pow^f\}$$

And the cost equation is the following:

$$c = \begin{cases} K & \text{if } Pow^f > P^F_{max}, \\ 0 & \text{otherwise} \end{cases}$$

The rationale behind this cost function is that the total transmission power of each femto cell does not exceed the allowed P^F_{max}.

Since the structure of this algorithm is very similar to the FQL-PMFB we only highlight the different parameters.

Layer 1: The term set of the input linguistic variables is defined by the following fuzzy set: $T(Pow^f) = \{$VL, L, M, H, VH$\}$. The Layer 1 of the fuzzy system is composed by $z = |T(Pow^f)| = 5$ term nodes.

Layer 2: This is composed by $n = |T(Pow^f)| = 5$ nodes.

- **FQL power and macro cell capacity-based (FQL-PMB)**: This algorithm objective is to maintain the total femto BS transmission power below the threshold and at the same time maximize the macro cell capacity. The input state variable is the following:

$$\vec{x} = \{Pow^f, C^m_r\}$$

and the cost equation:

$$c = \begin{cases} K & \text{if } Pow^f > P^F_{max} \text{ or } C^m_r < C^M_{Th}, \\ 0 & \text{otherwise} \end{cases}$$

The rationale behind this cost function is that, besides the total transmission power control of the femto cell, it guarantees that the macro cell capacity is above a desired threshold.

The layered structure is as for FQL-PMFB, but:

Layer 1: The linguistic variables of the system are defined by the following fuzzy sets: $T(Pow^f) = \{$VL, L, M, H, VH$\}$, $T(C^m_r) = \{$L, M, H$\}$, so that Layer 1 is composed by $z = |T(Pow^f)| + |T(C^m_r)| = 8$ term nodes, each one representing a fuzzy term of an input linguistic variable.

Layer 2: This is composed by $n = |T(Pow^f)| \times |T(C^m_r)| = 15$ nodes.

- **Smart power control**: In this algorithm the femto cell BS adjusts its RB transmission power based on the total received interference at the femto cell BS. The scheme is open loop, and it does not involve the femto users and signaling between network nodes [15]. The femto BS adjusts the RB maximum transmission power according to:

$$p^f_r = \max\left(\min\left(\eta \cdot (E_c + 10\log(R \cdot N_{sc})) + \beta, P^F_{max}\right), P^F_{min}\right)$$

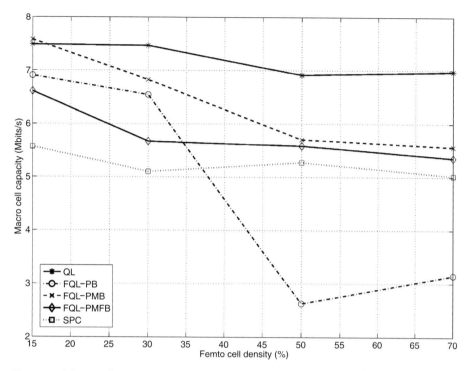

Figure 16.5 Macro cell capacity.

where η is a linear scalar that allows altering the slope of power control mapping curve and β is a parameter expressed in dB, both of which are femto cell configuration parameters. Furthermore, N_{sc} is the number of subcarriers, P_{min}^F is the minimum femto cell transmit power, and E_c is the reference signal received power per resource element present at the femto node.

16.1.5 Simulation results

In this subsection we evaluate the QL and fuzzy algorithms' behavior in terms of macro cell, average femto cell, total system capacity, and convergence speed. Figure 16.5 depicts the behavior in terms of macro cell capacity. Analyzing the fuzzy algorithms' behavior, we can see that the FQL-PMB and FQL-PMFB algorithms, differently from the FQL-PB algorithm, are required in their cost equations to maintain the macro cell capacity above the C_{Th}^M, and so they behave, contrarily to FQL-PB, which is not able to maintain this target. In addition, it is worth mentioning that the FQL-PMFB obtains lower values of macro cell capacity than QL and FQL-PMB, since due to its cost function definition, it also aims at maximizing the femto cells' capacity. QL maintains the macro cell capacity at higher values due to the discrete nature of its action space, while femto cells applying FQL, to fulfil the macro cell performance requirements, can transmit at higher power levels, which increases the femto capacity but decreases the macro one. Finally, the QL, FQL-PMB, and FQL-PMFB perform better than the benchmark algorithm, since

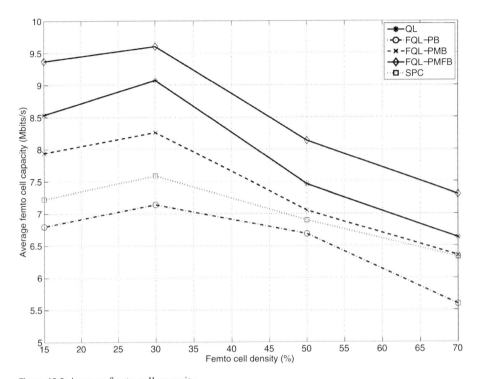

Figure 16.6 Average femto cell capacity.

when increasing the density of femto cells, it does not adaptively operate to maintain the interference below a threshold.

Figure 16.6 shows the system behavior in terms of average femto cell capacity. It can be observed that average femto cell capacity decreases with the femto cell density due to the increase of femto–femto interference in all cases. Since QL and FQL-PMFB include in their cost equation (16.5) the maximization of the femto cell capacity, they are able to increase the average femto cell capacity up to more than 0.5 and 1 Mbits/s with respect to FQL-PMB and 1.5 and 2 Mbits/s with respect to FQL-PB, respectively. As mentioned before, the FQL-PMFB is able to choose more precise actions, which allow femto cells to reach higher performances maintaining the macro cell system requirements. In addition, similarly to the macro cell capacity case, the QL, FQL-PMB, and FQL-PMFB algorithms outperform the SPC algorithm.

Figure 16.7 represents the system behavior in terms of total system capacity. As expected QL and FQL-PMFB have higher system capacity, outperforming the FQL-PMB and the SPC algorithm by up to 20 Mbits/s and the FQL-PB by up to 40 Mbits/s. The total system capacity increases with the femto cell occupation ratio since there are more nodes in the network. On the other hand, the femto cells' capacity decreases as a function of femto nodes' density due to the the interference between them.

Finally, to assess the speed of convergence, we compare the iterations before convergence required by the FQL algorithm and by the QL, in a scenario with 70% of femto

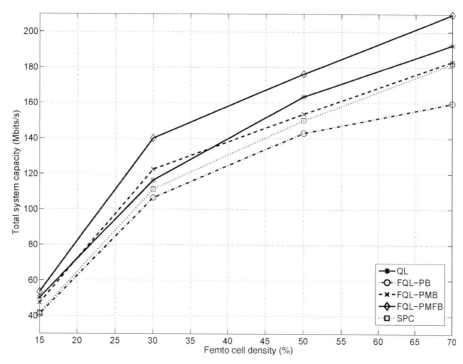

Figure 16.7 Total system capacity.

cell density. The FQL increases the speed of convergence up to 25% with respect to the QL algorithm, due to the previous knowledge that can be embedded in the rules.

16.2 Macro–femto interference minimization through self-organization of macro cell azimuth angles

In the second section of the chapter we turn our attention to the macro–femto interference problem, which remains a major challenge in heterogeneous networks. We particularly focus on interference minimization through self-organization of antenna parameters. In homogeneous macro cell networks BS antenna parameters are configured once in the planning and deployment phase and left unchanged for a long time. This approach works sufficiently well in homogeneous macro cells because the topology of macro BSs remains unchanged for substantially long periods of time. However, in heterogeneous small cell networks this is not the case. Small cells are deployed in impromptu manner. Consequently the density, locations, and activity levels of the small cells may change over space and time. Hence, there is need to revisit the macro cell planning from this perspective. However, in case of heterogeneous networks it might not be possible to determine macro BS azimuth angles during the planning and deployment phase, which remain optimal for a long period. Instead, the macro BSs need to be equipped with self-organizing (SO) mechanisms to update their azimuth angles continually based on the spatio-temporal dynamics of the heterogeneous network. To achieve this objective, in this section we propose and analyze a solution for self-optimization of the antenna azimuths of macro cells, while taking into account the current system topography, as well as user demography, in an online manner.

The rest of this section is organized as follows. Subsection 16.2.1 provides the background of the problem. In Subsection 16.2.2 we present the system model and the assumptions. Subsection 16.2.3 presents the problem formulation. In order to achieve an SO solution, in Subsection 16.2.4 we propose a low-complexity distributed method to solve the system-wide azimuth angle optimization problem. Subsection 16.2.5 presents numerical results.

16.2.1 Background

Azimuth orientation of sectors has a major impact on cellular system performance, both in homogeneous and heterogeneous networks, as it determines the interference, as well as the coverage. However, in the majority of academic literature on interference management, sectorization is overlooked altogether [16–18], for sake of simplicity, by assuming omnidirectional antennas. Some works that do incorporate sectorization generally assume ideally symmetric sector orientations, e.g., 30°, 150°, 270° [19]. In commercial cellular systems, on the other hand, sector azimuth angle optimization is a key component of the performance optimization process; and for that, azimuth angles have to be set optimally for each sector. This optimization needs to take into account not only the antenna heights and orientations of neighboring sites, but also the demographic and topographic factors in the vicinity of that site.

There are two possible approaches to solve this optimization problem. The first approach consists of two steps: First a simulation model of the cellular system is built to link optimization parameters such as azimuth angles, tilts, heights, and BS locations, with key performance indicators (KPIs) such as coverage or capacity (i.e., optimization objectives). Here the simulation model acts as a black box between the planning parameter and the KPIs of interest. Since such black box hardly provides a direct insight into systems behavior as a function of the optimization parameters, generally the optimal configuration has to be obtained by exploring the solution search space extensively. However, the search space of such optimization problems is tremendously large, even for reasonably small size networks. To get a quantitative perspective, consider that for only 19 base stations having 3 sectors each, whose antenna height, tilt, and sectors have to be optimally configured and can have 10 different values, each solution space can be as large as 10^{60}. With an optimistic assumption that one can have computation power to perform 10^{12} evaluations of the objective function per second, it may take longer than the known age of the universe to explore the whole search space for this commonly occurring problem. Note that for simulation-based evaluation engines, the evaluation time can be much longer than the one assumed above [20]. Given the non-polynomial hard nature of the problem [21] and the unfathomable search space, brute-force search is not computationally feasible. Therefore as a second step in this approach, meta-heuristics such as simulated annealing [22], genetic algorithms [23, 24], particle swarm

[25, 26], Taugchi's method [27], and ant colony optimization [28], are applied to selectively explore the search space and obtain near-optimal solutions. While this approach allows us to consider a large number of optimization parameters simultaneously in the optimization process – and therefore it is predominant in the academic literature in general and commercial planning tools in particular – it potentially suffers from the following drawbacks: (1) keeping a simulation model updated to capture the latest real topographic features is a daunting task; (2) metaheuristics generally generate suboptimal solutions for which the optimality gap may not be known; (3) the outcome of this approach largely depends on the practitioner's experience, i.e., the choice of parameters used to initially configure the metaheuristic techniques largely determines the quality of outcome. For example, in the case of simulated annealing – i.e., a widely used metaheuristic in cellular system planning and optimization – there are no general rules to set the *temperature* and *acceptance probability* relationships [29]. A detailed discussion on the use of metaheuristics in such problems can be found in [30]. Here it would suffice to comment that despite the prevailing use of metaheuristic techniques for offline planning, this approach is too time consuming to be implemented in an online manner. Therefore it is not suitable to cope with the acute dynamics of a heterogeneous network, mainly due to its ever-changing topology.

The second possible approach to solve the optimization problem is to resort to more robust mathematical optimization. Here, instead of relying on the black box of simulation model, as a first step a more transparent analytical model to link the system-level performance with the optimization parameter needs to be developed. In the second step, suitable optimization tools are invoked to obtain optimal solutions. Unlike simulation-based approaches this method is more transparent and less time consuming, and can lead to solutions whose quality can be verified easily. However, this approach suffers from one major drawback. It is not easy to mathematically model an amalgam of KPIs, such as coverage, capacity, and QoS, as tractable functions of a plethora of system optimization parameters, such as BS location, tilt, azimuth, frequency, heights, transmission powers, etc. It is because of this difficulty that works that leverage a purely mathematical approach to solve such system-level optimization problems are scarce in literature. A possible remedy to overcome this difficulty as advocated in [31] is to exploit the time-tested rule of divide and conquer. That is, instead of approaching the problem holistically, one or a few parameters at a time can be considered in the mathematical model and optimization process. This strategy to divide and conquer has been used even in the simulation and metaheuristic-based approach discussed above, to overcome the complexity of the cellular system parameter optimization problem.

In the following, we pursue this objective and develop an analytical model that can be used to optimize the sector azimuth angle, while taking into account the impromptu existence of small cells and realistic non-uniform user demographic distributions.

16.2.2 System model

We consider a heterogeneous deployment consisting of macro cells and small cells as shown in Figure 16.8. A universal frequency reuse of one is assumed. The term sector is

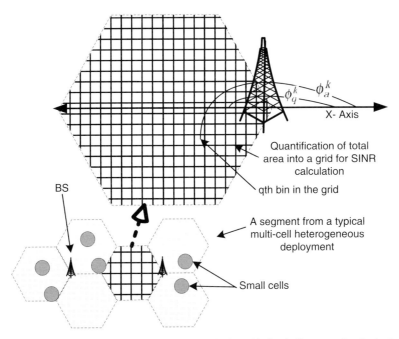

Figure 16.8 System model for problem formulation. Circles indicate randomly deployed small/femto cells.

used as synonymous of cell. Let $\mathcal{N}$ denote the set of points corresponding to the transmission antenna location of all cells, small or macro. We assume that the total area of interest A is divided into set of bins denoted by $\mathcal{Q}$ such that $\sum_{i=1}^{Q} q_i = A$, and $\frac{A}{Q} = q, \forall q \in \mathcal{Q}$ where $Q = |\mathcal{Q}|$. The area q is so small that long-term propagation conditions such as path loss and shadowing can be assumed to be constant in q. With this assumption, the downlink SINR perceived on the mth subcarrier in the qth bin served by the nth cell antenna, at time t, is given by

$$\gamma_q^{n,m}(t) = \frac{P^{n,m} G_q^n \alpha \left(d_q^n\right)^{-\beta} \sigma_q^n \epsilon_q^{n,m}(t)}{N_o B + \sum_{\forall \acute{n} \in \mathcal{N} \setminus n} \sum_{\forall \acute{m} \in \mathcal{M}} \left(P^{\acute{n},m} G_q^{\acute{n}} \alpha \left(d_q^{\acute{n}}\right)^{-\beta} \sigma_q^{\acute{n}} . u \left(\mu_q^{\acute{n},\acute{m}}\right) \epsilon^m(t) \right)} \quad (16.6)$$

where $P^{n,m}$ and $P^{\acute{n},m}$ are the transmission powers of the nth and $\acute{n}^{th}$ cells on m^{th} carrier, d_q^n and $d_q^{\acute{n}}$ are the distances of the qth bin from the nth and $\acute{n}^{th}$ cell respectively. G_q^n and $G_q^{\acute{n}}$ are the antenna gains from the nth and $\acute{n}^{th}$ cell to q^{th} bin, α and β are path loss model coefficient and exponents respectively. N_o is the thermal noise spectral density and B is carrier bandwidth.

σ_q^n and $\sigma_q^{\acute{n}}$ is the shadowing that the qth bin faces while receiving signal from the nth and $\acute{n}^{th}$ cells. $\epsilon^m(t)$ is the fast fading coefficient for the mth subcarrier at time t. Note that unlike fast fading, which is generally frequency selective and time dependent, shadowing mainly depends on the local clutter and angle of arrival. Since self-optimization to cope with topographic and demographic variations will run on a relatively larger time scale than that of fast fading, very short time-scale channel variations such as fast fading can be omitted.

Clutter-dependent shadowing, however, can have a relatively long-term impact and therefore cannot be neglected. We assume a worst-case scenario of full-system load. In that scenario, with frequency reuse of one, which is typical in LTE, the term $u(\mu_q^{m,\acute{m}})$ will always be 1. Another implication of full load and frequency reuse of 1 is that the downlink is generally interference limited. With these observations, for planning purposes, the SINR can be approximated with a time and subcarrier independent geometric signal to interference ratio (SIR).

$$\gamma_q^n = \frac{P^{n,m} G_q^n \alpha \left(d_q^n\right)^{-\beta} \sigma_q^n}{\sum_{\forall \acute{n} \in \mathcal{N}/n} \left(P^{\acute{n},m} G_q^{\acute{n}} \alpha \left(d_q^{\acute{n}}\right)^{-\beta} \sigma_q^{\acute{n}}\right)} \tag{16.7}$$

We take into account the fact that, in a given bin, the value of shadowing can be different for different BSs, i.e., generally $\sigma_q^n \neq \sigma_q^{\acute{n}}$. Since the relation between σ_q^n and $\sigma_q^{\acute{n}}$ plays an important role in the determination of the SIR, in order to make our framework more accurate we model this relationship using the shadowing cross-correlation models proposed in literature [32]. In this chapter we use the shadowing cross-correlation model used in [32], i.e., the cross-correlation coefficient to be used between values of σ_q^n and $\sigma_q^{\acute{n}}$ is determined as:

$$\eta_n^{\acute{n}} = \begin{cases} \sqrt{\dfrac{\min\left(d_q^n, d_n^{\acute{n}}\right)}{\max\left(d_q^n, d_n^{\acute{n}}\right)}} & 0 < \psi_{n,q,\acute{n}} < \psi_T \\ \left(\dfrac{\psi_T}{\psi_{n,q,\acute{n}}}\right)^{\mu} \sqrt{\dfrac{\min\left(d_q^n, d_n^{\acute{n}}\right)}{\max\left(d_q^n, d_n^{\acute{n}}\right)}} & \psi_{n,q,\acute{n}} < \psi_T < \pi \end{cases} \tag{16.8}$$

where $\psi_{n,q,\acute{n}}$ is the angle between the nth and $\acute{n}^{th}$ cell antennas, with respect to the qth bin.

The μ is a parameter that can be tuned to reflect the dependency of shadowing values on the angle of arrival, and can be determined in practice by the size and height of terrain and the height of the antenna. ψ_T is the threshold angle that can be given as

$$\psi_T = 2 \arctan\left(\frac{D_{cor}}{2 \min\left(d_q^n, d_q^{\acute{n}}\right)}\right) \tag{16.9}$$

where D_{cor} is the de-correlation distance that determines the degree of autocorrelation in the shadowing among the adjacent bins. This means that, if the shadowing in q^{th} bin from n^{th} cell is σ_q^n, the shadowing in the $\acute{q}^{th}$ bin that is at a distance $d_{\acute{q}}^q$ from the qth bin, while receiving a signal from the nth cell, is:

$$\sigma_{\acute{q}}^n = \left(e^{\frac{-d_{\acute{q}}^q}{D_{cor}}}\right) \sigma_q^n + \sigma \sqrt{1 - \left(e^{\frac{-d_{\acute{q}}^q}{D_{cor}}}\right)^2} q \tag{16.10}$$

where the term in parenthesis is the auto-correlation coefficient [32]. σ is a new instance of log normal random process. Note that D_{cor} can be used to tune the model according to the type of area under consideration. The tuning parameter can be derived from measurement campaigns or empirical values in literature. For example, for rural areas, the D_{cor} distance can be as high as 100 m, and may decrease to as low as 10 m in dense urban areas.

For 3GPP, LTE, and LTE-A, the antenna gain can be modeled as in [33] and written in dB as:

$$G_q^k = \lambda_v \left(G_{max} - \min\left(12\left(\frac{\theta_q^k - \theta_t^k}{B_v}\right)^2, A_{max} \right) \right)$$
$$+ \lambda_h \left(G_{max} - \min\left(12\left(\frac{\phi_q^k - \phi_a^k}{B_h}\right)^2, A_{max} \right) \right), k = n, m \quad (16.11)$$

where θ_q^k is the vertical angle in degrees from the qth bin to the kth cell and θ_{tilt}^k is the tilt angle of the kth cell antenna with respect to the horizon. The θ_q^k is the horizontal angle in degrees with similar meanings of subscript and postscript. Subscripts h and v denote the horizontal and vertical azimuth, respectively. Thus B_h and B_v represent horizontal and vertical beamwidths of the antenna, respectively, and λ_h and λ_h represent the weighting factors for the horizontal and vertical beam pattern of the antenna in a 3D antenna model [33], respectively. G_{max} and A_{max} denote the maximum antenna gain at the boresight of the antenna and maximum antenna attenuation at the sides and back of the boresight of the antenna respectively, in dB. For simplicity of expression, we can neglect the maximum attenuation factor A_{max} and assume maximum gain of 0 dB. Notice that these assumptions preserve the accuracy of the antenna model essential to the analysis in this chapter, but when evaluating the numerical results, we will remove them. The simplified antenna-gain model in linear form can be written as:

$$G_q^k = 10^{-1.2\left(\lambda_v \left(\frac{\theta_q^k - \theta_t^k}{B_v}\right)^2 + \lambda_h \left(\frac{\phi_q^k - \phi_a^k}{B_h}\right)^2 \right)}, \quad k = n, \acute{n} \quad (16.12)$$

For ease of expression we use the following substitutions:

$$\rho_q^k = \alpha \left(d_q^k\right)^{-\beta}; \varphi_q^k = \frac{B_h^2 \lambda_v}{\lambda_h} \left(\frac{\theta_q^k - \theta_t^k}{B_v}\right)^2; \tau = \frac{-1.2\lambda_h}{B_h^2}, \quad k = n, \acute{n} \quad (16.13)$$

Using equations (16.12) and (16.13), (16.14) can be written as:

$$\gamma_q^n = \frac{\rho_q^n \sigma_q^n 10^T \left(\varphi_q^n + \left(\phi_q^n - \phi_a^n\right)^2 \right)}{\sum_{\forall \acute{n} \in \mathcal{N}/n} \left(\rho_q^{\acute{n}} \sigma_q^{\acute{n}} 10^{T(\varphi_q^{\acute{n}} + (\phi_q^{\acute{n}} - \varphi_q^{\acute{n}})^2)} \right)} \quad (16.14)$$

Note that equation (16.14) provides a model for estimating SIR, and thus system KPIs such as capacity and spectral efficiency, as a function of cell-antenna azimuth angles. In case small cells are assumed to have omnidirectional antennas, the received signal from them will not be a function of the azimuth angle. Taking into account this scenario, the generalized geometric SINR can be written as

$$\gamma_q = \frac{\pi_q^n}{\sum_{\forall \acute{n} \in \mathcal{N} \setminus n} \pi_q^{\acute{n}}} \quad (16.15)$$

such that

$$\pi_q^k = \begin{cases} \rho_q^k \sigma_q^k 10^{T(\varphi_q^k + (\phi_q^k - \phi_{\acute{n}}^k)^2)} & k = n, \acute{n} \in \mathcal{M} \\ P^{k,k} C\alpha(d_q^k)^{-\beta} \sigma_q^k & k = n, \acute{n} \in \mathcal{S} \end{cases} \quad (16.16)$$

where $\mathcal{M}$ represents locations of macro cells that have directional antenna and thus azimuth angle dependent gain. $\mathcal{S}$ represents locations of small cells who have omnidirectional antennas with constant gain C in all directions. Note that $\mathcal{N} = \mathcal{M} \cup \mathcal{S}$.

16.2.3 Problem formulation

Our problem is to find and maintain the optimal azimuth angles for all cells in the system, i.e., ϕ_a^N that maximize system-wide user spectral efficiency for a given user demography and heterogeneous deployment. Mathematically our problem can be written as:

$$\phi_a^N = \arg\max_{\phi_a^N} \sum_{\forall q \in \mathcal{Q}} \xi_q \log_2(1 + \gamma_q(\phi_a^N)) \quad (16.17)$$

Here γ_q is the SIR in q^{th} bin perceived from the serving node. ξ_q is the weight associated with the qth bin, which can be set by the operator to reflect its significance in the optimization process, e.g., a bin in a dense hotspot area will have higher value of ξ than a bin in a less populated area. While ξ can be used to incorporate the quality of experience (QoE), pricing factors, and operators' policies etc., here we use a simple function to set the value of ξ as:

$$\xi_q = \frac{u_q}{U}, \quad \text{where} \quad U = \sum_{q=1}^{Q} u_q \quad (16.18)$$

where u_q is the expected number of active users in a qth bin, and U is the total number of average active users in the system.

Note in (16.17), that the SIR is a function of the vector of azimuth angles of all cells in the system i.e., ϕ_a^N, where $N = |\mathcal{N}|$. However, from (16.16), it can be observed that the signal received from the small cells is independent of the azimuth angles. Therefore, small cells can be excluded from the optimization process and the problem in (16.17) can be simplified as

$$\phi_a^M = \arg\max_{\phi_a^M} \sum_{\forall q \in \mathcal{Q}} \xi_q \log_2(1 + \gamma_q(\phi_a^M)) \quad (16.19)$$

where $M = |\mathcal{M}|$. Since, generally, the number of macro cells is much smaller than the number of small cells in the system, i.e., $M \ll N$, therefore compared to (16.17), the problem in (16.19) has lower complexity. However, despite the fact that (16.19) is much simpler than (16.17), still (16.19) is a large-scale non-convex optimization problem. Because of having no known polynomial time solution, problems similar to (16.19) have been solved using metaheuristics in the literature as explained above.

16.2.4 A low-complexity solution for online implementation

As discussed above, metaheuristic-based solutions are not suitable for online self-optimization of macro BS antennas. This is because the time needed to determine the optimal azimuth angle might be too long to make the solution agile enough to respond to the acute changes in the node topography and user demography in a heterogeneous network. To this end, in this subsection we present a low-complexity method to obtain an azimuth-angle optimization solution without resorting to the aforementioned conventionally used simulation-heuristics. We start by underpinning the key factor that makes the problem in (16.19) too complex to be tackled with analytical approach. That is, instead of looking at the problem in (16.19) from a top-to-bottom perspective, where azimuths of all macro cells are tried to be optimized together, while considering all the bins in the system, we exploit a bottom-to-top perspective that allows one cell azimuth to be optimized at a time. We observe that due to the sectorization, optimizing the azimuth in a single cell or sector affects only the numerator of the SIR in (16.15), for the bins that are being served by that macro cell. On the other hand, for the bins that are being served by small cells or other macro, only the denominator of SIR in (16.15) is affected by the azimuth angle adaptation of that cell. We further observe that, due to exponential path loss (see (16.7)), changes in azimuth angle of a sector strongly affect only the bins that are in close vicinity of that sector. With this observation, we infer that in order to enhance system performance without resorting to high complexity, one sector azimuth at a time can be optimized while considering only the bins it will affect. The number of bins to be considered in the optimization process can be further reduced by considering only the bins that will be strongly affected. That is, for optimizing azimuth of a sector, the bins associated with that sector and its immediate neighbor macro cells and small cells can be considered. Thus mathematically the problem can be rewritten as:

$$\max_{\phi_a^n} \left(\frac{\frac{1}{\sum_{\forall q \in \mathcal{Q}_M^n} \xi_q} \sum_{\forall q \in \mathcal{Q}_M^n} \left(\rho_q^n \sigma_q^n 10^{T(\varphi_q^n + (\phi_q^n - \phi_a^n)^2)} \right)}{\frac{1}{\sum_{\forall q \in \mathcal{Q}_S^n} \xi_q} \sum_{\forall q \in \mathcal{Q}_B \setminus \mathcal{Q}_M^n} \left(\rho_q^n \sigma_q^n 10^{T(\varphi_q^n + (\phi_q^n - \phi_a^n)^2)} \right)} \right), \forall n \in \mathcal{M} \quad (16.20)$$

where $\mathcal{Q}_M^n$ is the set of bins in the nth sector that are being served by the nth macro cell, and $\mathcal{Q}_B$ is the set of all bins within all the three sectors associated with the Bth BS site, to which the nth sector belongs (see Figure 16.8). Thus $\mathcal{Q}_B \setminus \mathcal{Q}_M^n$ denotes bins that are being served by small cells within all three sectors of the Bth BS site and the bins that are being served by the other two sectors of the Bth BS excluding the nth sector of the Bth BS. The problem in (16.20) effectively aims to optimize the azimuth angle of a given sector, so that it maximizes the average signal strength being received by bins that are being served by the macro sector, and minimizes the signal strength received in the bins that are being served by the adjacent two macro sectors or the small cells within that sector and adjacent two sectors. Even if the azimuth angles obtained through this method will not be globally optimal, as it takes into account the bins within a site only, its low complexity allows an agile and online implementable solution for enhancing system performance. A gain compared to currently rigid setting of regular azimuth angles, which does not change autonomously in response to topographic and demographic changes, can thus be expected.

The optimization problem in (16.20) can be solved with conventional derivative-based optimization methods. For the sake of brevity, we have skipped the full derivation and only the final result is presented here. By taking the first derivative, it can be shown that the critical points of the objective function in (16.20) correspond to azimuth angle given by the following optimality condition:

$$\frac{\sum_{\forall q \in \mathcal{Q}_M^n} \xi_q \vartheta_q^n (\phi_a^n - \phi_q^n)}{\sum_{\forall q \in \mathcal{Q}_B \backslash \mathcal{Q}_M^n} \xi_q \vartheta_q^n (\phi_a^n - \phi_q^n)} = \frac{\sum_{\forall q \in \mathcal{Q}_M^n} \xi_q \vartheta_q^n}{\sum_{\forall q \in \mathcal{Q}_B \backslash \mathcal{Q}_M^n} \xi_q \vartheta_q^n} \quad (16.21)$$

where $\vartheta = 10^{-1.2(\phi - \phi_q^n)^2} \times \frac{\sigma_q^n}{(d_q^n)^\beta}$ effectively denotes the received signal level (RSL) in the qth bin from the nth cell. The feasibility to obtain an analytical solution of (16.20) in the form of an easily tractable expression, given by (16.21), allows us to propose the following algorithm for implementation of proposed azimuth angle optimization framework:

Online implementation algorithm

1. Each macro BS divides its coverage area into virtual bins. The size of the bin is decided based on a tradeoff between the accuracy of the optimization process and the complexity.
2. Each macro BS cell/sector assigns weights to each bin in it, to reflect the significance of that bin in the optimization process. The weights can be determined, for example, based on the probability of containing active users, targeted QoS, or potential revenue expected from that fraction of the cell area, etc.
3. Calculate the total weight assigned to the bins for each macro cell.
4. Start with the BS site with the highest total weight and determine the azimuth angle of each of its three sectors independently, one by one according to the solution in (16.21).
5. Continue in descending order of weight of sites, until all sector azimuth angles are optimized.
6. Perform azimuth-angle optimization after pre-set intervals or on-need bases, by repeating steps 2 to 5.

16.2.5 Numerical results

In this subsection we present numerical results by considering a simple network scenario consisting of one BS with three sectors, all three of which are optimized one by one using the proposed solution. While these results do not fully predict the full-scale system-level gain of the proposed solution, their key significance is that they are readily obtainable from the analytical results presented in Subsection 16.2.4 using parameter values given in Table 16.1. Thus these results serve the purpose of validating the potential gain of the proposed solution.

Four sets of numerical results from Figures 16.9 to 16.12 are shown to represent different instances of user distribution and its split among macro and small cells and corresponding optimal azimuth angles obtained by the proposed solution. In each figure, each row of sub-figures represents one particular sector that is being optimized out of

Table 16.1 System-level simulation parameters.

Parameters	Values
System topology	1 BS × 3 sector, frequency reuse 1
BS transmission power	39 dBm
Cell radius, BS, and user height	600 m, 32 m, and 1.5 m respectively
User antenna gain	0 dB (omni directional)
B_h, B_v	70^0, 10^0
$\lambda_v = \lambda_h$	0.5
G_{max}, A_{max}	18, 20 dB
Frequency	2 GHz
Path-loss exponent β	4
Shadowing standard deviation	8 dB
Shadowing auto-correlation distance	20 m
Total user population	100 users
Small cell radius	100 m (located randomly)
Small cell location	random, three small cells per macro cell
% of users in small cells	20 to 80%

the three sectors in the site. The first column sub-figures show the objective function in (16.20) plotted as function of range of azimuth angles.

Note that the range has been confined to 30° above and below the nominal azimuth angles of [30°, 150°, 270°]. This constraint on the optimization range is optional. It has been placed to retain hexagonal-sectorized topology in order to minimize the negative impact of azimuth adaptation on the farther cells that are not incorporated in determining the optimal azimuth angle. The middle figure in each row shows the average RSL perceived by the users that are being served by the sector, and the average RSL (interference) perceived by the other users within that site. The last figure in each row shows the actual user distribution used in the given experiment. The circles indicate randomly deployed femto or small cells. Dark arrows in this figure show the nominal azimuth angles of the sectors i.e., [30°, 150°, 270°]. Light arrows show the optimal azimuth angle obtained from the proposed solution for that particular instance user distribution and femto cell locations. For example, in the result set given in Figure 16.9, the proposed solution yields optimal azimuth angles of [12°, 154°, 260°], which are plotted in the last column sub-figure as light arrows. Note that we start with optimization of the right top sector and move anticlockwise to optimize the next sector azimuth angle, while leaving the previously optimized sector azimuth unchanged. Thus each site is optimized in the further three steps, i.e., one sector at a time, and thus each row in a figure corresponds to one step.

Note, from the first column sub-figures in all result sets, that for different user distributions and split ratio between the macro and small cells, the objective function has a different shape but has a clear optimal point. This highlights the need for and benefit of the user demography-aware azimuth angle self-optimization scheme, able to respond to changing user locations and on-off status of small cells. An online and autonomous implementation of the proposed solution will obviously require the knowledge of the

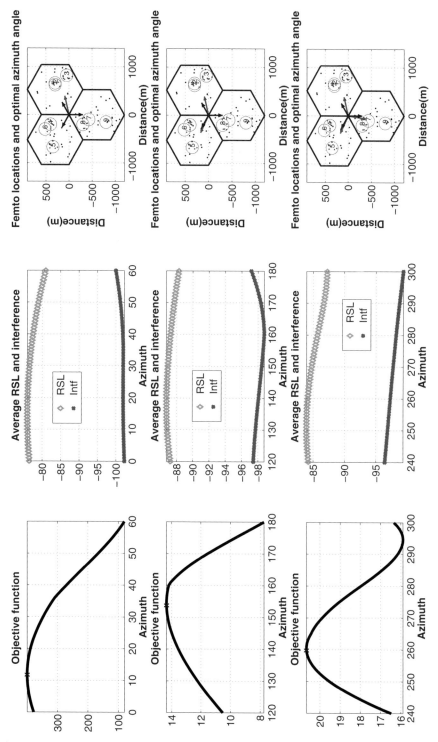

Figure 16.9 Optimal azimuth angles when 20% of users are in femto (small) cells.

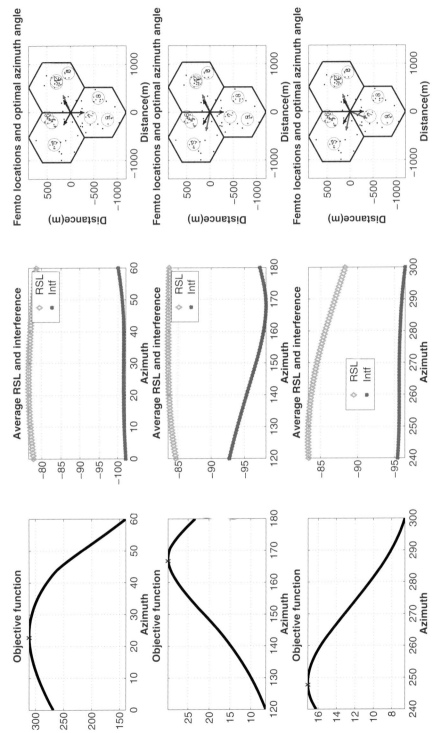

Figure 16.10 Optimal azimuth angles when 40% of users are in femto (small) cells.

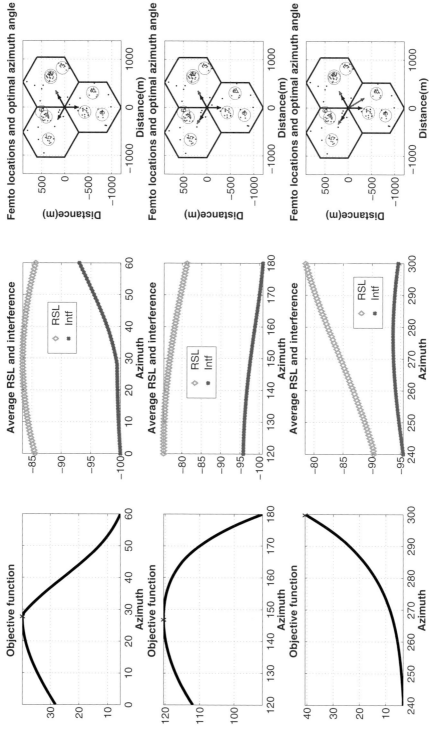

Figure 16.11 Optimal azimuth angles when 60% of users are in femto (small) cells.

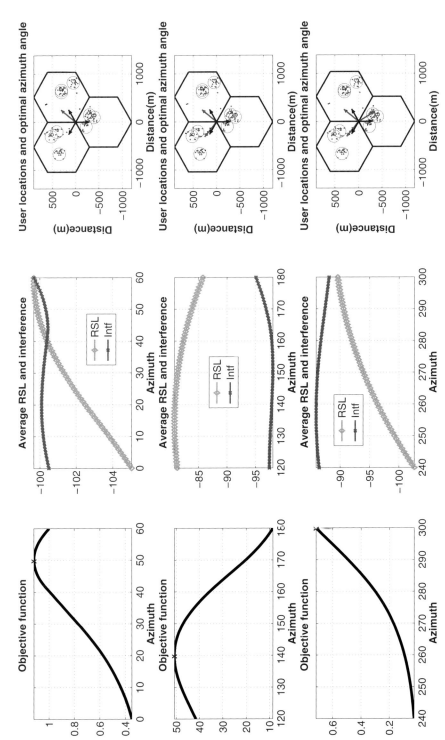

Figure 16.12 Optimal azimuth angles when 80% of users are in femto (small) cells.

locations of all small cells and macro users within a site. Location of small cells can be communicated to the BS through X2 interface. Notice that for scenarios where small or femto cells cover indoor spaces, only the users that are outdoors would be mostly served by the macro BS. In this case, since determining user location in an outdoor scenario, e.g., through GPS or triangulization, is not an issue, the presented solution is easy to implement.

To assess the gain of the proposed solution, we consider a scenario represented by the third row of the result set in Figure 16.11. Compared to regular azimuth angle, the proposed solution increases the desired average RSL of the users in that sector by $(-85 - (-90)) = 5$ dB, compared to that obtained with regular 270° azimuth. This azimuth adaptation also reduces the average interference received from this sector from the other cell users by $(-91 - (-90)) = -1$ dB. This effectively translates into 5 dB and 1 dB gain in SINR for users being served by those sectors and users being interfered by that sector, respectively. Note that this gain is indicative of optimization of one sector only. With optimization of all sector azimuths through the proposed algorithm, the gain is expected to be even higher. However, note from the second column sub-figures in all figure sets, that the exact gain of the proposed solution, compared to the regular azimuth angle, also depends on the user distribution and small cell locations. The full assessment of the overall gain requires the implementation of the proposed solution in a full-scale system-level simulator, which is beyond the scope of this chapter.

While the objective function remains convex in most of the scenarios investigated, for certain user demography and small cell location combinations it can become non-convex, as happens in the third row of Figure 16.13. However, a very small number of critical points and the presence of an analytically tractable solution means that the absolute optimal solution of (16.20) can be easily found.

16.3 Summary

Two approaches to the inter-cell interference coordination (ICIC) problem among different layers of a heterogeneous network have been presented in this chapter. In particular, we first propose a multi-agent RL approach to control the aggregated interference generated by a femto cell system onto a macro cell network. We propose a TD learning solution based on QL and then we identify its limits and propose novel improvements that combine fuzzy logic and learning. We show that both learning algorithms are able to learn a policy that allows them to correctly adapt to each realistic situation of the environment. In order to present the behavior of the proposed FQL algorithm, we compare the obtained results with two less-complex FQL approaches, with QL, and with a benchmark proposed by 3GPP. Results show that the proposed learning methods perform better than the SPC allowing a more efficient use of the network resources.

We then also study the macro–femto interference coordination problem. An analytical framework to model heterogeneous cellular system performance as function of sector azimuth angle is presented. An optimization problem is formulated to optimize azimuth angles while taking into account live user demography, small cell activity, and deployment scenarios. A low-complexity scalable solution is then analytically derived. A simple

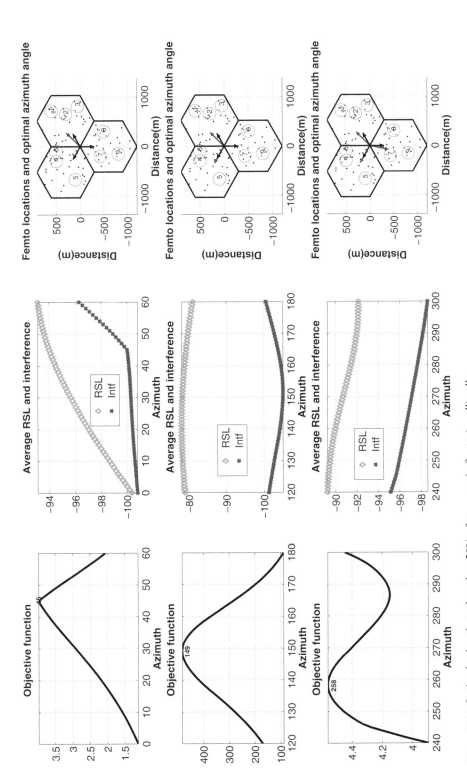

Figure 16.13 Optimal azimuth angles when 20% of users are in femto (small) cells.

algorithm for pragmatic implementation of the proposed solution in a real system is also presented. While full-scale system-level evaluation of the proposed solution is beyond the scope of this work, extensive numerical results demonstrate the significant gain of the proposed solution by self-optimizing macro cell azimuth angles to minimize both macro–femto and macro–macro interference. By compromising on global optimality, the presented solution is designed to work without relying on system-wide signaling, cumbersome planning tools, and metaheuristics, while taking into account a variety of factors of real systems, such as demography, small cell deployments, their on/off status and quality of service (QoS) constraints.

References

[1] Chandrasekhar, V., Andrews, J. G. & Gatherer, A. (2008), 'Femtocell networks: A survey', *IEEE Communication Magazine* 46(9), 59–67.

[2] Sycara, K. P. (1998), 'Multiagent systems', *AI Magazine* 19(2), 79–92.

[3] Harmon, M. E. & Harmon, S. S. (2000), 'Reinforcement learning: A tutorial'.

[4] Zadeh, L. (1965), 'Fuzzy sets', *Information and Control* 8, 338–353.

[5] Berenji, H. R. (1994), Fuzzy q-learning: a new approach for fuzzy dynamic programming, *in* 'IEEE World Congress on Computational Intelligence, Proceedings of the Third IEEE Conference on Fuzzy Systems', pp. 486–491.

[6] Jouffe, L. (1998), 'Fuzzy inference system learning by reinforcement methods', *IEEE Transactions on Systems, Man, and Cybernetics, Part C: Applications and Reviews* 28(3), 338–355.

[7] Glorennec, P. Y. & Jouffe, L. (1994), Fuzzy q-learning, *in* 'Proceedings of the Sixth IEEE International Conference on Fuzzy Systems', pp. 659–662.

[8] Glorennec, P. Y. & Jouffe, L. (1996), A reinforcement learning method for an autonomous robot, *in* 'Proceedings of Fourth European Congress on Intelligent Techniques and Soft Computing (EUFIT96)'.

[9] Bellman, R. (1957), *Dynamic Programming*, Princeton, NJ: Princeton Univ. Press.

[10] Panait, L. & Luke, S. (2005), 'Cooperative multi-agent learning: The state of the art', *Autonomous Agents and Multi-Agent Systems* 3(11), 383–434.

[11] Chen, Y.-H., Chang, C.-J. & Huang, C. Y. (2009), 'Fuzzy Q-learning admission control for WCDMA/WLAN heterogeneous networks with multimedia traffic', *IEEE Transactions on Mobile Computing* 8, 1469–1479.

[12] 3GPP (2009), 3GPP TSG RAN WG4 (Radio) Meeting 51: Simulation assumptions and parameters for FDD HeNB RF requirements, Technical report.

[13] Galindo-Serrano, A. & Giupponi, L. (2010), Distributed Q-learning for interference control in OFDMA-based femtocell networks, *in* 'Proc. of the IEEE 71th Vehicular Thechnology Conference, VTC Spring 2010', Taipei, Taiwan.

[14] Salo, J., Del Galdo, G., Salmi, J., Kysti, P., Milojevic, M., Laselva, D. & Schneider, C. (2005), 'MATLAB implementation of the 3GPP Spatial Channel Model (3GPP TR 25.996)', On-line.

[15] 3GPP (2010), 3GPP TR 36.921 evolved universal terrestrial radio access (E-UTRA); FDD home eNode B (HeNB) radio frequency (RF) requirements analysis, Technical report, 3GPP.

[16] Qi, Y., Imran, M. & Tafazolli, R. (2010), On the energy aware deployment strategy in cellular systems, *in* 'Personal, Indoor and Mobile Radio Communications Workshops (PIMRC Workshops), 2010 IEEE 21st International Symposium on', pp. 363–367.

[17] Hu, L., Kovacs, I., Mogensen, P., Klein, O. & Stormer, W. (2011), Optimal new site deployment algorithm for heterogeneous cellular networks, in 'Vehicular Technology Conference (VTC Fall), 2011 IEEE', pp. 1–5.

[18] Guruprasad, K. (2011), Generalized voronoi partition: A new tool for optimal placement of base stations, in 'Advanced Networks and Telecommunication Systems (ANTS), 2011 IEEE 5th International Conference on', pp. 1–3.

[19] Qi, Y., Imran, M. & Tafazolli, R. (2011), Energy-aware adaptive sectorisation in lte systems, in 'Personal Indoor and Mobile Radio Communications (PIMRC), 2011 IEEE 22nd International Symposium on', pp. 2402–2406.

[20] Ramiro, J. & Hameid, K., eds (2012), *Self Organizing Networks*, Wiley, ISBN 978-0470-97352-3.

[21] Amaldi, E., Capone, A. & Malucelli, F. (2003), 'Planning UMTS base station location: optimization models with power control and algorithms', *IEEE Transactions on Wireless Communications* 2(5), 939–952.

[22] Hurley, S. (2002), 'Planning effective cellular mobile radio networks', *IEEE Transactions on Vehicular Technology* 51(2), 243–253.

[23] Gu, F., Liu, H. & Li, M. (2009), Evolutionary algorithm for the radio planning and coverage optimization of 3G cellular networks, in 'International Conference on Computational Intelligence and Security, (CIS).', vol. 2, pp. 109–113.

[24] Yang, H., Wang, J., Song, X., Yang, Y. & Wang, M. (2011), Wireless base stations planning based on GIS and genetic algorithms, in '9th International Conference on Geoinformatics, 2011', pp. 1–5.

[25] Tsilimantos, D., Kaklamani, D. & Tsoulos, G. (2008), Particle swarm optimization for UMTS WCDMA network planning, in '3rd International Symposium on Wireless Pervasive Computing (ISWPC)', pp. 283–287.

[26] Elkamchouchi, H., Elragal, H. & Makar, M. (2007), Cellular radio network planning using particle swarm optimization, in 'National Radio Science Conference (NRSC)', pp. 1–8.

[27] Awada, A., Wegmann, B., Viering, I. & Klein, A. (2011), 'Optimizing the radio network parameters of the long term evolution system using taguchi's method', *IEEE Transactions on Vehicular Technology* 60(8), 3825–3839.

[28] Berrocal-Plaza, V., Vega-Rodriguez, M., Gomez-Pulido, J. & Sanchez-Perez, J. (2011), Artificial bee colony algorithm applied to wimax network planning problem, in '11th International Conference on Intelligent Systems Design and Applications (ISDA)', pp. 504–509.

[29] Kirkpatrick, S., Gelatt, C. D. & Vecchi, M. P. (1983), 'Optimization by simulated annealing', *Science* 220(4598), 671–680.

[30] Mendes, S., Molina, G., Vega-Rodriguez, M., Gomez-Pulido, J., Saez, Y., Miranda, G., Segura, C., Alba, E., Isasi, P., Leon, C. & Sanchez-Perez, J. (2009), 'Benchmarking a wide spectrum of metaheuristic techniques for the radio network design problem', *Evolutionary Computation, IEEE Transactions on* 13(5), 1133–1150.

[31] Eisenblatter, A. & Geerd, H.-F. (2006), 'Wireless network design: solution-oriented modeling and mathematical optimization', *IEEE Wireless Communications* 13(6), 8–14.

[32] Yang, X. & Tafazolli, R. (2003), A method of generating cross-correlated shadowing for dynamic system-level simulators, in 'Personal, Indoor and Mobile Radio Communications, 2003. PIMRC 2003. 14th IEEE Proceedings on', Vol. 1, pp. 638–642.

[33] Viering, I., Dottling, M. & Lobinger, A. (2009), 'A mathematical perspective of self-optimizing, wireless networks', *IEEE International Conference on Communications, 2009, (ICC '09)*, pp. 1–6.

17 Large-scale deployment and scalability

Iris Barcia, Simon Chapman, and Chris Beale

17.1 Introduction

Combined with a technology upgrade to LTE/LTE-A, small cells are presented by many industry players (from operators to equipment vendors to analysts) as the most cost-effective solution to the known increase in mobile data demand [1–3]. As small cells (e.g., femto and pico) become the broadly adopted solution for adding capacity to modern, smart phone-dominated cellular networks, their numbers, and the areas where they will be deployed, will increase dramatically (Figure 17.1).

This reduction in cell size and the growth in cell numbers has required many new approaches to be developed to ensure that the next generation of networks are built to exploit costly, limited spectrum resources while maximizing capacity.

Such new methods consider the network design process in a holistic manner and ensure sufficient computational power is available to remove any accuracy compromises inherent with the traditional design processes. The set of techniques used to accomplish this we call large-scale network design, or L-SND for short.

The US cellular market has many good examples of planned small cell deployments that are to occur at a national level [1]. Results and data from such small cell designs are included in this chapter to illustrate the L-SND accuracy and scalability difficulties that have been overcome when compared to the limitations found with traditional methods.

17.1.1 Large-scale network design

Large-scale network design groups advanced radio-planning algorithms and technical solutions to evaluate vast numbers of sites without overloading engineering resources. Big data, cloud computing, and optimized metaheuristic algorithms are basic parts of the pool of multiple components that have to be combined to solve the small cell challenge.

The methodology described in this chapter has been implemented in Overture[1] and the results shown have been shared by Keima Technologies.

Following model-based control theory [4, 5], L-SND aims to reduce the gap between business-case analysis and deployment. By utilizing the advantages of big data and

---

[1] Overture is a software platform developed and optimized for HetNet design. For more information visit http://keima.co.uk.

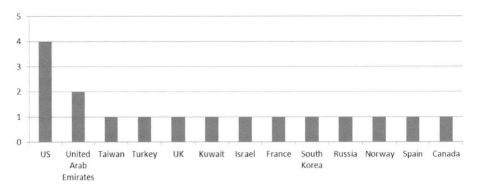

Figure 17.1 Small cell deployment commitments, 2013 data.

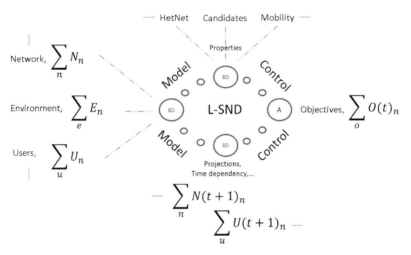

Figure 17.2 Simplified large-scale network design diagram.

cloud computing it is possible to seamlessly link marketing, planning, and deployment information. As the analysis evolves, robustness is built into the study by integration of feedback and new or updated big data components to review the objectives. In a real-life scenario, such systems will utilize practical and realistic information to evaluate outcomes and to maximize the advantages of economies of scale through a quicker (but more detailed) design phase.

To build a realistic model, all the multiple data sources must be integrated simultaneously. In a highly simplified case (Figure 17.2), this would be a discrete analysis at particular points in time and space; hence it could be represented as a summation. Under this simplification, each data source would equally add to its particular data type (note the sub-index under each data type: environment, usage, etc.) and would be directly injected into the system and evaluated according to project specifications. Control is executed as a discrete comparison of the model under different conditions, for example marketing projections. Each iteration has an effect on the properties that characterized previous iterations (take, for example, the candidate set) and that retro-feeds the system to add robustness to the final design.

A proper implementation of L-SND follows the same principles; however, it is characterized by a higher data processing complexity (BD in the diagram). The contribution of each big data component is weighted according to meaningful information that has been previously derived from the raw source. The qualities of the type of source drive the properties that can be extracted and how those are integrated into the design. For example, the contribution of social networks information can be correlated both with demographic profiles (social dimension) and clustering identification (spatial, since it is linked to location; and technical dimensions, since it can be associated to demand). The following section expands on the nature of the multiple big data sources.

Hence for its real implementation, L-SND explores the correlation (for multiple dimensions) between the different data sets, and simultaneously integrates their contributions into the objective functions. This maximizes the volume and quality of the inputs and the accuracy of the results; for each iteration only the data sets that add valuable information are part of the process.

The diverse constitution of the elements that influence network design of modern wireless networks leads to a complex interdependence between optimization objectives that hinders the ability to reach an absolute optimal solution. Even when detailed pre-processing of data inputs simplifies the definition of the objective functions by narrowing range and variety, L-SND requires meta-heuristic algorithms [6] to explore the optimization space in a multi-dimensional proposal. These algorithms help the selection of close-to-optimal solutions. One of the determining factors of the quality and success of meta-heuristic algorithms is the initial condition state. The more realistic the information available, the closer will be the result to the optimal solution. As the iterations of the algorithm advance, it is possible to inject energy into the system until no improvement is detected. Again, the more information we have about the conditions in the system, the higher the probability to reach the optimal solution. For that reason, big data brings important benefits to the process. It helps reduce the uncertainty and can be used to feed back to the system when it is in search of the solution.

For example, the case studies included in this chapter evaluate the deployment of small cells as a solution to capacity constraints in large operating mobile networks. The ultimate objective, however, is to create a small cell network (SCN) that maximizes the ROI of the network. That is only possible through a deep understanding of the demand and the resources of the network's operator.

From an engineering perspective, the design of the network has to consider:

- multiple radio access technologies (multi-RAT)
- multiple frequency bands (multi-band)
- multiple small cell types (i.e., metro, pico, micro, femto, atto)
- deep understanding of demand.

From a mathematical perspective, maximization of ROI is the top-level objective in Figure 17.3. This is influenced by the performance of the network (spectral efficiency)

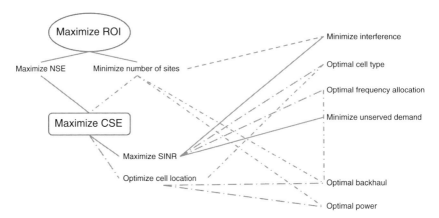

Figure 17.3 Interdependence between optimization objectives.

and the cost associated to the site (number of sites); spectral efficiency is dependent on frequency allocation, SINR, and usage; if we concentrate on usage (and within usage we only concentrate on the probability of access), this is dependent on interference and signal levels, which links back to the number of sites. The number of combinations possible to reach a solution for each function expands the more we break down the objectives. While it could be possible to adapt traditional methods for the planning of a few small cells in a simplified scenario, those are not scalable. Subsection 17.3.5 explores in more detail the consequences for network planning of this combinatorial explosion.

Cloud computing and automation are essential to achieve the performance requirements capable of processing this volume of information. Automation (A in Figure 17.2) facilitates, from the user perspective, the management, modeling, and control of the multiple sources. Cloud computing is, however, a key L-SND component quasi-transparent to the user. It has a direct effect on scalability since the ability to cluster processing resources allows flexible allocation of computational power to meet the immediate requirements of the analysis. It also facilitates data management and transferability, which provides an ideal framework for big data.

The same principle of flexible allocation of computing resources also applies to the ability of the optimization algorithms to adapt to different analysis conditions and objectives, helping to balance error vs complexity in extreme cases.

17.1.1.1 Big data

Integration of big data contributes at different levels to the accuracy and scalability objectives of L-SND: scalable because global sources of data are instantly available; accurate because valuable information can be inferred by correlating individual sources of partial information.

Big data groups a varied set of data types from a heterogeneous and unlimited set of sources that may not have any characteristics in common. There is no standard volume, unique gathering process, or unified processing method.

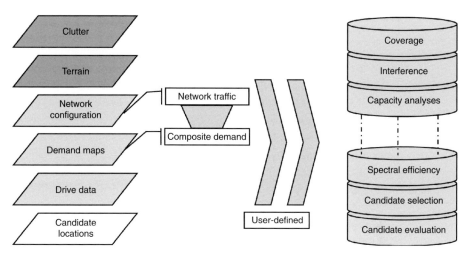

Figure 17.4 Automated network design: inputs/analyses.

Variety is one of the main characteristics of big data and some examples of the multi-dimensionality of the data have already been mentioned in the analysis of Figure 17.2 (social, spatial, and technical). From the perspective of value added to network planning, vector clutter and social geodata are specific practical cases of big data solutions to the L-SND challenges described in Subsections 17.3.1 and 17.3.2.

Clutter is an important input for network planning as it is used to model the behavior of the signal, but as Subsection 17.3.1 describes, it is one of the biggest contributors to error. Vector clutter is built from environmental big data and it is used to mimic entire cityscapes by recreating the elements that shape the landscape, and mainly those that the signal will encounter as it propagates (trees, water, roads, buildings, etc.). In general, environmental information is a set of many different objective public and private data sources. Interpretation and understanding of the data presents technical challenges, i.e., format compatibility, projections, etc., but the individual information sets add very little value. However, when aggregated, they provide the most accurate representation of the environment.

On the other hand, social networks information groups all those data sets gathered from social networks (a social network in this context is an application that facilitates any kind of content or idea sharing between a large population of individuals). It is characterized by high variety, volume, and availability. It also is content-rich and provides an invaluable source of knowledge about the life in the network. Individual data sources by themselves already provide in-depth objective (location, volume) and subjective (users' interactions with their networks) information. Social geodata specifically refers to the network planning input that provides location-aware social networks data: often users are able to "geo-tag" the information they share by granting the application access to the location system of their device. Full understanding of the qualities of the social network is vital in order to identify hidden usability or behavioral biases that could skew the network design. Some other aspects to consider are: demographics, uplink or downlink domination, locality, geographical scope, etc.

We will keep focusing on environmental and social information to describe in more detail the nature of big data in terms of variety, availability, volume, veracity, and value.

Variety. The social networks environment is very dynamic and the number of sources that fall in this classification expands and contracts easily: Twitter, Flickr, Facebook, Foursquare, Instagram, Formspring, Vine, Orkut, Tuenti, etc. The success and continuity of a networking site is linked to social trends and in some cases the data gathered is subject to a strong demographic bias. For example, Tuenti is a social site similar to Facebook very popular between the Spanish teenage community. For environmental information, data provision may be more spatially limited since common sources often are local authorities or governmental departments.

Availability. An article published by *The Washington Post* in 2013 [7] reports that 200 million active users generated 400 million tweets per day. Social data is strongly linked to personal information and the approach to data sharing and access to APIs differs between different social sources: open access (open APIs provided by Transport for London (TfL) to access public transport information; government and industry initiatives – mainly driven by the health sector); volume-restricted open access (Twitter); or paid access. Use of private information is highly controversial and it is a key requirement guaranteeing data anonymity or granting access only under user's consent. Taking again the environmental and social networks comparison: updates of environmental information are often irregular and limited in frequency. Focusing on Twitter for simplicity, we detect a totally different pattern: a continuous stream of raw data from the API is stored and automatically compartmentalized for processing. In both cases, though, the raw data is highly unstructured and it requires specific processing and analysis tools.

Volume. Both environmental and social networks information require the storage and processing of large data volumes. For the Twitter case, petabytes (PB) of data have been processed with useful results cached over the past three years. Data storage services facilitate the preservation of current volumes and provision exabytes (EB) of storage for coming years. Preserving raw data facilitates the exploitation of the data's original properties at any given time but the increasing volume of storage required fosters appearance of solutions that allows balancing that flexibility with the ability to perform partial analyses on pre-processed data.

Veracity and value. Value is given by the meaningful combination of big data with the different aspects of L-SND. Mobility patterns or user clustering are examples of value added by social network analyses to network demand profiles (Subsection 17.3.2). Veracity is essential to achieve value and it is brought into the system by statistical significance and error detection (model-based control theory). For example, height information in some social networks is captured from the location tags provided by the API. The accuracy of this height information depends on the sensor that has captured it and that varies from device to device. Veracity implies in this case a location-aware filtering algorithm detects spurious readings.

A concise summary of different big data solutions for storage, processing, visualization, etc., is provided in Pete Warden's *Big Data Glossary* [8].

17.2   L-SND for modern wireless networks

Smart phones and changes in user behavior have driven a transformation in the wireless world. From a mobile data traffic perspective, in 2012 users generated 50 times more traffic than the previous year [3]. Operators have to explore new business models and wireless network planning methodologies have to also be transformed to be effective.

17.2.1   Deployment-ready business-case analysis

It is only possible to accurately compute a business case for small cells if, and only if, that business case is deployment ready – taking account of interference factors, technology characteristics, etc.

At the time of writing this book, the variety of small cell solutions and the novelty of the devices also bring additional dimensions to network design, both from a deployment and a business-case analysis perspective. Many current planning solutions and processes struggle to capture the dependability between the multiple parameters and new requirements that small cells, and ultimately heterogeneous networks (HetNets), bring to modern wireless network planning. It is due to this heterogeneous nature, that cell location is the single most important factor for wireless planning and deployment. Small cell equipment can be easily installed at street level. It is cheaper, lighter, and versatile, which opens a new set of candidate locations that include a variety of street furniture assets: lamp posts, traffic lights, bus stops, etc. A clear advantage of small cells compared to the deployment limitations of macro cells, though it increases the complexity of factors that highly influence OPEX:

- Backhaul requirements for data-dominated networks bring substantial changes to traffic estimation [9, 10]. Small cell deployments will require the study of heterogeneous backhaul solutions, line of sight (LOS), near-line of sight (nLOS), non-line of sight (NLOS), availability of fiber, etc. Identification and provision of the appropriate backhaul solution is a critical requirement for any real small cell deployment.
- Provision of power will also be a decisive deployment driver. For example, for bank-switched street lighting power is delivered to an entire region simultaneously activating each light at the same time: if daytime power-saving systems are in place, additional cost considerations to provide continuous power have to be part of the deployment scenario.

But as the cell size shrinks, the detail in the analyses (accuracy) has to increase to be able to provide differential information about network performance. Big data facilitates data availability but automation is required if modern networks are to be planned understanding the return on investment (ROI) of each new site. An example of this successful pairing of data and automation is the identification of demand clustering from social geodata sources. For simplicity, demand distributions have traditionally been modeled as uniform or random; however, population distributions or user distributions are clustered by nature (see Figure 17.5).

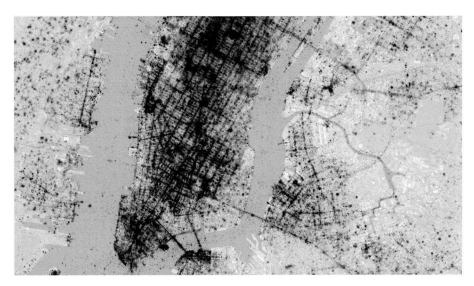

Figure 17.5 Visualization of demand clustering effects from social geodata for Manhattan.

From a deployment perspective, some of the changes that small cells bring to traditional macro design are:

- Lower radiating power reduces the cell's coverage area → analyses at a higher level of detail (accuracy) are required to deploy the new network element in a way that optimizes received signal and minimizes interference from more powerful network elements. From the data-processing perspective, accuracy implies:
  - if demand clustering ~ 20 m, modeled at <1 m; and
  - signal propagation resolution <1 m.
- New deployment locations → small cells can be installed at street level in any type of street furniture, on the side of buildings, or even inside the user's home, as in the case of femto cells. The number of candidate locations that has to be analyzed increases exponentially. A higher level of accuracy is required to select the location that optimizes the contribution of the new cell to the network.
- Multiple tools in the box → small cells group a variety of cell types that provide optimal solutions to different problems. Automatic HetNet cell planning (AHCP) takes into account the complex interplays between the different network elements.

The two following studies apply L-SND to large-scale scenarios for business-analysis purposes. Top-level results include the multiple requirements listed above, but present the targeted objectives as a decision support frame.

17.2.1.1 Spectrum allocation study

Figure 17.6 reflects the outcome of a spectrum allocation study. This tier 1 US operator faces the deployment of twice the number of sites currently operating in its network to cope with the increasing demand requirements. Spectrum resources are a key concern and have a direct input into current and future network strategies. Acquisition of more

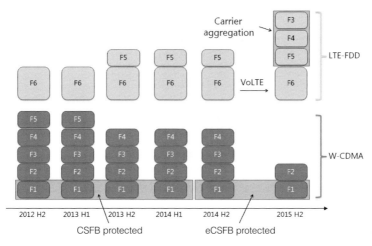

Figure 17.6 Spectrum allocation study.

spectrum or spectrum re-farming are examples of business decisions derived from this study. The graph summarizes available frequencies and how those frequencies are repurposed over time. These results are based on an extensive analysis of multiple variables to provide an accurate picture of the current status of the network and its future requirements: the performance of current and future technologies, network traffic information, market penetration, demand estimates and investment plans. This represents only one of the multiple analyses that can solve questions such as:

Network spectral efficiency?
Optimal frequency allocation for small cell deployments?
Future spectrum needs?
Performance of different UE devices and network equipment?
Optimal technology coexistence?
Would it be possible to benefit from economies of scale?
What is the schedule for a nation-wide deployment?

17.2.1.2 Deployment study

Large-scale network design facilitates the creation and rapid evaluation of multiple scenarios for business purposes, but providing the level of detail that engineering teams would use for deployment phases. Figure 17.7 is an example of support information provided for a frequency allocation exercise. The individual contributions of the existent sites are compared with a repurposed macro network where new small cells are deployed. The study includes accurate radio analysis (i.e., capacity increase due to MIMO), which helps to identify the impact of each site in the network and which macro sites can be decommissioned and substituted by a more spectrally efficient set of small cells. In this case, the results are shown in capacity terms, though it would also be possible to represent them according to different variables, for example dollars ($).

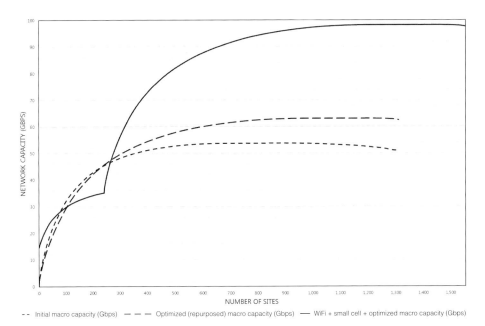

Figure 17.7 Deployment roll-out study.

Since L-SND targets accuracy and scalability, it is in general beneficial when there is a combination of multiple technical configurations (HetNets). However, one of the biggest advantages of L-SND applied to SCN is the ability to perform complex analysis without the requirement to subdivide the project into artificial regions or to accommodate a maximum number of sites.

By subdividing the project into smaller regions the interference continuum is broken. This is critical for small cell planning (independently of the scale) due to the high imbalance of radiated power levels between the different types of cells in the network. For example, the power of a macro cell and the power of a small cell could differ by 10 to 20 dB.

17.2.2 The interference continuum

Since L-SND targets accuracy and scalability, it is in general beneficial when there is a combination of multiple technical configurations (HetNets). However, one of the biggest advantages of L-SND applied to SCN is the ability to perform complex analysis without the requirement to subdivide the project into artificial regions or to accommodate a maximum number of sites.

By subdividing the project into smaller regions the interference continuum is broken. This is critical for small cell planning (independently of the scale) due to the high imbalance of radiated power levels between the different types of cells in the network. For example, the power of a macro cell and the power of a small cell could differ by

10 to 20 dB.

$$SINR = \frac{Signal}{Interference + Noise}$$

The higher the level of interference the more difficult it will be for the user to receive or send a clear message, and more redundancy and robust coding schemes will have to be selected to provide a quality communication.

For example, users at the cell edge will receive a lower level of signal coming from their server cell. At the same time, the level of signal received by them from the neighbor cells will be higher than for a user closer to the antenna. This means our cell-edge user will have to use lower throughput coding schemes, and more bandwidth will be dedicated to redundancy and to increase the robustness of the communication (with a higher power consumption also associated).

Overall, network performance is limited by interference, and SINR requirements are given by the technology, the UE characteristics, the environment, and the other active cells in the network. Use of different frequencies and careful management of the available spectrum could have a very positive effect, which stresses the importance of accurately computing the contribution of each network element.

As previously mentioned, one common solution to the issue of planning a growing number of sites is to cut down the project area into smaller regions limited by size or number of cells. By modeling entire cities, it is found that it is necessary to extend the signal predictions to much further distances in order to compute a site's capacity. This is because capacity is directly influenced by the interference generated by the other network components in the system, and due to the low nature of the power of the smaller entity in the HetNet stack; interference from a distant entity can have a great impact on the performance of the smaller cells.

For example, the power of a distant macro can often be at a similar value to the power of a femto cell's serving area. In addition to the interference problems that this situation can cause, if access to the femto and macro cells isn't controlled, users could be accessing the distant macro cell, even at a lower QoS. The capacity provided by the femto cell shouldn't be considered in isolation from the other elements of the network.

17.2.3 Meeting network capacity requirements

It is only possible to accurately compute network capacity requirements if, and only if, the analysis reflects the complex dependencies between network elements and parameters.

How many more cells are required is directly linked to network capacity, and ultimately, to wireless data demand.

$$\text{Network capacity (bps)} = \text{Quantity of spectrum (Hz)} \\ \times \text{Cell spectral efficiency (bps Hz}^{-1}) \times \text{Number of cells}$$

(17.1)

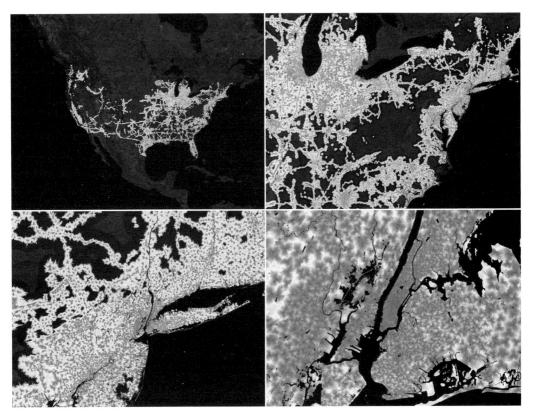

Figure 17.8 US nation-wide SINR analysis.

As the demand increases, more network capacity is required. If we analyse equation (17.1) above:

Acquiring spectrum to tackle initial capacity problems is a common approach. Spectrum is also considered as an asset and governments often auction it, in some cases opening "spectrum wars" between operators [11]. Without spectrum limitations it is possible to use a dedicated band for small cell deployment. This would provide the best SINR because it would avoid interference problems. However, acquiring additional spectrum is costly and, although it correlates directly to capacity, doubling bandwidth with the current growth rates can only double the capacity and does not keep up with the demand.

Cell spectral efficiency. Deploying 4G or upcoming new protocols and technologies can significantly improve the cell spectral efficiency. However, a technology-only solution is still failing to provide enough capacity increase to cope with the growth in data demand [3]. Furthermore, the maximum improvement in cell spectral efficiency is only possible when the design ensures that the most effective high-rate, high-order MCS and SINR improvement solutions, such as MIMO diversity/multiplexing features or beamforming, coincide with the demand. So it

Table 17.1 US macro sites projection.

$N_{1947} < 100$
$N_{2013} \sim 290{,}000$
$N_{2016} \sim 600{,}000$

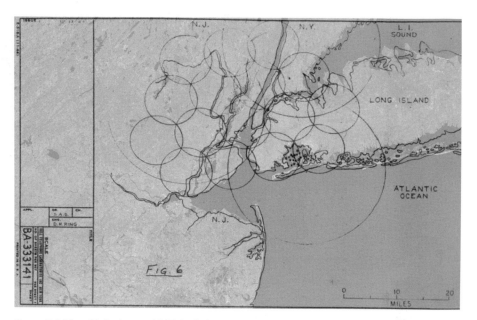

Figure 17.9 New York plan, c. 1947 "cellular" design in New York.

is important to ensure there is high SINR where there is demand, and this means that the cell location and orientation are critical.

Number of cells. The third element in the equation indicates that more cell sites are required to keep up with demand. This will be achieved by deploying additional, though smaller, network elements. The images below show how that has been the trend since the first cellular network was proposed in 1947. For the last 50 years, 97% of all mobile capacity has been delivered through an increase in cell numbers (spectrum re-use) [12].

Figure 17.9 is a hand-drawn network plan for Manhattan in 1947. From a total of 19 cells, we can easily differentiate 3 "macro" cells and 16 "small cells" that cover the whole area.

Figure 17.10 is that same initial network overlapped by the current network of a US mobile network operator (MNO). Every little symbol represents a macro cell. Small cells haven't been included in this representation but Table 17.1 shows the projection based on US analyses for site counts excluding femto cells and atto cells.

The nation-wide view in Figure 17.11 and the number of macro sites involved in the design may seem an extreme situation. However, Figure 17.12 shows a more realistic

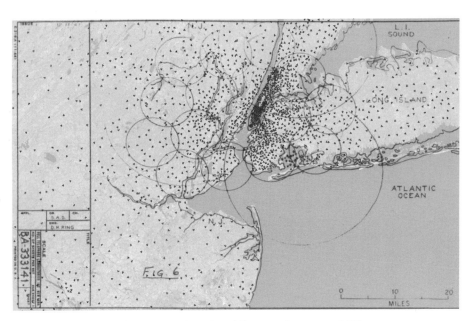

Figure 17.10 New York Plan, c. 1947 plus US mobile network operator, 2013.

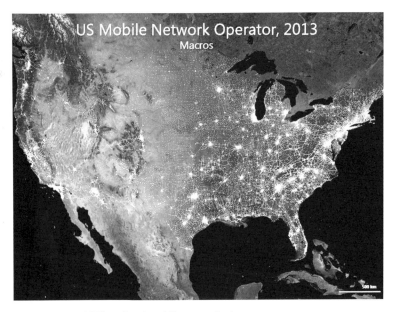

Figure 17.11 Real US-national mobile network view.

zoom in over Manhattan, this time including all the different elements of the HetNet stack: macro, pico, femto, and WiFi. Following Cooper's law [12], mobile operators are facing very large rollouts in the upcoming years and this is a clear example of the volume and variety of cells that are part of modern cellular networks. The complexity of cell

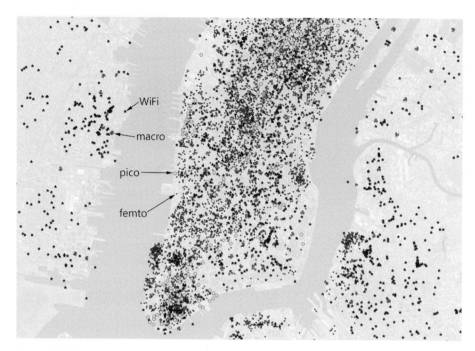

Figure 17.12 Modern heterogeneous network in Manhattan.

planning has increased not only due to the number of cells, but also because additional considerations have to be taken into account, such as frequency band assignment for the different cell types, power, etc. It is necessary to analyze as a whole the complex interplay between those network elements and parameters.

17.3 Large-scale challenges

Any engineering team dealing with a large-scale deployment (because of the number of sites, or because of the area, or because of both) will definitely face a combinatorial complex problem. To be effective, small cell planning has to be accurate and scalable, but special attention also has to be paid to the areas described in this section.

We have already discussed two key challenges that drive the use of L-SND:

- the need to analyze and manage large numbers of small cells
- the need to maintain the interference continuum to provide the required accuracy.

We have also discussed the need of a deployable business case for a small cell network. With large-scale deployments, this is only achievable by dedicating time to the project set-up phase. To make it possible to scale as to the point to reach the phase of deployment, key objectives have to be predefined and correctly specified. So we have to clearly understand those before we can automate the design.

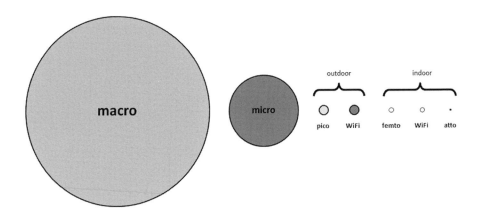

	macro	micro	pico	WiFi	femto	WiFi	atto
Urban radius	1,000 m	500 m	100 m	100 m	40 m	40 m	10 m
Power	40 – 20 W	15 – 5 W	5 – 1 W	0.25 W	0.25 W	0.25 W	0.1 W

Figure 17.13 Heterogeneous components in modern networks.

This section focuses on how L-SND helps to solve those network planning analyses already challenging for any common modern deployment: clutter accuracy, demand estimations, signal modeling, spectrum reuse, and dependency between parameters. The large scale just adds an extra level of complexity that would make the planning of a small cell network an impossible task.

17.3.1 Environment accuracy

Subsection 17.1.1 briefly described vector clutter and why an appropriate representation of the environment is critical: clutter data is often the primary decision input for significant calculations: signal propagation analyses (the accuracy of the interference calculations directly depends on the clutter resolution); optimization analyses (ACP and AHCP, number of sites); statistical analyses (indoor/outdoor capacity); etc.

As Figure 17.13 demonstrates, modern networks will be built as a mixture of a set of heterogeneous components.

Those HetNet components exist in forms that have very large power ranges from the smallest to the largest component. Planning for small cells means that a coarse modeling of the environment could provide a distorted representation of the effect that smaller components have at street level. For example, the components with more power can swamp the smaller network elements and render them inoperable. Careful consideration of location could, however, provide a different outcome. For example, if the smaller components are placed at street level in macro shadowing areas where the additional signal and capacity are really required.

Table 17.2 Maximum recommended bin size for different cell radius.

Type	R	ΔX
Macro	500 m	12.5 m
Small cell	100 m	2.5 m
Femto	40 m	1 m

For large-scale deployments, clutter has to meet the accuracy requirements of the smallest element but it also has to be scalable. Management of clutter files for big areas usually involves dedicated engineering time and memory resources. For example, the clutter required for small cells planning has to provide a better resolution than the clutter required for macro cells planning and small cells planning is more sensitive to changes in the clutter information, hence it needs to be updated more often (i.e., road changes, new developments, etc.).

Traditional clutter and environmental data management techniques present important scalability and accuracy limitations. To optimize the planning process, L-SND proposes the use of dynamic resolution clutter information. The objective is for engineering teams to be able to access the environmental information from different locations and, provided the data changes, they are also able to easily update the information they are working with.

17.3.1.1 Accurately modeling the environment

The environment is usually modeled based on a binned representation of it. The bin size is directly linked to the resolution we use to represent the data. By subdividing the data in bins we are quantizing the information and that generates a quantization error.

Low-resolution and low-accuracy clutter introduce an error in the results delivered. The study conducted by Bernadin and Manoj [13] establishes a practical view on the relationship between cell radius, R, and the quantization level, ΔX.

$$\Delta X < R/40$$

where R is the cell radius and ΔX is the bin size.

To maintain an acceptable level of accuracy, it is important to first model the environment to a high degree of accuracy. What this degree is for a large-scale deployment is given by the smallest element from the pool of HetNet components present in the network.

So if the cell radius is ~400 m, $\Delta X < 10$ m. As we can see in Table 17.2, macro cell planning requires a lower resolution than small cell or femto planning. However, in a situation where we are calculating traffic per cell to identify areas that require additional capacity (in particular co-channel small cell fill-in), it is necessary to work at the higher resolution limit or the results could be biased by the interaction of high-power macros hiding low-power cells. We recommend working at 1 m resolution to minimize the quantization error.

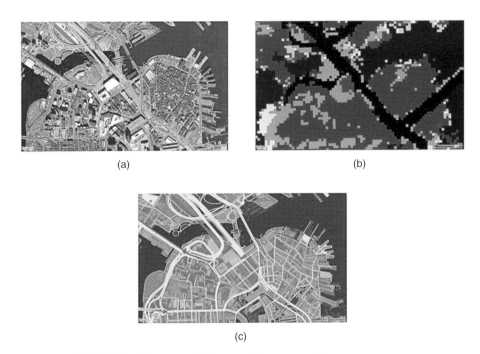

Figure 17.14 (a) Satellite, (b) raster, and (c) vector clutter comparison.

The use of vector clutter for L-SND provides the required level of accuracy but it also allows a dynamic calculation of the environment at a variable level of resolution. Different sources of information are dynamically integrated to provide a high-resolution picture of the environment. Every single element of the clutter is characterized independently to give a more realistic representation of the transformations that affect the radio waves when they encounter each element: buildings, roads, trees, pathways, etc.

Figure 17.14 shows a visual comparison of two different clutter types for the same area of Boston. The second image represents a very good traditional raster clutter (25 m resolution) sample used for the design of macro networks. The third image shows a sample of vector clutter (1 m resolution) for the same area calculated at 1 m resolution. In this last image, it is possible to distinguish clutter elements: building information, parks and greenery, different road types, and corrected information (compare with black lines and squares in the traditional raster clutter), etc.

The images in Figure 17.15 compare the signal propagation estimations for the same parameters when raster clutter is used and when vector clutter is used.

17.3.1.2 Scalability and environment modeling

The dynamic nature of the vector clutter eliminates the requirement to subdivide project files, avoiding file transfers or manual propagation of changes. This eases the data management burden of large-scale deployments. Scalability is then limited by the computational resources.

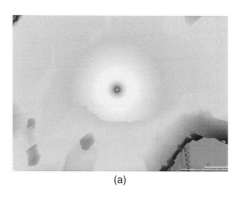 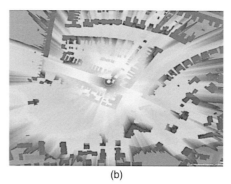

Figure 17.15 Signal propagation estimation using (a) raster clutter, and (b) vector clutter.

Cloud computing plays an important role in the possibility of an accurate and scalable modeling of the environment. Data inputs are stored as individual elements. This provides higher flexibility when any of the data sources are modified. Also, the cloud systems allow an instant distribution of changes, from any satellite location to the central storage point and from the central storage point to the rest of the satellite locations.

Petabytes of data can be easily shared, which reduces the limitation posed by the exchange of heavy files when high accuracy is required. Only the areas of interest are processed at any given time, saving local disk space but maintaining the availability of global high-definition data.

Aside from the data management challenge, scalability requires the optimization of the computational load to provide the level of accuracy demanded without overloading the system with unnecessary requirements. One of the principles of the L-SND methodology is the ability to automatically adapt calculations to the minimum resolution required. This speeds up run times, optimizes the use of computational resources, and simplifies the planning process by being transparent to engineers.

Figures 17.16 and 17.17 show a sample of scalability for the San Francisco Bay and San Francisco downtown areas and how different resolutions bias the modeling of the environment. The vector clutter has been dynamically built from a pool of multiple individual elements for the whole region providing a continuous layer of environmental information.

17.3.2 Understanding traffic and demand

In this subsection we will focus on the cell spectral efficiency component of equation (17.1).

A key challenge for an effective network design is to maximize spectral efficiency and ROI simultaneously. That is only possible if the network design ensures capacity is provided where the demand is located. As it is explained in Subsection 17.3.2, the location of a traffic hotspot determines whether there will be a return on investment or not, and maximizes profit margins.

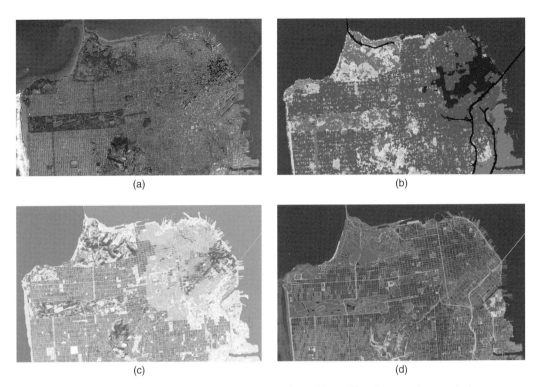

Figure 17.16 San Francisco Bay clutter comparison: (a) satellite, (b) raster low resolution, (c) raster high resolution, and (d) vector clutter.

Figure 17.17 San Francisco downtown clutter comparison: (a) satellite, (b) raster low resolution, (c) raster high resolution, and (d) vector clutter.

As per the formula above, the cell spectral efficiency is critical if there is to be an effective increase in network capacity. In the presence of traffic hotspots, an adaptive modulation and coding scheme brings the difference between users sharing 0.5 Mbps and 25 Mbps on a 5 MHz carrier; or 2 Mbps to 100 Mbps on a 20 MHz carrier: a dynamic range of ~50.

With carrier aggregation, this is carried through to each and every carrier, so badly placed cells never achieve a return on investment. More importantly, SON-style algorithms have limited effect. A hotspot at the cell edge is almost always likely to remain at the cell edge no matter what improvements are made.

So, this adaptive modulation and coding scheme means that the users' locations actually determine the capacity of the serving cell. For a commuting, dynamic population the use of population estimations from static residential or business census information is very limiting.

Also, HetNet components are sensitive to the mobility of the demand and the environment. As is described in more detail in Subsection 17.3.2, fast-moving demand cannot be served by small cells such as picos. Different small cell components are suited to different environments. Femto cells, for example, are designed for indoor locations; macros and picos work effectively outdoors.

Big data brings to network planning a variety of new information sets that can be inter-connected to achieve a better understanding of users and networks. For example, social geodata provides a solution to the location and demand accuracy problem. Tracking demand through traditional sources such as mobile data (PCMD/MMR) and census information provides some understanding of the live network. Public data from social networks such as Twitter, Facebook, Instagram, Foursquare, and flickr provide information about the life in the network. Associated benefits are the higher accuracy of the user-location information or the ability to easily identify and predict user clustering, for example for special events.

17.3.2.1 Modulation schemes

There are fundamental differences between network design for data-dominated networks compared to network design for voice networks. With voice, it is simply enough to provide a sufficient SNR (signal-to-noise ratio) such that a circuit-switched connection can be maintained. Data, however, is highly dependent on interference and load. Network performance and user perception varies strongly with those parameters.

Automatic modulation techniques adapt to the performance of the link. Higher SNR means higher, more spectrally efficient modulation schemes may be deployed [14].

Figure 17.18 explains why, ultimately, locating the sites where users congregate can improve the cell spectral efficiency by an order of magnitude. It also shows why, in order to maximize ROI and the network's spectral efficiency, capacity has to be provided where the demand is. Careful study of location must be considered to ensure high-rate, high-order MCS zones and MIMO spatial diversity/multiplexing features are in the areas where there is high, clustered demand. Additionally, the performance

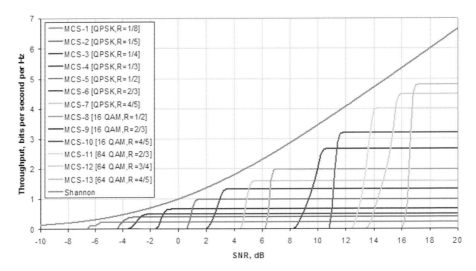

Figure 17.18 Throughput vs SNR for LTE modulation and coding schemes.

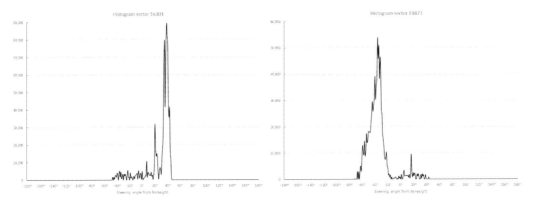

Figure 17.19 Steering beam histogram for two sample radios; time spent away from boresite (0 on X-axis).

of LTE-TDD beamforming antennas can be transformed by understanding the most optimal orientation of the steered traffic beam, see Figure 17.19.

We can also see the benefits of optimal cell location in Figure 17.20. It compares two SCN deployment scenarios in terms of the percentage of area that is covered by the coding schemes available for LTE. The bars to the right represent the MCS values when the small cells are placed at every two crossroads without any consideration to the demand profile, resulting in a grid-shape deployment. The bars to the left represent the MCS achieved when the location of the small cells is defined following demand hotspots.

Optimally placed small cells and hotzones will have the added benefit of reducing the overheads on larger macro cells, increase their overall cell spectral efficiency, and reduce ICIC/eICIC scheduling conflicts.

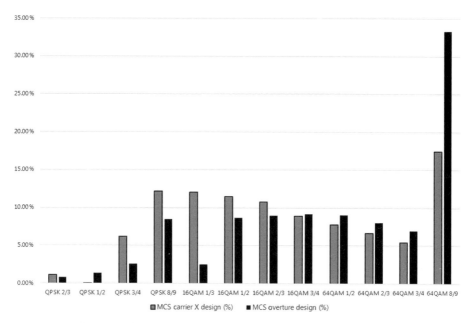

Figure 17.20 MCS comparison for gridded or demand-driven SCN deployments.

The statement above looks very simple and straightforward as long as there is information about where the users are and what are they doing (what is their data consumption profile). And this is a really big challenge for large-scale small cell networks.

Planning to deploy an enterprise network often offers the advantage to be able to include known parameters; this could be a floor plan and/or an employee distribution (even when mobility considerations have to be taken into account). However, for a commercial network, working with an accurate demand profile presents one of the biggest challenges for L-SND. Factors such as mobility or user clustering have a huge effect on the network planning and, ultimately, on user experience. Outdoor small cell deployments are initially targeting highly developed areas, though there is growing interest in providing a better mobile data experience for areas with a lower population density. User mobility is one of the drivers of that change, since users expect the same data service everywhere they go. We will explore the different demand profiles in Subsection 17.3.2.

17.3.2.2 Traffic

Small cells are often considered as an offload solution to solve the capacity problems of the congested areas in the network. Traffic is the actual number of bytes that circulate through the network; hence, the network operator is often aware of the areas that present a higher load and can act accordingly.

Together with other network parameters such as congestion, blocked or dropped calls, busy hour traffic, throughput etc., traffic data is a critical input to understand macro demand, therefore it is very useful for an offload small cell design. In many cases,

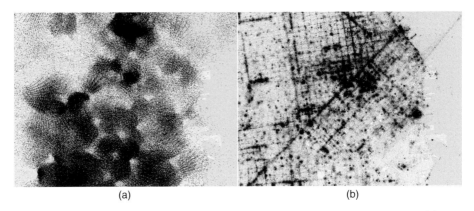

Figure 17.21 (a) Visual representation of mobile measurement reports (MMR) and (b) social networks geotagged events for the same area: downtown San Francisco.

marketing projections for future technologies and network evolution can be applied to the current traffic information to estimate future network capacity requirements. Also, traffic values can be spread over the serving area of the network entities to estimate the effect on the different demand profiles. For example, a served demand map is a reflection of the association between network traffic and serving radio and provides a more accurate estimation of the areas and the capacity that has to be injected into the system to cope with the real total demand generated by the users.

However, network traffic is a reactive source of information. The wireless operator will only have information about its current network and it doesn't necessarily present a realistic picture of demand due to:

- Limitations on the understanding of the information, i.e. inaccurate location estimations (Figure 17.21), busy hour, network status...
- Traffic profiles may be skewed by the limitations of the current network and by the profile of the current users or the available cellular technology: an MNO with reduced network share (~15%) would be in that situation if the objective is to expand into areas where there is limited or no infrastructure: existing traffic in this case would provide a partial view of the network.

For a greenfield design, use of adapted traffic information should be even more careful, since even the user profile may differ from the targets of the deployment.

17.3.2.3 Demand

The demand profile is a very important input for network planning. If accurate clutter helps provide a realistic modeling of the environment and the signal propagation, accurate demand has a direct effect on a realistic evaluation of the network. This is because demand is non-uniform and to evaluate the real impact of a site or the real performance of the network it is necessary to identify where users are and understand their behavior. Techniques that cannot identify usage clustering at the scale of the cell radius will have limited use.

Large-scale deployment and scalability 449

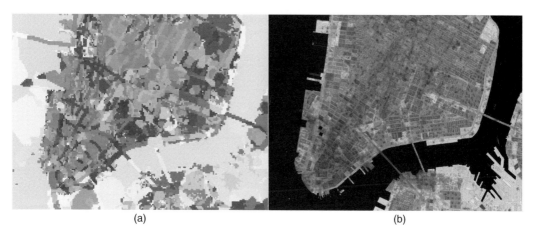

(a)　　　　　　　　　　　　　　　　　　　(b)

Figure 17.22 Demand maps comparison: low resolution (a) vs. high resolution (b).

Same as the vector clutter, for L-SND and in particular due to the characteristics of small cells, the demand maps require a new approach that overcomes the limitations of current methodologies. For that reason, demand profiles are also dynamically built from a set of layers, which include new big data sources of information such as social networks traffic information. This provides a high level of flexibility and the ability to scale the analysis to the project requirements.

The images in Figure 17.22 represent a traditional demand map used for macro design for the Lower Manhattan area (a), compared to a high-resolution small cells demand map generated using L-SND capabilities (b). The high-resolution image clearly shows a higher depth of information. For example, it is possible to identify the different demand profiles for indoor/outdoor, user clustering effects, etc.

Demand profiles and social analytics
But the benefits of combining new sources of information with L-SND are broader. Big data provides additional dimensions of information. For example, we can see in Figure 17.23 data gathered during the period corresponding with the London 2012 Olympics. It clearly reflects the increase in traffic due to specific events, such as the opening and closing ceremonies or the day that Usain Bolt won the men's 100 m final. The graph also reveals clear usage variations depending on the time of the day.

Daytime changes and weekday changes are also clear in the following analysis of Flickr information (Figure 17.24).

The depth of information provided by the analysis of social networks opens totally new aspects that enhance network planning. The images in Figure 17.25 are a sample for the United States of the extent of the raw data. By simply utilizing the location information, we can zoom up to meter level into the Manhattan area, where clustering effects are easily visualized. But parallel data analysis can be used for prediction, mobility, customer satisfaction (sentiment analysis), etc.

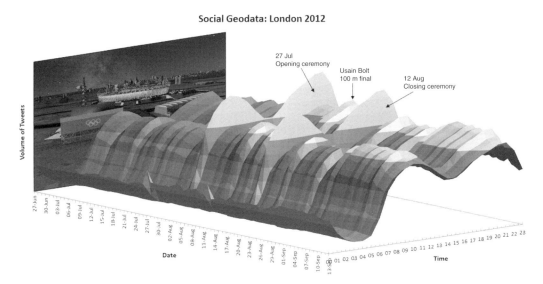

Figure 17.23 Twitter analysis for the London 2012 Olympics period.

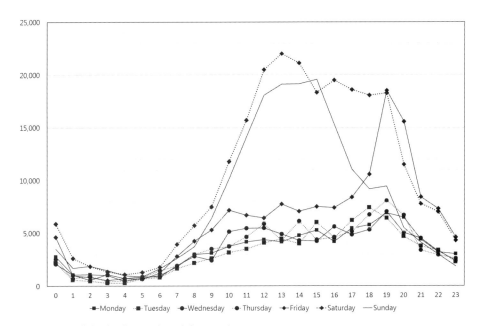

Figure 17.24 Flickr daytime and weekday trends.

Combining information from social networks offers a dimensional view of user clustering and network agnostic demand. This helps create valuable demand profiles that help tailor the objectives for the network design and the automation of the process leading to a simplified decision support report for the engineering team.

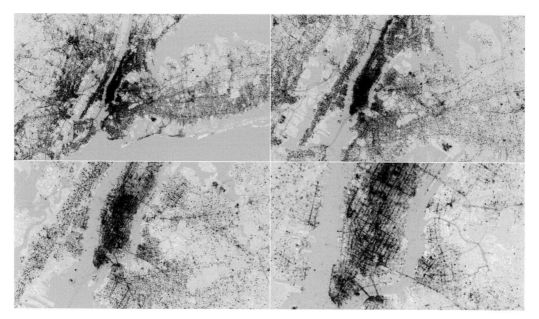

Figure 17.25 Social geodata zoom in samples for the United States.

Together with the network and traffic information, the demand map helps to automatically define which areas are "commercially" interesting, that is, where the network is under stress and an action is required. The evaluation and solution criteria are very flexible and may change depending on the project objectives. Demand profiles can integrate marketing projections or target particular deployment objectives. For example, the criteria may prioritize outdoor capacity vs. indoor, or it may concentrate the deployment efforts according to mobility (specific criteria for low-mobility roads vs. highways, or special coverage efforts for railway routes).

When traffic information is available, it is also possible to directly link the effect that the current operating network has in the demand estimations and automate the process for analysis of larger areas. For example, subtracting the served demand from the offered demand, we can work with a map of unserved demand, which automatically shows the areas where more demand has to be added to the network:

$$\text{Unserved Demand} = \text{Offered Demand} - \text{Served Demand}$$

- Offered demand. The given demand in an area. That is, if the network was able to process all the user requests with no limitations, this would be the value of the offered demand.
- Served demand. The demand that the current network is already processing.
- Unserved demand. The demand that the current network is unable to process.

From a business-case analysis to a deployment scenario, the objective of a SCN can be to complement the current network in order to eliminate the unserved demand. New or candidate sites are evaluated and selected based on this specific demand profile. The

Table 17.3 The dynamic population of Manhattan.

	Week days	Weekends	Week nights
Local residents	1.46 million	1.54 million	1.58 million
Commuting workers	1.61 million	0.57 million	0.23 million
Visitors	0.78 million	0.78 million	0.24 million
Others	0.09 million	0.02 million	0.02 million
Total	3.94 million	2.91 million	2.07 million

L-SND algorithms will automatically identify and quantify: which areas of the network require additional capacity; if that is enough to cope with the demand; how much capacity is required; how many sites have to be deployed in order to meet the objective, etc.

Dynamic population
An interesting effect clearly highlighted by big data is the variability of the demand profiles according to user behavior. Daily commuters have an important effect on network load. The population of most modern cities grows and shrinks during the working day or at the weekends and for that reason it is not okay to work with just the residential census information. Transient demand is vitally important to providing adequate capacity.

Table 17.3 shows the values of the dynamic population of Manhattan. With a US Census-measured residential population of 1.6 million, this value more than doubles to a daytime population of 3.9 million. Peak population events, or "day-trip" events, often push Manhattan's daytime population to 5 million, depending on the conditions and circumstances [15].

Manhattan can be considered as an extreme example of dynamic population, but worldwide data demonstrate that big population variations are also observed in cities with different profiles. Table 17.4 shows the changes observed in the UK for cities of different sizes.

Big data methodologies ensure multiple inputs of MMR/PCMD (per call measurement data), social data, mobility events, census, etc., make it possible to get an accurate picture of where and what people do. Use of smart phone geotagged social data allows for a more accurate idea of where people congregate, and hence where there are traffic hotspots and hotzones. Locating smaller HetNet components near to these hotspots/hotzones ensures they achieve a high spectral efficiency since most users will be in high-order 64-QAM modulation schemes.

The moving demand
Mobility is another factor associated with daily commuters, but the effect of this on the network design is different to the effect of the dynamic population.

Each cell type from the small cell portfolio is more appropriate for different problem areas. For example, the coverage area of a pico cell is relatively small. The handoff burden as a moving user in a vehicle would be ineffective, as that type of user handsoffs every few seconds.

Table 17.4 Population changes during the working day for UK cities [16].

	Population	Population changes during the working day: 2011 Census – % increase
York	198,051	2.8
Coventry	316,900	5
Bristol	428,100	8.1
Birmingham	1,074,000	8.4
Peterborough	116,570	9.9
Leeds	757,700	9.7
Cardiff	324,800	15.4
Oxford	150,200	25
Manchester	2,553,379	27.6
Cambridge	123,900	35.3
Islington	206,300	36.5
Tower Hamlets	234,800	57.9
Westminster	236,000	266

Figure 17.26 (a) Indoor and (b) outdoor demand maps.

Estimating whether the user is moving and at what speed is critical to ensure picos are not used to cover fast moving traffic, which is the domain of the macro.

Using big data methodologies it is also possible to extract information from multiple sources such as GPS tracks and dense RF scanner measurements to establish the mobility of users to select the optimal HetNet component. Automatic HetNet cell planning (AHCP) has to include these mobility profiles into the decision-making process.

Indoor/outdoor demand

A large-scale small cell deployment will be coexisting with different technological solutions. For example, for a small cell deployment targeting capacity problems in the network, indoor and outdoor considerations are also important (Figure 17.26). According to the statistics and the research published by several industry sources, for example Cisco [3] [17], close to 80% of the traffic is conducted indoors.

Different areas need different solutions. Macros are designed to give coverage over larger distances, but in a situation where a big number of users congregate in an area, they are more limited to provide data capacity. Pico cells are perfect for provisioning highly efficient capacity across small outdoor areas, but they are ineffective solutions for indoor demand. WiFi, femto, and atto can provide indoor concentrated capacity.

Estimating whether data capacity is required indoors or outdoors is a critical step in the choice of HetNet component and an accurate demand profile is the foundation of the analysis. From the perspective of the challenge posted by network capacity, all the different elements have to be considered simultaneously when evaluating a small cell design. This will offer a more realistic view on the spectral efficiency of the resulting network.

The vector clutter allows us to establish if the demand is originated indoors or outdoors and the AHCP automatically optimizes the selection of the best network element. Generally that will be macro and pico cells for outdoor capacity and femto and atto cells to provide indoor capacity.

17.3.3 Signal accuracy

Signal accuracy isn't an exclusive challenge for large-scale deployments. Much research has been conducted about this topic, including signal prediction modeling for the different elements that we can find in the networks. How to tune the models, how to transform the theoretical formula into a more realistic model, has also been subject to extensive research.

Traditional signal analyses lack the accuracy and scalability characteristics required for large-scale small cell deployments. While the theoretical base may be quite similar, the implementation of signal modeling processes for large-scale rollouts requires a substantial new approach. The use of big data and cloud computing is critical to overcome those limitations. For the examples in this chapter and the images shown below (Figures 17.27 and 17.28), new algorithms that maximize new and extensive sources of information have been implemented. The benefits and the use, though, are not restricted to large-scale deployments.

For example, by using scanner data, it is now possible to model the signal of each single radio in a sector with a lower root mean square (RMS) and standard deviation (SD). The higher accuracy in the signal estimation is given by the high-definition modeling of the environment. By combining this with big data it is possible to scale the analysis to include all the radios on a site, and by multiplying the scalability the analysis can be extended to all the sites in the network.

That is, each and every radio, be it an LTE radio or a W-CDMA radio, has its own propagation model at its exact operational frequency.

In particular, for complex city environments, it is preferable to use deep scanner information to create a model for every radio element. This means for an area such as

Large-scale deployment and scalability 455

Figure 17.27 Sample of drive data for Manhattan.

Figure 17.28 Sample of tuned signal prediction.

Manhattan, there may be around 6,000+ individually tuned models across all bands and radio-access technologies.

Deep measurements from scanners such as PCTEL's MX scanner typically detect up to 16 PN codes or 16 PCI codes at a given location. The measurements are gathered at a sampling of 50 ms and a signal depth of 64 dB is possible.

Automatic disaggregation of these codes and allocation to the relevant radio ensures that each radio in the network has enough measurements to ensure accurate propagation models at each band/RAT. Every sector, every band, and every technology should have a tuned prediction model. High-resolution vector clutter improves the accuracy, and large classifications ensure minimum difference between predicted and measured.

The signal of a site is interference for the neighbor site so the interference analyses will be as accurate as the signal analyses are. The combination of per-radio path-loss models, vector clutter, technology profiles, etc., provides a more accurate network analysis.

17.3.4 Co-channel and spectral utilization

Management and optimal utilization of the spectrum is one of the most challenging tasks for most nation-wide deployments. The fragmentation of the spectrum differs according to country-specific regulations, but in many cases we will find MNOs are subject to spectrum limitations.

While for a small deployment, for example in the case of a trial, it is easy to allocate a chunk of spectrum that can be used in that particular time and for a limited number of sites, when dealing with large deployments, optimizing the use of the spectrum requires a deep understanding of:

- the legacy network
- marketing considerations
- technology considerations and strategy.

A disconnect between the marketing and the technology sides, causes a "nightmare" large-scale deployment. An unrealistic business-case analysis leads to artificially balanced engineering budgets and ultimately to a poorer network both in technical and ROI terms.

The concept of "protected zones" is an example of how an integrated approach to network and technology strategy provides an immediate direct solution to saturated areas, while allowing a long-term network evolution strategy. For a design that targets hotzones of demand traditionally covered by macros:

– the future-proof solution is to deploy picos, but the existing macros have significantly higher power
– if the available spectrum is enough, an entire channel could be dedicated to the picos, but this reduces the spectral utilization.

So "protected zones" are necessary: where the multi-band nature can be used to allow zones of locally deployed picos to have a locally "dedicated" channel, which then blends into the surrounding macro-deployed regions (Figure 17.29). As the technology

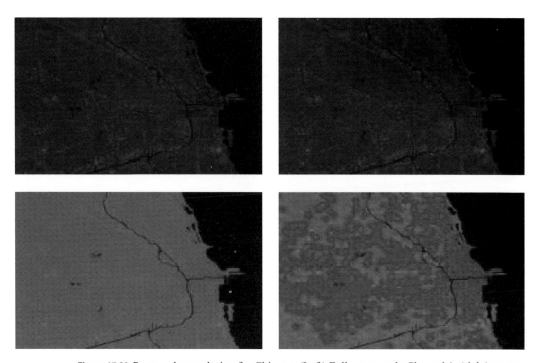

Figure 17.29 Protected zone design for Chicago. (Left) Full macro-only Channel 1; (right) macro and pico co-channel for Channel 2.

evolves, this channel distribution can be reconfigured to follow demand requirements. For example, macros operating on the pico channel in that area could be switched off to avoid strong interference problems but as inter-cell interference coordination has evolved with the different 3GPP releases from ICIC, to eICIC for Release 10, and FeICIC for Release 11, the channel allocation can be redistributed.

Hence it is very important to produce a holistic view of the network, understanding what is it going to be used for and what are the available resources in terms of spectrum and technology tools.

17.3.5 Combinatorial explosion

Planning of modern networks is a complex interplay of dependent objectives and holistic meta-heuristics are required to seek the optimal solution to the unknowns of such a large pool of considerations. Automation and big data are again important elements to solve this combinatorial explosive problem.

The previous section was focused on the set of issues that we could face in a large-scale deployment. Some of the issues described are linked to the larger nature of the area subject to analysis. For example: computing limitations, running times, management of data files, etc.

However, many of the issues described are due to the volume of sites that a small cell network deployment may involve. Furthermore, design of modern networks has a larger set of requirements and considerations: location, configuration and technology parameters; backhaul proximity and wireless clearance; rental costs; latency, X2; etc. Because location is the most important point when it comes to the design of the network (SON could solve other later optimization considerations), all those additional requirements play an important role in the design and deployment.

The objective for the deployments will be to look for areas where there is a positive return on investment and deploying small cells makes sense. This will often be analyzed in terms of ROI:

- sites that pay back the investment by being located near high-demand "hotspots"
- sites with manageable interference impact
- backhaul of rental costs are affordable.

There is a long list of objectives to consider but only by considering *all* of them can we maximize return on investment.

17.3.5.1 New York case study

The New York case study is an example of how those variables are interconnected. Some of the points considered in the analysis were:

– optimal traditional towers
– optimal utility poles
– optimal wall mounting
– suggesting search ring
– backhaul costs
– fiber routes
– rental costs
– etc.

The scope of the study was to design a small cell network for Manhattan, maximizing spectral efficiency. The equipment to be deployed was low-powered 1 W small cells. These should be installed in street furniture elements such as lighting fixtures, kiosks, power lines, etc. (Figure 17.30). All of them should have available backhaul (wireless or fiber).

From a total candidate set of 470,000 locations, there were selected 1,852 primary locations based on the pool of interconnected objectives. As part of the analysis, secondary locations were also selected to provide an alternative deployment option to the primary location.

Figure 17.30 demonstrates how the number of selected candidates follows the demand. The higher demand in Lower Manhattan concentrates a higher number of primary locations and the corresponding secondary option.

Candidate locations were also evaluated according to their backhaul characteristics: availability of wireless links at 10 GHz and 60 GHz and distance to fiber. There were cost considerations associated to this evaluation as well, since a fiber node presents accessibility advantages compared to the cost of initiating works that involve digging trenches or laying new pavement. Power was also taken into account to guarantee the

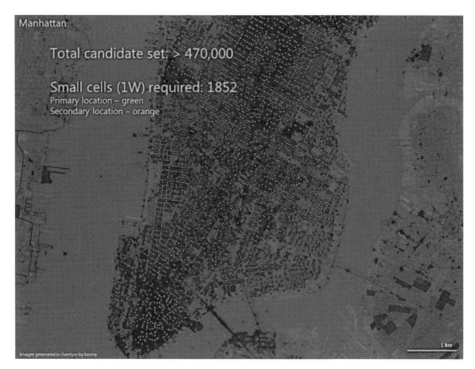

Figure 17.30 Selected street furniture locations for a Manhattan SCN deployment case study.

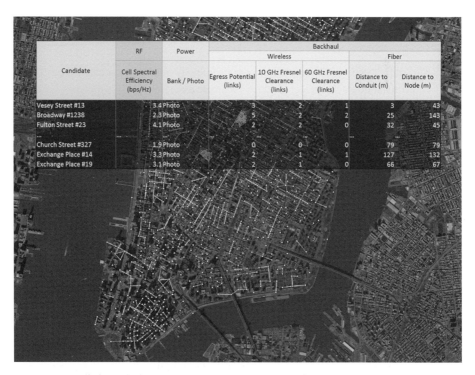

Figure 17.31 Holistic analysis output for a Manhattan SCN deployment case study.

Figure 17.32 3D view of selected sites and signal propagation estimations.

small cell would be operating continuously and independently from the power system installed on the selected street furniture asset.

The table included in Figure 17.31 is a sample of intermediate information generated during the selection process. The outcomes of the study included detailed interference and signal analyses at 1 m resolution. Figure 17.32 is a sample of the signal output.

References

[1] Small Cell Forum Org., 2013. *Small Cell Market Status – Informa February 2013*. www.smallcellforum.org/resources-reports [Accessed 18 January 2014].

[2] Alcatel-Lucent, 2011. *Metro Cells, A Cost Effective Option for Meeting Growing Capacity Demands*. www.alcatel-lucent.com/solutions/small-cells [Accessed 18 January 2014].

[3] Cisco, 2013. *Cisco Visual Networking Index: Global Mobile Data Traffic Forecast Update, 2012*. www.cisco.com/en/US/solutions/collateral/ns341/ns525/ns537/ns705/ns827/white_paper_c11-520862.pdf [Accessed 18 January 2014].

[4] Pistikopoulos, E., Galindo, E., and Dua, V. (eds). *Multi-Parametric Model-Based Control: Theory and Applications, Volume 2*.

[5] Rossiter, J. A., 2003. *Model-based Predictive Control: a Practical Approach*. Boca Raton: CRC Press.

[6] Doerner, K. F., Gendreau, M., Greistorfer, P., Gutjahr, W., Hartl, R. F., and Reimann, M. (eds.) 2007. *Metaheuristics: Progress in Complex Systems Optimization*. USA: Springer.

[7] Tsukayama, H., 2013. "Twitter turns 7: users send over 400 million tweets per day," *The Washington Post,* 21 March. www.washingtonpost.com/business/technology/twitter-turns-7-users-send-over-400-million-tweets-per-day/2013/03/21/2925ef60-9222-11e2-bdea-e32ad90da239_story.html [Accessed 23 January 2014].

[8] Warden, P., 2011. *Big Data Glossary.* Sebastopol: O'Reilly Media.

[9] NGMN Alliance, 2011. *Guidelines for LTE Backhaul Traffic Estimation.* www.ngmn.org/uploads/media/NGMN_Whitepaper_Guideline_for_LTE_Backhaul_Traffic_Estimation.pdf [Accessed 23 January 2014].

[10] Robson, J. 2012. *NGMN Requirements for Small Cell Backhaul.*www.cambridgewireless.co.uk/Presentation/Julius%20Robson%20Presentation%20031012.pdf [Accessed 23 January 2014].

[11] Ofcom, 2013. *Ofcom announces winners of the 4G mobile auction.* [press release] 20 February 2013. http://media.ofcom.org.uk/2013/02/20/ofcom-announces-winners-of-the-4g-mobile-auction [Accessed 18 January 2014].

[12] ArrayComm LLC. *Cooper's Law.* www.arraycomm.com/index.php/technology/coopers-law [Accessed 21 January 2014].

[13] Bernadin, P. and Manoj, K., 2000. "The post-processing resolution required for accurate RF coverage validation and prediction," *IEEE Transactions on Vehicular Technology*, Vol. 49, No. 5, pp. 1516–1521

[14] 3GPP, TR 36.942 version 10.2.0 Release 10, Annex A, 2011. www.3gpp.org/ftp/Specs/html-info/36942.htm [Accessed 22 January 2014].

[15] Moss, M. L. and Qing, C., 2012. *The Dynamic Population of Manhattan.* Wagner School of Public Service, New York University.

[16] Office for National Statistics, 2013. *The Workday Population of England and Wales: An Alternative 2011 Census Output Base.* www.ons.gov.uk/ons/dcp171776_333420.pdf [Accessed 23 January 2014].

[17] Cisco, 2011. *Cisco IBSG Connected Life Market Watch, 2011.* www.cisco.com/web/about/ac79/docs/clmw/CLMW_Service_Delivery_US_Short.pdf [Accessed 23 January 2014].

18 Energy efficient heterogeneous networks

Y. Qi, M. A. Imran, M. Z. Shakir, and K. A. Qaraqe

18.1 Introduction

This chapter introduces novel approaches in heterogeneous networks (HetNets) where both large and small cells are deployed in a mixed manner to satisfy the increasing traffic demand and, at the same time, to improve the energy efficiency (EE) of future cellular networks.

In recent years, there has been a tremendous increase in the number of mobile handsets, in particular smart phones, supporting a wide range of applications, such as image and video transfer, cloud services, and cloud storage. The average smart phone usage rate has nearly been tripled and the overall amount of mobile data traffic demand grew 2.3 times in 2011 [1]. Furthermore, the amount of mobile data traffic is expected to increase dramatically in the coming years; recent forecasts are expecting the data traffic to increase more than 500 times in the next ten years [2, 3]. The current cellular systems would not be able to cope with the expected traffic demand increase. This huge amount of traffic demand leads to the need for further densification of the networks, for example in hotspot areas where traffic demand is concentrated as seen in Figure 18.1.

However, traffic load varies from time to time because of the typical night–day behavior due to the users' daily activities in offices and being back to residential areas during the night [4]. In the current cellular networks, the power consumption of the radio access network (RAN) does not effectively scale with the traffic variations as shown in Figure 18.2. The traffic variations create the opportunities for the design of an adaptive network paradigm that can dynamically scale its power consumption according to the traffic variations.

Generally speaking, the power consumption of the RAN scales with the number of deployed base stations (BSs), each with offset power consumption. In cellular networks, only 10% of the overall power consumption stems from the user equipments (UEs) whereas nearly 90% of power consumption is incurred by the operator networks [5]. Figure 18.3 gives an idea on how the power consumption is distributed across the different parts of a typical cellular network. It is obvious that the RAN and the operation of data centers that provide computations, storage, applications, and data transfer are the most energy intensive parts of the entire network. It has also been identified that a large portion of the 90% power consumption is actually consumed by the BSs of the RAN while the backbone and aggregation networks present a much lower energy

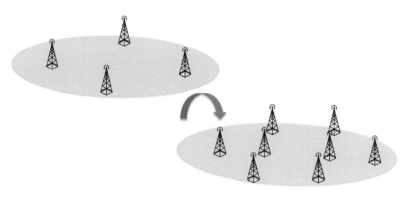

Figure 18.1 Continous densification of cellular networks.

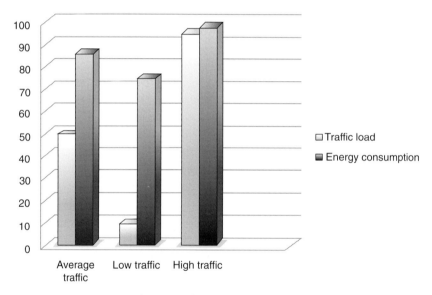

Figure 18.2 Power consumption vs. traffic load.

demand. Therefore improving the EE is mainly focused on design of an EE RAN. In the design of an EE RAN, in one aspect, overprovisioning of BSs will cause excessive power consumption and in this regard, larger cells are preferable. However, in another aspect smaller cells, which in nature are designed for lower transmit power in the first place, are more EE than large cells in the sense that they come with much less overhead power consumption. Small cells also bridge less path loss in the downlink connections to the UEs. An EE deployment methodology has to be able to intentionally balance the offset power consumptions of the network elements and the user experience in terms of throughput and quality of service (QoS) to avoid both under- and overprovisioning.

The state of the art EE cellular network design approaches can be categorized into three classes:

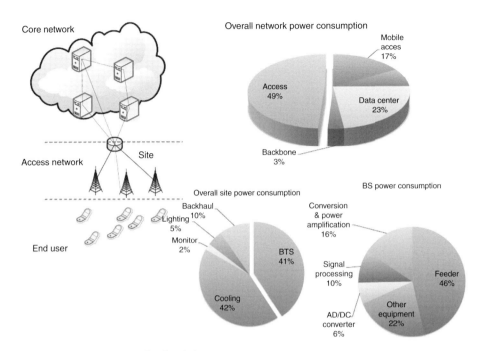

Figure 18.3 Power consumption breakdown.

1. HetNet deployment
2. network management
3. coordinated multi-point (CoMP).

Firstly, the EE HetNet deployment is investigated in the next subsection. The data traffic experienced by users can be increased with more small cells deployed to boost the data rates in a given area [6]. It is widely believed that using a mixed topology of large cells with small cells could lower the energy consumption and increase the capacity of the network. It is also well known that the radio signals are subject to various channel attenuations. For a large cell only network, the wireless connections could suffer from high losses in a hostile radio environment and these connections are the most expensive connections in terms of allocated resources. For example, one of the major losses is building propagation loss: the indoor users suffer, compared to outdoor users, from penetration loss. By deploying indoor small cells to serve the indoor users, outdoor large cell resources can be saved as well as the energy consumption. The benefit of offloading some users to small cells is discussed in [6], where traffic load of certain users is offloaded from large cells to the small cells adjacent to those users. Another advantage of deploying small cells is that the small cells can easily scale their power consumption according to their activity level [7]. The deployment of small cells along with large cells achieves considerable power savings compared to conventional large cell only homogeneous deployment without sacrificing spectrum efficiency and throughput [7–10].

Network management is then discussed. The future cellular networks are expected to guarantee an "always connected" experience for the users. However, "always connected" does not necessarily require all the network elements to be activated at the same time. A promising approach would be dynamically switching off the BSs. However, when a BS is switched off, its radio coverage and service quality must still be guaranteed by neighboring BSs or by other means [11–15]. In this regard, various network management algorithms are provided to effectively choose those network elements that should be activated to connect the users to the network. The self-organizing network (SON) is proposed to manage both the on/off operation of the network elements and the dynamic usage of the network resources while at the same time reduce the human intervention to a minimal level. Network densification by deploying small cells is also an important subject in SON [15]. In this chapter, a cloud-based network paradigm with self-organizing capability based on fuzzy logic is introduced.

Moreover, CoMP has drawn considerable attention from both academia and industry [16, 17] because one UE can be connected to more than one cell to achieve enhanced performance in terms of network throughput while at the same time reduce the energy expenditure. Recently, 3GPP has embarked with more focus on solutions containing small cell enhancements with CoMP [34]. The strength of received signals for the users at the cell edge is expected to be weak due to propagation loss; however, the interference from adjacent cells could be significant when the same carrier frequency is used. The concept of CoMP is to allow one UE to communicate with adjacent multiple accessing points. In the downlink, multiple cells form a distributed MIMO system with precoding, thus the interference from neighboring cells is turned into useful information. In the uplink, the received multiple copies of the same transmitted signals are forwarded to the central processing cell and combined before final decoding. We will focus on the uplink part in this chapter and show how the cell-edge user can benefit from receive diversity by efficient backhaul link optimization and how to achieve improved EE.

The rest of the chapter is organized as follows: EE HetNet deployment is discussed in Section 18.2. In Section 18.3, a cloud-based HetNet architecture as well as the on-demand network management strategies is introduced. CoMP and its EE performance is analyzed in Section 18.4 and, finally, Section 18.5 concludes the chapter.

18.2 Conventional HetNet

There are several options for the heterogeneous deployment of large cells and small cells and one of them is to place the small cells at the cell edge where the users are experiencing weak radio channel due to the path loss and high level of interference. The small cells deployed at the cell edge make it viable to increase the inter-cell distance for large cells while at the same time maintain satisfactory user experience. Given constant large cell power consumption, the increased inter-cell distance is able to lower the area power consumption in terms of Watts/km^2 for large cells. The power consumption reduction might be so significant that the added offset power required by the newly deployed small cells is compensated and the overall area power consumption is reduced.

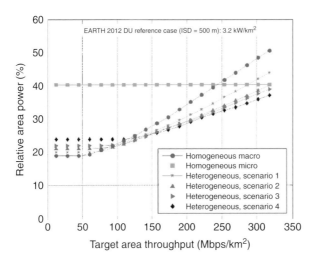

Figure 18.4 Relative area power consumption.

A system level simulation is conducted in [18] to demonstrate the relative area power consumption against the full buffer case in dense urban (DU) environment defined in [19–20]. Four heterogeneous scenarios are considered with 1, 2, 3, and 4 micro cells deployed at the macro cell edges. Homogeneous deployment with macro cells and micro cells, respectively, is shown as benchmark scenarios. As shown in Figure 18.4, with low target area throughput, homogeneous macro cell deployment achieves the lowest area power consumption because such low throughput target can be satisfied with large inter-cell distance. Thus the area power consumption is small. When the target throughput is increased, homogeneous macro cell deployment is densified and the area power consumption is increased. However, as we mentioned before, deployment of micro cells to help cell-edge users enlarges the inter-cell distance of macro cells, which reduces the area power consumption. The added power consumption offset of micro cells can be more than compensated for by the saved power from macro cells. In such a case, the heterogeneous deployment becomes more efficient.

An alternative to the aforementioned heterogeneous deployment is to deploy the small cells in the traffic hotspot where high traffic demand happens. In [18], small cells are dropped in the covered area of macro cells and the traffic in their adjacent area is assumed to be concentrated to the small cells to form traffic hotspots. Figure 18.5 shows the area power consumption for two homogeneous deployment scenarios (three and six-sector macro cells with 500 meters inter-cell distance) and four heterogeneous deployment scenarios (3, 10, 20, and 30 pico cells deployed per macro cell). Given a fixed inter-cell distance, the maximal traffic that can be served by homogeneous macro cell deployment is limited and thus for the case where the number of micro cells is zero, the curves are saturated. Actually, every curve including the heterogeneous ones will eventually be saturated. We can draw similar conclusions from Figure 18.5 that in low traffic the macro only deployment is more energy efficient. Heterogeneous deployment can only outperform homogeneous deployment when the traffic is not in low level. Furthermore,

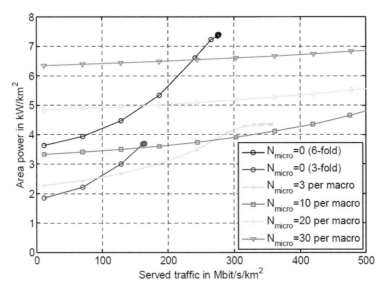

Figure 18.5 Area power consumption.

Figure 18.5 discloses an interesting observation that the optimal number of small cells increases with the traffic level.

18.3 HetNet based on cloud architecture

In the previous section, dynamics is assumed for macro cells by allowing the macro cells to adapt the transmission power to the traffic variations. Because the inter-cell distance cannot be easily adjusted once the infrastructure rollout is complete, this adaptation capability is crucial to improve the EE when the traffic is varying from time to time. As pointed out in [18], the sectors of macro cells and even the entire macro cell can be switched off and turned into sleep mode to save energy when traffic is at a low level. However, no dynamics is assumed for small cells in the previous results, which is the reason why homogeneous small cell deployment is only EE at very high traffic level.

In this section, we consider a cloud-based heterogeneous deployment. The cloud-based heterogeneous deployment originates from the idea of integrating cloud computing into the mobile environment. In this cloud-based architecture, the small cells and macro cells play different roles. The responsibility of small cells is mainly to boost the data traffic, but macro cells are more focused on control and management. A typical scenario is that multiple small cells are deployed in the same area of a macro cell. However, unlike the previous case, the operation of these small cells is dynamically controlled by the macro cell. They have the capability of turning into sleep mode and are only activated by the macro cell by demand, i.e., whenever traffic demand happens in their adjacent areas. The decision of which small cells are activated can be made following different rules and in this chapter a fuzzy logic-based approach is introduced. This approach aims to adapt the cell structure to serve users better while taking into account their conditions and demands. The advantages of implementing this concept are described in the section. The suggested algorithm is flexible yet capable of evaluating current cell status and providing decisions to enhance its performance level.

The concept of cloud heterogeneous deployment stems from two facts: with low traffic demand and sparse user distribution, deploying large cells is not only more EE but it also minimizes the handover operation; on the other hand, in a high-traffic scenario with dense user distribution, the availability of small cells is more efficient because they provide enhanced traffic-serving capability. Once a user establishes a connection with a macro cell, the macro cell will evaluate the user's location, traffic demand, etc., to decide whether or not to activate certain small cells to enhance the user experience. In another word, the small cells are only activated on a demand basis. For example, if a large number of users gather in a certain area of the macro cell, the small cells in that area are activated while other small cells are still in sleep mode to save energy.

In order to achieve higher EE, the cloud operation should be optimized, but this optimization should be subject to certain constraints. Moving users connected to small cells require handover operation when they move out from the coverage area of one small cell to another. It is also possible that the traffic request is rejected by the activated small cells because no resources are available, so that extra small cells need to be activated. Such handover probability and blocking probability are the constraints to be considered in the cloud-operation optimization. Here we propose a framework performing dynamic cloud operation for future cellular systems. This framework is based on the self-organization concept aiming at providing means to autonomously adapt the cell density in terms of activated small cells as well as the user association based on user locations and spatial traffic profile with a given performance objective. This framework adopts multiple optimization objectives and optimizes the system based on a fuzzy logic approach.

In the short-term approach each macro cell is responsible for making a decision on which cloud small cells are activated. Cloud small cells are responsible for serving the users in their coverage areas once they are activated. Figure 18.6 illustrates the main self-organization concept in the proposed cloud-coverage approach. In the observation and analysis stages, the system determines, via measurement, whether the current deployment is insufficient to provide the desired performance in terms of key performance indicators (KPIs). If not, a request to find an alternative deployment approach is automatically triggered and the optimization algorithm is executed. Three main KPIs are considered here: blocking probability indicating the possibility of a user request being rejected, handover probability indicating user mobility, and the area power consumption.

18.3.1 Key performance indicators

In order to evaluate the blocking probability, we use the multi-class MMPP/M/1/D-PS queue model (a single server processor sharing queue, with Markov-modulated Poisson arrival process, Markovian service time, and finite capacity), which has been widely used to study and dimension the telecommunication systems for more than a century [21, 22]. The assumptions made in this model are as follows.

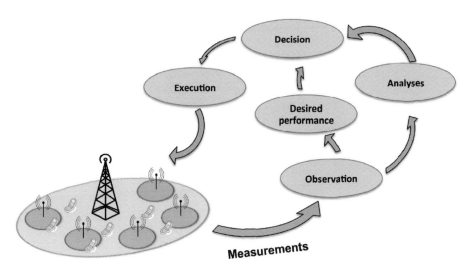

Figure 18.6 Self-organization process.

1. No more than U users and D data connections can be admitted in a cell at the same time. The session arrival rate is λ_u and with u users and the data connection arrival rate is $u\lambda_D$.
2. The user service time is exponentially distributed with mean value T, hence the user service rate is exponentially distributed with mean value $\mu_u = 1/T$.
3. In each data connection, the amount of information transferred is exponentially distributed with mean value G, hence the data connection service time is exponentially distributed with mean value $\mu_d = T_{out}/G$, where T_{out} is the user throughput.

  T_{out} depends on the channel condition as well as the allocated resources to a particular user. Here we use a simple approximation. Using the adaptive modulation and coding (AMC) scheme, we can provided an analytical formula for spectrum efficiency ξ_u in terms of bits/s/Hz for a target bit error rate (BER) in Rayleigh fading channel [23]

$$\xi_\mu = \log_2\left(1 + \frac{-1.5\gamma_\mu}{\ln(5\,\text{BER})}\right), \tag{18.1}$$

where γ_u is the received SINR of user u. From the achievable spectral efficiency we can calculate the user throughput as follows:

$$T_{out} = \frac{B^*\xi\mu}{U_i}, \tag{18.2}$$

where B denotes the system bandwidth and U_i is the total number of users associated with the ith cell. The blocking probability p_{blk} is given in [22].

The handover probability is calculated in [23], where, as the cloud small cells are activated dynamically, several handover types are considered.

1. Intra macro cell handover representing the scenario of a user handing over from one macro cell to a neighboring macro cell.
2. Intra macro and small cell handover represents a user handing over from a cloud small cell to a neighboring macro cell.

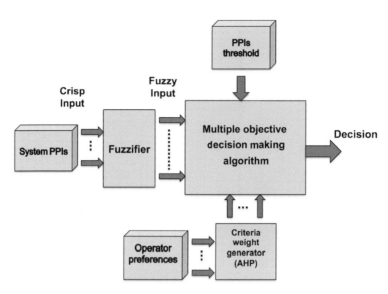

Figure 18.7 Weighted fuzzy logic.

3. Inter macro and small cell handover representing a user handing over from a macro cell to a cloud small cell within its coverage area.
4. Inter small cell and macro cell handover.
5. Inter small cell handover representing a user handing over from a cloud small cell to another cloud small cell.

The last and the most important KPI is the area power consumption, which can be easily obtained based on the power models established in [19, 20].

18.3.2 Optimization based on fuzzy logic

The weighted fuzzy logic approach is employed for multiple objective decision making (MODM), where all the KPIs are evaluated to generate a decision on how to activate the cloud small cells. There are a number of different MODM methods adopted throughout the literature and here we employ simple additive weighting (SAW): the weighted sum of all the attribute values determines a given network score level. Fuzzy logic has the ability to deal with imprecise data and evaluate multiple criteria simultaneously and provides a robust decision-making framework.

The block diagram of the weighted fuzzy logic system is presented in Figure 18.7, where the system KPIs are first normalized based on the desired performance objective. Then the fuzzifier converts the crisp KPIs to fuzzy values. The elements of a fuzzy set are given membership values ranging from zero to one depending on the defined membership function. In a fuzzy set an element can be a member to more than one class. The membership functions are set in a way that larger membership values are given for the outcome closer to the desired objectives. Assuming n KPIs $C_1, C_2, \ldots, C_n$, importance values are assigned for each KPI through a pairwise comparison. The weightings are

Table 18.1 Simulation assumptions.

Parameters	Values
Base station ICL	ICU = 600 m
Shadow fading	Log-normal. 6 dB standard deviation
Path loss	PL = 131.1 + 41.8 log(d) (dB); d = distance in km
Cell structure	Hexagonal grid of three-sector sites
Micro base station	$P_0 = 56.0$ W
Micro base station	$\Delta p = 2.6$
Micro base station	$P_{max} = 6.3\ 11$
Micro base station	$N_{TEX} = 3$
Channel bandwidths	10 MHz
Maximum users that can be admitted	$U = 20$
Maximum data connections per user	$D = 10$
User data connections arriving rate	$\lambda_d = 12$ connection/sec
User service time mean value	$T_h = 0.1017$ sec
Information transferred mean value	$R = 2$ Mbits
Maximum cell load	30 Mbits/km^2
Blocking probability weight χ	$w_{Pb,\chi} = 0.35$
Handover probability weight χ	$w_{Ph,\chi} = 0.43$
Power consumption weight χ	$w_{Pc,\chi} = 0.22$
Blocking probability weight β	$w_{Pb,\beta} = 0.2$
Handover probability weight β	$w_{Ph,\beta} = 0.6$
Power consumption weight β	$w_{Pc,\beta} = 0.2$
Blocking probability weight ς	$w_{Pb,\varsigma} = 0.6$
Handover probability weight ς	$w_{Ph,\varsigma} = 0.2$
Power consumption weight ς	$w_{Pc,\varsigma} = 0.2$
Measurement internal	T 15 sec
User terminal maximum speed	v_{max} 1.4 m/s
User terminal location	X random distribution in macro cell area

then applied to each KPI for decision making. If we assume N KPIs and M possible decisions to be made, then we have:

$$F_1 = w_{1,1} C_{1,1} + w_{1,2} C_{1,2} + \cdots + w_{1,N} C_{1,N},$$
$$F_2 = w_{2,1} C_{2,1} + w_{2,2} C_{2,2} + \cdots + w_{2,N} C_{2,N},$$
$$F_M = w_{M,1} C_{M,1} + w_{M,2} C_{M,2} + \cdots + w_{M,N} C_{M,N},$$

where w is the weight. Then fuzzy rules are applied to defuzzify outputs into crisp values. There are different methods to define the fuzzy rules including the maximum membership principle, centroid method, weighted average method, and mean maximum membership.

18.3.3 Simulation results and discussions

System-level simulations were conducted to evaluate the performance of the proposed approach with given assumptions in Table 18.1. Three sets of weights are considered. The blocking probability and handover probability are illustrated in Figures 18.8 and

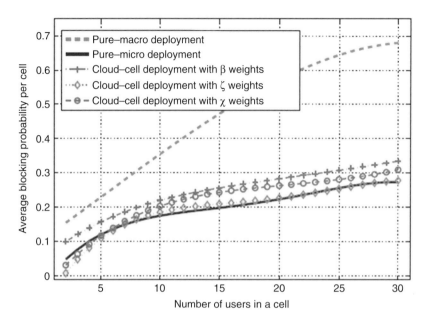

Figure 18.8 Blocking probability 0.5.

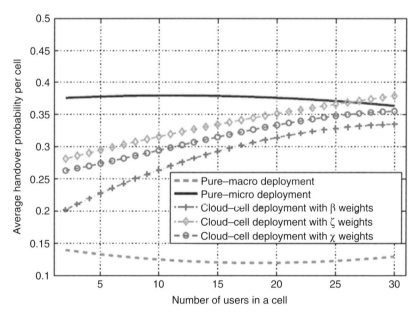

Figure 18.9 Handover probability.

18.9, respectively. We first compare the performance of homogeneous macro cell and micro cell deployment. The deployment of cloud small cells achieves better blocking probability at the expense of higher handover probability. However, it is still better than homogeneous small cell deployment in terms of handover while maintaining low blocking probability by intelligently allocating users to the most appropriate cloud small

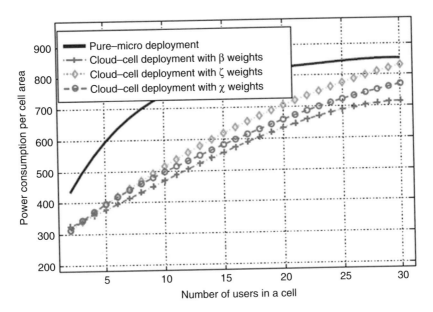

Figure 18.10 Area power consumption.

cells. The area power consumption is depicted in Figure 18.10. Pure macro deployment is not demonstrated because its area power consumption is obviously the lowest without added small cells power offset. However, it should be emphasized that this is at the expense of high blocking probability, i.e., the energy efficiency is obtained by sacrificing user experience. It is envisaged that in a high-traffic scenario with a large number of users, to achieve a given blocking probability, the inter-cell distance of macro cells should be shortened and the area power consumption of homogeneous macro deployment will eventually be higher than cloud deployment. The cloud small cell deployment outperforms homogeneous small cell deployment in terms of EE because the small cells are only activated when needed to save energy.

We can reach the conclusion that the cloud nodes concept provides a cost-effective solution to lower the network power consumption while at the same time keeping a satisfactory level of user experience. It has flexible capability compared to static approaches.

18.4 Multi-point coordination

In the previous sections, the user can only be connected to one BS, which can be either a large/macro cell or a small/micro cell. The overlaid network with dual connection is considered in this section. The network is of two layers: the coverage layer and the traffic-boosting layer. The coverage layer consists of macro cells maintaining coverage in a large area to ensure the UEs are always connected to the network. The traffic-boosting layer consists of small cells overlaid in the same coverage area and these small cells are deployed to boost the traffic. The small boosting cells (BCs) are connected

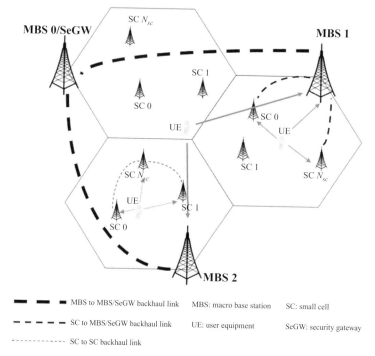

Figure 18.11 Uplink CoMP scenarios.

to the coverage cell (CC) via X2 interface (wired or wireless). Considering the highly densified deployment of BCs, it is more appropriate and cost efficient to use a wireless X2 interface, which is also referred to the wireless either fronthaul or backhaul in downlink and uplink CoMP, respectively [24]. We focus on uplink CoMP and refer this wireless X2 interface as backhaul links hereafter. Once the UEs are connected to boosting cells because of high traffic demand, the mobility management is of paramount importance because the BCs are normally very small so that frequency handover is expected even with very low moving velocity. CoMP is envisaged to be able to tackle this problem because the moving UE is connected to multiple adjacent BCs simultaneously and no handover request occurs as long as the UE is inside the CoMP area. There are three coordination strategies in this dual-layer structure as follows.

1. Single layer: one processing small BC + multiple coordinating small BCs, and the final processing happens in the processing BC.
2. Single layer: one processing CC + multiple coordinating CCs, and the final processing happens in the processing CC.
3. Dual layer: one processing CC + multiple coordinating small BCs, and the final processing happens in the processing CC.

The focus of our work is on the last strategy. As shown in Figure 18.11, UEs are simultaneously connected with N_{BC} coordinating BCs, which are then connected to the large CC via wireless backhaul links. The channel between the UE and the jth BC is denoted as $h_{UC,j}$ and the channel between the jth BC and CC is $h_{CC,j}$. These channels are assumed to be quasi-static, i.e., they keep constant within one frame and change from frame to frame. The following notations and definitions are employed in this work: capital letters e.g., **X**, stand for random variables and lower case letters e.g., x, represent realization of these variables.

Vectors are represented by bold letters, e.g., **X**. Equalities of vectors are element-wise, i.e., $\mathbf{X} > \mathbf{Y}$ means $X_i > Y_i \; \forall i$. A subscript, e.g., $\mathbf{X}_n$ represents the vector $(X_i, \ldots, X_n)$. Calligraphic letters denote a set, e.g., $\mathcal{T} \equiv \{1, \ldots, T\}$, whose cardinality is $|\mathcal{T}|$. A subset $\mathcal{S}$, which fulfils $\mathcal{S} \subseteq \mathcal{T}$, has a complementary set $\mathcal{S}^C$ where $\mathcal{S} \cup \mathcal{S}^C = \mathcal{T}$ and $\mathcal{S} \cap \mathcal{S}^C = \phi$. A subset with element t removed is denoted as $\mathcal{S} \setminus t$ and x_s refers to $\{x_1, \ldots, x_i, \ldots, x_{|\mathcal{S}|}\}$. Mutual information and entropy are denoted as $I(\cdot)$ and $H(\cdot)$, respectively. All the logarithms are in base 2. $\mathbf{E}\{\cdot\}$ denotes expectation.

18.4.1 System model

Assuming synchronized transmission across the entire network, the UE broadcasts a message w to the BCs and CC by sending a frame consisting of n symbols. At the jth BC and the CC, the received signals are

$$y_{BC,j}[t] = h_{UB,j} x_{UE}[t] + n_j[t],$$
$$y_{CC}[t] = h_{UC} x_{UE}[t] + n_C[t],$$

where n_j and n_c are the additive Gaussian noise at BC j and CC, respectively, following circularly symmetric complex Gaussian (CSCG) distribution with zero mean and variance σ_j and σ_{cc}. Here we assume that $\sigma_j = \sigma_{BC} = \sigma_{CC} = \sigma_C$.

Once the BCs receive the signal, they do not intend to decode the message because the decoding is likely to fail due to the low signal-to-noise ratio (SNR). Instead, the received continuous signals will be quantized and forwarded to the CC and coherently combined there with enhanced SNR. In this chapter, we consider a more efficient source coding scheme – a distributed source coding, where each BC does not only quantize the received signal but also compresses the quantization index and then forwards the compressed quantization index to the CC. By doing this, the number of information bits after compression is smaller than normal source coding schemes without compression, thus the transmission in the backhaul links can be more effective. The whole process of compression/decompression can be re-sorted to the Wyner–Ziv lossy distributed source coding [25, 26].

The wireless backhaul links operate at a different carrier frequency from the UE to BC links. In some of the future-generation cellular network proposals, they are designed to operate at a very high carrier frequency, e.g., 60 GHz, while the UEs operate at 2.5 GHz. Thus there is no interference between these backhaul links. Since the backhaul transmission is assumed to operate at a different frequency band from the UEs, the coordinating BCs can be regarded as full-duplex nodes because they can forward the compressed index of the ($n-1$)th received frame to the CC and at the same time receive the nth frame from the UEs as long as the duration of index transmission is shorter than the UE frame duration, which is normally true because of the relatively high capacity of backhaul links. Assuming the duration of frames from the UE and BCs is T and αT, respectively, the total duration of n frames is $nT + \alpha T$. If $n \gg 1$, it can be approximated as nT. When all the BCs forward the Wyner–Ziv bin index to the CC simultaneously, the received signals at the CC are assumed not to interfere with each other. This is because for 60 GHz wireless backhaul links, the wavelength is very small thus the transmission is highly directional. One independent directional receiving antenna and a following RF chain are assumed to be installed for each BC at the CC. The power leakage from one backhaul link to another is small enough to be negligible. The received signal from the jth boosting cell is

$$y_{CC,j}[t] = h_{BC,j} x_{BC,j}[t] + n_C[t], \qquad (18.3)$$

where n_C is the additive Gaussian noise at the CC following CSCG distribution with zero mean and variance σ_C. At the CC, the compression indices are decompressed to reconstruct the received signals at the BCs. Then the reconstructed signals are coherently combined to decode the messages from the UEs.

18.4.2 Achievable rate analysis

The achievable rate Φ for such a system can be derived according to [24, 29] and can be expressed as a function of $\sigma_{w\mathcal{L}}$. With Gaussian distributed input to the channel assumed, the achievable rate optimization problem can be formed as

$$\Phi(\sigma_{w\mathcal{L}}) = \log\left(1 + \frac{\gamma_0 P_U}{\sigma_C} + \sum_{i \in \mathcal{L}} \frac{\gamma_i P_U}{\sigma_C + \sigma_{wi}}\right). \qquad (18.4)$$

Here $\gamma_0 = |h_{UC}|^2$, $\gamma_i = |h_{UB,i}|^2$, and σ_{wi} is compression noise of the reconstructed observation at BC i. The achievable rate can be maximized by optimizing the compression noise σ_{wi} jointly subject to some constraints [24, 29]. Since the maximization objective function is not in standard concave form, we resort to its dual problem by forming its Lagrangian dual $L(\sigma_{w\mathcal{L}}, \lambda^c)$, where λ is the Lagrange multiplier. The dual function is then defined as

$$\phi(\lambda) = \max_{\sigma_{w\mathcal{L}}} L(\sigma_{w\mathcal{L}}, \lambda^c). \qquad (18.5)$$

The dual problem takes the following form: minimize

$$\phi(\lambda) \qquad (18.6)$$

subject to

$$\lambda^c \geq 0^c \qquad (18.7)$$

The dual objective function λ^c is a convex function regardless of the concavity of the primal function [27]. It is proved that the duality gap of the primal problem and the dual problem is zero in [24] and [29]. It means that the primal problem can be solved by searching for the solution of the dual problem. At first, we need to find the optimal

$\sigma_{w\mathcal{L}}$ to maximize the Lagrangian function. Due to its high complexity, it is difficult to achieve a closed-form solution. A successive optimization algorithm, i.e., non-linear Gauss–Seidel algorithm, is applied, where only one σ_{wi} is optimized at one time while other $\sigma_{w\mathcal{L}/i}$ are kept constant [28]. The algorithm optimizes σ_{w1} to $\sigma_{w\mathcal{L}}$ in one iteration and repeats the same procedure until $\sigma_{w\mathcal{L}}$ converges. The optimization algorithm for $\sigma_{w\mathcal{L}}$ is derived in [24] and [29]. In the nth iteration, the elements of $\sigma_{w\mathcal{L}}$ will be updated one after another until $\sigma_{w\mathcal{L}}$ eventually converges to the optimal value. Once the optimal $\sigma_{w\mathcal{L}}$ is obtained, the dual minimization problem can be solved by successively optimizing elements of λ^c. The searching domain of λ^c is proven to be $\lambda^c < 1^c$ in [24]. With this feasible range, the searching domain of λ^c is greatly reduced to $0^c \leq \lambda^c < 1^c$. Due to the convexity of f (A), a subgradient method can be used for solving the minimization problem [30]. The searching direction of $\lambda_{S_{ji}}$ is given as

$$g_{S_j^i}(\sigma_{w\mathcal{L}}) \bigg| = \frac{\partial \phi(\lambda)}{\partial \lambda S_{ji}} = C_{S_j^i} - f_{S_j^i}(\sigma_{w\mathcal{L}}) \qquad (18.8)$$

The searching criterion is: if $g_{S_j^i}(\sigma_{w\mathcal{L}}) \leq 0$, increase $\lambda_{S_j^i}$; otherwise decrease $\lambda_{S_j^i}$. The overall algorithm is given as

1. $\lambda_{min}^c = 0^c$ and $\lambda_{max}^c = 1^c$;
2. $\lambda^c = (\lambda_{min}^c + 1^c \lambda_{max}^c)/2$;
3. Solve the optimal $\sigma_{w\mathcal{L}}$;
4. Update λ_{min}^c or λ_{max}^c based on $g_{S_j^i}(\sigma_{w\mathcal{L}})$;
5. If A converges, stop the algorithm; otherwise, go back to step 2.

18.4.3 EE analysis

The total energy consumption of a CoMP system is composed of not only the transmission/reception power but also the circuitry energy consumption of all involved nodes. During the first frame, the UE broadcasts to the BCs and CC, the consumed energy can be expressed as

$$E^1 = T(P_U + P_{U,c} + P_{BC,rc} + P_{CC,sm}), \qquad (18.9)$$

where $P_{U,c}$ is the circuitry power consumption of the UE while transmitting, $P_{BC,rc}$ is the receiving circuitry power consumptions of the BCs, and $P_{CC,sm}$ is the sleep mode power consumption of CC, which is depending on the dynamics of the circuitry. From the second frame, the BCs transmit to the CC simultaneously and, at the same time, the UE transmits to the BCs at a different carrier frequency. The BCs operate in full-duplex mode. The consumed energy is

$$E^n = T(P_U + P_{U,c} + P_{bh} + P_{BC,tc} + P_{BC,rc} + P_{CC,c}), \qquad (18.10)$$

where P_{bh} is the transmission power of BCs, $P_{BC,tc}$ is the circuitry power consumption of a BC while transmitting, and $P_{BC,rc}$ is the receiving circuitry power consumption of the CC. Since the transmission and the reception of the BCs are simultaneous, two circuitry power consumptions $P_{BC,tc}$ and $P_{BC,rc}$ should be taken into consideration only once. Therefore we assume that $P_{BC,rc} = 0$. The backhaul link transmission power of BCs P_{BH} is assumed to be linearly related with link capacity in [31] as

$$P_{BH} = \frac{R_{bh}}{R_{mw-link}} P_{mw-link}, \qquad (18.11)$$

where $P_{mw-link} = 100$ Mbps and $P_{mw-link} = 50$ Watts.

In the last frame, the BCs transmit to the CC and the UE is silent. The consumed energy is

$$E^n = T(P_{U,sm} + P_{BC,tc} + P_{bh} + P_{CC,c}), \qquad (18.12)$$

where $P_{U,sm}$ is the sleep mode power consumption of the UE. As long as the number of frames N is large enough, the overall EE for N frames in terms of joules/bit is

$$E = \frac{P_U + P_{U,tc} + P_{BC,tc} + P_{bh} + P_{CC,c}}{BR}, \qquad (18.13)$$

where B is the system bandwidth.

Well-defined macro and micro BS power models based on measurement are presented in [19–20] and [32], where a linear approximation is justified:

$$P = N_{TRX}(P_0 + \Delta_p P_{out}), \qquad (18.14)$$

where N_{TRX} is the number of RF chains (for simplicity, N_{TRX} is assumed to be 1 in this chapter), P_0 is the linear model parameter representing consumed power when RF output power P_{out} is zero, Δ_p is the slope of the load-dependent power consumption. The circuitry power consumption of the BCs $P_{BC,tc}$ can be regarded as the difference between overall power consumption and the RF output power, i.e., $P_0 + \Delta_p P_{out} - P_{out}$. The receiving circuitry power consumption of the CC $P_{CC,c}$ is assumed to be equal to P_0. The power consumption of the UE is discussed in [33], where three states are identified: active, sleep, and idle. For simplicity, we assume that the transmission of the UE is continuous thus only the active state energy is considered. As indicated in [33], the power consumption of the UE is related to data transfer rate. For the considered uplink scenario, a similar linear power model can be formed.

18.4.4 Simulation results and discussions

In cellular networks, uplink CoMP is normally used to help the cell-edge users. In this dual-layer network architecture, we still focus on the users at the macro cell edge, where micro cells are deployed and the edge users' throughput is enhanced by connecting to multiple micro cells and a macro cell at the same time. Here we assume a cellular network as shown in Figure 18.12. The macro cell radius is denoted as r_c. The micro cells are deployed close to the cell edge. The UE to BC distance is assumed to r_{UBC} and the macro BS to micro cells distances are approximately assumed to be $\sqrt{3}r_c$. The UEs within the same cell are orthogonal in the frequency domain. We assume that the interfering UEs

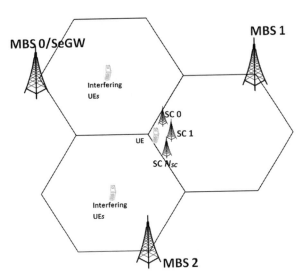

Figure 18.12 Uplink CoMP.

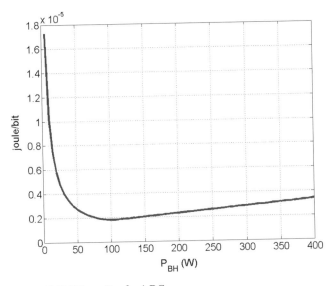

Figure 18.13 EE vs. P_{BH} for 1 BC.

in the other cells are located in the macro cell central and transmit with $P_U = 3$ Watts. The path loss is given as

$$P = N_{TRX}(P_0 + \Delta_p P_{out}), \quad (18.15)$$

where d (m) denotes distance. The thermal noise power density $N_0 = -171$ dBm/Hz and the bandwidth is 10 MHz, i.e., 50 RBs in total. Here we assume the UE is allocated with all 50 RBs. The circuitry power consumptions are assumed as $P_{BC,tc} = 23$ Watts, $P_{CC,c} = 204$ Watts, and $P_{U,tc} = 1.8$ Watts [19, 20, 33]. Figure 18.13 shows the EE with increased

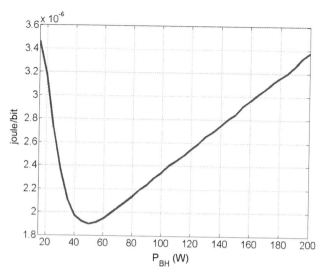

Figure 18.14 EE vs. P_{BH} for 2 BC.

backhaul links power P_{bh} for 1, 2, and 3 small BCs, respectively, when $r_c = 250$ m and $r_{UBC} = 10$ m. As seen from the figures, with increased P_{bh}, the backhaul links capacity is increased, resulting in smaller compression noise as well as better re-construction of the BCs' received signals at the CC, and the compression noise becomes smaller. The denominator of (18.13) increases according to (18.14) but the nominator increases as well and its pace is faster than the denominator, leading to decreasing of overall joules per bit. However, when the backhaul power increases further so that the compression noise approaches zero, the achievable rates saturate and the denominator keeps almost unchanged. It means that the backhaul power consumption is still increasing as well as the denominator of (18.13) but the nominator almost keeps unchanged. In such a case, the overall energy consumption in terms of joules per bit is increased.

Figures 18.14 and 18.15 show the EE with increased number of BCs. Introducing new BC increases the achievable rate but at the same time the energy expenditure is also increased. At first, the increased energy can be more than compensated by the improved capacity and the EE is improved until four BCs are deployed. After that, a new BC only brings in very small capacity improvement but the added energy consumption becomes significant and the EE performance degrades.

18.5 Conclusions

This chapter sheds light on a pathway in which future traffic demands can be satisfied by implementing HetNets. We first discuss the HetNet where small cells are deployed at the cell edge or traffic hotspots. The deployment of micro cells adds power consumption, which can be more than compensated for by the saved power from macro cells, and eventually the HetNet deployment becomes more energy efficient. It is also demonstrated

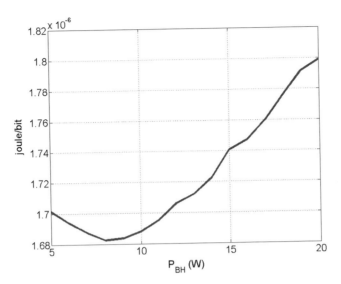

Figure 18.15 EE vs. P_{BH} for 3 BC.

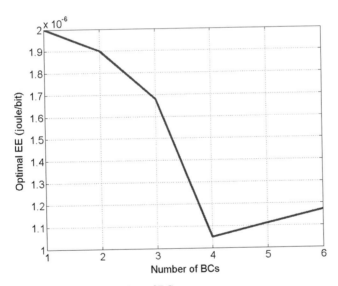

Figure 18.16 EE vs. number of BCs.

that small cells are able to provide very high capacity at hotspots at reasonable power and can serve the traffic demand at much lower joules per bit.

In addition, we consider a cloud-based architecture. The cloud-nodes concept is not only a cost and energy-effective solution but it also enhances network capability in terms of capacity and throughput. By using a simple weighting approach and a fuzzy logic, a group of system-performance metrics are employed as optimization objectives to flexibly promote the most important parameters in a balanced manner, thus avoiding oversacrificing any operational parameter in the expense of achieving a goal beyond the desired. Compared with conventional static approaches, which require human intervention, the dynamics of the SON network provides a flexible solution yet capable of providing a high level of performance with minimal human–network interactions and the impacts of such a self-organizing network on current networks are significant.

We also investigate CoMP from the EE perspective. It is found that in cellular networks, the backhaul link power consumption plays an important role in determining the overall EE of the entire network. For any number of BCs, there is always an optimal backhaul power to achieve the lowest joules per bit. The number of small BCs also has impact on the EE (Figure 18.16). The number of BCs should be optimized to achieve the best EE.

References

[1] CISCO, "Cisco Visual Networking Index: global mobile data traffic forecast update, 2012–2017," White Paper. pp. 1–34, Feb. 2013.

[2] H. Ishii, Y. Kishiyama, and H. Takahashi, "A novel architecture for LTE-B:C- plane/U-plane split and phantom cell concept," *Globecom Workshops, 2012*, pp. 624–630, 2012.

[3] T. Nakamura, S. Nagata, A. Benjebbour, *et al.* "Trends in small cell enhancements in LTE advanced," *IEEE Communications Magazine*, vol. 51, no. 2, pp. 98–105, 2013.

[4] M. Marsan, L. Chiaraviglio, D. Ciullo, and M. Meo, "Optimal energy savings in cellular access networks," *ICC Communications Workshops*, pp. 1–5, 2009.

[5] P. D. G. Koutitas, "A review of energy efficiency in telecommunication network," *Telfor Journal*, vol. 2, no. 1, 2010.

[6] H. Claussen, "Performance of macro- and co-channel femtocells in a hierarchical cell structure," *Proc. of PIMRC 2007*, pp. 1–5, 2007.

[7] F. Richter, A. J. Fehske, and G. P. Fettweis, "Energy efficiency aspects of base station deployment strategies for cellular networks," *Proc. of Vehicular Technology Conference Fall (VTC 2009-Fall)*, pp. 1–5, 2009.

[8] A. J. Fehske, F. Richter, and G. P. Fettweis, "Energy efficiency improvements through micro sites in cellular mobile radio networks," *Proc. of Globecom Workshops*, pp. 1–5, 2009.

[9] F. Richter, A. Fehske, P. Marsch, and G. Fettweis, "Traffic demand and energy efficiency in heterogeneous cellular mobile radio networks," *Proc. of Vehicular Technology Conference (VTC 2010-Spring)*, pp. 1–6, 2010.

[10] F. Richter and G. Fettweis, "Cellular mobile network densification utilizing micro base stations," *Proc. of ICC 2010*, pp. 1–6, 2010.

[11] E. Oh and B. Krishnamachari, "Energy savings through dynamic base station switching in cellular wireless access networks," *Proc. of Globecom 2010*, vol. 2010, pp. 1–5.

[12] M. A. Marsan, L. Chiaraviglio, D. Ciullo, and M. C. Meo, "Multiple daily base station switch-offs in cellular networks," *Proc. of Communications and Electronics (ICCE)*, 2012.

[13] G. Micallef, "Methods for reducing the energy consumption of mobile broadband networks," *Telektronikk*, Jan 2010.

[14] 3GPP, "3rd Generation Partnership Project; Technical Specification Group Radio Access Network; evolved universal terrestrial radio access (E-UTRA); base station (BS) radio transmission and reception, Release 9," 2011.

[15] M. Simsek and A. Czylwik, "Improved decentralized fuzzy Q-learning for interference reduction in heterogenous LTE networks," *Proc. of InOWo '12*, pp. 1–6, 2012.

[16] I. Bahceci, "Distributed CoMP set selection for heterogenous networks," *Proc. of SIU 2013*, pp. 1–4.
[17] D. Matsuo et al., "Shared remote radio head architecture to realize semi-dynamic clustering in CoMP cellular networks," *Proc. of Globecom Workshops*, pp. 1145–1149, 2012.
[18] I. Godorv (ed.), "D3.3: Final report on green network technologies," *Tech. Rep.*, Jun. 2012.
[19] M. A. Imran (ed.), "D2.3: Energy efficiency analysis of the reference systems, areas of improvements and target breakdown," *Tech. Rep.*, Nov. 2010.
[20] G. Auer et al, "How much energy is needed to run a wireless network?" *IEEE Wireless Communications*, October 2011.
[21] L. N. Singh and G. R. Dattatreya, "A novel approach to parameter estimation in Markov-modulated Poisson processes," *Proc. of IEEE Emerging Technologies Conference (ETC)*, Oct. 2004, Richardson, Texas.
[22] Y. Qi, M. Imran, and R. Tafazolli, "Energy-aware adaptive sectorisation in LTE systems," *Proc. of Personal Indoor and Mobile Radio Communications (PIMRC)*, 2011.
[23] T. Alsediary, Y. Qi, M. A. Imran, and B. Evans, "Energy-efficient dynamic deployment architecture for future cellular systems," *Proc. of PIMRC 2013*, 2013.
[24] Y. Qi, M. A. Imran, A. Quddus, and R. Tafazolli, "Achievable rate optimization for coordinated multi-point transmission (CoMP) in cloud-Based RAN architecture," *ICC* 2014.
[25] A. D. Wyner and J. Ziv, "The rate-distortion function for source coding with side information at the decoder," *IEEE Trans. Inform. Theory*, vol. 22, no. 1, Jan. 1976.
[26] A. D. Wyner, "The rate-distortion function for source coding with side information at the decoder-II: general sources," *Inform. and Control*, pp. 60–80, 1978.
[27] C. A. Floudas, *Non-linear and Mixed-integer Optimization*, Oxford University Press, 1995.
[28] T. A. Porsching, "Jacobi and Gauss–Seidel methods for nonlinear network problems," *SIAM Journal*, vol. 6, no. 3, 1969.
[29] A. del Coso, and S. Simoens, "Distributed compression for MIMO coordinated networks with a backhaul constraint," *IEEE Trans. Wireless Commun.*, vol. 8, no. 9, pp. 4698–4709, Sep. 2009.
[30] D. P. Bertsekas, *Nonlinear Programming*, 2nd edition. Athena Scientific, 1999.
[31] J. Wu, S. Rangan, and H. Zhang, *Green Communications: Theoretical Fundamentals, Algorithms and Applications*, 1st edition. CRC Press, 2012.
[32] B. Debaillie et al., "Flexible power modeling of LTE base stations," *Proc. of IEEE WCNC 2012*, Paris, Apr. 2011.
[33] J. Huang et al., "A close examination of performance and power characteristics of 4G LTE networks," *Proc. of MObiSys'12*, UK, Jun. 2012.
[34] 3GPP, "3rd Generation Partnership Project;Technical Specification Group Radio Access Network; Scenarios and requirements for small cell enhancements for E-UTRA and E-UTRAN, Release 12," 2013.

19 Time- and frequency-domain e-ICIC with single- and multi-flow carrier aggregation in HetNets

Meryem Simsek, Mehdi Bennis, and Ismail Guvenc

Multi-layer heterogeneous network (HetNet) deployments including small cell base stations (BSs) are considered to be the key to further enhancements of the spectral efficiency achieved in mobile communication networks [1]. Besides the capacity enhancement due to frequency reuse, a limiting factor in HetNets has been identified as inter-cell interference. The 3rd Generation Partnership Project (3GPP) discussed inter-cell interference coordination (ICIC) mechanisms in long term evolution (LTE) Release 8/9 [2]. LTE Release 8/9 ICIC techniques were introduced to primarily save cell-edge user equipments (UEs). They are based on limited frequency domain interference information exchange via the X2 interface, whereby ICIC related X2 messages are defined in the 3GPP standard [3]. In LTE Release 8/9 ICIC, a BS provides information about set of frequency resources in which it is likely to schedule DL transmissions to cell-edge UEs, for the benefit of a neighboring BS. The neighboring BS in turn avoids scheduling its UEs on these frequency resources.

With the growing demand for data services and the introduction of HetNets it has become increasingly difficult to meet a UE's quality of service (QoS) requirements with these mechanisms. To cope with the QoS requirements and growing demand for data services, enhanced ICIC (e-ICIC) solutions have been proposed in LTE Release 10 and further e-ICIC (Fe-ICIC) solutions to reduce cell reference signal (CRS) interference in e-ICIC techniques are discussed in LTE Release 11 [4].

In LTE Release 10 e-ICIC techniques, the focus is on time- and frequency-domain techniques and power-control techniques. While in time-domain techniques, the transmissions of the victim UEs are coordinated in time-domain resources, in frequency-domain techniques, e-ICIC is mainly achieved by frequency-domain orthogonalization. The power-control techniques have been intensively discussed in 3GPP. Hereby, power control is performed by the aggressor cell to reduce inter-cell interference to victim UEs. In 3GPP studies, e-ICIC mechanisms with adaptive resource partitioning, cell range expansion (CRE), and interference coordination/cancellation take a central stage [5].

In the following, the inter-cell interference problem in HetNets is introduced and time- and frequency-domain e-ICIC techniques are discussed based on 3GPP specifications. In addition, single- and multi-flow transmission techniques for e-ICIC and system capacity improvement are described.

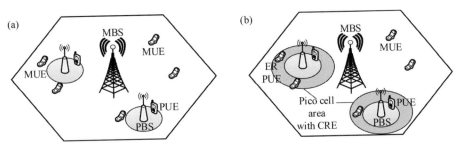

Figure 19.1 Macro- and pico cell based HetNet: (a) without CRE and (b) with CRE.

19.1 Inter-cell interference in HetNets

One of the major features extensively studied for LTE Release 10, also known as LTE-Advanced, is the HetNet coverage and capacity optimization, e.g., through the use of cell-range expansion (CRE) techniques. CRE has been considered jointly with advanced interference coordination and cancellation methods to mitigate resulting interference problems.

As illustrated in Figure 19.1a, a pico cell's coverage area is quite small due to its low transmission power and the strong interference from the macro cell. Hence, many UEs connect to the macro cell due to their large transmit power, rather than to pico cells at shorter distance with lesser number of UEs, so that the traffic load is unevenly distributed in the network. As a result, the macro cell will be overloaded whereas pico cells will be underutilized [6–8]. To increase the efficiency of a HetNet containing macro and pico cells, the CRE technique has been introduced, which enables pico cells to *virtually* increase their coverage area in order to offload more UEs from the macro cellular network, see Figure 19.1b. Hereby, a UE adds a positive bias value to its reference signal received power (RSRP) from the pico cell. A UE gets access from the pico cell if this RSRP plus bias value is larger than its RSRP from the macro cell. These UEs are called expanded-range UEs (ERUEs).

Cell-range expansion allows offloading more UEs from the macro cell to the pico cell and thus achieving performance improvement, but causes extra downlink interference, which becomes severe for higher bias values and may degrade the throughput of ERUEs. Extra downlink interference is caused, because ERUEs get access from cells with lower RSRP, so that the larger RSRP from the macro cell becomes the interfering link. As a result, ERUEs suffer from low downlink signal-to-interference-plus-noise ratios (SINRs) [9]. Therefore CRE should be jointly designed with e-ICIC schemes to boost network capacity by smartly offloading traffic to open access pico cells, thereby achieving cell splitting gains.

The major challenge in HetNets is to decide *when* to handover *which* UE to which BS and *how* the remaining BSs in the network should behave in this case. For this reason, pico cells need to self-organize in a way that maximizes the throughput of their UEs, as well as alleviating the macro cell's traffic. With this in mind, *intelligent* and flexible CRE techniques across time and frequency must be devised for macro and pico cells, to mitigate excessive DL inter-cell interference suffered by ERUEs, while at the same time not jeopardizing pico UE (PUE) QoS requirements.

19.2 Time domain e-ICIC in HetNets

Dense HetNet deployment in LTE is a big challenge especially for co-channel use cases. With the deployment of small cells, more cells with intra-frequency interference are introduced into the macro cellular network, in which coverage areas overlap between small cells and macro cells. Additionally, the concept of CRE allows UEs to access pico cells with weak RSRPs, so that low received power and enhanced downlink interference yield lower SINR. As a result, both service channels and control channels in LTE are greatly affected by inter-cell interference.

To suppress the inter-cell interference in a HetNet and improve the overall system capacity and UE experience, e-ICIC is introduced in LTE-Advanced Release 10. The 3GPP and redefines the ICIC concepts of Release 8/9 and introduces an additional time dimension for e-ICIC techniques, which allows signals in different cells to be orthogonal in the time domain [10, 11]. The basic idea of time-domain ICIC is that an aggressor node creates protected subframes for a victim node by reducing its transmission power in certain subframes. These subframes are called almost blank subframes (ABS). Notably, in co-channel deployments, ABSs are used to reduce interference created by transmitting nodes while providing full legacy support. During ABS, the aggressor cell transmits only the following control signals: primary/secondary synchronization signal (PSS/SSS), physical broadcast channel (PBCH) CRS, paging, and system information block 1 (SIB1). This signaling is required to ensure backward compatibility to UEs in LTE Release 8/9.

19.2.1 Almost blank subframes

The concept of ABS is introduced in time-domain e-ICIC to coordinate inter-cell interference in the time domain. Time-domain e-ICIC configures ABSs in the interfering cell. These ABSs are used by the interfered cell to provide service for its (especially ER) UEs, which previously experienced strong interference. In this way, inter-cell interference is coordinated in the time domain.

Figure 19.2 depicts an ABS example with a duty cycle of 50%. During ABS, the aggressor cell/macro cell does not transmit data but may transmit reference signals, critical control channels, and broadcast information. If the pico BS schedules its (especially ER) UEs in subframes 1, 3, 5, 7, 9, it protects its UEs from strong inter-cell interference. In other words, there is no interference in these subframes from the macro cell, and therefore, data transmission can be much faster. When several pico cells are used in the coverage area of a single macro cell, overall system capacity is increased as each pico cell uses the ABSs for data transmission of ERUEs and pico cell UEs. The drawback is of course that the macro cell capacity is diminished due to scarcer resources to support its

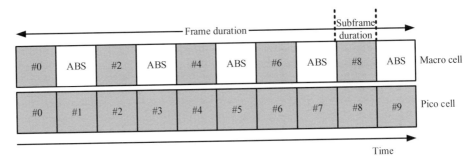

Figure 19.2 Time-domain e-ICIC in LTE-A with ABS structure.

own users. Therefore methods have to be put in place to find out the optimum ABS pattern in a network to reduce inter-cell interference, while at the same time not decreasing the aggressor cell's performance.

The configuration of ABSs is transferred via the X2 interface or operation and maintenance (OAM). In general, a pico cell transfers the configuration via the X2 interface, and a femto cell transfers the configuration manually via OAM.

19.2.2 Almost blank subframes with MBSFN configuration

The ABS patterns can be constructed by configuring so-called multi-cast/broadcast over single-frequency network (MBSFN) subframes or/and by not scheduling unicast traffic (or by reducing transmit powers) in certain subframes. Unlike the pure ABS approach as described in Subsection 19.2.1, the ABS with multi-cast/broadcast over single-frequency network (MBSFN) method does not require backhaul signaling for subframe partitioning. Instead, it is only necessary to convey MBSFN configurations between coordinating BSs. Remaining aspects are the same as for pure ABSs, whereby ABSs with MBSFN configuration are fully backward compatible with LTE Release 8/9.

19.2.3 Almost blank subframes with OFDM symbol shift

In this approach, in addition to pure ABSs, orthogonal frequency division multiplexing (OFDM) symbol shift is configured to avoid CRS interference to the physical downlink control channel (PDCCH) region. Data resource element muting is utilized to prevent radio link failure (RLF) of LTE Release 8/9 UEs. The remaining aspects are the same as for pure ABSs, whereby ABSs with OFDM symbol shift are fully backward compatible with LTE Release 8/9.

19.2.4 Blank subframes for TDD only

Within ABSs, the blank subframes for time-division multiplexing (TDD) approach is the last approach discussed in 3GPP [12]. This approach utilizes different downlink and uplink partitioning at different BSs to create blank downlink subframes. In these subframes no data transmission is performed by the aggressor cell. This approach requires backhaul signaling for downlink and uplink partitioning coordination and is fully backward compatible.

19.3 Frequency-domain e-ICIC in HetNets

In 3GPP LTE Release 10 and beyond, frequency-domain e-ICIC is performed through the concept of carrier aggregation (CA) [13, 14]. CA is one of the most important features of LTE Release 10. It bonds multiple carriers together and enables a UE to connect to several carriers simultaneously.

In frequency-domain e-ICIC solutions, control channels (e.g., PDCCH, PHICH, PCFICH) and physical signals (i.e., synchronization signals and CRSs) of different cells are scheduled in reduced bandwidths in order to have totally orthogonal transmission of these signals at different cells. Frequency-domain orthogonalization may be achieved in a static or dynamic manner through victim UE detection. According to [15], an LTE Release 8/9 UE accesses and operates on the reduced bandwidth while an LTE Release 10 UE accesses the reduced bandwidth, but may be scheduled over the entire bandwidth. In other words, in LTE Release 10 the orthogonality in frequency domain is for control channels only while the data channels may be scheduled over the entire bandwidth.

To ensure considerably high data rates, the entire bandwidth is increased in 3GPP LTE Release 10 and beyond. The method proposed by 3GPP is called carrier aggregation and is described in Subsection 19.3.1.

19.3.1 Carrier aggregation

Carrier aggregation is used in LTE Release 10 in order to increase the bandwidth, and thereby enhance the overall network performance. Allowing concurrent utilization of different frequency carriers, the bandwidth that can be allocated to a UE is efficiently increased through CA. Hereby, each aggregated carrier is referred to as a component carrier (CC). The CC can have a bandwidth of 1.4, 3, 5, 10, 15, or 20 MHz, which are the bandwidths defined in LTE Release 8. This is essential to keep backward compatibility with LTE Release 8/9 UEs. These UEs are able to connect only to one CC while LTE Release 10 UEs can connect to several CCs simultaneously, whereby the maximum number of aggregated CCs is five, so that the maximum aggregated bandwidth is 100 MHz.

Carrier aggregation can be applied to frequency-division duplex (FDD) and TDD. The major difference is that in FDD mode the number of aggregated carriers and their bandwidths can be different in the downlink and uplink while they are the same in TDD mode.

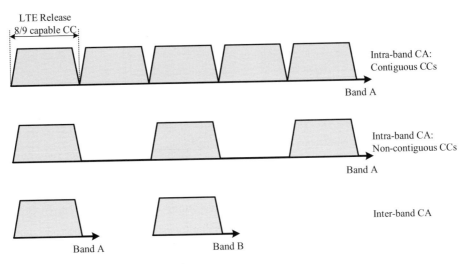

Figure 19.3 Types of carrier aggregation.

19.3.2 Types and deployment strategies of carrier aggregation

According to 3GPP LTE-Advanced, the following CA scenarios shall be considered, as illustrated in Figure 19.3:

- **Intra-band contiguous CA**. This type of CA uses a single band A, in which the carriers are adjacent to each other. It is the easiest LTE-Advanced CA to implement, because in intra-band contiguous CA, the aggregated CCs can be considered as a single enlarged channel by the BS. In this case only one transceiver is required within the BS or UE.
- **Intra-band non-contiguous CA**. This type of CA is more complex than the intra-band contiguous CA. The multi-carrier signal cannot be treated as one single signal, so that more transceivers are required. This adds significantly more complexity in terms of space, power, and cost.
- **Inter-band contiguous CA**. In this type of CA different bands A and B are used as depicted in Figure 19.3, which requires two transceivers. The inter-band contiguous CA has been introduced because of the fragmentation of frequency bands.

19.3.3 Carrier aggregation based e-ICIC

In CA based e-ICIC, an aggressor cell (macro cell) creates protected CCs to enable reception of downlink physical signals on PDCCH, system information, and control channels at victim cells (pico cells). At the same time, data can be received on any CC at physical downlink shared channels (PDSCH) by using the concept of cross-carrier scheduling. In macro and pico cell based HetNets, the macro and pico cells operate downlink signaling on different CCs while data transmission may be operated on same CCs when using cross-carrier scheduling. Hence, using cross-carrier scheduling control

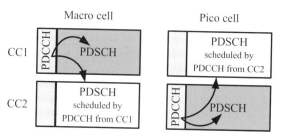

Figure 19.4 Illustration of cross-carrier scheduling in HetNets.

channel interference is avoided as illustrated in Figure 19.4. In this illustration, CC1 of the macro cell would cause high interference to CC1 of the pico cell. Therefore the pico cell uses CC2 for PDCCH transmission to schedule PDSCH transmission on CC1 and vice versa for the macro cell.

19.4 Single-flow and multi-flow transmission

The 3GPP standardization body has introduced notable changes in LTE resulting in a number of important enhancements for better flexibility and improved performance during the past years.

One of the main improvements introduced in the LTE technology is the possibility of CA as described in Subsection 19.3.1. The CA enables flexible spectrum adaptation, while at the same time taking full benefit of multiple carriers that may reside in one or several bands. 3GPP introduced HetNets along with the need to cope with challenges caused by high traffic demand especially in so-called hotspot and indoor environments. Operators are seeking an opportunity to combine CA solutions with the HetNet environment to take full advantage of the resulting architecture defined by 3GPP. This section describes such a solution, which is known as multi-flow CA or dual connectivity [16–18].

In [19], CA is studied as a function of bias values and frequency band deployment, in which CA enables UEs to connect to several carriers simultaneously. Two different methods are considered, namely single- and multi-flow CA. While in single-flow CA, UEs associate with only one of the available tiers at one time, in multi-flow CA based HetNets, UEs can be served by both macro and pico cells at the same time. This mandates a smart mechanism in which the different tiers coordinate their transmission through adaptive CRE across different CCs.

In Subsections 19.4.1 and 19.4.2 the concepts of single- and multi-flow transmissions are presented, respectively. Considering these two concepts, Subsection 19.4.3 presents learning based frequency-domain e-ICIC algorithms. Q-learning based CA techniques are presented that consider a heterogeneity, in which BSs individually select CRE bias values across different CCs in a self-organizing manner. The dynamic learning based CA approach is applied to both single- and multi-flow CA.

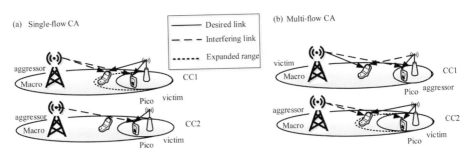

Figure 19.5 (a) Single-flow CA technique; (b) multi-flow CA technique.

19.4.1 Single-flow transmission

The single-flow CA technique is a basic CA approach as introduced by 3GPP LTE Release 10. In single-flow CA, the macro cell is the aggressor cell and the pico cell is the victim cell as depicted in Figure 19.5a. While a pico cell performs CRE to offload the macro cellular network, and to transmit UE data and UE control signaling on CC1 in Figure 19.5a, the macro cell causes inter-cell interference if no e-ICIC techniques are applied. On CC2, the ERPUE does not receive any data while regular PUEs are served on CC2.

19.4.2 Multi-flow transmission

A recent feature in 3GPP LTE Release 12, referred to as multi-flow CA [20], enables a better use of resources and improves system capacity, where multiple BSs (from different tiers) simultaneously transmit data to a UE on different CCs as depicted in Figure 19.5b. Dual connectivity or multi-flow CA, in which users are served by different layers on different CCs, has recently emerged as a key research challenge in 3GPP Release-12 [21]. Besides the multi-flow CA technique, another related approach to provide an efficient and flexible network performance improvement is to split the control and user plane (C- and U-plane). This concept was introduced and discussed in [22, 23] whereby, the C-plane is always provided at a low-frequency band to maintain good connectivity and mobility. On the other hand, the U-plane is provided by both the macro cells and the small cells (deployed at higher frequency bands) for data transfer. Since small cells are not configured with cell-specific signals and channels, they are named *phantom cells* [23]. The *phantom cell* concept is introduced because of the lack of capacity solutions for high-traffic outdoor environments that can also support feasible mobility and connectivity. Thus the concept of macro cell assisted pico cells has been introduced as a capacity solution that offers mobility support while capitalizing on the existing macro cellular network. Hereby, the C-plane of a pico cell UE is provided by a macro cell in a lower frequency band and the U-plane is provided by the pico cell at higher frequency bands. For a macro cell UE, both the C-plane and U-plane are provided by the serving macro cell in the same way as in a conventional LTE system. However, a network architecture that supports the C/U-plane split, and interworking between the macro cell and *phantom cell* is still required.

19.4.3 Q-learning based e-ICIC for single-flow transmission

In contrast to existing frequency-domain ICIC solutions, such as single-flow CA where pico BSs select one CC and apply a fixed CRE bias, a heterogeneous case where different CRE bias values are used across different CCs in a self-organizing manner is considered in this subsection.

Aiming at optimality, a Q-learning formulation is proposed, which consists of a set $\mathcal{P}$ of pico BSs and a set $\mathcal{M}$ of macro BSs, denoted as the players/agents. We define a set of states $\mathcal{S}$ and actions $\mathcal{A}$ aiming at finding a policy that minimizes the observed costs over the interaction time of the players. In detail, every player explores its environment, observes its current state s and takes a subsequent action a according to its decision policy $\pi : s \to a$. For all players, individual Q-tables maintain their knowledge of the environment with which autonomous decisions are made based on local and limited information. The Q-learning has been shown to converge to optimal values in Markov decision process (MDP) environments [24], where the goal of a player is to find an *optimal* policy $\pi^*(s)$ for each state s, that minimizes the cumulative costs over time. This Q-learning procedure is applied to HetNets to enable e-ICIC at the BSs for single-flow transmission.

Hereby the single-flow CA problem is divided into primary CC selection, bias value selection, and power allocation sub-problems. These three sub-problems are inter-related in which the pico BS and macro BS (as players) learn their optimal e-ICIC strategy. The pico BS first selects its optimal CC to perform CRE, then the bias value for CRE in the selected CC, after which the transmit power is allocated accordingly. Hence, a three-stage decision-making process is considered, in which the macro BS is informed about the pico BS's primary CC via the X2 interface. The macro BS selects the pico BS's secondary CC as its primary CC and learns its optimal power allocation strategy. While the macro BS selects low power levels on its secondary (pico BS's primary) CC, it selects higher power levels on its primary CC. The rationale behind considering two different power levels for the macro BS's primary and secondary CC is to reduce interference on ER PUEs, which are served on pico BS's primary CC.

The player, state, action, and perceived cost associated to the Q-learning procedure are defined as:

- **Player**. Pico BS p, $\forall 1 \leq p \leq P$ and macro BS m, $\forall 1 \leq m \leq M$.
- **State**. The state representation of player n at time t in RB r is given by the vector state $\vec{s}_r^n = \{I_r^{u(p)}, I_r^{u(m)}\}$.

$$I_r^{u(n)} = \begin{cases} 0, & \text{if } \Gamma_r^{u(n)} < \Gamma_{\text{target}} - 2\,\text{dB} \\ 1, & \text{if } \Gamma_{\text{target}} - 2\,\text{dB} \leq \Gamma_r^{u(n)} \leq \Gamma_{\text{target}} + 2\,\text{dB} \\ 2, & \text{otherwise} \end{cases} \quad (19.1)$$

with $n = \{p, m\}$, $\Gamma_r^{u(n)} = 10\log(\gamma_r^{u(n)})$ is the instantaneous SINR of UE u in dB in RB r and $\Gamma_{\text{target}} = 20$ dB is the selected target value.

- **Action.** For player pico BS p the action set is defined as $A^p = \{C_i^p, \beta^p, a_r^p\}_{r \in \{1,\dots,R\}}$, where $C_{i \in \{1,2\}}^p$ is the selected component carrier to perform CRE on the selected CC, $\beta^p \in \{0, 6, 12\}$ dB is the bias value for CRE on selected C_i^p of pico BS p, and a_r^p is the transmit power level of pico BS p over a set of RBs$\{1, \dots, R\}$. Hence, the pico BSs will independently learn on which CC it performs range expansion with which bias value and how to optimally perform power allocation.

  For player macro BS m the action set is defined as, $A^m = \{a_{r,C_i}^m\}_{r \in \{1,\dots,R\}}$, where a_{r,C_i}^m is the transmit power level of macro BS m over a set of RBs$\{1, \dots, R\}$ on CC C_i. Different power levels are defined for macro BS's primary and secondary CCs.

- **Cost.** The considered cost in RB r of player n is

$$c_r^n = \begin{cases} 500, & \text{if } P_{\text{tot}}^n > P_{\text{max}} \\ \left(\Gamma_r^{u(n)} - \Gamma_{\text{target}}\right)^2, & \text{if otherwise} \end{cases} \quad (19.2)$$

The rationale behind this cost function is that the Q-learning aims to minimize its cost, so that the SINR at UE u is close to a selected target value Γ_{target}. Considering a cost function with a minimum as target SINR as in (19.2) will enable the player to develop a strategy that leads to SINR values close to the target SINR. The target SINR is set to be 20 dB, and this corresponds to a maximum CQI of 15 in our look-up tables. Since the Q-learning approach aims at that maximization, we select this target SINR. Furthermore, the Q-learning equation is updated as follows:

$$Q^n(s, a) \leftarrow (1 - \alpha)Q^n(s, a) + \alpha \left[c + \lambda \min_a Q^n(s', a)\right], \quad (19.3)$$

where $\alpha = 0.5$ is the player's willingness to learn from its environment, $\lambda = 0.9$ is the discount factor and s' is the next state. This is an iterative procedure in which the previous knowledge ($Q^n(s, a)$) is updated by considering the newly obtained cost value (c) and estimation of future costs ($\min_a Q^n(s', a)$).

Algorithm 19.1 Dynamic Q-learning based e-ICIC algorithm for single- and multi-flow CA

```
1: loop
2:     for player p do
3:         Select primary CC Cᵖ ∈ {1, 2}
4:         Select bias value bᵖ for primary CC Cᵖ
5:         Select power level aᵣᵖ according to argmin_{a∈Aᵖ} Qᵖ(s, a) on both CCs
6:     end for
7:     Inform player m about primary CC Cᵖ
8:     for player m do
9:         Select player p's secondary CC as primary CC Cᵐ
10:        if multiflow CA then
```

```
11:        Select bias value b^m for primary CC C^p
12:     end if
13:     Select power level a_r^m ∈ A^m according to arg min_{a∈A^m} Q^m(s, a)
14:   end for
15:   Receive an immediate cost c
16:   Observe the next state s'
17:   Update the table entry according to equation (1.3)
18:   s = s'
19: end loop
```

A pseudo code of the dynamic Q-learning based e-ICIC algorithm for single-flow CA is presented in Algorithm 19.1.

19.4.4 Q-learning based e-ICIC for multi-flow transmission

In contrast to the single-flow CA in which the macro BS is always the aggressor cell, in multi-flow CA either the macro BS or pico BS is the aggressor cell. This is because both macro BS and pico BS perform CRE on their primary CCs, so that a UE can be served on different CCs by different BSs based on its biased received power. Similar to the single-flow CA learning algorithm in Subsection 19.4.3, the multi-flow CA based e-ICIC learning algorithm considers pico BS and macro BS as players. Both pico BS and macro BS learn their optimal primary CC, CRE bias values, and power levels. Algorithm 19.1 summarized the steps of the Q-learning based e-ICIC for multi-flow transmission.

The main difference of the multi-flow CA and the single-flow CA based e-ICIC learning algorithm is the action definition. It is redefined for multi-flow CA based e-ICIC as follows.

- **Action**. For player pico BS p the action set is defined as $A^p = \{C_i^p, \beta^p, a_r^p\}_{r \in \{1,...,R\}}$, and for player macro BS m the action set is defined as $A^m = \{C_i^m, \beta^m, a_r^m\}_{r \in \{1,...,R\}}$, where $C_{i, i \in \{1,2\}}$ is the component carrier index that can be selected in order to perform CRE on the selected CC, $\beta \in \{0, 6, 12\}$ dB is the bias value for CRE on selected CC C_i, and a_r is the transmit power level over a set of RBs $\{1, \ldots, R\}$. Hence the pico BSs and macro BS will independently learn the CCs on which they perform range expansion, with which bias value, and how to optimally perform the power allocation. Since both pico BS and macro BS can be aggressor cells, different power levels are considered for CCs on which the BSs perform CRE, and the regular CCs, which do not have CRE.

In addition, the one player formulation case is considered, in which the pico BS is the player. In this case, the pico BS carries out the multi-flow CA based Q-learning procedure and informs the macro BS about its primary CC and the macro BS uses reduced power levels on this CC and higher power levels on the pico BS's secondary CC. A uniform power allocation is considered in this case. However, even if no CRE is performed by the macro BS, a UE can be served by both pico BS and macro BS

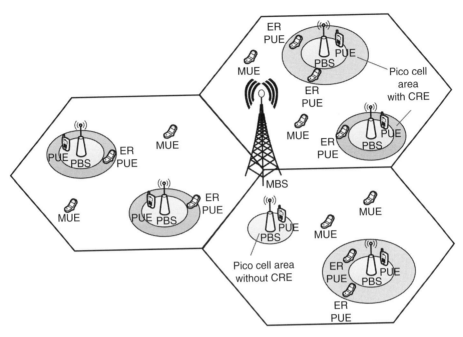

Figure 19.6 A HetNet scenario with cell-range expansion (CRE).

on different CCs at the same time based on its biased received power. In Subsection 19.4.5, this learning algorithm will be coined as *MF static QL*, while the two player algorithm is named *MF dynamic QL*.

19.4.5 Simulation results

In this subsection the system model and simulation results for the presented Q-learning based e-ICIC algorithms in Subsections 19.4.3 and 19.4.4 are described.

19.4.5.1 System model

In this chapter the focus is on a network deployment with multiple pico cells overlaying a macro cellular network consisting of three sectors per macro cell. A network consisting of a set of $\mathcal{M} = \{1, \ldots, M\}$ macro cells and a set of $\mathcal{P} = \{1, \ldots, P\}$ uniformly randomly distributed co-channel pico cells per macro sector, as depicted in Figure 19.6, is considered. We consider that the total bandwidth (BW) is divided into subchannels with bandwidth $\Delta f = 180$ kHz. OFDM symbols are grouped into resource blocks (RBs). Both macro and pico cells operate in the same frequency band and have the same number of available RBs, denoted by R. We consider that all transmitters and receivers have single antennas. A set of UEs $\mathcal{U} = \{1, \ldots, U\}$ is defined, whereby the UEs are dropped according to scenario #4b in [25]. We denote by $u(m)$ a macro cell UE, while $u(p)$ refers to a pico cell UE. We denote by p_r^m and p_r^p the downlink transmit power of macro BS m

and pico BS p in RB r, respectively. The SINR at macro UE (MUE) u allocated in RB r of macro cell m is:

$$\gamma_r^{u(m)} = \frac{p_r^{m(u),\mathrm{M}} g_{m,u,r}^{\mathrm{MM}}}{\underbrace{\sum_{j=1,j\neq m}^{M} p_r^{j(u),\mathrm{M}} g_{j,u,r}^{\mathrm{MM}}}_{I^{\mathrm{M}}} + \underbrace{\sum_{p=1}^{P} p_r^{m(u),\mathrm{P}} g_{p,u,r}^{\mathrm{PM}}}_{I^{\mathrm{P}}} + \sigma^2}. \tag{19.4}$$

In (19.4), $g_{m,u,r}^{\mathrm{MM}}$ indicates the channel gain between the transmitting macro BS m and its MUE u; $g_{j,u,r}^{\mathrm{MM}}$ indicates the link gain between the transmitting macro BS j and MUE u in the macro cell at BS m; $g_{l,u,r}^{\mathrm{PM}}$ indicates the link gain between the transmitting pico BS l and MUE u of macro cell m; and σ^2 is the noise power. I^{M} and I^{P} are the interference terms caused by the macro BSs and the pico BSs, respectively.

The SINR at PUE u allocated in RB r of pico cell p is:

$$\gamma_r^{u(p)} = \frac{p_r^{p(u),\mathrm{P}} g_{p,u,r}^{\mathrm{PP}}}{\underbrace{\sum_{j=1,j\neq p}^{P} p_r^{j(u),\mathrm{P}} g_{j,u,r}^{\mathrm{PP}}}_{I^{\mathrm{P}}} + \underbrace{\sum_{k=1}^{K} p_r^{p(u),\mathrm{M}} g_{m,u,r}^{\mathrm{MP}}}_{I^{\mathrm{M}}} + \sigma^2}. \tag{19.5}$$

Here $g_{p,u,r}^{\mathrm{PP}}$ indicates the link gain between the transmitting pico BS p and its PUE u; $g_{j,u,r}^{\mathrm{PP}}$ indicates the link gain between the transmitting pico BS j and PUE u in the pico cell at pico BS p; $g_{m,u,r}^{\mathrm{MP}}$ indicates the link gain between the transmitting macro BS m and PUE u of pico BS p.

19.4.5.2 Simulation parameters

The scenario used in the system-level simulations is based on configuration #4b in [25]. We consider a macro cell consisting of three sectors and P {2, 4, 8} pico BSs per macro sector, uniformly randomly distributed within the macro cellular environment. $N_{\mathrm{UE}} = 30$ mobile users are generated within each macro sector from which $N_{\mathrm{hotspot}} = \lceil \frac{2}{3} \cdot N_{\mathrm{UE}}/P \rceil$ are randomly and uniformly dropped within a 40 m radius of each pico BS. The remaining UEs are uniformly distributed within the macro cellular area. All UEs have an average speed of 3 km/h. A full buffer model is assumed for the traffic of users. Further details about the system level simulation parameters are provided in Table 19.1.

19.4.5.3 System-level simulation results

As an evaluation metric, the average UE throughput is considered, which is defined as the ratio of the number of information bits that the user successfully receives divided by the total simulation time T. If UE u has V DL packet calls with $W_{v,u}$ packets for the v-th DL packet call and $b_{w,v,u}$ bits for the w-th packet, the average UE throughput for UE u is:

$$R_u = \frac{\sum_{v=1}^{V} \sum_{w}^{W_{v,u}} b_{w,v,u}}{T}. \tag{19.6}$$

Table 19.1 Simulation parameters.

Parameter	Value
Cellular layout	Hexagonal grid, 3 sectors per cell, reuse 1
Carrier frequency	2 GHz
System bandwidth	10 MHz
Subframe duration	1 ms
Number of RBs	50
Number of macro cells	1
Number of pico BSs per macro cell P	{2, 4, 8}
Max. macro (pico) BS transmit power	$P_{max}^{M} = 46$ dBm ($P_{max}^{P} = 30$ dBm)
Number of UEs per sector N_{UE}	30
Number of hotspot UEs $N_{hotspot}$	$2/3 \cdot N_{UE}/Pe$
Macro path-loss model	$128.1 + 37.6\log_{10}(R)$ dB (R[km])
Pico path-loss model	$140.7 + 36.7\log_{10}(R)$ dB (R[km])
Traffic model	Full buffer
Scheduling algorithm	Proportional fair
Transmission mode	Transmit diversity
Macro UE speed	$3 \frac{km}{h}$
Min. dist. macro BS–pico BS	75 m
Min. dist. pico BS–pico BS	40 m
Min. dist. macro BS–macro UE	35 m
Min. dist. pico BS–pico UE	10 m
PUE radius	40 m
Thermal noise density	−174 dBm
Macro antenna gain	14 dBi
Pico antenna gain	5 dBi

For the proposed frequency-domain e-ICIC algorithms an analysis of the tradeoffs for single-flow CA (*SF QL*) and multi-flow CA (*MF static QL* and *MF dynamic QL*) is performed. Figure 19.7 plots the UE throughput for two active pico cells per macro cell. While the *SF QL* and *MF static QL* algorithms are in average very close to each other, the *MF dynamic QL* algorithm shows a performance improvement of 47% on average. A close-up view of the cell-edge UE throughput shows that the multi-flow CA algorithms outperform the single-flow case. This is because in multi-flow CA, cell-edge UEs are served by macro and pico cell at the same time.

The behavior of the learning-based frequency domain e-ICIC algorithms when increasing the number of pico cells per macro cell is depicted in Figure 19.8. Here the solid curves belong to the left ordinate showing the total throughput and the dashed curves refer to the right ordinate reflecting the cell-edge UE throughput. It can be observed that the *MF dynamic QL* algorithm outperforms the other algorithms in terms of total throughput, while the *SF QL* algorithm is slightly better than the *MF static QL* algorithm for less number of pico cells (and vice versa for large numbers). The *SF QL* algorithm shows the lowest performance for cell-edge UE throughput. It can be concluded that cell-edge UEs benefit more from multi-flow CA than from single-flow

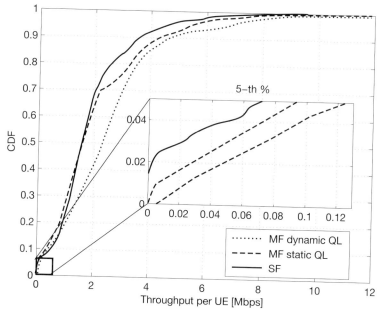

Figure 19.7 Cumulative distribution function of the UE throughput for the single- and multi-flow e-ICIC learning algorithm for two pico cells per macro cell.

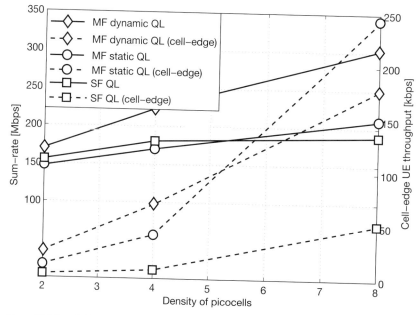

Figure 19.8 Total-throughput and cell-edge throughput vs. the number of pico cells in frequency-domain e-ICIC.

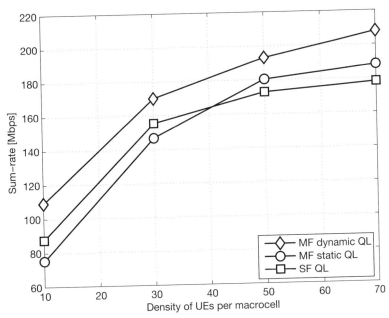

Figure 19.9 Sum-rate vs. the number of UEs per macro cell in the frequency-domain e-ICIC for two pico BSs per macro cell and two CCs.

CA. Interestingly, it can be observed that the *MF static QL* algorithm outperforms the *MF dynamic QL* for larger number of pico cells. This is because in the two-player case, the macro BS cannot fully adapt to the ICIC strategies of all pico BSs in the system, when the number of pico BSs is large.

In Figure 19.9 the sum-rate of the proposed frequency domain e-ICIC algorithms for different number of UEs in the case of two pico cells per macro cell is depicted. While the *MF dynamic QL* algorithm outperforms the other two algorithms, a tradeoff between the *SF QL* and *MF static QL* algorithm can be seen. For more than 40 UEs per macro cell the *MF static QL* algorithm outperforms the single-flow case. Hence, the multi-flow CA technique does not only protect cell-edge UEs, but also improves the total sum-rate at high loads.

References

[1] A. Ghosh, R. Ratasuk, B. Mondal, N. Mangalvedhe, and T. Thomas, "LTE-Advanced: next-generation wireless broadband technology," *IEEE Commun. Mag.*, vol. 17, no. 3, pp. 10–22, Jun. 2010.

[2] 3GPP, "Evolved universal terrestrial radio access (E-UTRA); FDD Home eNode B (HeNB) radio frequency (RF) requirements analysis," Technical Report (3GPP TR 36.921) v.9.0.0, 2010.

[3] S. Sesia, I. Toufik, and M. Baker, (eds.), *LTE: The UMTS Long Term Evolution from Theory to Practice*. John Wiley & Sons, 2011.

[4] 3GPP, "Requirements for further advancements for evolved universal terrestrial radio access (E-UTRA)," Technical Report (3GPP TR 36.913) v.10.0.0, 2011.
[5] A. Damnjanovic, J. Montojo, W. Yongbin, J. Tingfang, and M. Vajapeyam, "A survey on 3GPP heterogeneous networks," *IEEE Commun. Mag.*, vol. 18, no. 3, pp. 10–21, Jun. 2011.
[6] D. Lopez-Perez, I. Guvenc, G. de la Roche, M. Kountouris, T. Q. S. Quek, and J. Zhang, "Enhanced inter-cell interference coordination challenges in heterogeneous networks," *IEEE Commun. Mag.*, vol. 18, no. 3, pp. 22–30, Jun. 2011.
[7] D. Lopez-Perez, X. Chu, and I. Guvenc, "On the expanded region of picocells in heterogeneous networks," *IEEE J. Select. Topics in Signal Processing*, vol. 6, no. 3, pp. 281–294, Mar. 2012.
[8] I. Guvenc, J. Moo-Ryong, I. Demirdogen, B. Kecicioglu, and F. Watanabe, "Range expansion and inter-cell interference coordination (ICIC) for picocell networks," *Proc. IEEE Vehic. Technol. Conf. (VTC)*, San Francisco, CA, Jun. 2011.
[9] R. Madan, J. Borran, A. Sampath, N. Bhushan, A. Khandekar, and J. Tingfang, "Cell association and interference coordination in heterogeneous LTE-A cellular networks," *IEEE J. Select. Areas Commun.*, vol. 28, no. 9, pp. 1479–1489, Dec. 2010.
[10] LG Electronics, "Comparison of time-domain eICIC solutions," 3GPP Work Item Description (R1-104661), Spain, Aug. 2010.
[11] NTT DOCOMO, "Views on eICIC schemes for Rel-10," 3GPP Work Item Description (R1-105442), China, Oct. 2010.
[12] Qualcomm Incorporated, "eICIC Solutions Details," 3GPP Work Item Description, Jun. 2010.
[13] 3GPP, "Carrier aggregation for LTE," Technical Report v0.0.7, 2013.
[14] 3GPP, "Evolved universal terrestrial radio access (E-UTRA); user equipment (UE) radio transmission and reception," Technical Report (3GPP TR 36.807) v10.0.0, 2012.
[15] CMCC, "Summary of the description of candidate eICIC solutions," 3GPP Work Item Description (R1-104968), Spain, Aug. 2010.
[16] Qualcomm Incorporated, "Range expansion and multiflow enhancements in heterogeneous networks," 3GPP Work Item Description (R2-130423), Malta, Jan. 2013.
[17] Samsung, "Discussion on small cell enhancement and dual connectivity," 3GPP Work Item Description (R2-130100), Malta, Jan. 2013.
[18] CMCC, "Discussion on some issues of dual connectivity," 3GPP Work Item Description (R2-130055), Malta, Jan. 2013.
[19] X. Lin, J. G. Andrews, and A. Ghosh, "Modeling, analysis and design for carrier aggregation in heterogeneous cellular networks," *CoRR*, vol. abs/1211.4041, 2012. http://arxiv.org/abs/1211.4041.
[20] Huawei, HiSilicon, "Discussion on range expansion techniques with multiflow," 3GPP Work Item Description (R2-130158), Malta, Jan. 2013.
[21] Nokia Siemens Networks, "On advanced UE MMSE receiver modeling in system simulations," 3GPP Work Item Description (R1-111031), Taiwan, Feb. 2011.
[22] NTT DOCOMO, Inc., "Requirements, candidate solutions, and technology roadmap for LTE Rel. 12 onward," 3GPP Workshop on Release 12 and Onwards, Jun. 2012. www.3gpp.org/ftp/workshop/2012-06-1112RANREL12/Docs/RWS-120010.zip.
[23] H. Ishii, Y. Kishiyama, and H. Takahashi, "A novel architecture for LTE-B:C-plane/U-plane split and phantom cell concept," *Proc. IEEE Global Commun. Conf.: International*

Workshop on Emerging Technologies for LTE-Advanced and Beyond – 4G, Anaheim, USA, Dec. 2012.

[24] M. E. Harmon and S. S. Harmon, "Reinforcement learning: a tutorial," 2000.

[25] 3GPP, "Evolved universal terrestrial radio access (EUTRA); further advancements for E-UTRA physical layer aspects," Technical Report (3GPP TR 36.814), 2010.

Index

3GPP, v, vi, 1–4, 10, 16, 18, 19, 20, 22, 29, 30, 33, 34, 36–40, 46, 47, 50, 54, 56, 57, 59, 64, 66, 73–77, 79, 80, 81, 83–86, 88, 90–93, 95, 96, 107, 120, 121, 125, 128, 130–134, 136, 138, 145, 146, 147, 167, 193, 195, 200, 208, 210, 211, 213–219, 223, 229, 230, 235, 240, 241, 243, 244, 245, 262, 279, 280, 282, 287, 290, 307, 310, 313, 315, 321, 323, 328, 329, 333, 334, 336, 337, 361, 369, 384, 395, 402, 403, 404, 412, 421, 423, 457, 461, 465, 482, 483, 484, 486–491, 499, 500, 501
4G, 120, 129, 147, 170
5G, i, 31, 54, 56, 94, 180, 242, 279

ABS, 3, 4, 19, 20, 22, 29, 47, 78, 79, 125, 224–230, 233, 234, 326, 367, 486, 487
access node, 213
active antenna systems, 363, 369
adaptation, 237
ANDSF, 36, 122, 132, 134–137, 144, 147, 334
APs, 31, 36, 47, 52, 138–142, 148–152, 154–158
asymmetric, 58, 60, 72, 73

backhaul, 368, 390
big data, 425, 427–430, 449, 452, 453, 454, 457
BS, vii, 16, 58, 59, 60, 63, 64, 67, 70, 72, 96, 158, 160, 161, 164, 181–185, 223, 224, 284–293, 296–306, 393, 394, 396, 397, 402, 403, 404, 407, 408, 409, 414, 415, 416, 421, 465, 473, 478, 482, 484, 485, 486, 489, 492–497, 499
BSSID, 335

capacity, 338, 339
carrier aggregation, ix, 79, 82, 86, 89, 90, 91, 239, 240, 445, 488, 489
CDF, 101, 104, 151, 162, 209, 319, 344, 353, 359, 360, 498
clustering, 69–72, 200, 308, 427, 430, 431, 432, 445, 447–450, 483
co-channel, 1, 495
cognitive small cell networks, vii, 167, 191, 210
coordination, 3
coverage, 3
CPLANE, 219, 231

CQI, 4, 12, 22, 47, 78, 85, 95, 229, 325, 493
CRE, cell range expansion, 1, 42, 78, 79, 223, 224, 229, 230, 326, 327, 484, 485, 486, 490–495
CSI, channel state information, 79, 228, 249

D2D, device-to-device, 92, 93, 94, 96
density, 83, 93
discovery, 76, 88, 89
downlink, 2, 4, 5, 11, 12, 41, 42, 48, 50, 58, 59, 63, 77, 80, 81, 85, 87, 91, 94, 95, 113, 146, 148, 149, 150, 155, 156, 158, 166, 167, 172, 173, 178, 179, 182, 183, 185, 189, 202, 213, 219, 221–225, 228, 229, 233–238, 240, 258, 259, 260, 264, 266, 270, 271, 272, 275, 313, 317, 319, 320, 325, 326, 327, 342, 357, 389, 390, 395, 396, 410, 411, 430, 463, 465, 474, 485–489, 495

EESM, exponential effective SIR mapping, 317
eICIC, enhanced inter-cell interference coordination, 1, 3, 4, 18, 19, 77, 78, 79, 81, 83, 90, 95, 202, 211, 225, 229, 230, 239, 323, 326, 327, 335, 365, 367, 369, 386–389, 446, 457, 500
eNBs, 2, 16, 18, 23, 83, 96, 97, 99, 115, 119, 192, 196, 197, 200, 202, 203, 320, 326, 329, 366, 385, 386
energy efficiency, i, 55, 113, 138, 144, 284, 286, 288, 370, 462, 473, 482

FDD, frequency division duplex 85–6, 94
femto cell, 138, 145
field trial, viii, 338, 361
filtering, 3, 4, 8, 9, 12, 16, 129, 312, 317, 370, 375, 430

game theory, 242, 250, 251, 258
GBR, guaranteed bit rate, 315
green communications, 306, 483
GSM, global system for mobile communications, 131, 147, 309, 389
guard period, 59

HARQ, hybrid automatic repeat request, 47, 225, 227, 332
heterogeneous networks, 307, 309

Index

HO, handover, 1–4, 7–15, 17–23, 26, 29, 75, 83, 85, 89, 91, 92, 95, 125, 130, 132–135, 138, 144, 172, 179, 223, 226, 232, 245, 246, 249, 292, 309–315, 317, 318, 321–332, 334–337, 468–472, 474, 485
HO CMD, handover command, 313, 320
HO CONF, handover confirm, 313
hysteresis, 309, 310

ICI, inter-cell interference, 3, 19, 61, 77, 82, 87, 118, 176, 180, 223, 225, 227, 229, 240, 247, 287, 326, 368, 369, 389, 484, 486
IIR, infinite impulse response, 8, 312
interference management, mitigation, viii, 73, 78, 80, 82, 85, 86, 95, 107, 108, 112–115, 120, 122, 123, 125, 144, 145, 231, 243, 258, 264, 270, 282, 287, 363–366, 371, 390, 394, 408
Internet of Things, i, 33, 56, 74

large-scale deployment, 425, 439, 441, 456, 457
learning techniques, vii, 249, 270, 277, 278, 282
LOS, line-of-sight, 102, 344
LPN, low power nodes, 75, 76, 87, 113, 126
LTE, 1, 2, 3, 34, 46
LTE-A, viii, 73, 79, 90, 91, 213, 214, 218, 219, 223, 229, 236, 243, 244, 363, 365, 369, 385, 412, 425, 487, 500

M2M, machine-to-machine, 96, 317–321, 325, 326, 327
MAC, medium access control, 6, 40, 41, 46, 62, 140, 147, 216, 315, 317
macro cell, 80, 141
MBSFN, multicast-broadcast single frequency network, 218, 219, 221, 222, 230, 487
MDP, Q-learning, 274, 394, 492
measurement, 218, 228
micro cells, 286
MME, mobility management entity, 23, 130, 219, 231, 312, 314
mobility robustness optimization, 336
MRC, maximum rate combining, 317
MSE, mobility state estimation, 18
MTBH, mean time between handovers, 318, 319, 326, 331
multi-objective optimization, 378, 389
multi-radio, v, 31–34, 36, 40, 55, 57

NCH, network controlled handover, 324
non-line-of-sight, 60

PAN, personal area network, 32
PBS, pico BS, 42, 43, 321
PDCCH, physical downlink control channel, 5, 12, 20, 78, 81, 82, 91, 94, 95, 201, 202, 218, 221, 222, 225, 229, 233, 314, 317, 487–490
PeNBs, pico cell eNBs, 326

performance analysis, i, v, 61, 167, 181, 188, 189
pico cells, 1, 2, 3, 8, 18, 20, 23–27, 78, 83, 310, 317, 319, 321, 326, 338, 339, 341, 342, 344, 347, 360, 364, 365, 366, 371, 383–386, 454, 466
power control, 146, 175, 234
PRB, physical resource blocks, 233, 317
PUCCH, physical uplink control channel, 81, 94, 95, 221, 234, 325
PUEs, pico cell UEs, 326, 491, 492

RACH, random access channel, 11, 12, 313, 315, 317
RAN, radio access network, vi, 9, 33, 34, 36–40, 50, 51, 52, 54–57, 74–77, 92, 93, 128, 131, 134, 211, 240, 310, 329, 334, 363, 402, 423, 462, 463, 483
RAT, radio access technology, 2, 278, 309
RCPI, received channel power indicator, 335
reinforcement learning, 271, 393, 501
RLC, radio link control, 6, 11, 79, 315, 317
RLF, radio link failure, 1, 4, 6, 7, 29, 310, 487
RLM, radio link monitoring, 4, 5, 12, 21, 228, 229, 230, 328
RRC, radio resource control, 4–7, 23, 24, 29, 62, 74, 79, 81, 83, 216, 218, 219, 223, 240, 312, 315, 317, 324, 328, 329, 331, 334, 335, 336
RRH, remote radio head, 483
RS, reference symbol/signal, 81, 90, 94, 201, 209, 218, 221, 235, 237, 238, 239, 312
RSNI, received signal-to-noise indicator, 335
RSRP, reference symbol received power, 2, 3, 4, 7, 8, 9, 12, 16, 18, 37, 39, 47, 51, 52, 78, 88, 115, 209, 228, 232, 235, 312, 313, 326, 336, 339, 342–353, 366, 485
RSRQ, reference symbol received quality, 2, 7, 8, 18, 52, 114, 119, 228, 232, 235, 312, 313, 325

s-cell, source cell, 12, 23, 90, 239, 313, 315, 317, 322, 324, 325, 329, 331
Scells, small cells, vi, viii, 1, 2, 4, 7, 8, 9, 18, 21, 23, 27, 31, 33, 36, 37, 39–42, 44, 46, 48, 49, 56, 60, 61, 63, 64, 67, 72, 75, 78, 83, 85, 88, 93, 94, 97, 100, 110, 111, 116, 117, 118, 124–127, 137, 138, 139, 141, 144, 148, 149, 168, 169, 171, 172, 173, 175, 177–182, 184–188, 192, 196, 197, 213, 214, 232, 235–239, 242, 243, 246, 249, 255, 257–262, 264, 265, 267, 270, 274, 277, 278, 279, 286, 287, 288, 307, 310, 311, 323, 328, 332, 335, 338, 339, 340, 344, 351, 354–361, 393, 407, 409, 412–416, 421, 425, 427, 428, 431, 432, 433, 437, 439, 440, 441, 445, 446, 449, 458, 462, 464–470, 472, 473, 480, 481, 486, 491
scheduling, 366, 367, 385
self-optimization, viii, 247, 249, 280, 287, 363, 364, 383, 388, 390, 416

s-eNB, source eNB, 10, 11, 23, 324, 325
SGW, serving gateway, 312, 314
signaling, 404, 423, 488
SINR, signal to interference plus noise ratio, 20, 38, 43, 60, 61, 63, 64, 67–73, 75, 78, 85, 99–103, 106, 108, 110, 115, 126, 127, 138, 140, 141, 148, 151, 181, 183, 184, 189, 197, 198, 201, 209, 224, 229, 247, 251, 256, 257, 258, 260, 261, 262, 265, 267, 268, 272, 275, 277, 315, 317, 319, 320, 324, 325, 358, 359, 366, 367, 386, 396, 410, 411, 412, 421, 428, 435, 436, 437, 469, 486, 493, 496
sleep mode, vii, 284, 285, 288, 289, 302, 467, 468, 477, 478
social networks, 427, 429, 430, 445, 449, 450
SON, self-organised network, 328
SRB, signaling radio bearer, 328
SSID, service set identifier, 137, 334, 335
standardization, 80–85, 98, 124, 138, 195, 213, 217, 223, 230, 235, 239, 243, 244, 245, 490
SToS, short time-of-stay, 1, 11, 13, 14, 16, 17, 19, 20, 29

t-cell, target cell, 311, 313, 329
technical specification, 29, 30, 145, 210, 240, 307, 402, 482, 483
t-eNB, target eNB, 5, 6, 10, 11, 23, 324
traffic offloading, 25, 122–125, 132, 141, 144, 333
TTT, time to trigger, 3, 9, 10, 13, 18, 19, 20, 23, 93, 312, 314, 317, 319–322, 325, 329, 330, 331

UE, user equipment, 36, 133, 173, 214, 284
UMTS, universal mobile terrestrial system, 90, 92, 93, 95, 123, 188, 190, 215, 307, 323, 334, 336, 361, 389, 424, 499
U-plane, user plane, 83, 219, 238, 309, 312, 328, 491
uplink, 430

WANs, wide area networks, 32
WCDMA, wideband code division multiple access, 170, 173, 174, 175, 178, 180, 183, 309, 311, 423, 424
WLAN, wireless local area network, 33, 39, 49, 51, 56, 57, 85, 86, 87, 91, 95, 136, 146, 333–337, 423